Contributions in Mathematical and Computational Sciences

Volume 9

More information about this series at http://www.springer.com/series/8861

Thomas Carraro • Michael Geiger •
Stefan Körkel • Rolf Rannacher
Editors

Multiple Shooting and Time Domain Decomposition Methods

MuS-TDD, Heidelberg, May 6-8, 2013

Springer

Editors

Thomas Carraro
Heidelberg University
Institute for Applied Mathematics
Heidelberg, Germany

Michael Geiger
Heidelberg University
Institute for Applied Mathematics
Heidelberg, Germany

Stefan Körkel
Heidelberg University
Interdisciplinary Center for Scientific Computing
Heidelberg, Germany

Rolf Rannacher
Heidelberg University
Institute for Applied Mathematics
Heidelberg, Germany

ISSN 2191-303X ISSN 2191-3048 (electronic)
Contributions in Mathematical and Computational Sciences
ISBN 978-3-319-23320-8 ISBN 978-3-319-23321-5 (eBook)
DOI 10.1007/978-3-319-23321-5

Library of Congress Control Number: 2015954722

Mathematics Subject Classification (2010): 34-XX, 34Bxx, 34B08k, 34B15, 34B60, 34Hxx, 34H05, 35-XX, 35D30, 35Jxx, 35Kxx, 35Q30, 35Q40, 35Q92, 49-XX, 49Mxx, 49M05, 49M15, 49M37, 62-XX, 62Kxx, 62K05, 62K25, 65-XX, 65Kxx, 65K05, 65K10, 65Lxx, 65L04, 65L05, 65L06, 5L09, 65L10, 65L80, 65Mxx, 65M20, 65M32, 65M60, 68-XX, 68Uxx, 68U20, 90-XX, 90Cxx, 90C30, 90C53, 90C55

Springer Cham Heidelberg New York Dordrecht London

Printed on acid-free paper

Springer International Publishing AG Switzerland is part of Springer Science+Business Media (www.springer.com)

Preface to the Series

Contributions to Mathematical and Computational Sciences

Mathematical theories and methods and effective computational algorithms are crucial in coping with the challenges arising in the sciences and in many areas of their application. New concepts and approaches are necessary in order to overcome the complexity barriers particularly created by nonlinearity, high-dimensionality, multiple scales and uncertainty. Combining advanced mathematical and computational methods and computer technology is an essential key to achieving progress, often even in purely theoretical research.

The term mathematical sciences refers to mathematics and its genuine sub-fields, as well as to scientific disciplines that are based on mathematical concepts and methods, including sub-fields of the natural and life sciences, the engineering and social sciences and recently also of the humanities. It is a major aim of this series to integrate the different sub-fields within mathematics and the computational sciences and to build bridges to all academic disciplines, to industry and to other fields of society, where mathematical and computational methods are necessary tools for progress. Fundamental and application-oriented research will be covered in proper balance.

The series will further offer contributions on areas at the frontier of research, providing both detailed information on topical research and surveys of the state of the art in a manner not usually possible in standard journal publications. Its volumes are intended to cover themes involving more than just a single "spectral line" of the rich spectrum of mathematical and computational research.

The Mathematics Center Heidelberg (MATCH) and the Interdisciplinary Center for Scientific Computing (IWR) with its Heidelberg Graduate School of Mathematical and Computational Methods for the Sciences (HGS) are in charge of providing and preparing the material for publication. A substantial part of the material will be acquired in workshops and symposia organized by these institutions in topical areas of research. The resulting volumes should be more than just proceedings collecting papers submitted in advance. The exchange of information and the discussions during the meetings should also have a substantial influence on the contributions.

Starting this series is a venture posing challenges to all partners involved. A unique style attracting a larger audience beyond the group experts in the subject areas of specific volumes will have to be developed.

The first volume covers the mathematics of knots in theory and application, a field that appears excellently suited for the start of the series. Furthermore, due to the role that famous mathematicians in Heidelberg like Herbert Seifert (1907–1996) played in the development of topology in general and knot theory in particular, Heidelberg seemed a fitting place to host the special activities underlying this volume.

Springer Verlag deserves our special appreciation for its most efficient support in structuring and initiating this series.

Heidelberg, Germany

Hans Georg Bock
Willi Jäger
Hans Knüpfer
Otmar Venjakob

Preface

The point of origin of this book was an international workshop with the same title (*Multiple Shooting and Time Domain Decomposition Methods—MuSTDD 2013*) that took place at the Interdisciplinary Center of Scientific Computing (IWR) at Heidelberg University in late spring 2013. Most of the chapters presented here are based on topics exposed in the talks given during this workshop.

The leading motivation for realizing this book project was its potential to fill a gap in the existing literature on time domain decomposition methods. So far, in contrast to domain decomposition methods for the spatial variables, which have found broad interest in the past two decades, the decomposition of the time domain still constitutes a niche. There is no comparable compendium on this subject, although an increasing amount of journal articles proves a growing need for these methods. Therefore, we firmly believe that this volume provides a useful overview over the state-of-the-art knowledge on the subject and offers a strong incentive for further research.

The book at hand is divided into two parts, which roughly reflect a classification of the articles into theoretical and application-oriented contributions:

- The first part comprises methodical, algorithmic, and implementational aspects of time domain decomposition methods. Although the context is often given by optimization problems (optimal control and parameter estimation with nonstationary differential equations), the covered topics are also accessible and crucial for researchers who intend to utilize time decomposition in a modeling and simulation framework. The topics covered in this theoretical part range from a historical survey of time domain decomposition methods via state-of-the-art environments for multiple shooting (such as ODE parameter estimation or DAE problems) up to recent research results, e.g. on different multiple shooting approaches for PDE, on multiple shooting in the optimal experimental design (OED) or the nonlinear model predictive control (NMPC) frameworks or on parareal methods as preconditioners.
- The second part is concerned with applications in different scientific areas that can potentially benefit from multiple shooting schemes and the related

parareal methods. In the application fields covered in this volume (amongst them fluid dynamics, data compression, image processing, computational biology, and fluid structure interaction problems), the two essential features of time domain decomposition methods, namely the stabilization of the solution process and its parallelizability, display their full potential.

Overall, we are convinced that this volume constitutes a unique compilation of methodical and application-oriented aspects of time domain decomposition useful for mathematicians, computer scientists, and researchers working in different application areas. Although it does not claim to be exhaustive, it provides a comprehensive accumulation of material that can both serve as a starting point for researchers who are interested in the subject and extend the horizon of experienced scientists who intend to deepen their knowledge.

We would like to acknowledge the support of several sponsors who made the MuSTDD workshop possible: the Priority Program 1253 of the German Research Association (DFG), the Mathematics Center Heidelberg (MATCH), and the Heidelberg Graduate School of Mathematical and Computational Methods for the Sciences (HGS MathComp). Furthermore, we thank all the authors for their precious contributions. The cooperation with Springer, MATCH, and IWR should not be left unmentioned: it was a pleasure to work with them, and we thank all the people who rendered this 9th volume of *Contributions in Mathematical and Computational Sciences* possible by quietly and efficiently acting behind the scenes.

Heidelberg, Germany

Thomas Carraro
Michael Geiger
Stefan Körkel
Rolf Rannacher

Contents

Direct Multiple Shooting and Generalized Gauss-Newton Method for Parameter Estimation Problems in ODE Models

Hans Georg Bock, Ekaterina Kostina, and Johannes P. Schlöder

Abstract The paper presents a boundary value problem approach for optimization problems in nonlinear ordinary differential equations, in particular for parameter estimation, based on multiple shooting as originally introduced by Bock in the 1970s. A special emphasis is placed on the theoretical analysis including numerical stability, grid condition and a posteriori error analysis. The paper discusses advantages of multiple shooting versus single shooting which are illustrated by numerical examples.

1 Introduction

The history of shooting methods for optimization problems in differential equations goes back to the 1950s when shooting methods were first used to solve two point boundary value problems (TPBVP) resulting from application of the Pontryagin maximum principle to optimal control problems, see e.g. [26, 31, 38, 40]. A first versatile algorithm capable to treat TPBVP with switching points has been developed by Bulirsch and Stoer [18] giving start to numerous theses in the Bulirsch group extending optimal control theory and multiple shooting algorithms for TPBVP. The drawbacks of this "indirect" approach are that the boundary value problems resulting from the maximum principle for optimal control problems are difficult to derive, moreover, they are usually ill-conditioned and highly nonlinear in terms of state and adjoint variables, they have jumps and switching conditions.

Another type of methods for optimal control problems—"direct" single shooting methods—appeared in the 1960s. Instead of using the maximum principle and adjoint variables, in these methods the controls were discretized, such that solving differential equations resulted in finite nonlinear programming problems. Numerous

H.G. Bock • J.P. Schlöder
IWR, Heidelberg University, Heidelberg, Germany
e-mail: bock@iwr.uni-heidelberg.de; schloeder@iwr.uni-heidelberg.de

E. Kostina (✉)
IAM, Heidelberg University, Heidelberg, Germany
e-mail: kostina@uni-heidelberg.de

T. Carraro et al. (eds.), *Multiple Shooting and Time Domain Decomposition Methods*, Contributions in Mathematical and Computational Sciences 9,
DOI 10.1007/978-3-319-23321-5_1

direct algorithms were developed, usually as feasible step gradient type methods, e.g. [17, 36, 37]. The reason for this was that effective non-feasible step algorithms for constrained nonlinear optimization emerged only in the 1970s, i.e. SQP [25, 41].

In this paper we consider a "direct" multiple shooting method for optimization problems with differential equations, also known nowadays as "optimization boundary value problem", "all-at-once", or "simultaneous" approaches. This approach is based on constrained nonlinear optimization instead of the maximum principle, and the discretized BVP is treated as an equality constraint in the optimization problem which is then solved by non-feasible step methods. This approach became a standard tool for solving optimization problems for differential equation models, see [9, 10, 12, 13].

The multiple shooting combined with the generalized Gauss-Newton is especially suitable for parameter estimation.

In this paper we focus on the multiple shooting for parameter estimation for differential equations. A special emphasis is placed on the theoretical analysis including numerical stability, grid condition and a posteriori error analysis. We discuss also advantages of multiple shooting versus single shooting which are illustrated by numerical examples.

2 Parameter Estimation Problem in ODE Systems

We consider parameter estimation problems that are characterized by a system of ordinary differential equations for state variables $x(t)$

$$\dot{x} = f(t, x, p),$$

the right hand side of which depends on a parameter vector p. Furthermore, measurements η_{ij} for the state variables or more general for functions in the states are given

$$\eta_{ij} = g_{ij}(x(t_j), p) + \varepsilon_{ij},$$

which are collected at measurement times $t_j, j = 1, \ldots, k$,

$$t_0 \leq t_1 < \ldots < t_k \leq t_f,$$

over a period $[t_0, t_f]$, and are assumed to be affected by a measurement error ε_{ij}.

The unknown parameters p have to be determined such that the measured (observed) process is "optimally" reproduced by the model. If the measurement errors ε_{ij} are independent, Gaussian with zero mean value and variances σ_{ij}^2 an appropriate objective function is given by a weighted l_2-norm of the measurement errors

$$l_2(x, p) := \sum_{i,j} \sigma_{ij}^{-2} \varepsilon_{ij}^2 = \sum_{i,j} \sigma_{ij}^{-2} (\eta_{ij} - g_{ij}(x(t_j), p))^2. \tag{1}$$

In this case the minimization of the weighted squared errors provides a maximum-likelihood estimator for the unknown parameter vector.

In a more general case where the measurement errors are correlated with a known (positive definite) covariance matrix C, we have to use C^{-1} as a weight for the definition of a scalar product replacing (1) to receive a maximum-likelihood estimator.

In many problems additional (point-wise) equality and/or inequality constraints on parameters and state variables arise as restrictions onto the model

$$e(t_j, x(t_j), p) = 0, \; u(t_j, x(t_j), p) \geq 0, j = 1, \ldots, k.$$

For notation simplicity we assume the constraints are stated at the same points as measurements.

A rather general parameter estimation problem for ordinary differential equations can be summarized as follows:

Problem [PE1] Find a *parameter vector* $p \in \mathbb{R}^{n_p}$ and a trajectory $x : [t_0, t_f] \to \mathbb{R}^{n_d}$, that minimize the objective function (describing a weighted norm of measurement errors)

$$l_2 = \|r_1(x(t_1), \ldots, x(t_k), p)\|_2, r_1 \in \mathbb{R}^{n_1}, \tag{2}$$

where n_1 is a number of all measurements. The solution has to satisfy the *system of ordinary differential equations* of dimension n_d

$$\dot{x} = f(t, x, p), \quad t \in [t_0, t_f], \tag{3}$$

the n_2 *equality constraints*

$$r_2(x(t_1), \ldots, x(t_k), p) = 0 \tag{4}$$

and the n_3 *inequality constraints*

$$r_3(x(t_1), \ldots, x(t_k), p) \geq 0. \tag{5}$$

Other functionals than the l_2-norm can be used, like the general l_q-functional

$$l_q(x, p) = \sum_{i,j} \left|\beta_{ij}(\eta_{ij} - g_{ij}(x(t_j), p))\right|^q, \; 1 \leq q < \infty$$

with the sum of absolute values of errors for $q = 1$ as the most significant special case besides $q = 2$, and the limit case of the Chebyshev or minimax problem

$$l_\infty(x, p) = \max_{i,j} \left|\gamma_{ij}^{-1}(\eta_{ij} - g_{ij}(x(t_j), p))\right|.$$

l_1- and l_∞-estimators have some specific properties which make them interesting. Under certain regularity assumptions l_∞-optimization leads to a solution in which exactly $n+1$ of the weighted measurement errors take the maximum value, whereas exactly n errors vanish in case of the l_1-optimization so that the l_1-optimal solution interpolates n measurement values. Here $n = n_v - n_{\text{con}}$ is the number of the remaining degrees of freedom, where $n_v = n_d + n_p$ counts the differential equations and parameters, n_{con} counts the equality and active inequality constraints at the solution. Hence, the optimal solution is only specified by few "best" (l_1) and "worst" (l_∞) measurement values, respectively. From statistical point of view l_∞-based estimator provides a maximum-likelihood estimation for uniformly distributed errors with maximum γ_{ij}, l_q-based estimator is related to distributions of the type $\alpha_{ij} \exp(-|\beta_{ij}\varepsilon_{ij}|^q)$. For further discussion see [2, 15, 32].

3 Initial Value Problem Approach

The simplest—and maybe most obvious—approach for the numerical treatment of parameter estimation problems in differential equations is the repeated *solution of the initial value problem* (IVP) for fixed parameter values in framework of an iterative procedure for *refinement of the parameters* to improve the parameter estimates and to fulfill possible constraints on states and parameters. Thus, the inverse problem is lead back to a sequence of IVPs.

Besides the undeniable advantage of a simple implementability the IVP approach has two severe fundamental disadvantages which are clearly shown in numerical practice and are verified by theoretical analysis.

On the one hand the state variables $x(t)$ are eliminated—by means of differential equation (3)—in favor of the unknown parameters p by the re-inversion of the inverse problem. As a consequence any information during the solution process that is especially characteristic for the inverse problem is disregarded. Consequently, the structure of the inverse problem is destroyed.

On the other hand the elimination of the state variables can cause a drastical loss of stability of the numerical procedure. At least for bad initial guesses of the parameters, which always have to be expected in practice, the (non-linear) initial value problem can be ill-conditioned and difficult to solve or can be not solvable at all even if the inverse problem is well-conditioned. As a consequence the IVP approach places high demands on the used iterative method or on the quality of the initial values.

Let us illustrate these properties of the IVP approach by two examples.

3.1 Test Problem 1

This parameter estimation problem is a modification of a two-point boundary value problem originally given in Bulirsch and Stoer [18]. A system of ODEs with two states and one unknown parameter p is given by

$$\begin{aligned} \dot{x}_1 &= x_2, & x_1(0) &= 0 \\ \dot{x}_2 &= \mu^2 x_1 - (\mu^2 + p^2)\sin pt, & x_2(0) &= \pi \end{aligned} \qquad t \in [0,1] \qquad (6)$$

For the true value of parameter $p = \pi$ the solution is

$$x_1(t) = \sin \pi t, \; x_2(t) = \pi \cos \pi t.$$

In this example measurement data is generated by adding a normally distributed pseudo random noise $(N(0, \sigma^2), \sigma = 0.05)$ to the solution $x(t), t \in [0, 1]$ at measurement points $t_j = 0.1j, j = 1, \ldots, 10$. The results for the IVP approach shown in Fig. 1 are disappointing. Choosing $p^{(0)} = 1.$ as an initial value, the numerical integration routine stops at about $t \approx 0.4$ (in case $\mu = 60$) and even for the true value of the parameter $p^{(0)} = \pi$ (up to 16 decimals) the solution is correctly reproduced even less than for the first half of the interval. Not even the objective function is (numerically) defined for these trajectories. Hence, the problem is not numerically solvable with the IVP approach. The reason for this does not lie in the

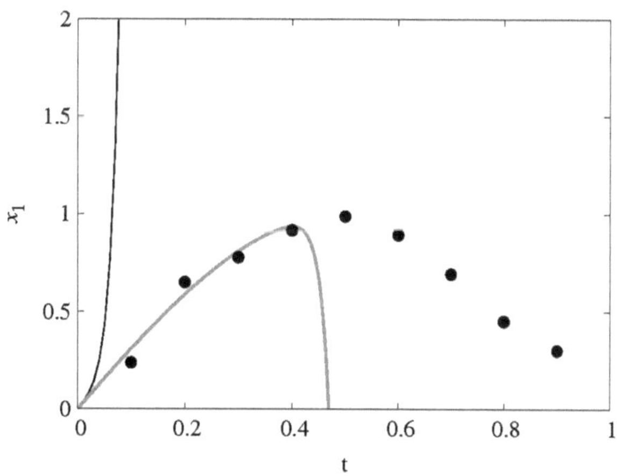

Fig. 1 Test problem 1: IVP approach, initial trajectories $x_1(\cdot)$ for $p^{(0)} = 1$ (*black line*), $p^{(0)} = \pi$ (*grey line*), measurements (*black dots*)

integration method: the eigenvalues of the matrix $f_x(t, x(t), p) := \frac{\partial f(t, x(t), p)}{\partial x}$ are $\lambda_{1,2} = \pm\mu$ and the general solution of the ODE has the form

$$\begin{aligned} x_1(t) &= \sin \pi t + \varepsilon_1 \sinh \mu t + \varepsilon_2 \cosh \mu t; \qquad \varepsilon_1 = \frac{x_2(0) - p}{\mu}, \\ x_2(t) &\equiv \pi \cos \pi t + \varepsilon_1 \cosh \mu t + \varepsilon_2 \sinh \mu t; \;\; \varepsilon_2 = x_1(0). \end{aligned} \tag{7}$$

Even smallest errors of the initial values and the parameter and the (unavoidable) discretization and rounding errors, respectively, are drastically intensified, as a result the exact solution is dominated by highly increasing parasitic solution components.

3.2 Test Problem 2

The performance of an ecological system consisting of one predator and one prey can be described by the following model of Lotka and Volterra

$$\dot{x}_1 = -k_1 x_1 + k_2 x_1 x_2, \quad \dot{x}_2 = k_3 x_2 - k_4 x_1 x_2, \quad t \in [0, 10].$$

The measurements in the points $t_j = j, j = 1, \ldots, 10$, are simulated by numerical integration of ODE for the "true" values $k_i = 1, i = 1, \ldots, 4$, and $x_1(0) = 0.4$, $x_2(0) = 1.0$ on $[0, 10]$, disturbed by normally distributed pseudo random noise $(N(0, \sigma^2), \sigma = 0.05)$.

The measurement data and the initial trajectory of the IVP approach for $k = (0.5, 0.5, 0.5, -0.2)^T$ are shown in Fig. 2. The solution has a singularity at $\hat{t} = 3.3$, therefore the IVP approach must fail.

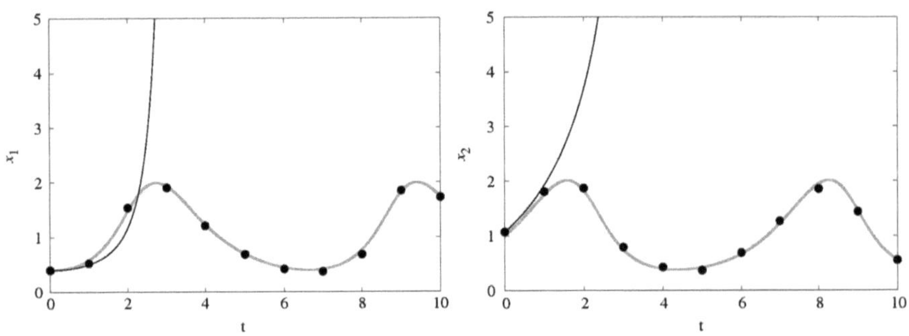

Fig. 2 Test problem 2. Initial trajectory of the IVP approach (*black line*), true solution (*grey line*) and measurements (*black dots*)

4 A Multiple Shooting Method for Parameter Estimation

Alternatively to the IVP approach, the inverse problem [PE1] can be interpreted as an over-determined, constrained multi-point *boundary value problem* (BVP). Therefore, a stable and efficient solution method has been developed as an appropriate generalization of the method for the treatment of multi-point boundary value problems described in Bock [7], based on the multiple shooting approach (cf. Bulirsch and Stoer [18], Stoer and Bulirsch[47], and Deuflhard [19]).

A BVP method for parameter estimation problem [PE1] has been first presented by Bock in [6] and implemented in the program package PARFIT.

4.1 *Description of the Multiple Shooting Algorithm*

For an appropriate *grid* of *m nodes* τ_j

$$\tau_1 < \tau_2 < \ldots < \tau_m, \quad \Delta\tau_j := \tau_{j+1} - \tau_j \qquad j = 1, \ldots, m-1 \tag{8}$$

that covers the sampling interval ($[\tau_1, \tau_m] = [t_0, t_f]$), we introduce the *discrete trajectory* $\{s_j := x(\tau_j)\}$ as optimization variables in addition to the unknown parameters p. For a given estimate of the extended vector of variables $(s_1, \ldots, s_m, p)$ we compute the solutions $x(t; s_j, p)$ of $m-1$ *independent* initial value problems

$$\dot{x} = f(t, x, p),\ x(\tau_j) = s_j, \quad t \in I_j := [\tau_j, \tau_{j+1}]$$

on each subinterval I_j and in this way obtain a (discontinuous) parameterization of the ODE solution $x(t), t \in [t_0, t_f]$. Formally inserting the solutions at the measurements points t_i

$$x(t_i; s_{j(i)}, p) \qquad \text{for} \quad t_i \in [\tau_{j(i)}, \tau_{j(i)+1}[$$

into the objective function and constraints (2), (4), and (5)

$$R_l(s_1, \ldots, s_m, p) := r_l(x(t_1; s_{j(1)}, p), \ldots, x(t_k; s_{j(k)}, p), p),\ l = 1, 2, 3,$$

we get a constrained *finite-dimensional* estimation problem the solution of which is equivalent to the solution of [PE1].

Problem [PE2] (Discretized Parameter Estimation Boundary Value Problem)
Find a parameter vector $p \in \mathbb{R}^{n_p}$ and a discrete trajectory $(s_1^T, \ldots, s_m^T)^T \in \mathbb{R}^{n_d \times m}$, that minimize the objective function

$$\|R_1(s_1, \ldots, s_m, p)\| \tag{9}$$

subject to the constraints

$$R_2(s_1,\ldots,s_m,p)=0, \quad R_3(s_1,\ldots,s_m,p)\geq 0, \tag{10}$$

and the additional *continuity conditions*

$$h_j(s_j,s_{j+1},p):=x(\tau_{j+1};s_j,p)-s_{j+1}=0, \tag{11}$$

which ensure the continuity of the solution.

4.2 Well-Definedness

By an appropriate choice of the multiple shooting nodes we can ensure that the parameterized trajectory of the state variables exists on the whole interval and remains in a specified region. Well-definedness and differentiability properties of problem [PE2] then follow from existence and differentiability of the trajectories $x(t;s_j,p), j=1,\ldots,m$, in the multiple shooting intervals, under certain assumptions on functions r_1, r_2, r_3, see e.g. [4, 5]. Other advantages of the multiple shooting approach will be discussed in the following.

The following lemma provides a qualitative statement about the choice of multiple shooting nodes.

Lemma 1 *Let $\eta(t)\in C^0[t_0,t_f]^{n_d}$ be piecewise continuously differentiable, $\hat{p}\in\mathbb{R}^{n_p}$, $\delta>\varepsilon_0>0$, $\varepsilon_p>0$ arbitrary, $f\in C^r(D)$, D a domain, such that $S:=\{(t,x,p)\mid t\in[t_0,t_f], \|x-\eta(t)\|\leq\delta, \|p-\hat{p}\|\leq\varepsilon_p\}\subset D$. Then there exists a grid such that for all s_j, p with*

$$\|s_j-\eta(\tau_j)\|\leq\varepsilon_0, \qquad \|p-\hat{p}\|\leq\varepsilon_p$$

the solution $x(\tau;x_j,s_j,p)$ of

$$\dot{x}=f(t,x,p), \qquad x(\tau_j)=s_j, \qquad t\in[\tau_j,\tau_{j+1}]$$

exists, is r times continuously differentiable in (t,τ_j,s_j,p) and does not deviate from $\eta(t)$ more than δ

$$\|x(t;\tau_j,s_j,p)-\eta(t)\|\leq\delta. \tag{12}$$

Proof The proof is based on classical arguments. The central statement follows from the elementary estimate for $\delta x(t):=x(t;\tau,s,p)-\tau(t)$, $\delta p:=p-\hat{p}$

$$\|\delta x(t)\|\leq\|\delta x(\tau)\|\cdot e^{\mu|t-\tau|}+\mu^{-1}(\nu\|\delta p\|+\beta)(e^{\mu|t-\tau|}-1) \tag{13}$$

where $\|f_x(t,x,p)\| \leq \mu$, $\|f_p(t,x,p)\| \leq \nu$ in S and $\|\dot{\eta}(t) - f(t,\eta(t),\hat{p})\| \leq \beta$. From (13) we get

$$|\Delta\tau_j| \leq \mu^{-1}\log((\delta+\theta)/(\varepsilon_0+\theta)), \qquad \theta := \mu^{-1}(\varepsilon_p\nu+\beta) \tag{14}$$

as a bound for the grid size. □

Let us note, that the criterion for well-definedness (12) can be ensured even without knowing the constants μ, ν, β by using e.g. a node strategy suggested by Bulirsch and Stoer [18]: Starting with an initial function $\eta(t)$ that has to be given, the integration of the initial value problem is stopped as soon as the distance from $\eta(t)$ exceeds a bound δ and a node τ_j with $s_j = \eta(\tau_j)$ is inserted at this position.

For a differential equation with a perturbed right hand side

$$\dot{x} = f(t,x,p) + \varphi(t) \tag{15}$$

the relations (13) and (14) also hold with $\bar{\beta} = \beta + \varepsilon_d$, $\varepsilon_d \geq \|\varphi(t)\|$. With this presentation the discretization error of a numerical integrator can easier be taken into account.

In the following let $(\eta(t),\hat{p})$ be a solution of [PE1] (i.e. $\beta = 0$). From (13) it follows that a “drift off the solution” as a result of bad initial values, which in IVP approaches can lead to an exponential error growth near the solution or even to singularities in the non-linear case, can be avoided by a suitable choice of the nodes.

Therefore, in multiple shooting the domain and the convergence region of the parameter estimation problem are enlarged, the demands on the initial values for a solution procedure can be significantly reduced.

Note, that particularly the information available for the inverse problem allows an excellent choice of the initial values s_j in the state space for which $\delta x(\tau_j)$ is small. Hence, the evaluation of the objective function and constraints can be done near the optimal solution trajectory right from the beginning of the iterative solution procedure. As models in non-trivial parameter estimation problems are almost always non-linear, the efficiency of the solution scheme is thus seriously improved.

Hence, better initial guesses improve global convergence of the iterative procedure, help to avoid local minima and reduce nonlinearity of the optimization problem leading to improved local convergence, e.g. even up to one step if parameters enter the differential equations linearly.

Figure 3 shows for the test problem 2 how the consideration of the given measurement data damps the influence of bad parameter estimates in the multiple shooting approach. Though the initial guesses for the constants are chosen rather poor with $k_i^{(0)} = (0.5, 0.5, 0.5, -0.2)^T$ (the optimized values are $k_i = 1$, $i = 1,\ldots,4$), the trajectories lie relatively “near” to the optimal solution.

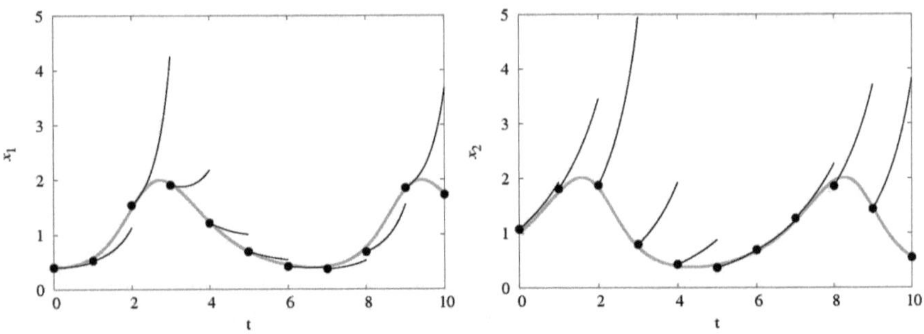

Fig. 3 Test problem 2: measurements (*black dots*), initial (*black line*) and optimal (*grey line*) trajectories of the multiple shooting approach, $k_i^{(0)} = (0.5, 0.5, 0.5, -0.2)^T$

4.3 Stability

Besides the advantages of the multiple shooting method concerning applicability, generation of a favorable initial trajectory and the involved increase of efficiency compared to the IVP approach, which are illustrated by the numerical examples (see Sect. 7), the multiple shooting approach particularly has the advantage of numerical stability. With the notations

$$\bar{\gamma}_j := e^{\mu|\Delta\tau_j|}, \qquad \bar{\gamma}_j^p := \mu^{-1}\nu\bar{\gamma}_j$$

and $\varepsilon_x \geq \varepsilon_0 + |\Delta\tau_j|\varepsilon_d \geq \|\delta x(\tau_j)\| + |\Delta\tau_j|\,\|\varphi(s)\|$ Eqs. (13) and (15) lead to the bound

$$\|\delta x(t)\| \leq \bar{\gamma}_j \cdot \varepsilon_x + \bar{\gamma}_j^p \cdot \varepsilon_p.$$

By the choice of the nodes the propagation of inevitable rounding and discretization errors can be limited sufficiently.

But for a numerically reasonable grid choice strategy the estimations from Lemma 1 are useless. The bounds μ, ν are hardly accessible in practice, furthermore they usually provide an overestimation of the true error propagation.

Realistic and quantitatively accessible estimations of the error propagation result from a linear perturbation analysis for IVP (15) with perturbed initial values and parameters. As a global error on the interval I_j we get as a first-order approximation

$$\delta x(t) \doteq G(t,\tau_j)\delta x(\tau_j) + \int_{\tau_j}^{t} G(t,s)\varphi(s)\,ds + G^p(t,\tau_j)\delta p,$$

where the sensitivity matrices G, G^p are the solution of the variational differential equations

$$\begin{aligned} \frac{d}{dt}(G,G^p)(t,\tau) &= f_x(t,y(t),p)(G,G^p)(t,\tau) + (0,f_p(t,y(t),p)) \\ (G,G^p)(\tau,\tau) &= (I,0). \end{aligned} \tag{16}$$

Inserting discretization and rounding error and estimating via

$$\gamma_j := \max_{t_2 \geq t_1 \in I_j} \|G(t_2,t_1)\| \geq 1, \quad \gamma_j^p := \max_{t_1 \geq \tau_j \in I_j} \|G^p(t_1,\tau_j)\| \geq 0,$$

we get as a minimum requirement for the choice of nodes the grid condition [GC]

$$[GC] \qquad \gamma_j \varepsilon_x + \gamma_j^p \varepsilon_p \leq \delta_j. \tag{17}$$

Here, δ_j is the (global) accuracy of the solution of the IVP problem on the multiple shooting interval I_j, which bounds the error appearing in the computation of the objective and constraint functions $R_i, i = 1,2,3$ and the continuity functions h_j and which depends on the condition of the problem and the desired accuracy of the solution.

4.4 Grid Determination

Knowing the matrices G, G^p, which may be computed in a very efficient way, see e.g. [10], the bounds γ_j, γ_j^p can be easily approximated.

For a fixed grid, condition [GC] can be satisfied by reduction of the discretization error if necessary. This also my be done by a somewhat finer estimation than (17) by adapting the discretization error.

If limit accuracy is reached, the grid has to be refined in order to reduce the constants of γ_j, γ_j^p or the stopping criterion and thus the tolerance δ_j has to be damped.

Example In case of test problem 1 one can easily verify that γ_j, $\gamma_j^p \approx 10^{26}$ for the IVP approach. Computing with 16 digits precision, [GC] is not satisfiable for reasonable accuracies δ_j. In contrast, for the multiple shooting approach an equidistant grid with $\Delta\tau_j = 0.1$ (11 nodes) already suffices for γ_j, $\gamma_j^p \approx 10^4$. With integration accuracies around $10^{-4} \cdot \delta_j$ one can achieve practically sufficient errors δ_j which are global on I_j.

Basically, the grid condition [GC] is a stability condition and hence also allows relatively gross grids for most problems with moderate instabilities of the differential equations. Thus, the node choice can be used itself with the aim of a good initial trajectory and low memory space complexity at the same time.

In practical applications it turns out that the number of required nodes often is significantly smaller than the number of measurement points.

Let us note, that condition [GC] provides the possibility of an automatic grid choice for which the nodes are distributed in order to reach an overall low computational effort.

5 A Generalized Gauss-Newton Method for Constrained Non-linear Least Squares Problems

The discretized parameter estimation boundary value problem [PE2] is a non-linear constrained least squares problem of the general form

Problem PN (Non-linear Constrained Least Squares Problem)

$$\min_x \|F_1(x)\|_2^2, \text{ subject to the constraints } F_2(x) = 0,\ F_3(x) \geq 0.$$

Let the mappings $F_i : D \subset \mathbb{R}^n \to \mathbb{R}^{m_i} \in C^3(D)$ on a domain D. The derivatives are denoted as $J_i(x) := F_i'(x)$.

The basic algorithm can be formulated as follows:

Basic Algorithm A given estimate x_0 is iteratively improved by

$$x_{k+1} = x_k + [t_k]\Delta x_k, \quad t_k \in]0, 1], \tag{18}$$

where the *increment* Δx_k is a solution of the problem [PL(x_k)] linearized in x_k:

Problem [PL(x)] (Linearized Constrained Least Squares Problem)

$$\min_{\Delta x} \|F_1(x) + J_1(x)\Delta x\|_2^2, \text{ s.t. } F_2(x) + J_2(x)\Delta x = 0,\ F_3(x) + J_3(x)\Delta x \geq 0.$$

There is numerous literature about Gauss-Newton methods for the unconstrained problem. The general constrained case was first implemented in the context of the parameter estimation in the algorithm PARFIT, including also inequality constraints. The properties of the generalized Gauss-Newton method are investigated also for the l_1-case, see e.g. [10, 15, 16, 32]. We briefly sketch here the facts that are necessary for further analysis of multiple shooting approach.

5.1 *Characterization of an Optimal Solution*

Let the inequality constraints that are active in a point $z \in D$ be defined by the index set

$$I(z) := \{i \mid F_{3i}(z) = 0,\ i = 1, \ldots, m_3\},$$

let these components and their derivatives be combined in $\tilde{F}_3$, $\tilde{J}_3$. Furthermore, denote by

$$F_c := \begin{pmatrix} F_2 \\ \tilde{F}_3 \end{pmatrix}, \qquad J_c := \begin{pmatrix} J_2 \\ \tilde{J}_3 \end{pmatrix}$$

the mapping $F_c : D \subset \mathbb{R}^n \to \mathbb{R}^{m_c}$, $m_c := m_2 + \tilde{m}_3$, $\tilde{m}_3 = |I(z)|$ and its derivative. The point z is called *regular* if the constraint qualification

$$\text{Rang } J_c(z) = m_c \tag{19}$$

is satisfied.

With the *Lagrange function* $L(x, \lambda, \mu)$,

$$L(x, \lambda, \mu) = \frac{1}{2}\|F_1(x)\|_2^2 - \lambda^T F_2(x) - \mu^T F_3(x),$$

we get the following optimality conditions as a consequence of standard statements of non-linear optimization:

Theorem 2 (Necessary Optimality Conditions) *Let $x^* \in D$ be regular solution of the non-linear problem* PN. *Then x^* is feasible,*

$$F_2(x^*) = 0, \qquad F_3(x^*) \geq 0,$$

and the necessary conditions of the first order hold: There exist adjoint variables λ^*, μ^* such that conditions hold

$$\frac{\partial}{\partial x}L(x^*, \lambda^*, \mu^*) = F_1(x^*)^T J_1(x^*) - \lambda^{*T} J_2(x^*) - \mu^{*T} J_3(x^*) = 0 \tag{20a}$$

$$\mu^* \geq 0, \quad i \notin I(x^*) \Rightarrow \mu_i^* = 0 \quad (i.e.\ \mu^{*T} F_3(x^*) = 0). \tag{20b}$$

Furthermore, the second-order necessary conditions hold: the Hessian of the Lagrange function is positive semi-definite

$$p^T H(x^*, \lambda^*, \mu^*) p \geq 0, \qquad H(x^*, \lambda^*, \mu^*) := \frac{\partial^2}{\partial x^2} L(x^*, \lambda^*, \mu^*).$$

for all directions $p \in T(x^*)$

$$T(x^*) := \{p \neq 0 \mid J_2(x^*)p = 0, \tilde{\mu}_i^* \tilde{J}_{3i}(x^*)p = 0, \tilde{J}_3(x^*)p \geq 0\}.$$

Theorem 3 (Sufficient Conditions) *Let (x^*, λ^*, μ^*) be a Kuhn-Tucker point of [PN] and let the Hessian of the Lagrange function be positive definite for all directions $p \in T(x^*)$*

$$p^T H(x^*, \lambda^*, \mu^*)p > 0.$$

Then x^* is a strict local minimum of [PN].

Lemma 4 *Let a Kuhn-Tucker triple $\{x^*, \lambda^*, \mu^*\}$ satisfy the constraint qualification (19), the first-order necessary conditions, the strict complementarity $\mu_i^* > 0, i \in I(x^*)$, and let the Hessian H be positive definite for all directions $p \in \{d \neq 0 | J_c(x^*)d = 0\}$. Then x^* is a strict local minimum of [PN].*

Applying Theorem 2 to the linear problem [PL(x^*)] we obtain that, if the triple $\{x^*, \lambda^*, \mu^*\}$ satisfies the assumptions of the theorem, then x^* is a fixed point of the Gauss-Newton iterations. Using Lemma 4 we may conclude that as the iterates of the Gauss-Newton method approach a local minimum satisfying the assumptions of Lemma 4, namely constraint qualification, positive definiteness and strict complementarity, the active set remains fixed and hence in a neighbourhood D of the solution x^* we may ignore inequalities not belonging to $I_c(x^*)$ and may consider (locally) only equality constrained problems in the form

$$\min_x \|F_1(x)\|_2^2, \text{ s.t. } F_c(x) = 0.$$

If the Jacobians $J_1(x)$ and $J_c(x)$ satisfy two regularity assumptions in the domain D

$$\begin{aligned} &[\text{CQ}]\ \operatorname{Rang} J_c(x) = m_c, \\ &[\text{PD}]\ \operatorname{Rang} J(x) = m, \quad \text{where } J(x) := \begin{pmatrix} J_1(x) \\ J_c(x) \end{pmatrix}. \end{aligned} \tag{21}$$

then the linearized constrained least squares problem

$$\min_{\Delta x} \|J_1 \Delta x + F_1\|_2^2, \text{ s.t. } J_c \Delta x + F_c = 0,$$

has a unique solution Δx^* and a unique Lagrange vector λ_c^* satisfying the following optimality conditions

$$\begin{aligned} &J_1^T J_1 \Delta x - J_c^T \lambda_c + J_1^T F_1 = 0, \\ &J_c \Delta x + F_c = 0. \end{aligned} \tag{22}$$

Under regularity assumptions (21) this system has a unique solution

$$\begin{pmatrix} \Delta x^* \\ -\lambda_c^* \end{pmatrix} = -\begin{pmatrix} J_1^T J_1 & J_c^T \\ J_c & 0 \end{pmatrix}^{-1} \begin{pmatrix} J_1^T & 0 \\ 0 & I \end{pmatrix} \begin{pmatrix} F_1 \\ F_c \end{pmatrix}. \tag{23}$$

Using a linear mapping $J^+ : \mathbb{R}^{m_1+m_c} \to \mathbb{R}^n$:

$$J^+ := (I, 0) \begin{pmatrix} J_1^T J_1 & J_c^T \\ J_c & 0 \end{pmatrix}^{-1} \begin{pmatrix} J_1^T & 0 \\ 0 & I \end{pmatrix}$$

the solution Δx^* can be formally written as

$$\Delta x = -J^+ F, \quad F := \begin{pmatrix} F_1 \\ F_c \end{pmatrix}.$$

Let us note that the solution operator J^+ is a generalized inverse, i.e. it satisfies the condition $J^+ J J^+ = J^+$.

6 Aspects of Numerical Implementation

In this section we concentrate ourselves on numerical aspects of the basic algorithm resulting from multiple shooting.

6.1 *Solution of Linearized Problems*

The linearization of the discretized parameter least squares boundary value problem [PE2] according to (9)–(11)

$$\begin{aligned} \|R_1(s_1, \ldots, s_m, p)\| &\to \min_{s_1,\ldots,s_m,p}, \\ R_2(s_1, \ldots, s_m, p) &= 0, \qquad R_3(s_1, \ldots, s_m, p) \geq 0, \\ h_j(s_j, s_{j+1}, p) &= 0, \qquad j = 1, \ldots, m-1, \end{aligned}$$

leads to a sparse Jacobian which is large compared to the initial value approach and has the special structure

$$J = \left[\begin{array}{ccccc|c} D_1^1 & D_1^2 & \cdots & & D_1^m & D_1^p \\ \hline D_2^1 & D_2^2 & \cdots & & D_2^m & D_2^p \\ \hline D_3^1 & D_3^2 & \cdots & & D_3^m & D_3^p \\ \hline G_1^l & -G_1^r & & & & G_1^p \\ & \ddots & \ddots & 0 & & \vdots \\ & & \ddots & \ddots & & \vdots \\ & 0 & & \ddots & \ddots & \vdots \\ & & & G_{m-1}^l & -G_{m-1}^r & G_{m-1}^p \end{array}\right], \quad F = \left[\begin{array}{c} R_1 \\ \hline R_2 \\ \hline R_3 \\ \hline h_1 \\ \vdots \\ \vdots \\ \vdots \\ h_{m-1} \end{array}\right] \tag{24}$$

with the notations

$$D_i^j := dR_i/ds_j, \quad D_i^p := dR_i/dp,$$
$$G_j^l := dh_j/ds_j, \quad G_j^r := dh_j/ds_{j+1}, \quad G_j^P := dh_j/dp.$$

6.1.1 The Classical Condensing Algorithm

For the multiple shooting method according to Sect. 4 we get for the derivatives of the continuity conditions h_j

$$G_j^r = I, \qquad G_j^l = G(\tau_{j+1}, \tau_j) =: G_j,$$
$$G_j^p = G^p(\tau_{j+1}, \tau_j),$$

where (G, G^p) are the sensitivity matrices which solve the variational differential equation (16). The constrained linear least squares problem characterized by the structured Jacobian (24) can be transformed by *backward recursion*

$$\text{Start}: \; u_i^m := R_i, P_i^m := D_i^p, E_i^m := D_i^m,$$
$$u_i^{j-1} := u_i^j + E_i^j h_j,$$
$$P_i^{j-1} := P_i^j + E_i^j, G_{j-1}^p$$
$$E_i^{j-1} := D_i^{j-1} + E_i^j G_{j-1}, \quad j = m, \ldots, 2, \quad i = 1, 2, 3,$$

to a *condensed problem*

$$\min_{\Delta s_1, \Delta p} \; \|u_1^1 + E_1^1 \Delta s_1 + P_1^1 \Delta p\|, \tag{25}$$
$$u_2^1 + E_2^1 \Delta s_1 + P_2^1 \Delta p = 0, \quad u_3^1 + E_3^1 \Delta s_1 + P_3^1 \Delta p \geq 0,$$

which only depends on the increments of the initial values Δs_1 and of the parameters Δp and thus has just dimension $(m_1 + m_2 + m_3) \times n_v$, $n_v := n_d + n_p$. Having solved the condensed problem , the increments of the remaining variables can be computed via *forward recursion*

$$\Delta s_{j+1} := G_j \Delta s_j + G_j^p \Delta p + h_j, \quad j = 1, \ldots, m-1.$$

Further variants of the condensing algorithm based on other matrix decompositions as block-Gauss are described in [16].

Numerical solution of the condensed problem can be performed by combination of active set strategy with various numerical algorithms for solving resulting equality constrained problems. One of such algorithms based on QR-decompositions of problem matrices can be found e.g. in Stoer [46].

6.1.2 Computation of the Adjoint Variables of the Complete System

The adjoint variables λ_c for equality and active inequality constraints λ_c can be computed together with the solution of the condensed problem.

With the definitions

$$
\begin{aligned}
r &:= R_1 + \sum_{j=1}^{m} D_1^j \Delta s_j + D_1^p \Delta p, \\
d^j &:= D_1^{j^\top} r, \quad d^p := D_1^{p^\top} r,
\end{aligned}
\tag{26}
$$

the Kuhn-Tucker conditions (20) for the complete system take the form

$$
\begin{pmatrix}
D_c^{1^\top} & G_1^{l^\top} & & & \\
D_c^{2^\top} & -G_1^{r^\top} & G_2^{l^\top} & & 0 \\
\vdots & & -G_2^{r^\top} & \ddots & \\
\vdots & 0 & & \ddots & G_{m-1}^{l^\top} \\
D_c^{m^\top} & & & & -G_{m-1}^{r^\top} \\
D_c^{p^\top} & G_1^{p^\top} & G_2^{p^\top} & \cdots & G_{m-1}^{p^\top}
\end{pmatrix}
\cdot
\begin{pmatrix}
\lambda_c \\ \mu_1 \\ \vdots \\ \vdots \\ \mu_{m-1}
\end{pmatrix}
=
\begin{pmatrix}
d^1 \\ d^2 \\ \vdots \\ \vdots \\ d^m \\ d^p
\end{pmatrix}.
\tag{27}
$$

This in general overdetermined problem has a unique solution according to Theorem 2. Therefore the adjoint variables μ_i to the continuity conditions can be efficiently computed using the decompositions of the matrices of the condensing algorithm by *backward recursion*

$$
\begin{aligned}
\text{Start: } \mu_{m-1} &:= (G_{m-1}^{r^\top})^{-1}(D_c^{m^\top}\lambda_c - d_m), \\
\mu_{j-1} &:= (G_{j-1}^{r^\top})^{-1}(D_c^{j^\top}\lambda_c - d^j + G_j^{l^\top}\mu_j), \quad j = m-1, \ldots, 2.
\end{aligned}
\tag{28}
$$

6.2 *Condition of the Discretized Parameter Estimation Boundary Value Problem*

In this section based on a generalization of the condensing algorithm, we give an explicit representation of the generalized inverse of the Jacobian J of the complete system and derive *grid independent bounds for its norm* which can be used to estimate the condition of the problem. For simplification we assume that the Jacobian along a solution

$$
(t, x(t), p), \qquad t \in I = [t_0, t_f] \tag{29}
$$

of the parameter estimation boundary value problem has been determined for which the assumptions of Lemma 1 hold with $\eta(t) = x(t)$, $\hat{p} = p$. Then for arbitrary $t_1, t_2 \in I$ the solutions $(G, G^p)(t_1, t_2)$ of the variational differential equation (16) exist, are nonsingular, and there exist bounds γ, γ^p with

$$\|G(t_1, t_2)\| \leq \gamma, \quad \|G^p(t_1, t_2)\| \leq \gamma^p, \quad \forall t_1, t_2 \in I. \tag{30}$$

With the notations

$$D^k = \begin{pmatrix} D_1^k \\ D_c^k \end{pmatrix}, \quad B^l = \sum_{k=l}^{m} D^k G(\tau_k, \tau_l), \quad A^l = -\sum_{k=1}^{l-1} D^k G(\tau_k, \tau_l),$$

$$E^l = B^l - A^l, \quad P^l = D^p + \sum_{j=1}^{l-1} A^{j+1} G_j^p + \sum_{j+l}^{m-1} B^{j+1} G_j^p$$

we get a condensed problem in the node τ_l

$$\begin{aligned} &\|E_1^l \Delta s_l + P_1^l \Delta p + R_1 + \sum_{j=1}^{l-1} A_1^{j+1} h_j + \sum_{j=l}^{m-1} B_1^{j+1} h_j\|_2^2 \to \min_{(\Delta s_l, \Delta p)}, \\ &E_c^l \Delta s_l + P_c^l \Delta p + R_c + \sum_{j=1}^{l-1} A_c^{j+1} h_j + \sum_{j=l}^{m-1} B_c^{j+1} h_j = 0, \end{aligned} \tag{31}$$

with a corresponding generalized inverse

$$V^l = \begin{bmatrix} E_1^l & P_1^l \\ E_c^l & P_c^l \end{bmatrix}^+ = [V_1^l \quad V_c^l]. \tag{32}$$

Due to (30) we have, that if the matrix

$$[E^l \; P^l] = \begin{bmatrix} E_1^l & P_1^l \\ E_c^l & P_c^l \end{bmatrix}$$

satisfies the regularity conditions (21) for a node τ_l then this property holds true for all other nodes as well.

Due to the uniqueness of the solution of the complete system the part of the matrix V^l corresponding to the parameters is apparently independent of the node index l. Defining

$$M_{lj} := \begin{cases} V^l A^{j+1} & (j < l) \\ V^l B^{j+1} & (j \geq l) \end{cases}$$

we get the explicit representation of the solution

$$-\begin{pmatrix} \Delta s_l \\ \Delta p \end{pmatrix} = V_1^l R_1 + V_c^l R_c + \sum_{j=1}^{m-1} M_{lj} h_j.$$

Theorem 5 (Condition of the Discretized Parameter Estimation Boundary Value Problem) *Under assumption (30) it holds:*

1. *There exist constants k_1^l, K_1, k_c^l, K_c, a_j^l, b_j^l, k_M^l, K_M such that*

$$\|V_1^l\|_\infty \le k_1^l \le K_1 < \infty \qquad l = 1, \dots, m,$$
$$j = 1, \dots, m-1,$$
$$\|V_c^l\|_\infty \le k_c^l \le K_c < \infty$$
$$\|M_{lj}\| \le \begin{cases} a_j^l \ (j < l) \\ b_j^l \ (j \ge l) \end{cases} \le k_M^l \le K_M < \infty$$

2. *The following bounds hold for $\Delta S = (\Delta s_1^T, \dots, \Delta s_m^T, \Delta p^T)^T$:*

$$(a) \quad \|\Delta S\|_\infty \le \max_l \left\| \begin{pmatrix} \Delta s_l \\ \Delta p \end{pmatrix} \right\|_\infty$$
$$\le \max_l \left(k_1^l \|R_1\|_\infty + k_c^l \|R_c\|_\infty + \sum_{j=1}^{l-1} a_j^l \|h_j\| + \sum_{j=l}^{m-1} b_j^l \|h_j\| \right)$$
$$\le \max_l \left(k_1^l \|R_1\|_\infty + k_c^l \|R_c\|_\infty + k_M^l (m-1) \|h\|_\infty \right)$$
$$\le K_1 \|R_1\|_\infty + K_c \|R_c\|_\infty + (m-1) K_M \|h\|_\infty$$
$$(b) \quad \|J^+\|_\infty \le K_1 + K_c + (m-1) K_M.$$

3. (a) *The bounds K_1, K_c, K_M are independent of position and number of the nodes.*
   (b) *The bounds k_1^l, k_c^l, k_M^l, a_j^l, b_j^l depending on the nodes are grid independent insofar as they hold for any grid which includes the nodes τ_l or (τ_l, τ_j), respectively.*

Proof Consider the variation system of the parameter estimation boundary value problem

$$\min_{\varepsilon, \hat{p}} \left\| \sum_{i=1}^{k} \bar{D}_1^i \xi(t_i) + D_1^p \hat{p} + \varepsilon_1 \right\|_2, \tag{33a}$$

$$\dot{\xi}(t) - X(t)\xi(t) - Y(t)\hat{p} - \varphi(t) = 0, \tag{33b}$$

$$\sum_{i=1}^{k} \bar{D}_c^i \xi(t_i) + D_c^p \hat{p} + \varepsilon_c = 0, \tag{33c}$$

where $(X, Y)(t) := (f_x, f_p)(t, x(t), p)$, $\bar{D}^i_{c,1} := \frac{\partial r_{c,1}}{\partial x_i}(x(t_1), \ldots, x(t_k), p)$. Here we have assumed (without loss of generality) that the time points $t_1 < t_2 < \ldots < t_k \subset [t_0, t_f]$ are fixed.

For a solution of the linear system (33b) the relation holds

$$\xi(s) = G(s,\tau)\xi(\tau) + \int_\tau^s G(s,t)(Y(t)\hat{p} + \varphi(t))\,dt,\ s, \tau \in I. \tag{34}$$

After inserting (34) into (33c), (33a) we get

$$\sum_{i=1}^{k} \bar{D}^i G(t_i,\tau)\xi(\tau) + D^p\hat{p} + \sum_{l=1}^{k} \bar{D}^i \int_\tau^{t_i} G(t_i,t)Y(t)\hat{p}\,dt$$
$$+\varepsilon + \sum_{i=1}^{k} \bar{D}^i \int_\tau^{t_i} G(t_i,t)\varphi(t)\,dt.$$

Using the matrices

$$B(\tau) := \sum_{\substack{i=1\\(t_i \geq \tau)}}^{k} \bar{D}^i G(t_i,\tau), \qquad A(\tau) := -\sum_{\substack{i=1\\(t_i < \tau)}}^{k} \bar{D}^i G(t_i,\tau),$$
$$E(\tau) := B(\tau) - A(\tau). \quad \tau \in [t_0, t_f]$$

we get the representation

$$\sum_{i=1}^{k} \bar{D}^i \int_\tau^{t_i} G(t_i,t)g(t)\,dt = \int_\tau^{t_f} B(t)g(t)\,dt + \int_{t_0}^{\tau} A(t)g(t)\,dt.$$

Setting

$$P(\tau) = D^p + \int_{t_0}^{\tau} A(t)Y(t)\,dt + \int_\tau^{t_f} B(t)Y(t)\,dt$$

we get the condensed at the node τ least squares problem:

$$\left\| E_1(\tau)\xi(\tau) + P_1(\tau)\hat{p} + \varepsilon_1 + \int_{t_0}^{\tau} A_1(t)\varphi(t)\,dt + \int_\tau^{t_f} B_1(t)\varphi(t)\,dt \right\|_2 \to \min_{\xi(\tau),\hat{p}},$$
$$E_c(\tau)\xi(\tau) + P_c(\tau)\hat{p} + \varepsilon_c + \int_{t_0}^{\tau} A_c(t)\varphi(t)\,dt + \int_\tau^{t_f} B_c(t)\varphi(t)\,dt = 0.$$

Under the assumption (30) a bounded generalized inverse $(E(\tau)P(\tau))^+$ exists for a specific value τ as well as a bounded generalized inverse $(E(t)P(t))^+$ exists for all $t \in [t_0, t_f]$, and the relation holds

$$[E(t)\ P(t)]^+ = \begin{bmatrix} G(t,\tau) & G^p(t,\tau) \\ 0 & I \end{bmatrix} [E(\tau)\ P(\tau)]^+ =: [V_1(t),\ V_c(t)]$$

since

$$\int_\tau^s G(s,t)Y(t)\hat{p}\,dt \equiv G^p(s,\tau)\hat{p}.$$

With the notation

$$M(\tau,t) = \begin{cases} V(\tau)A(t) & \tau > t \\ V(\tau)B(t) & \tau \le t \end{cases}$$

we explicitly get the solution of the linearized parameter estimation boundary value problem as

$$-\begin{pmatrix} \xi(\tau) \\ \hat{p} \end{pmatrix} = V_1(\tau)\varepsilon_1 + V_c(\tau)\varepsilon_c + \int_{t_0}^{t_f} M(\tau,t)\varphi(t)\,dt.$$

Therefore, the bounds

$$k_1(t) := \|V_1(t)\|, \quad K_1 := \sup_{t\in[t_0,t_f]} k_1(t),$$

$$k_c(t) := \|V_c(t)\|, \quad K_c := \sup_{t\in[t_0,t_f]} k_c(t),$$

$$\|M(t,\tau)\| =: \begin{cases} a(t,\tau) & (t > \tau) \\ b(t,\tau) & (t \le \tau) \end{cases},$$

$$k_M(t) := \sup_{\tau\in[t_0,t_f]} \|M(t,\tau)\|, \quad K_M := \sup_{t\in[t_0,t_f]} k_M(t)$$

exist and consequently the continuous estimates are true

$$\begin{aligned} \sup_{t\in I} \left\| \begin{matrix} \xi(t) \\ \hat{p} \end{matrix} \right\| &\le \sup_{t\in I} \left(k_1(t)\|\varepsilon_1\| + k_c(t)\|\varepsilon_c\| + k_M(t) \int_{t_0}^{t_f} \|\varphi(\tau)\|\,d\tau \right) \\ &\le K_1\|\varepsilon_1\| + K_c\|\varepsilon_c\| + K_M \int_{t_0}^{t_f} \|\varphi(\tau)\|\,d\tau. \end{aligned}$$

The statements for the discretized case follow immediately after verifying

$$B^l = \sum_{k=l}^{m} D^k G(\tau_k, \tau_l) = \sum_{\substack{i=1 \\ (t_i \geq \tau_l)}}^{k} \bar{D}^i G(t_i, \tau_l) = B(\tau_l)$$

for a given grid and analogously $A^l = A(\tau_l)$, $E^l = E(\tau_l)$, $P^l = P(\tau_l)$, $M_{lj} = M(\tau_l, \tau_j)$. □

The bounds derived in Theorem 5 describe the sensitivity of the parameter estimation boundary value problem with respect to perturbations in the continuity conditions, constraints and the right hand side of the differential equations system. They also allow to estimate the influence of the discretization error on the computed solution (see next section), thus proving the numerical stability of the multiple shooting approach.

6.3 Influence of Discretization Errors on the Global Error of the Solution

In the following explicit relations between the local errors and the resulting global perturbation of the optimal solution are derived.

Consider the perturbed equality constrained linear parameter least squares problem

$$\begin{aligned} &\|(J_1 + \delta J_1)(x + \delta x) + (F_1 + \delta F_1)\|_2 \to \min_{\delta x} \\ &(J_c + \delta J_c)(x + \delta x) + (F_c + \delta F_c) = 0. \end{aligned} \tag{35}$$

Denote

$$\bar{J} = J + \delta J, \quad \operatorname{cov}(\bar{J}) := \bar{J}^+ \begin{pmatrix} I & 0 \\ 0 & 0 \end{pmatrix} (\bar{J}^+)^T.$$

Lemma 6 (Solution of a Perturbed Least Squares Problem) *Assume that J_c satisfies (19). Let x be a stationary point of the unperturbed problem ($\delta J = 0$, $\delta F = 0$), $R = J_1 x + F_1$ be the residual and λ be the corresponding adjoint variable with $J_1^T R - J_c^T \lambda = 0$. Then the solution $\bar{x}$ of the perturbed problem (35) is equal to $\bar{x} = x + \delta x$ where*

$$\begin{aligned} \delta x &= -\bar{J}^+ \delta F - \operatorname{cov}(\bar{J}) \delta J^T \begin{pmatrix} R \\ -\lambda \end{pmatrix} - \bar{J}^+ \delta J x \\ &=: \delta x_1 + \delta x_2 + \delta x_3. \end{aligned}$$

Proof The problem (35) can be rewritten as

$$\begin{aligned} \|\bar{J}_1\delta x + \delta F_1 + \delta J_1 x + R\| &\to \min_{\delta x} \\ \bar{J}_c\delta x + \delta F_c + \delta J_c x &= 0. \end{aligned}$$

Hence, the solution δx can be presented with the help of a generalized inverse:

$$\begin{aligned} \delta x &= -\bar{J}^+ \begin{pmatrix} \delta F_1 + \delta J_1 x + R \\ \delta F_2 + \delta J_2 x \end{pmatrix} = -\bar{J}^+(\delta F + \delta J x) - \bar{J}^+ \begin{pmatrix} R \\ 0 \end{pmatrix} \\ &= \delta x_1 + \delta x_3 - \bar{J}^+ \begin{pmatrix} R \\ 0 \end{pmatrix}. \end{aligned}$$

Using properties of generalized inverse and the equality $J_1^T R - J_c^T \lambda = 0$ we may reformulate the third term

$$\begin{aligned} \bar{J}^+ \begin{pmatrix} R \\ 0 \end{pmatrix} &= \bar{J}^+ \begin{pmatrix} I & 0 \\ 0 & 0 \end{pmatrix} \begin{pmatrix} R \\ -\lambda \end{pmatrix} = \bar{J}^+ \begin{pmatrix} I & 0 \\ 0 & 0 \end{pmatrix} (\bar{J}^+)^T \bar{J}^T \begin{pmatrix} R \\ -\lambda \end{pmatrix} \\ &= \operatorname{cov}(\bar{J})(J^T + \delta J^T) \begin{pmatrix} R \\ -\lambda \end{pmatrix} = \operatorname{cov}(\bar{J})\delta J^T \begin{pmatrix} R \\ -\lambda \end{pmatrix}. \end{aligned}$$

□

Let us note, that the perturbation estimations by Lawson and Hanson for the unconstrained problem [34] immediately follow from this lemma as special cases. Furthermore, Lemma 6 has the advantage of explicitly giving the presentation of the global error of the solution.

For applications particularly the terms δx_1, δx_2 are important as long as the inequality $\alpha := \|\bar{J}^+\delta J\| \ll 1$ holds, namely in a fixed point of the generalized Gauss-Newton method we get $\bar{x} = 0$, and hence $\|\delta x_3\| \leq \alpha\|\delta x\| \leq \alpha/(1 - \alpha)\|\delta x_1 + \delta x_2\|$. Consequently, though the term δx_3 affects the convergence rate of the generalized Gauss-Newton method, this term is of minor importance for the perturbation of the fixed point.

Lemma 6 allows to interpret the perturbation of the derivative matrix δJ as an additional perturbation of the measurement data, which is weighted by the residual R (and the adjoint variables λ, respectively) and by the norm of the covariance matrix $\|\operatorname{cov}(\bar{J})\|$.

For an estimation of the global error δx by means of the errors δF, δJ we assume that the bounds

$$\|\bar{J}^+ \begin{pmatrix} I & 0 \\ 0 & 0 \end{pmatrix}\| \leq \bar{K}_1, \qquad \|\bar{J}^+ \begin{pmatrix} 0 & 0 \\ 0 & I \end{pmatrix}\| \leq \bar{K}_C$$

or their estimates are available e.g. by the repeated solution of the linear systems during the Gauss-Newton process. Furthermore, we assume that the vectors $\tilde{R}$ and $\tilde{\lambda}$ exist, such that the component-wise inequalities hold

$$|R_i| \leq \tilde{R}_i, \qquad |\lambda_j| \leq \tilde{\lambda}_j, \; i = 1, \ldots, n_1, \, j = 1, \ldots, n_c,$$

Using the notation

$$\tilde{y} := \left[|\delta J_{1\,ij}|\right]^T \tilde{R} + \left[|\delta J_{cij}|\right]^T \tilde{\lambda} \tag{36}$$

we get the estimation for δx

$$\|\delta x\| \leq \bar{K}_C \|\delta F_C\| + \bar{K}_1 \|\delta F_1\| + \bar{K}_1^2 \|\tilde{y}\|.$$

Now let us apply this analysis to the linearized discretized parameter estimation boundary value problem in the case of the multiple shooting method. For simplification we assume that all errors except the discretization errors can be neglected and that these only occur in the right hand sides of the continuity conditions h_j and the matrices G_j, G_j^p. With the notations (26)–(28) we have

$$\tilde{y}_j = \left[|\delta G_{j_{ik}}|\right]^T \tilde{\mu}_j, \qquad j = 1, \ldots, m-1,$$

$$\tilde{y}^p = \textstyle\sum_{j=1}^{m-1} \tilde{y}_j^p, \qquad \tilde{y}_j^p = \left[|\delta G_{j\,ik}^p|\right]^T \tilde{\mu}_j.$$

If we require in the node τ_{j+1} (we assume without loss of generality that $\bar{K}_C \neq 0,\; \bar{K}_1 \neq 0$), that

$$\|\delta h_j\| \qquad \leq \frac{|\Delta\tau_j|}{2\bar{K}_C|\tau_m - \tau_1|} e =: TOL_j^X,$$

$$\|\tilde{y}_j\| + \|\tilde{y}_j^p\| \leq \frac{|\Delta\tau_j|}{2\bar{K}_1^2|\tau_m - \tau_1|} e =: TOL_j^G,$$

then the following bound holds for the global solution error $\|\delta x\|$

$$\|\delta x\| \leq \bar{K}_C \sum_{j=1}^{m-1} \|\delta h_j\| + \bar{K}_1^2 \sum_{j=1}^{m-1} (\|\tilde{y}_j\| + \|\tilde{y}_j^p\|) \leq e.$$

Let us note that, the constants $\bar{K}_C$ and $\bar{K}_1$ can be bounded by the grid independent estimates $(m-1)K_M$, K_C and K_1 from Theorem 5. This analysis may be used to tailor accuracies of integration and computing of constraint Jacobians in order to reduce computational efforts.

7 Numerical Examples

The aim of this section is the illustration of the numerical analysis presented in this paper. The choice of the problems is not necessarily representative for the large class of treatable problems but still allows an examination of the numerical properties of the algorithm in practical use under aspects of stability, reliability, efficiency, accuracy, and last but not least general applicability.

For further challenging problems of parameter estimation and optimal control and their treatment with BVP methods see e.g. [3, 8, 11, 14, 20, 21, 27, 28, 30, 33, 35, 43–45, 48].

7.1 Test Problem 1

Test problem 1 was introduced in Sect. 3 to demonstrate the effect of instabilities in the initial value problem approach (single shooting). In contrast, with the multiple shooting method the problem remains solvable also for significantly larger values of the constants μ and thus of the eigenvalues $\lambda_{1,2} = \pm\mu$ of the Jacobians of the differential equations.

The numerical results for $10 \leq \mu \leq 120$ are summarized in Table 1. The IVP approach still provides results for $\mu = 10, 20$, though only with great integration effort. For $\mu = 40$ error propagation (fourth row) over the whole interval $[0, 1]$ is already at 10^{19}, for computation with 16 digits precision the required local errors δ_j can no longer be maintained, see grid condition (17).

The multiple shooting approach with 11 nodes (all measurement points) still satisfies the grid condition up to $\mu = 120$ with $\delta_j^{\min} \leq 10^{-8}$ (third row)

Figure 4 shows the initial and the solution trajectory for the medium high value $\mu = 60$ and the parameter initial value $p^{(0)} = 1$ like in Fig. 1. Despite the extreme variations convergence is achieved within 4 iterations with a relative accuracy of $\varepsilon = 10^{-3}$.

Table 1 Test problem 1 (6)–(7)

μ	10	20	40	60	80	100	120
Number of iterations	4	4	4	4	4	4	4
Min. local error $\delta_j^{\min} = \gamma_j \varepsilon_{\text{mach}}$	10^{-14}	10^{-13}	10^{-12}	10^{-11}	10^{-10}	10^{-9}	10^{-8}
Error propagation $\|G(0,1)\|$	10^{5}	10^{10}	10^{19}	10^{27}	10^{36}	10^{45}	10^{54}

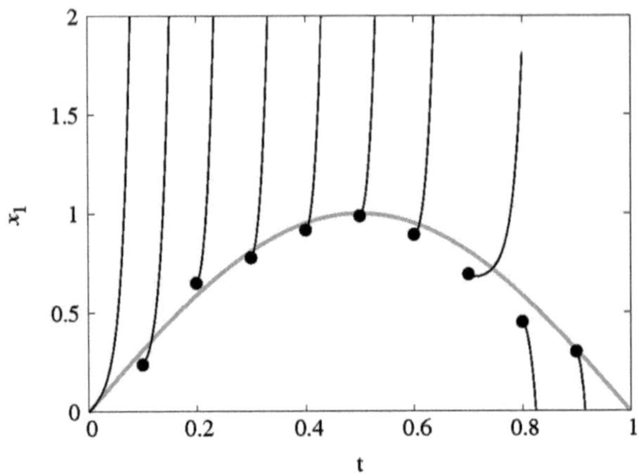

Fig. 4 Test problem 1, multiple shooting approach ($\mu = 60$, $p^{(0)} = 1$), measurements (*black dots*), initial (*black line*) and optimal (*grey line*) trajectories

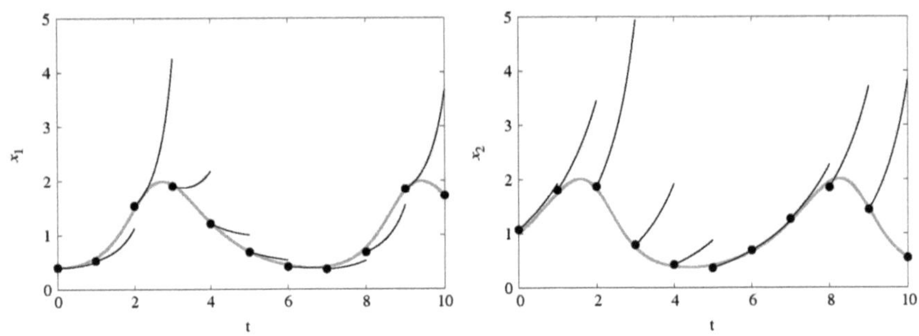

Fig. 5 Test problem 2, multiple shooting approach (10 multiple shooting nodes), initial trajectory (*black line*), measurements (*black dots*) , solution (*grey line*), *left*: $x_1(\cdot)$, *right*: $x_2(\cdot)$

Table 2 Solution and estimated error, test problem 2

	k_1	k_2	k_3	k_4
Solution	0.923	0.934	1.106	1.110
Standard error	±0.040	±0.044	±0.045	±0.080

7.2 Test Problem 2

Also the predator prey problem (Sect. 3.2) can be treated with multiple shooting without difficulties. Figure 5 shows the initial trajectory for the set of parameters $k = (0.5, 0.5, 0.5, -0.2)$. The singularity at $t = 3.3$ apparently does not show up. The generalized Gauss-Newton method needs 8 iterations until convergence ($\varepsilon = 10^{-3}$) (Table 2).

This test problem also clearly demonstrates that the multiple shooting combined with generalized Gauss-Newton method reduces the nonlinearity of the problem

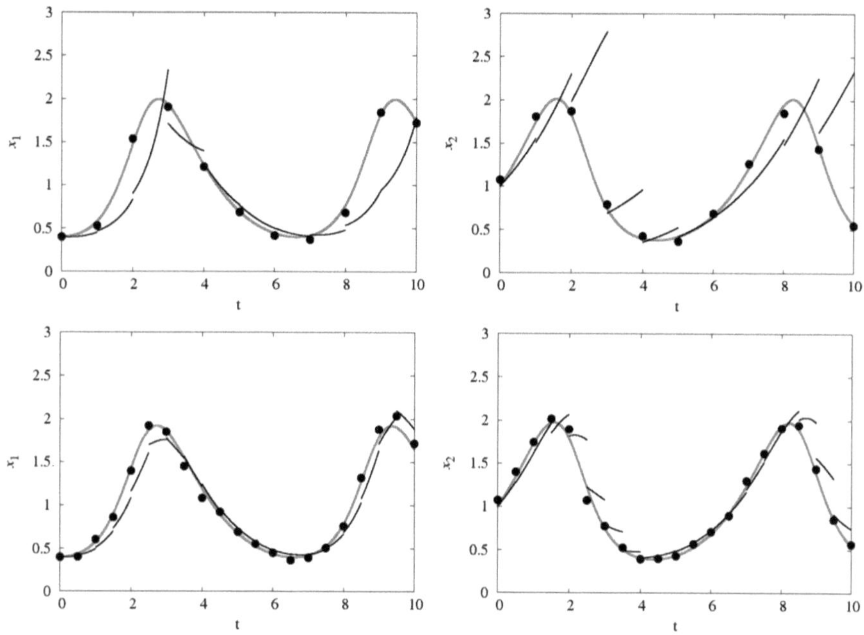

Fig. 6 Test problem 2, multiple shooting approach, trajectory after first iteration (*black line*), measurements (*black dots*), solution (*grey line*), *left*: $x_1(\cdot)$, *right*: $x_2(\cdot)$, *first row*: 10 multiple shooting nodes, *right*: 20 multiple shooting nodes

thus improving the convergence from arbitrary initial parameter guesses up to one step convergence for problems with dense not-noisy measurements for all states and parameters entering linearly the differential equations. Figure 6 show the multiple shooting trajectories after the first iteration. Obviously, the more measurements we have the nearer the multiple shooting trajectories are to the optimal solution.

7.3 FitzHugh-Nagumo Test Problem

We consider now a test problem developed by FitzHugh [24] and Nagumo et al. [39] as simplifications of the Hodgkin and Huxley model

$$\dot{V} = c(V - \frac{V^3}{3} + R), \quad \dot{R} = -\frac{1}{c}(V - a + bR), \quad t \in [0, 20],$$
$$V(0) = -1, \quad R(0) = 1$$

This test problem was used by Ramsay et al. [42] and characterized as complicated. The "true" parameter set is $a = 0.2$, $b = 0.2$ and $c = 3.0$. Measurement data is

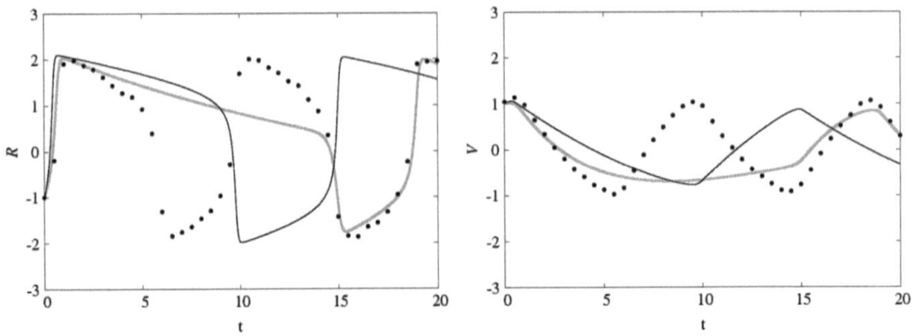

Fig. 7 FitzHugh-Nagumo test problem, single shooting approach, initial trajectory (*black line*), measurements (*black dots*), solution (*grey line*), starting guesses $a^{(0)} = 0.4$, $b^{(0)} = 0.1$ and $c^{(0)} = 6.0$, convergence to $a^* = 0.3709$, $b^* = 0.9153$ and $c^* = 4.3204$, *left*: *R*, *right*: *V*

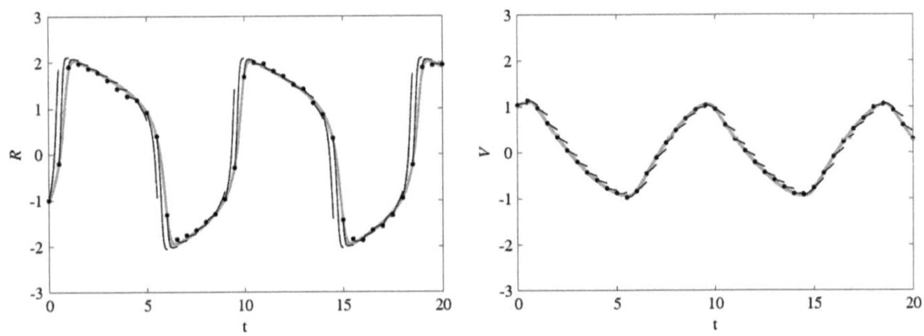

Fig. 8 FitzHugh-Nagumo test problem, multiple shooting approach, initial trajectory (*black line*), measurements (*black dots*), solution (*grey line*), starting guesses $a^{(0)} = 0.4$, $b^{(0)} = 0.1$ and $c^{(0)} = 6.0$, *left*: *R*, *right*: *V*

Table 3 Solution (after 5 iterations) and estimated error, FitzHugh-Nagumo test problem

	a	*b*	*c*
Solution	0.1998	0.2002	2.9990
Standard error	±0.0001	±0.0014	±0.0017

generated by adding a normally distributed pseudo random noise ($N(0, \sigma^2), \sigma = 0.01$) to the solution $x(t), t \in [0, 20]$ at measurement points $t_j = 0.5j, j = 1, \ldots, 10$. This test problem demonstrates that IVP approaches tend to converge to wrong local minima for poorly chosen parameter initial values. In the multiple shooting method this is widely avoided due to advantages of the multiple shooting and last but not least also due to the applied generalized Gauss-Newton method (Figs. 7 and 8; Table 3).

7.4 *Oscillating Chemical Reactions: The Oregonator Model*

The next test problem is the so-called oregonator [22, 23], which is regarded as the simplest realistic model of the chemical dynamics of the oscillatory Belousov-Zhabotinsky reaction and can be written in the scaled form

$$\begin{aligned}\varepsilon\frac{dx_1}{dt} &= qx_2 - x_1x_2 + x_1(1-x_1),\\ \varepsilon^1\frac{dx_2}{dt} &= -qx_2 - x_1x_2 + fx_3,\\ \frac{dx_3}{dt} &= x_1 - x_3.\end{aligned} \tag{37}$$

Defining the parameters

$$\begin{aligned}&p_1 = q/\varepsilon, \quad p_2 = -1/\varepsilon, \quad p_3 = 1/\varepsilon, \quad p_4 = -1/\varepsilon,\\ &p_5 = -q/\varepsilon^1, \quad p_6 = -1/\varepsilon^1, \quad p_7 = f/\varepsilon^1, \quad p_8 = 1, \quad p_9 = -1,\end{aligned}$$

we rewrite (37) as

$$\begin{aligned}\dot{x}_1 &= p_1x_2 + p_2x_1x_2 + p_3x_1 + p_4x_1^2,\\ \dot{x}_2 &= p_5x_2 + p_6x_1x_2 + p_7x_3,\\ \dot{x}_3 &= p_8x_1 + p_9x_3.\end{aligned} \tag{38}$$

Our aim is now to estimate the parameters p_1–p_9 of the ODE model (38). As in previous tests the measurements have been generated using "true" solution at time points $t_i = 0.5i$, $i = 1, \ldots, 20$ with adding normally distributed noise $(N(0, \sigma^2), \sigma = 0.01)$. The "true" parameter values have been calculated using values presented in [23]. The initial guesses for parameter have been taken as $p_i^{(0)} = 2p_i^{\text{“true”}}$, $i = 1, .., 9$. The initial values for the states have been taken as $x_1(0) = x_2(0) = x_3(0) = 1$. The parameter estimates resulted after 4 iterations using the multiple shooting are presented in Table 4. Figure 9 shows the solution

Table 4 Oregonator problem: "True" parameter, estimated values and error

	p_1	p_2	p_3	p_4	p_5
"True"	0.007697	−10.101010	10.101010	−10.101010	−3.848485
Solution	7.778×10^{-1}	-1.020×10^1	1.013×10^1	-0.101×10^2	-0.383×10^2
Error	$\pm 1.911 \times 10^{-1}$	$\pm 0.01131 \times 10^1$	$\pm 0.01268 \times 10^1$	$\pm 0.01973 \times 10^1$	$\pm 0.02590 \times 10^1$
	p_6	p_7	p_8	p_9	
"True"	−5050.505002	5050.505002	1.000000	1.000000	
Solution	-5.136×10^3	5.088×10^3	1.004	1.001	
Error	$\pm 0.01017 \times 10^3$	$\pm 0.01005 \times 10^3$	±0.0663	±0.0231	

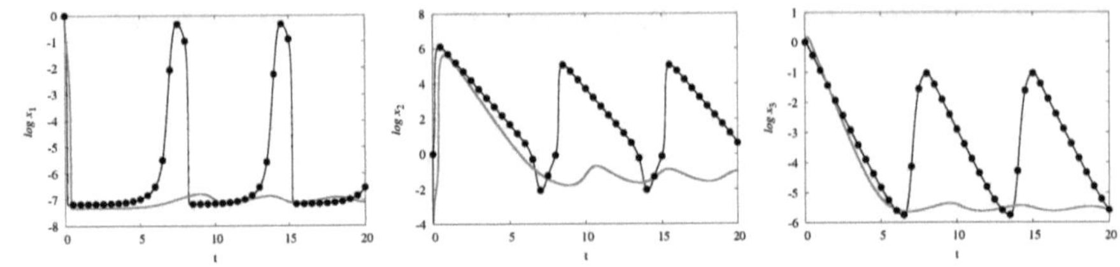

Fig. 9 Oregonator model, measurements (*black dots*), solution using the single (*black line*) and the multiple shooting (*grey line*), *left*: $\log x_1$, *middle*: $\log x_2$, *right*: $\log x_3$

trajectories for the single and multiple shooting approaches. Clearly, also in this test the single shooting converges to a wrong minimum.

7.5 Chaotic Test Problem

The last test problem is a parameter estimation problem with chaotic data. We consider an ODE model

$$\begin{aligned} \dot{x_1} &= \sum_{j=1}^{3}\sum_{k\geq j}^{3} a^1_{j,k} x_j x_k + \sum_{j_1}^{3} b^1_j x_j + c^1 \\ \dot{x_2} &= \sum_{j=1}^{3}\sum_{k\geq j}^{3} a^2_{j,k} x_j x_k x_j x_k + \sum_{j_1}^{3} b^2_j x_j + c^2 \\ \dot{x_3} &= \sum_{j=1}^{3}\sum_{k\geq j}^{3} a^3_{j,k} x_j x_k + \sum_{j_1}^{3} b^3_j x_j + c^3 \end{aligned} \tag{39}$$

with 30 unknown parameters $a^1_{j,k}$, b^1_j, c^1, $a^2_{j,k}$, b^2_j, c^2, $a^3_{j,k}$, b^3_j, c^3 and initial values $x_1(0), x_2(0), x_3(0)$. Note that the Lorenz model

$$\begin{aligned} \dot{x_1} &= 10x_2 - 10x_1, & x_1(0) &= 5.76540 \\ \dot{x_2} &= -x_1x_3 + 46x_1 - x_2, & x_2(0) &= 10.50547 \\ \dot{x_3} &= x_1x_2 - \tfrac{8}{3}x_3, & x_3(0) &= 30.58941 \end{aligned} \tag{40}$$

is a special case of the model (39). Our aim is to identify all 30 parameters from the data simulated by the Lorenz model (40) with adding normally distributed noise $(N(0,\sigma^2), \sigma = 0.01)$. The multiple shooting is initiated with parameters equal to 0. Results of the numerical experiments are presented in Fig. 10 and Table 5. We can

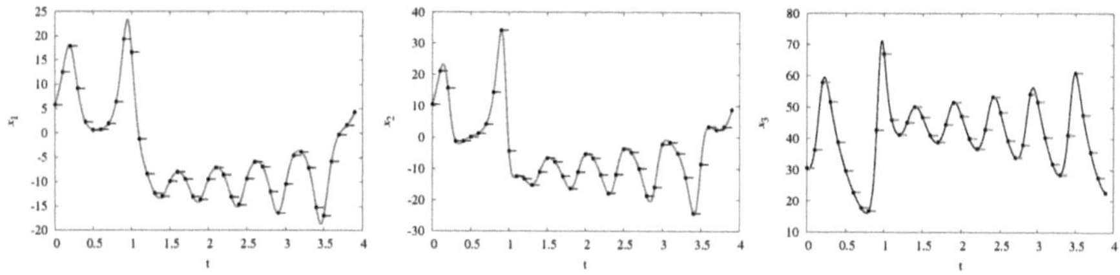

Fig. 10 Chaotic model, measurements (*black dots*), initial (*black line*) and solution (*grey line*) trajectories, multiple shooting approach, *left*: x_1, *middle*: x_2, *right*: x_3

Table 5 Chaotic test problem: "True" parameter, estimated values and error

	a_{11}^1	a_{12}^1	a_{13}^1	a_{22}^1	a_{23}^1
"True"	0	0	0	0	0
Solution	−0.0015	0.0014	−0.0014	−0.0004	0.0008
Standard error	± 0.0055	± 0.0029	± 0.0029	± 0.0024	± 0.0024
	a_{33}^1	b_1^1	b_2^1	b_3^1	c^1
"True"	0	−10	10	0	0
Solution	0.0004	−9.9242	9.9585	−0.0206	0.2817
Standard error	± 0.0024	± 0.0211	± 0.0169	± 0.0180	± 0.0690
	a_{11}^2	a_{12}^2	a_{13}^2	a_{22}^2	a_{23}^2
"True"	0	0	−1	0	0
Solution	−0.0006	−0.0000	−0.9993	0.0001	−0.0003
Standard error	± 0.0052	± 0.0047	± 0.0030	± 0.0025	± 0.0026
	a_{33}^2	b_1^2	b_2^2	b_3^2	c^2
"True"	0	46	−1	0	0
Solution	0.0000	45.9551	−0.9754	−0.0046	0.1129
Standard error	± 0.0023	± 0.0215	± 0.0177	± 0.0185	± 0.0764
	a_{11}^3	a_{12}^3	a_{13}^3	a_{22}^3	a_{23}^3
"True"	0	1	0	0	0
Solution	−0.0017	1.0015	0.0008	−0.0005	−0.0006
Standard error	± 0.0055	± 0.0048	± 0.0029	± 0.0024	± 0.0025
	a_{33}^3	b_1^3	b_2^3	b_3^3	c^3
"True"	0	0	0	$\frac{8}{3}$	0
Solution	0.0001	−0.0394	0.0283	−2.6694	0.0487
Standard error	± 0.0023	± 0.0207	± 0.0167	± 0.0176	± 0.0670

see that even with chaotic noisy data the inverse problem is well-posed and allows to identify all parameters relatively good. For more details and other examples see [1, 29].

8 Conclusions

We have presented a boundary value problem approach for parameter estimation problems in nonlinear ordinary differential equations based on multiple shooting as originally introduced by Bock in the 1970s. A special emphasis is placed on the theoretical analysis including numerical stability, grid condition and a posteriori error analysis. The advantages of multiple shooting versus single shooting have been illustrated by numerical examples.

References

1. Baake, E., Baake, M., Bock, H.G., Briggs, K.M.: Fitting ordinary differential equations to chaotic data. Phys. Rev. A **45**(8), 5524–5529 (1992)
2. Binder, T., Kostina, E.A.: Gauss-Newton methods for robust parameter estimation. In: Bock H.G., Carraro T., Jäger W., Körkel S., Phu H.X., Rannacher R., Schlöder J.P. (eds.) Model Based Parameter Estimation: Theory and Applications, pp. 55–87. Springer, New York (2013)
3. Binder, M., Sulaimanov, N., Clausznitzer, D., Schulze, M., Hber, C.M., Lenz, S.M., Schlöder, J.P., Trippler, M., Bartenschlager, R., Lohmann, V., Kaderali, L.: Replication vesicles are load- and choke-points in the hepatitis C virus lifecycle. PLoS Pathog. **9**(8), e1003561 (2013)
4. Bock, H.G.: Numerische Optimierung zustandsbeschränkter parameterabhängiger Prozesse, Diplomarbeit, Köln (1974)
5. Bock, H.G.: Zur numerischen Behandlung zustandsbeschränkter Steuerungsprobleme mit Mehrzielmethode und Homotopieverfahren. Z. Angew. Math. Mech. **57**, 266 (1977)
6. Bock, H.G.: A Multiple Shooting Method for Parameter Identification in Nonlinear Differential Equations. GAMM Tagung, Brüssel (1978)
7. Bock, H.G.: Numerical solution of nonlinear multipoint boundary value problems with applications to optimal control. Z. Angew. Math. Mech. **58**, 407 (1978)
8. Bock, H.G.: Numerical treatment of inverse problems in chemical reaction kinetics. In: Ebert, K.H., Deuflhard, P., Jäger, W. (eds.) Modelling of Chemical Reaction Systems. Springer Series Chemical Physics, vol. 18, p. 102. Springer, Heidelberg (1981)
9. Bock, H.G.: Recent advances in parameter identification techniques for ODE. In: Deuflhard, P., Hairer, E. (eds.) Progress in Scientific Computing, vol. 2. Birkhäuser, Boston (1983)
10. Bock, H.G.: Randwertproblemmethoden zur Parameteridentifizierung in Systemen nichtlinearer Differentialgleichungen, vol. 183. Bonner Mathematische Schriften, Bonn (1987)
11. Bock, H.G., Longman, R.W.: Computation of optimal controls on disjoint control setsastronomy societyration. In: Proceeding of the American Astronomy Society. Symposium on Engineering Science and Mechanics, Taiwan (1981)
12. Bock, H.G., Plitt, K.J.: A multiple shooting algorithm for direct solution of constrained optimal control problems. In: Proceeding of the 9th IFAC World Congress Automatic Control. Pergamon, Oxford (1984)
13. Bock, H.G., Schlöder, J.: Numerical solution of retarded differential equations with state-dependent time lags. Z. Angew. Math. Mech. **61**, 269 (1981)
14. Bock, H.G., Schlöder, J.P.: Recent progress in the development of algorithm and software for large-scale parameter estimation problems in chemical reaction systems. In: Kotobh, P. (ed.) Automatic Control in Petrol, Petrochemical and Desalination Industries. IFAC Congress. Pergamon, Oxford (1986)
15. Bock, H.G., Kostina, E.A., Schlöder, J.P., Gienger, G., Ziegler, G., Pallaschke, S.: Robust parameter estimation for identifying satellite orbits. In: Bock, H.G., Kostina, E.A., Phu, H.X., Rannacher, R. (eds.) Modeling, Simulation and Optimization of Complex Processes.

Proceeding of the International Conference on High Performance Scientific Computing. Springer, New York (2005)
16. Bock, H.G., Kostina, E., Schlöder, J.P.: Numerical methods for parameter estimation in nonlinear differential algebraic equations. GAMM Mitteilungen **30**(2), 376–408 (2007)
17. Bryson, A.E., Jr., Ho, Y.C.: Applied Optimal Control. Blaisdell, New York (1969)
18. Bulirsch, R., Stoer, J.: Die Mehrzielmethode zur numerischen Lösung von nichtlinearen Randwertproblemen und Aufgaben der optimalen Steuerung, Carl-Cranz-Gesellschaft, Technical Report (1971)
19. Deuflhard, P.: Recent advances in multiple shooting techniques. In: Gladwell, I., Sayers, D.K. (eds.) Computational Techniques for Ordinary Differential Equations. Academic, London (1980)
20. Dieses, A.E., Schlöder, J.P., Bock, H.G., Richter, O., Aden, K., Gottesbüren, B.: A parameter estimation tool for nonlinear transport and degradation processes of xenobiotics in soil. In: Human and Environmental Exposure to Xenobiotics. Proceeding of the XI Symposium Pesticide Chemistry, 12–15 September, Cremona (1999)
21. Dieses, A.E., Schlöder, J.P., Bock, H.G., Richter, O.: Parameter estimation for nonlinear transport and degradation processes of xenobiotica in soil. In: Keil, F., et al. (eds.) Scientific Computing in Chemical Engineering II, vol. 2, pp. 290–297. Springer, New York (1999)
22. Field, R.J., Noyes, R.M.: Oscillations in chemical systems. IV. Limit cycle behavior in a model of a real chemical reaction. J. Chem. Phys. **60**(5), 1877–1884 (1974)
23. Field, R.J.: Oregonator. Scholarpedia **2**(5), 1386 (2007)
24. FitzHugh, R.: Impulses and physiological states in models of nerve membrane. Biophys. J. **1**, 445–466 (1961)
25. Han, S.P.: A globally convergent method for nonlinear programming. J. Optim. Theory Appl. **22**, 297–309 (1977)
26. Haselgrove, C.B., Hoyle, F.: A mathematical discussion of the problem of stellar evolution. Mon. Not. R. Astron. Soc. **116**(5), 515–526 (1956). With reference to the use of an automatic digital computer
27. Hatz, K.: Efficient numerical methods for hierarchical dynamic optimization with application to cerebral palsy gait modeling. Dissertation, University of Heidelberg (2014)
28. Hatz, K., Schlöder, J.P., Bock, H.G.: Estimating parameters in optimal control problems. SIAM J. Sci. Comput. **34**(3), 1707–1728 (2012)
29. Kallrath, J., Bock, H.G., Schlöder, J.P.: Least squares parameter estimation in chaotic differential equations. Celest. Mech. Dyn. Astron. **56**, 353–371 (1993)
30. Kallrath, J., Altstädt, V., Schlöder, J.P., Bock, H.G.: Analysis of fatigue crack growth behaviour in polymers and their composites based on ordinary differential equation parameter estimations. Polym. Test. **18**, 11–35 (1999)
31. Keller, H.B.: Numerical Methods for Two-Point Boundary-Value Problems. Blaisdell, Waltham (1968)
32. Kostina, E.A.: Robust parameter estimation in dynamic dystems. Optim. Eng. **5**(4), 461–484 (2004)
33. Krämer-Eis, P.: Numerische Berechnung optimaler Feedback-Steuerungen bei nichtlinearen Prozessen, Diplomarbeit, Bonn (1980)
34. Lawson, C.L., Hanson, R.J.: Solving Least Squares Problems. Prentice Hall, Englewood Cliffs (1974)
35. Lenz, S.M., Bock, H.G., Schlöder, J.P., Kostina, E.A., Gienger, G., Ziegler, G.: Multiple shooting method for initial satellite orbit determination. AIAA J. Guid. Control. Dyn. **33**(5), 1334–1346 (2010)
36. Miele, A.: Gradient methods in control theory, part 6. Combined Gradient-Restoration Algorithm Rice University, Aero-Astronautics Report No. 74 (1970)
37. Miele, A.: Recent advances in gradient algorithms for optimal control problems. J. Optim. Theory Appl. **17**(5–6), 361–430 (1975)
38. Morrison, D.D., Riley, J.D., Zancanaro, J.F.: Multiple shooting method for two-point boundary value problems Commun. ACM **5**(2), 613–614 (1962)

39. Nagumo, J.S., Arimoto, S., Yoshizawa, S.: An active pulse transmission line simulating a nerve axon. Proc. IRE **50**, 2061–2070 (1962)
40. Osborne, M.R.: On shooting methods for boundary value problems. J. Math. Anal. Appl. **27**, 417–433 (1969)
41. Powell, M.J.D.: A method for nonlinear constraints in minimization problems. In: Fletcher, R. (ed.) Optimization. Academic, London (1969)
42. Ramsay, J.O., Hooker, G., Campbell, D., Cao, J.: Parameter estimation for differential equations: a generalized smoothing approach. J. R. Stat. Soc. B **69**(5), 741–796 (2007)
43. Scheer, E., Bock, H.G., Platt, U., Rannacher, R.: Retrieval of height profiles of trace gases by optimal parameter estimation. In: Varotsos, C. (ed.) NATO ASI Series I: Global Environmental Change, vol. 53, pp. 285–291. Springer, Berlin Heidelberg (1997)
44. Schlöder, J.: Zeitoptimale Orbit-Orbit-Transfers und Designstudien für Sonnensegelsatelliten, Diplomarbeit, Bonn (1980)
45. Schlöder, J., Conrads, A., Frank, T.: Neuere Verfahren zur Parameterschätzung dargestellt am Beispiel der Modellierung von Rübenwachstum. In: Möller, D.P.F. (ed.) Simulationstechnik, Proc. 3. Symposium Simulationstechnik, number 109 in Iformatik-Fachberichte, pp. 304–308. Springer, Berlin (1985)
46. Stoer, J.: On the numerical solution of constrained least squares problems. SIAM J. Numer. Anal. **8**, 382 (1971)
47. Stoer, J., Bulirsch, R.: Einführung in die Numerische Mathematik II. Springer, Berlin (1973)
48. von Schwerin, R., Winckler, M., Schulz, V.: Parameter estimation in discontinuous descriptor models. In: Bestle, D., Schiehlen, W. (eds.) IUTAM Symposium on Optimization of Mechanical Systems, pp. 269–276. Kluwer Academic, Heidelberg (1996)

Direct and Indirect Multiple Shooting for Parabolic Optimal Control Problems

Thomas Carraro and Michael Geiger

Abstract We present two multiple shooting approaches for optimal control problems (OCP) governed by parabolic partial differential equations (PDE). In the context of ordinary differential equations, shooting techniques have become a state-of-the-art solver component, whereas their application in the PDE case is still in an early phase of development. We derive both direct (DMS) and indirect (IMS) multiple shooting for PDE optimal control from the same extended problem formulation. This approach shows that they are algebraically equivalent on an abstract function space level. However, discussing their respective algorithmic realizations, we underline differences between DMS and IMS. In the numerical examples, we cover both linear and nonlinear parabolic side conditions.

1 Introduction

Multiple shooting methods have been extensively used during the past four decades to solve both ODE boundary value problems (BVP) and optimal control problems (OCP), and for the latter problem class different shooting techniques were developed, according to the general dichotomy of indirect and direct solution methods for OCP. In the context of PDE constrained OCP, where the past 15 years have seen a rapid development of both theoretical insights and solution algorithms, shooting methods are up to now rarely used despite their success in the ODE framework. Direct multiple shooting (DMS) or related time-domain decomposition methods are treated, e.g., in [13, 29], and an indirect shooting approach was introduced in [14]. All these articles employ shooting techniques focusing on specific aspects such as adaptivity or preconditioning, while our scope in this article is to show the derivation of DMS and IMS from the same extended problem formulation to underline their peculiarities in the PDE framework.

Shooting algorithms for parabolic OCP have to overcome additional difficulties caused mainly by the spatial variables. On the continuous level, the functional

T. Carraro (✉) • M. Geiger
Institute of Applied Mathematics, Heidelberg University, Heidelberg, Germany
e-mail: thomas.carraro@iwr.uni-heidelberg.de; michael.geiger@iwr.uni-heidelberg.de

T. Carraro et al. (eds.), *Multiple Shooting and Time Domain Decomposition Methods*, Contributions in Mathematical and Computational Sciences 9,
DOI 10.1007/978-3-319-23321-5_2

analytic setting (solution spaces, weak formulations etc.) has to be developed carefully; e.g., the initial values for the subinterval solutions are L^2 functions rather than $\mathbb{R}^n$ vectors as in the ODE case. On the discrete level, the usually high-dimensional spatial discretization leads to large-scale optimization problems and strongly advises the development of suitable adaptive techniques.

The two most attractive features of multiple shooting are its intrinsic stabilizing effect and its potential for parallelization. The former enables the solution of ill-conditioned problems where other methods fail; such instabilities are mirrored by local stability estimates such as

$$\|u(t;s_1) - u(t;s_2)\| \leq ce^{L(t-t_0)}\|s_1 - s_2\|$$

that are common in the analysis of initial value problems (IVP). Here, t is the independent variable normally interpreted as time, t_0 is the initial time-point, s_1 and s_2 are two parameter values (in multiple shooting methods, they denote initial values) and $u(t;s)$ is the solution depending on the parameter s. Furthermore, L is a fixed Lipschitz constant inherent to the problem; the value of the exponential factor can be controlled by splitting the solution interval into smaller parts as is common in multiple shooting.

The parallelizability of shooting techniques has been addressed and exploited in the PDE context by the so-called parareal method (developed in [22]) which has been shown to be equivalent to multiple shooting in [11]. Despite the results mentioned so far, there are many aspects that have not yet been systematically examined, which is why shooting techniques for PDE problems still constitute an interesting and promising subject.

A detailed presentation of indirect multiple shooting (IMS) for nonlinear parabolic OCP with additional control constraints has recently been given in [9]. In the current publication, we continue this work by comparing IMS and DMS techniques for the mentioned problem class, but without considering additional control or state constraints. The latter simplification prevents us from losing track of our main objective, namely to show the equivalence of IMS and DMS on an abstract function space level and the differences of their respective algorithmic concretization. Furthermore, as in the well studied ODE case, we expect IMS and DMS to behave differently in the presence of control and/or state constraints. In fact, in the ODE context the presence of state and control constraints has led to prefer DMS to IMS. An accurate performance comparison between DMS and IMS in PDE context is left as an interesting topic for further research.

In the literature of multiple shooting methods, the distinction between direct and indirect approaches is done according to two aspects: how the method is derived and at which stage the underlying system is discretized. According to the first aspect, the distinction is done between methods that are derived by certain optimality conditions (indirect approach) and methods that are not obtained by optimality conditions (direct approach). Considering the second aspect, the same methods are classified as indirect if they use the 'first optimize then discretize' approach or direct if they follow the 'first discretize, then optimize' approach.

Formerly in the ODE context, to develop indirect methods a Hamiltonian functional was introduced and the optimality conditions were derived following the Pontryagin Maximum Principle [25] and introducing adjoint (or co-state) variables. Numerical methods for this approach lead to a boundary value problem for the state and adjoint variables. As a result, in indirect methods the control is not present in the shooting system. On the contrary, in direct methods the control variable is included in the shooting system, leading to a procedure that in the multiple shooting context is also called 'all-at-once' approach.

Following the other distinction between indirect and direct methods, namely the differentiation between the approaches that optimize first and approaches that discretize first, in indirect methods first the optimality conditions are derived at the continuous level, and then a discretization method is used to derive a finite dimensional system that can be solved numerically. On the contrary, in direct methods the state and control variables are discretized first and typically the discrete control space has low dimension. Then, a method for nonlinear programming problems is applied to the resulting discrete system [18, 28]. According to this distinction, direct methods are not derived in a function space setting. This has the advantage of avoiding the definition of adjoint variables in function spaces, especially if state constraints are included in the optimization problem.

We do not want to discuss the several arguments that indicate advantages and disadvantages of either one method or the other, we will rather discuss the derivation of the two methods starting from the same formulation at the continuum level. Therefore, we give a unique argumentation to define whether a method is 'direct' or 'indirect' allowing to distinguish the two approaches without recurring to specific discretization methods. Proceeding like this has practical consequences because keeping the derivation at the continuous level will allow for example to derive error estimation methods in PDE context for the specific discretization of choice. This opens up new directions for future research.

The remainder of this contribution is oriented along the following outline: In the next section, we recapitulate the notational framework for PDE optimal control. Section 3 presents the OCP in a slightly modified but equivalent form which provides the suitable context for shooting methods. The KKT conditions of this extended OCP formulation are the starting point for both IMS and DMS methods on a function space level, which are described in separate sections. We start Sect. 4 by introducing a second variant of DMS which is common in ODE and DAE governed optimal control (here called 'classical' DMS); the connection between these two DMS approaches constitutes the main result of the section. After some brief remarks on discretization in Sect. 5, we discuss the effects of the differences between IMS and DMS in Sect. 6 by considering two concrete numerical examples. Some concluding remarks as well as an outlook toward possible further research constitute the final section.

2 Preliminaries

The general structure of an OCP requires the minimization of an objective functional, where the minimum is sought in a set of functions u given as solutions of a differential equation which depends on a control quantity q:

$$\min_{(q,u)} J(q,u) \text{ subject to } e(q,u) = 0. \tag{1}$$

In many problems, one has to deal with additional constraints to the control and/or state variables of the form $c(q,u) \leq 0$ which make the problems more difficult to solve. We skip them in order to avoid an excessive notation and in order not to lose track of our actual objective, namely comparing IMS and DMS methods. Concerning multiple shooting for parabolic OCP with control constraints $c(q) \leq 0$, we refer to [9].

We now describe in detail the parabolic OCP which is considered throughout this article. In this context, we have to deal with the following theoretical setting. The OCP reads in detail:

$$\min_{(q,u)} J(q,u) = \kappa_1 J_1(u) + \kappa_2 J_2(u(T)) + \frac{\alpha}{2}\|q\|_Q^2, \tag{2}$$

subject to the parabolic PDE

$$\partial_t u(x,t) + \mathscr{A}(u(x,t)) + \mathscr{B}(q(x,t)) = f(x,t) \quad \text{in } \Omega \times I, \tag{3a}$$

$$u(x,0) = u_0(x) \quad \text{in } \Omega \tag{3b}$$

We discuss the constituent parts of this formulation separately. Therefore, we assume that $V \hookrightarrow H = H^* \hookrightarrow V^*$ is a Gelfand triple of Hilbert spaces of functions on Ω (where the superscript $*$ denotes duality of spaces) and R is a suchlike Banach space. In the objective functional $J(q,u)$, we eliminate either $J_1(u)$, which we always assume to be of tracking type $\int_I \|u(t) - \hat{u}(t)\|_V^2\, dt$, or the end-time matching term $J_2(u(T)) := \|u(T) - \hat{u}_T\|_H^2$ by imposing the conditions $\kappa_i \in \{0, \frac{1}{2}\}, \kappa_1 \neq \kappa_2$. The term $\frac{\alpha}{2}\|q\|_Q^2$ serves as a regularization term, and α is the usual regularization parameter.

The computational domain of our problem is a space-time cylinder $\Omega \times I$ with a bounded convex polygonal or polyhedral spatial domain $\Omega \subset \mathbb{R}^d$ with $d \in \{1,2,3\}$ and a finite time interval $I = (0,T)$. The function spaces for state and control variables are usually Bochner spaces of the type $W(I;Y)$ where the time variable t is mapped into a Banach space Y.

In the above function space framework, the natural setting for the parabolic PDE (3a) is the following: For given $q(x,t) \in Q := L^2(I;R)$ and righthand side $f(x,t) \in L^2(I;V^*)$, find a state function $u(x,t)$ that satisfies (3) obeying additionally imposed suitable boundary conditions. Under these structural assumptions, the solution space for $u(x,t)$,

$$X := \{v(x,t) \in L^2(I;V) \mid \partial_t v(x,t) \in L^2(I;V^*)\}, \tag{4}$$

is known to be continuously embedded into the space $C(\bar{I};H)$ of temporally continuous functions with values in H (see, e.g., [10]), which means that an initial condition $u_0(x) \in H$ is well-defined. The differential operator $\mathscr{A} : X \to L^2(I;V^*)$ may be linear or nonlinear, whereas $\mathscr{B} : L^2(I;R) \to L^2(I;V^*)$ is usually linear and often simply an injection operator, given $R \hookrightarrow V^*$.

For a weak formulation of (3) we need some preparatory definitions. If $\overline{\mathscr{A}} : V \to V^*$ and $\overline{\mathscr{B}} : R \to V^*$ are pointwise-in-time operators corresponding to $\mathscr{A}$ and $\mathscr{B}$, respectively, we assume the elliptic operator $\overline{\mathscr{A}}$ to be coercive and define the following scalar products and semilinear forms:

$$((u,\varphi))_I := \int_I (u(t),\varphi(t))_H \, dt, \qquad a_I(u)(\varphi) := \int_I \langle \overline{\mathscr{A}}(u(t)), \varphi(t)\rangle_{V^* \times V} \, dt,$$

$$b_I(q)(\varphi) := \int_I \langle \overline{\mathscr{B}}(q(t)), \varphi(t)\rangle_{V^* \times V} \, dt.$$

We normally omit the index I denoting the integration interval if it is evident from the context. In this notational framework, the weak formulation of (3) reads: Find $u \in X$, so that for all $\varphi \in X$

$$((\partial_t u, \varphi)) + a(u)(\varphi) + b(q)(\varphi) + (u(0), \varphi(0)) = ((f,\varphi)) + (u_0, \varphi(0)), \tag{6}$$

where we included the initial condition (3b) weakly.

Remark 1 Conditions which guarantee that linear parabolic equations of type (3) or (6) possess a unique solution are provided in [16]. This framework is based on the underlying elliptic case which was presented, e.g., in [32]. We emphasize that both our theoretical observations for the linear case in the current and next sections and our linear example Sect. 6.1 fulfil the mentioned conditions and therefore allow for unique solutions.

The solution of the OCP is known to be among the stationary points of the Lagrange functional

$$\mathscr{L}(q,u,z) := J(q,u) + e(q,u;z) \tag{7}$$

where $e(q,u;z)$ is an abbreviation for the weakly formulated PDE side condition (6),

$$e(q,u;z) := ((\partial_t u, z)) + a(u)(z) + b(q)(z) - ((f,z)) + (u(0) - u_0, z(0)),$$

which sometimes enables a compact presentation on a more abstract level, see the discussion in Sect. 4. The Lagrange multiplier $z \in X$ denotes the solution of the adjoint equation $\mathscr{L}_u'(\delta u) = 0$ which naturally arises as part of the following optimality conditions:

$$\mathscr{L}_z'(\delta z) = ((\partial_t u, \delta z)) + a(u)(\delta z) + b(q)(\delta z) \\ -((f, \delta z)) + (u(0) - u_0, \delta z(0)) = 0, \tag{8a}$$

$$\mathscr{L}_u'(\delta u) = J_u'(q, u)(\delta u) - ((\partial_t z, \delta u)) + a_u'(u)(\delta u, z) + (z(T), \delta u(T)) = 0, \tag{8b}$$

$$\mathscr{L}_q'(\delta q) = J_q'(q, u)(\delta q) + b_q'(q)(\delta q, z) = 0. \tag{8c}$$

This so-called KKT system consists of the derivatives of (7) that form the state, adjoint and control equations. In the next section, we rewrite the above OCP in a form that is more suited to the derivation of multiple shooting algorithms.

Remark 2 The regularity of the Lagrange multiplier z is in general a delicate matter. Due to the structure of our objective functional, which may either be a temporally distributed L^2-term or an L^2-term at the final timepoint T, in our case the adjoint variable z lies in the same space X as the state variable u. A similar argument reveals the regularity of the adjoint variables $\lambda \in H$ in Sect. 3.1.

3 Multiple Shooting Methods for Parabolic OCP

As a suitable starting point for our observations we have introduced a general parabolic OCP which we extend, in Sect. 3.1, to a formulation tailored to the derivation of multiple shooting. The remainder of this section will be concerned with two variants of multiple shooting that are common in ODE optimal control. We will embed them into the context of parabolic OCP, thereby taking into account the additional challenges arising during the transfer from ODE to PDE (see Sect. 3.2 for IMS and Sect. 3.3 for DMS). Instead of only emphasizing the differences between these multiple shooting approaches, it is our objective to also show that they are rooted in one single common problem formulation and merely constitute different algorithmic realizations that may render either IMS or DMS preferable in concrete situations, depending on the problem at hand. We compare both shooting techniques in the concluding Sect. 3.4.

Remark 3 For simplicity, we assume the PDE side conditions of the OCP occurring in the following sections to be uniquely solvable, which has to be justified separately in every concrete problem. Furthermore, we assume also the OCP themselves to be uniquely solvable, which is guaranteed in a linear-quadratic framework on a convex domain but in general has to be verified. By these assumptions we avoid a detailed discussion of theoretical issues; for more information, we refer to the textbooks [16, 30].

3.1 The Modified Formulation of the Optimal Control Problem

The modification of the OCP (2) subject to (6) relies on a decomposition of the closure $\bar{I}$ of the interval I,

$$\bar{I} = \{\tau_0\} \cup \bigcup_{j=0}^{M-1} I_j, \quad I_j = (\tau_j, \tau_{j+1}], \tag{9}$$

where $\tau_0 = 0$ and $\tau_M = T$, and the subsequent redefinition of the OCP in terms of local control and state functions q^j, u^j on the subintervals I_j, which lie in the spaces $Q^j := L^2(I_j; R)$ and $X^j := \{v \in L^2(I_j; V) \mid \partial_t v \in L^2(I_j; V^*)\}$, respectively. For a more global view on these intervalwise problems, we define the compositions $\mathbf{u} = ((u^j)_{j=0}^{M-1})$ and $\mathbf{q} = ((q^j)_{j=0}^{M-1})$ as well as the corresponding spaces

$$\mathbf{X} := \bigtimes_{j=0}^{M-1} X^j, \quad \mathbf{Q} := \bigtimes_{j=0}^{M-1} Q^j.$$

We note that $\mathbf{X} = \{v \in L^2(I; V) \mid v|_{I_j} \in X^j\}$ and $\mathbf{Q} = \{q \in L^2(I; R) \mid q|_{I_j} \in Q^j\}$, which implies $X \subsetneq \mathbf{X}$ and $Q = \mathbf{Q}$. With these notations, the modified control problem reads:

$$\begin{aligned} \min_{(\mathbf{q},\mathbf{u})} \bar{J}(\mathbf{q}, \mathbf{u}) :&= \sum_{j=0}^{M-1} J^j(q^j, u^j) = \kappa_1 \sum_{j=0}^{M-1} \int_{I_j} \|u^j - \hat{u}|_{I_j}\|_V^2 \, dt \\ &+ \kappa_2 \|u^{M-1}(\tau_M) - \hat{u}_T\|_H^2 + \frac{\alpha}{2} \sum_{j=0}^{M-1} \int_{I_j} \|q^j\|_Q^2 \, dt \end{aligned} \tag{10a}$$

$$\begin{aligned} \text{s. t.} \quad & ((\partial_t u^j, \varphi)) + a(u^j)(\varphi) + b(q^j)(\varphi) - ((f|_{I_j}, \varphi)) \\ & + (u^j(\tau_j) - s^j, \varphi(\tau_j)) = 0 \ \text{ for } j \in \{0, \dots, M-1\}. \end{aligned} \tag{10b}$$

In this formulation, Eq. (10b) represent IVP on the subintervals I_j. However, we do not know the exact values $u(\tau_j)$ and therefore have to impose artificial initial values $\mathbf{s} = (s^j)_{j=0}^{M} \in H^{M+1}$. This leads to jumps in the global solution $\mathbf{u}$ composed of the interval solutions (i.e. $\mathbf{u}|_{I_j} \equiv u^j$). Therefore, problem (10) cannot be equivalent to the original OCP, because $\mathbf{u} \notin C(\bar{I}; H)$, whereas the solution $u \in X$ of the global OCP has to be continuous on I due to the above mentioned embedding $X \hookrightarrow C(\bar{I}; H)$. This matches the fact that X is a proper subset of $\mathbf{X}$. We therefore have to enforce the global continuity of the solution $\mathbf{u}$ of (10) by imposing the following additional continuity conditions:

$$(s^0 - u_0, \phi) = 0 \quad \forall \phi \in H, \tag{11a}$$

$$(s^{j+1} - u^j(\tau_{j+1}), \phi) = 0 \quad \forall \phi \in H, j \in \{0, \dots, M-1\}. \tag{11b}$$

Before we prove the equivalence of the original OCP and the extended problem formulation (10)–(11), we state the following preparatory lemma.

Lemma 1 *The objective functionals $J(q,u)$ and $\overline{J}(\mathbf{q},\mathbf{u})$ coincide for $\mathbf{u} = ((u^j)_{j=0}^{M-1})$ with $u^j = u|_{I_j}$, i.e. for globally continuous intervalwise defined functions* $\mathbf{u}$.

Proof Due to the additivity of integration on subintervals, we obtain

$$J(q,u) = \kappa_1 \int_I \|u(t) - \hat{u}(t)\|_V^2 \, dt + \kappa_2 \|u(T) - \hat{u}_T\|_H^2 + \frac{\alpha}{2} \int_I \|q(t)\|_R^2 \, dt$$

$$= \kappa_1 \sum_{j=0}^{M-1} \int_{I_j} \|u^j(t) - \hat{u}|_{I_j}(t)\|_V^2 \, dt + \kappa_2 \|u^{M-1}(\tau_M) - \hat{u}_T\|_H^2 + \frac{\alpha}{2} \sum_{j=0}^{M-1} \int_{I_j} \|q^j(t)\|_R^2 \, dt.$$

This corresponds to $\overline{J}(\mathbf{q},\mathbf{u}) = \sum_{j=0}^{M-1} J^j(q^j, u^j)$. □

The following theorem states the equivalence of the original and the modified OCP.

Theorem 1

(*a*) *Let $(q,u) \in Q \times X$ be a solution to the original OCP* (2) *subject to* (6)*. Then $(\mathbf{q},\mathbf{u}) \in \mathbf{Q} \times \mathbf{X}$, defined by $q^j := q|_{I_j}$ and $u^j := u|_{I_j}$, is a solution to the modified OCP* (10)–(11).

(*b*) *Let, on the other hand, $(\mathbf{q},\mathbf{u}) \in \mathbf{Q} \times \mathbf{X}$ solve the modified problem* (10)–(11)*. If we define q by $q|_{I_j} := q^j$ and u by $u|_{I_j} := u^j$, then $(q,u) \in Q \times X$ solves the original OCP* (2) *subject to* (6).

Proof

(a) Since $u \in X$ is globally continuous in time, we have $s^0 = u_0$ as well as $s^{j+1} = u^{j+1}(\tau_{j+1}) = u(\tau_{j+1})$ and $u^j(\tau_{j+1}) = u(\tau_{j+1})$, which means in turn $s^{j+1} = u^j(\tau_{j+1})$. Thus, the matching conditions (11) are fulfilled. Let now $(\tilde{\mathbf{q}}, \tilde{\mathbf{u}}) = ((\tilde{q}^j, \tilde{u}^j)_{j=0}^{M-1}) \in \mathbf{Q} \times \mathbf{X}$ such that $\overline{J}(\tilde{\mathbf{q}}, \tilde{\mathbf{u}}) < \overline{J}(\mathbf{q},\mathbf{u})$ and the continuity conditions (11) are fulfilled. The latter assumption immediately implies $\tilde{\mathbf{u}} \in X$, i.e. $(\tilde{q}, \tilde{u}) := (\tilde{\mathbf{q}}, \tilde{\mathbf{u}}) \in Q \times X$ due to $\mathbf{Q} = Q$. Lemma 1 now yields

$$J(\tilde{q}, \tilde{u}) = \overline{J}(\tilde{\mathbf{q}}, \tilde{\mathbf{u}}) < \overline{J}(\mathbf{q},\mathbf{u}) = J(q,u)$$

which is a contradiction to the assumed optimality of (q,u).

(b) Since $\mathbf{u}$ is part of a solution of the modified OCP, especially (11), we know that $s^0 = u_0$ and $s^{j+1} = u^j(\tau_{j+1})$. The initial value s^{j+1} on I_{j+1} clearly fulfils $s^{j+1} = u^{j+1}(\tau_{j+1})$. From $\mathbf{u} \in \mathbf{X}$ we know that $u^j \in C(\overline{I}_j; H)$, and together with the global continuity we know $\mathbf{u} \in C(\bar{I}; H)$. Considering $\partial_t u^j \in L^2(I_j; V^*)$, the corresponding global property $\partial_t u \in L^2(I; V^*)$ now directly follows. This means that u, defined by $u|_{I_j} := u^j$, lies in X, and together with q (analogously defined by $q|_{I_j} := q^j$) we obtain $(q, u) \in Q \times X$. Assuming that there is $(\tilde{q}, \tilde{u}) \in Q \times X$ with $J(\tilde{q}, \tilde{u}) < J(q, u)$, we get

$$\bar{J}(\tilde{\mathbf{q}}, \tilde{\mathbf{u}}) = J(\tilde{q}, \tilde{u}) < J(q, u) = \bar{J}(\mathbf{q}, \mathbf{u})$$

by Lemma 1, which is a contradiction to the optimality of $(\mathbf{q}, \mathbf{u})$. □

Agreement *To avoid a cumbersome case-by-case analysis, we assume for the rest of this contribution that all considerations are based on a distributed objective functional corresponding to* $\kappa_1 = \frac{1}{2}$ *and* $\kappa_2 = 0$. *The necessary modifications in case of an end-time functional (where* $\kappa_1 = 0$ *and* $\kappa_2 = \frac{1}{2}$*) are straightforward and will be covered in brief remarks.*

The reformulated problem (10)–(11) is the starting point for multiple shooting algorithms. In order to state the IMS and DMS methods properly, we have to derive the first order necessary optimality conditions of the modified OCP. Therefore, we first define the corresponding Lagrange functional, which is an extended version of (7) where the equality constraints (11) are considered in addition. We have in detail:

$$\begin{aligned}\mathscr{L}((q^j, u^j, z^j)_{j=0}^{M-1}, (s^j, \lambda^j)_{j=0}^{M}) := \sum_{j=0}^{M-1} J^j(q^j, u^j)&\\ + \sum_{j=0}^{M-1} [((\partial_t u^j, z^j)) + a(u^j)(z^j) + b(q^j)(z^j) - ((f|_{I_j}, z^j))]&\qquad (13)\\ + \sum_{j=0}^{M-1} (u^j(\tau_j) - s^j, z^j(\tau_j)) + \sum_{j=0}^{M-1} (s^{j+1} - u^j(\tau_{j+1}), \lambda^{j+1}) + (s^0 - u_0, \lambda^0)&\end{aligned}$$

The cost functional (10a) has been rearranged in an intervalwise fashion where all addends are structured alike. We have two kinds of Lagrange multipliers, the adjoint variables $\mathbf{z} = ((z^j)_{j=0}^{M-1}) \in \mathbf{X}$ corresponding to the intervalwise PDE side condition, and, newly, the spatial functions $\boldsymbol{\lambda} = (\lambda^j)_{j=0}^{M} \in H^{M+1}$ as multipliers for the equality constraints (11). We are now able to derive the first order optimality conditions, the so-called KKT system, by differentiating the above Lagrangian w.r.t. all its arguments. This yields, for all test functions $(\delta z, \delta u, \delta q, \delta\lambda, \delta s) \in X^j \times X^j \times Q^j \times H \times H$ and for all $j \in \{0, \cdots, M-1\}$, the intervalwise equations

$$\begin{aligned}\mathscr{L}'_{z^j}(\delta z) = ((\partial_t u^j, \delta z)) + a(u^j)(\delta z) + b(q^j)(\delta z)&\\ -((f|_{I_j}, \delta z)) + (u^j(\tau_j) - s^j, \delta z(\tau_j)) = 0,&\qquad (14a)\end{aligned}$$

$$\mathscr{L}'_{u^j}(\delta u) = J^{j'}_u(q^j, u^j)(\delta u) - ((\partial_t z^j, \delta u)) + a'_u(u^j)(\delta u, z^j)$$
$$+(z^j(\tau_{j+1}) - \lambda^{j+1}, \delta u(\tau_{j+1})) = 0, \quad (14b)$$
$$\mathscr{L}'_{q^j}(\delta q) = J^{j'}_q(q^j, u^j)(\delta q) + b'_q(q^j)(\delta q, z^j) = 0, \quad (14c)$$
$$\mathscr{L}'_{\lambda^0}(\delta\lambda) = (s^0 - u_0, \delta\lambda) = 0, \quad (14d)$$
$$\mathscr{L}'_{\lambda^j}(\delta\lambda) = ((s^{j+1} - u^j(\tau_{j+1}), \delta\lambda) = 0, \quad (14e)$$
$$\mathscr{L}'_{s^j}(\delta s) = ((\lambda^j - z^j(\tau_j), \delta s) = 0, \quad (14f)$$
$$\mathscr{L}'_{s^M}(\delta s) = (\lambda^M, \delta s) = 0. \quad (14g)$$

Remark 4 The last equation (14g) reflects a homogeneous initial condition at the final time-point $\tau_M = T$ [see also the original adjoint equation (8b)], whereas the terms $J^{j'}_u(q^j, u^j)(\delta u)$ serve as righthand sides for the intervalwise adjoints. In case of an end-time functional, (14g) comprises an extra term describing the initial condition for the adjoint equation, whereas the derivatives of the distributed functional terms w.r.t. u in (14b) vanish.

This system of equations can be split into two parts. The first one, Eqs. (14a)–(14c), correspond to the KKT system of the original problem (2) subject to (6), but restricted to a subinterval I_j [compare these equations to (8)]. The corresponding unknowns u^j, z^j and q^j are functions depending on spatial variables and time. The second part consists of Eqs. (14d)–(14g) and appears only in our problem reformulation. The unknowns s^j and λ^j are spatial functions in the isolated time-points τ_j.

Stationary points of the Lagrangian, i.e. solutions of (14), are solution candidates for the modified OCP. The KKT system constitutes a root-finding problem, which can, e.g., be handled by Newton's method. For this purpose, we need the second derivatives of the extended Lagrange functional (13), which is the Jacobian of the optimality conditions (14). We recall that Newton's method for solving a nonlinear but continuously differentiable problem $f(x) = 0$ consists in the iteration

$$x_{k+1} = x_k - J_f(x_k)^{-1} f(x_k)$$

initialized by a suitable starting point x_0. To avoid inverting the Jacobian J_f, this is usually rewritten in the two-step form

$$J_f(x_k) \cdot \delta x = -f(x_k), \quad (15a)$$
$$x_{k+1} = x_k + \delta x. \quad (15b)$$

The linear system displayed in the following is merely a formal representation of (15a) transferred to our context. We therefore rearranged Eq. (14) in a way that facilitates the illustration of IMS and DMS concepts in the subsequent sections but is not performed in practice:

$$\begin{pmatrix} 0 & \mathcal{L}''_{uz} & \mathcal{L}''_{qz} & \mathcal{L}''_{sz} & 0 \\ \mathcal{L}''_{zu} & \mathcal{L}''_{uu} & 0 & 0 & \mathcal{L}''_{\lambda u} \\ \mathcal{L}''_{zq} & 0 & \mathcal{L}''_{qq} & 0 & 0 \\ \mathcal{L}''_{zs} & 0 & 0 & 0 & \mathcal{L}''_{\lambda s} \\ 0 & \mathcal{L}''_{u\lambda} & 0 & \mathcal{L}''_{s\lambda} & 0 \end{pmatrix} \begin{pmatrix} \delta z \\ \delta u \\ \delta q \\ \delta s \\ \delta \lambda \end{pmatrix} = - \begin{pmatrix} \mathcal{L}'_{z} \\ \mathcal{L}'_{u} \\ \mathcal{L}'_{q} \\ \mathcal{L}'_{s} \\ \mathcal{L}'_{\lambda} \end{pmatrix}. \tag{16}$$

The righthand side of (16) consists of block vectors, i.e., the components of $\mathcal{L}'_z$, e.g., are the subinterval state equations, i.e. $\mathcal{L}'_z = (\mathcal{L}'_{z^0}, \cdots, \mathcal{L}'_{z^{M-1}})^\top$. Analogously, each of the solution variables is a block vector consisting of subinterval update values (e.g., $\delta q = (\delta q^{(0)}, \cdots, \delta q^{(M-1)})^\top$). The blocks of the matrix are either zero submatrices in case the equation to be differentiated does not depend on the variable w.r.t. which we differentiate, or they are sparse (often diagonal, or, after discretization, block diagonal) matrices due to the decoupling of the component equations of (14) between different subintervals. We underline that in the context of large scale parabolic OCP this system is never assembled explicitly due to its huge size. The multiple shooting techniques derived in the following sections rely on different splittings of system (16) which reduce its size significantly. Nevertheless, we will still not assemble the corresponding smaller matrices, but employ Krylov-Newton methods that allow to solve the respective Newton equations in a matrix-free manner.

Remark 5 For a discussion of the size of system (16), we exemplarily describe the matrix in more detail. The upper left 3×3 block consists of nine quadratic $M \times M$ blocks, whereas the lower right 2×2 block comprises four $(M+1) \times (M+1)$ blocks. The remaining submatrices are rectangular matrices of appropriate dimension. Summarizing, the Newton matrix is of size $(5M + 2) \times (5M + 2)$. Assuming the number M of subintervals I_j to be of moderate size (from $M = 1$ in the case of simple shooting up to $M \approx 10{-}100$), the system appears to be small. We emphasize, however, that so far we are still situated in a function space environment, i.e. up to now we considered neither time nor space discretization. We will see later in Sect. 5 that especially the discretization of the spatial variables leads to a huge enlargement of the systems that have to be solved numerically (cf. also Remark 12).

3.2 Indirect Multiple Shooting

We start with indirect shooting and describe first the overall structure of the method. As stated above, we seek ways of splitting the solution process of the linear system (16) (respectively, of its discrete counterpart). One such splitting leads to IMS, which is structured like a two-step (fixed-point) iteration. Furthermore, we discuss some algorithmic details that can also be found in [9]; therefore, we keep the presentation rather short.

3.2.1 Structure

In the Newton system (16), all variables u^j, z^j, q^j, s^j and λ^j are independent. We regroup them according to the following scheme,

$$\left(\begin{array}{ccc|cc} 0 & \mathscr{L}''_{uz} & \mathscr{L}''_{qz} & \mathscr{L}''_{sz} & 0 \\ \mathscr{L}''_{zu} & \mathscr{L}''_{uu} & 0 & 0 & \mathscr{L}''_{\lambda u} \\ \mathscr{L}''_{zq} & 0 & \mathscr{L}''_{qq} & 0 & 0 \\ \hline \mathscr{L}''_{zs} & 0 & 0 & 0 & \mathscr{L}''_{\lambda s} \\ 0 & \mathscr{L}''_{u\lambda} & 0 & \mathscr{L}''_{s\lambda} & 0 \end{array}\right) \begin{pmatrix} \delta z \\ \delta u \\ \delta q \\ \hline \delta s \\ \delta\lambda \end{pmatrix} = - \begin{pmatrix} \mathscr{L}'_z \\ \mathscr{L}'_u \\ \mathscr{L}'_q \\ \hline \mathscr{L}'_s \\ \mathscr{L}'_\lambda \end{pmatrix},$$

thus creating two subsystems and introducing inherent dependencies between the variables. In a **first solution step**, we fix $\mathbf{s} = (s^j)_{j=0}^M$ and $\boldsymbol{\lambda} = (\lambda^j)_{j=0}^M$ and solve the intervalwise boundary value problems

$$((\partial_t u^j, \delta z)) + a(u^j)(\delta z) + b(q^j)(\delta z) - ((f|_{I_j}, \delta z)) \\ + (u^j(\tau_j) - s^j, \delta z(\tau_j)) = 0, \tag{17a}$$

$$J_u^{j'}(q^j, u^j)(\delta u) - ((\partial_t z^j, \delta u)) + a'_u(u^j)(\delta u, z^j) \\ + (z^j(\tau_{j+1}) - \lambda^{j+1}, \delta u(\tau_{j+1})) = 0, \tag{17b}$$

$$J_q^{j'}(q^j, u^j)(\delta q) + b'_q(q^j)(\delta q, z^j) = 0. \tag{17c}$$

These equations correspond to $\mathscr{L}'_{z^j} = 0, \mathscr{L}'_{u^j} = 0$ and $\mathscr{L}'_{q^j} = 0$ in (14). The variables u^j, z^j and q^j do now depend on s^j and λ^{j+1}. The BVP character of the intervalwise problems results from the forward-backward structure of the state and adjoint equations: s^j is the initial value for u^j at τ_j, and λ^{j+1} is the initial value for z^j at the subinterval endpoint τ_{j+1}.

Remark 6 It is an interesting problem in its own right how to choose s^j and λ^{j+1}, as the quality of the initial choice certainly influences the convergence of Newton's method. For ODE problems, there have been several suggestions; e.g., additional information on the solution, if available, could improve the initial guesses (for an example, see [8]). Alternatively, one could employ homotopy methods as in [23].

The states $u^j(s^j, \lambda^{j+1})$ and $z^j(s^j, \lambda^{j+1})$ are coupled via the control equation, and together the three equations yield the same structure on each subinterval as the global KKT system (8). However, the intervalwise solutions do not fit together at the subinterval endpoints, thus the solution is globally discontinuous due to the artificially chosen initial values s^j and λ^{j+1}. This contradicts the embedding $X \hookrightarrow C(\bar{I}; H)$. Therefore, we use the local solutions $u^j(s^j, \lambda^{j+1}), z^j(s^j, \lambda^{j+1})$ and $q^j(s^j, \lambda^{j+1})$ in order to update s^j and λ^{j+1} in the **second solution step**. This consists in solving the following system (corresponding to $\mathscr{L}'_{\lambda^j} = 0, \mathscr{L}'_{s^j} = 0$):

Algorithm 1 Indirect multiple shooting for PDE governed OCP

Require: Decomposition $I = \{\tau_0\} \cup \bigcup_{j=0}^{M-1} (\tau_j, \tau_{j+1}]$, initial values $\{(s_0^j, \lambda_0^{j+1})_{j=0}^{M-1}\}$.
1: Set $k = 1$.
2: **while** Shooting conditions (18) not fulfilled **do**
3:     **for** $j = 0$ to $M - 1$ **do**
4:         Solve intervalwise boundary value problems (17).
5:     **end for**
6:     Solve (19), compute update $\{(s_k^j, \lambda_k^{j+1})_{j=0}^{M-1}\}$ of initial values, set $k \leftarrow k + 1$.
7: **end while**

$$(s^0 - u_0, \delta\lambda) = 0, \tag{18a}$$

$$(\lambda^j - z^j(\tau_j; s^j, \lambda^{j+1}), \delta s) = 0, \quad (j = 0, \dots, M - 1) \tag{18b}$$

$$(s^j - u^{j-1}(\tau_j; s^{j-1}, \lambda^j), \delta\lambda) = 0, \quad (j = 1, \dots, M) \tag{18c}$$

$$(\lambda^M, \delta s) = 0. \tag{18d}$$

These equations constitute the shooting system, which is the part of (14) we actually solve by Newton's method. Abbreviating the above shooting equations by $F(\mathbf{s}, \boldsymbol{\lambda}) = 0$, we thus have to solve

$$\nabla_{(\mathbf{s},\boldsymbol{\lambda})} F(\mathbf{s}, \boldsymbol{\lambda}) \cdot \begin{pmatrix} \delta\mathbf{s} \\ \delta\boldsymbol{\lambda} \end{pmatrix} = -F(\mathbf{s}, \boldsymbol{\lambda}). \tag{19}$$

This leads to improved initial values $\mathbf{s}^{\text{new}} = \mathbf{s} + \delta\mathbf{s}$, $\boldsymbol{\lambda}^{\text{new}} = \boldsymbol{\lambda} + \delta\boldsymbol{\lambda}$, with which we restart from step one described above. We see now the asserted structure of a two-step fixed-point iteration where we alternate between computing (u^j, z^j, q^j) and updating (s^j, λ^j). The whole process is resumed in the following Algorithm 1.

3.2.2 Algorithms for the Subproblems

We now focus on the two essential Steps 4 and 6 of Algorithm 1, namely the solution of the intervalwise BVP (17) (the first part of our two stage problem) and the solution of the shooting system (19) (the second part, correspondingly).

There are several possibilities how to solve the intervalwise BVP. In Algorithm 2, we present a so-called reduced approach which has also been implemented for our numerical examples. The important feature is the reduction of the set of independent variables from (q^j, u^j) to the control q^j alone, meaning that the interval state $u^j = u^j(q^j)$ is interpreted in terms of the interval control. To clarify the notation of Algorithm 2, we define the reduced cost functional (on subinterval I_j)

$$j(q^j) := J^j(q^j, u^j(q^j)). \tag{20}$$

Algorithm 2 Solution of the intervalwise BVP (reduced approach)

Require: Set $\nu = 0$, prescribe tolerance TOL_1 and initial control q_0^j.

while $\|\nabla j(q_\nu^j)\| > TOL_1$ **do**
  Solve state equation (17a).
  Solve adjoint equation (17b).
  Compute gradient $\nabla j(q_\nu^j)$ of reduced cost functional.
  Set $i = 0$, prescribe tolerance TOL_2 and $\delta q_{\nu,0}^j$.
  while $\|\delta q_{\nu,i+1}^j - \delta q_{\nu,i}^j\| > TOL_2$ **do**
    Compute matrix-vector product $\nabla^2 j(q_\nu^j)\delta q_{\nu,i}^j$.
    Solve system $\nabla^2 j(q_\nu^j)\delta q_{\nu,i}^j = -\nabla j(q_\nu^j)$ by a Newton-CG type method (this requires the solution of two additional equations per iteration; these so-called *tangent* and *additional adjoint equations* are obtained by linearization of (17a) and (17b)).
  end while
  Set $\nu \leftarrow \nu + 1$ and $q_{\nu+1}^j = q_\nu^j + \delta q_{\nu,end}^j$.
end while

Algorithm 3 Solution of the IMS shooting system (matrix-free approach)

Require: Shooting variables $(\mathbf{s}_k, \boldsymbol{\lambda}_k)$, intervalwise OCP solutions u^j, z^j

1: Build up residual $-F(\mathbf{s}_k, \boldsymbol{\lambda}_k)$.
2: Set $i = 0$, prescribe tolerance TOL and choose $(\delta\mathbf{s}_k^{(0)}, \delta\boldsymbol{\lambda}_k^{(0)})$.
3: **while** $\|\nabla F(\mathbf{s}_k, \boldsymbol{\lambda}_k)(\delta\mathbf{s}_k^{(i)}, \delta\boldsymbol{\lambda}_k^{(i)}) + F(\mathbf{s}_k, \boldsymbol{\lambda}_k)\| > TOL$ **do**
4:   Compute matrix-vector product $\nabla F(\mathbf{s}_k, \boldsymbol{\lambda}_k)(\delta\mathbf{s}_k^{(i)}, \delta\boldsymbol{\lambda}_k^{(i)})$.
5:   Solve system $\nabla F(\mathbf{s}_k, \boldsymbol{\lambda}_k)(\delta\mathbf{s}_k^{(i)}, \delta\boldsymbol{\lambda}_k^{(i)}) = -F(\mathbf{s}_k, \boldsymbol{\lambda}_k)$ by a Newton-GMRES type method (this requires the solution of two additional BVP, the linearizations of (17) w.r.t. $\mathbf{s}$ resp. $\boldsymbol{\lambda}$, in each iteration).
6: **end while**
7: Set $k \leftarrow k + 1$ and $\mathbf{s}_{k+1} = \mathbf{s}_k + \delta\mathbf{s}_k^{end}$, $\boldsymbol{\lambda}_{k+1} = \boldsymbol{\lambda}_k + \delta\boldsymbol{\lambda}_k^{end}$.

For a more detailed presentation we refer to [9],where IMS for parabolic OCP has already been described thoroughly. We will meet the concept of reduced control problems again in Sect. 4 in the context of different DMS techniques.

The solution of (19) by Newton's method involves the Jacobian matrix $\nabla_{(\mathbf{s},\boldsymbol{\lambda})} F(\mathbf{s}, \boldsymbol{\lambda})$ of the shooting conditions (18). Despite having substantially reduced the size of the Newton system [in this regard, (19) must be compared to (16)], the effort for explicitly assembling this Jacobian is still not manageable. Therefore, we choose a matrix-free method (in our case a Newton-GMRES approach) to solve (19). Algorithm 3 comprises the essential steps. Details of the Krylov-Newton method addressed in Step 5 can again be found in [9]. The proceeding in the case of DMS is quite similar in many respects and will be discussed in more detail in the next section.

3.3 Direct Multiple Shooting

We have already seen in the introduction that almost all literature on multiple shooting for PDE governed OCP concentrates on DMS methods (see [13, 29, 31]). Furthermore, the 'classical' method mostly used for ODE optimal control introduced in [3–5] is a direct method. Many algorithmic and implementational details can be found, e.g., in [20]. Further developments of this method, especially considering specific condensing techniques, can be found for example in [6, 17–19, 27]. The application of the 'classical' DMS method to large scale systems can be found for example in [21, 26]. The DMS method presented here is substantially equivalent to this 'classical' approach, the difference consists in the reduced and unreduced strategies to solve the problem as will be shown in Sect. 4.

We will now introduce direct shooting focusing on algorithmic details that we have skipped in the IMS context.

3.3.1 Structure

We derive DMS along the very same lines as IMS by splitting the solution process of system (16) into two parts. However, DMS relies on a different regrouping of the variables which is illustrated by the following scheme:

$$\left(\begin{array}{cc|ccc} 0 & \mathcal{L}''_{uz} & \mathcal{L}''_{qz} & \mathcal{L}''_{sz} & 0 \\ \mathcal{L}''_{zu} & \mathcal{L}''_{uu} & 0 & 0 & \mathcal{L}''_{\lambda u} \\ \hline \mathcal{L}''_{zq} & 0 & \mathcal{L}''_{qq} & 0 & 0 \\ \mathcal{L}''_{zs} & 0 & 0 & 0 & \mathcal{L}''_{\lambda s} \\ 0 & \mathcal{L}''_{u\lambda} & 0 & \mathcal{L}''_{s\lambda} & 0 \end{array}\right) \begin{pmatrix} \delta z \\ \delta u \\ \hline \delta q \\ \delta s \\ \delta\lambda \end{pmatrix} = - \begin{pmatrix} \mathcal{L}'_z \\ \mathcal{L}'_u \\ \hline \mathcal{L}'_q \\ \mathcal{L}'_s \\ \mathcal{L}'_\lambda \end{pmatrix}.$$

Thus, we now fix s^j, λ^j, and here also the controls q^j, and compute in the **first solution step** only the state and adjoint variables $u^j = u^j(q^j, s^j)$ and $z^j(q^j, s^j, \lambda^{j+1})$, which have now become dependent variables. This leads to the following initial value problems:

$$\begin{aligned} ((\partial_t u^j, \delta z)) + a(u^j)(\delta z) + b(q^j)(\delta z) \; - \; ((f|_{I_j}, \delta z)) & \\ + (u^j(\tau_j) \; - \; s^j, \delta z(\tau_j)) = 0, & \qquad (21a) \\ J_u^{j\prime}(q^j, u^j)(\delta u) - ((\partial_t z^j, \delta u)) + a'_u(u^j)(\delta u, z^j) & \\ + (z^j(\tau_{j+1}) \; - \; \lambda^{j+1}, \delta u(\tau_{j+1})) = 0. & \qquad (21b) \end{aligned}$$

Contrarily to the IMS case, these equations bear no BVP structure, because they are not fully coupled. We may in a first step compute the state solutions $u^j(q^j; s^j)$ and then use the result to compute the adjoint solutions $z^j(u^j(q^j; s^j); \lambda^{j+1})$, albeit backward in time. The local IVP (21) correspond to the first two equation blocks of (14), $\mathscr{L}'_{z^j} = 0$ and $\mathscr{L}'_{u^j} = 0$.

In the current situation, we still have to solve the matching conditions $\mathscr{L}'_{s^j} = 0, \mathscr{L}'_{\lambda^j} = 0$ and the control equation $\mathscr{L}'_{q^j} = 0$, together constituting the **second solution step**. The resulting system that has to be solved by Newton's method reads

$$(s^0 - u_0, \delta\lambda) = 0, \tag{22a}$$

$$J_q^{0\prime}(q^0, u^0)(\delta q) + b'_q(q^0)(\delta q, z^0(\tau_j; q^0, s^0, \lambda^1)) = 0, \tag{22b}$$

$$(\lambda^j - z^j(\tau_j; q^j, s^j, \lambda^{j+1}), \delta s) = 0, \quad (j = 0, \ldots, M-1) \tag{22c}$$

$$(s^j - u^{j-1}(\tau_j; q^{j-1}, s^{j-1}), \delta\lambda) = 0, \quad (j = 1, \ldots, M) \tag{22d}$$

$$J_q^{j\prime}(q^j, u^j)(\delta q) + b'_q(q^j)(\delta q, z^j(\tau_j; q^j, s^j, \lambda^{j+1})) = 0, \quad (j = 0, \ldots, M-1) \tag{22e}$$

$$(\lambda^M, \delta s) = 0. \tag{22f}$$

Thus we have again reduced the size of the original Newton system (16), but here the resulting system is larger than in the IMS framework. Therefore, the first step in DMS consists of only solving the IVP (21) in contrast to the far more complicated BVP (17). Abbreviating (22) by $F(\mathbf{q}, \mathbf{s}, \boldsymbol{\lambda}) = 0$, we end up with Newton's equation

$$\nabla_{(\mathbf{q},\mathbf{s},\boldsymbol{\lambda})} F(\mathbf{q}, \mathbf{s}, \boldsymbol{\lambda}) \cdot \begin{pmatrix} \delta\mathbf{q} \\ \delta\mathbf{s} \\ \delta\boldsymbol{\lambda} \end{pmatrix} = -F(\mathbf{q}, \mathbf{s}, \boldsymbol{\lambda}). \tag{23}$$

Altogether, the structure of DMS is again a two-step fixed-point iteration, where we first keep the controls and initial values fixed and compute (u^j, z^j) before updating (q^j, s^j, λ^j). We resume the solution process in the following Algorithm 4.

Algorithm 4 Direct multiple shooting for PDE governed OCP

Require: Decomp. $I = \{\tau_0\} \cup \bigcup_{j=0}^{M-1} (\tau_j, \tau_{j+1}]$, initial values and controls $\{(q_0^j, s_0^j, \lambda_0^{j+1})_{j=0}^{M-1}\}$.
1: Set $k = 1$.
2: **while** Shooting conditions (22) not fulfilled **do**
3:     **for** $j = 0$ to $M - 1$ **do**
4:         Solve intervalwise initial value problems (21).
5:     **end for**
6:     Solve (23), comp. update $\{(q_k^j, s_k^j, \lambda_k^{j+1})_{j=0}^{M-1}\}$ of initial values and controls, set $k \leftarrow k + 1$.
7: **end while**

3.3.2 Algorithm for Newton's Method

In contrast to the IMS case above, where we presented the details of the solution of both the intervalwise BVP (17) and the system (19) of shooting conditions, only Step 6 of Algorithm 4, i.e. the realization of Newton's method, is worth being discussed more thoroughly. The solution of the IVP in Step 4 is straightforward; the only important feature is that, on each subinterval, we have to first solve the state equation, because u^j is needed for solving the adjoint equation. With this restriction in mind, we turn our attention to Newton's system (23), the solution of which is different from that of the corresponding system in IMS and has not been presented elsewhere; therefore, we elaborate the following presentation in detail.

In Sect. 3.2, we stated that application of a matrix-free Krylov-Newton method is desirable due to the size of problem (19). We did not go into the details and referred to [9] instead. As the system (23) is even larger than (19) (due to the presence of the controls, see also Remark 12 below), a direct solver is even less advisable here. We will now discuss a Newton-GMRES method that has not been addressed in the DMS context before.

We see that the Jacobian $\nabla_{(\mathbf{q},\mathbf{s},\boldsymbol{\lambda})}F(\mathbf{q},\mathbf{s},\boldsymbol{\lambda})$ of (22) involves derivatives u_s^j, u_q^j of u^j w.r.t. s^j and q^j as well as derivatives $z_s^j, z_\lambda^j, z_q^j$ of z^j w.r.t. s^j, λ^{j+1} and q^j. These derivatives, the so-called sensitivities, are obtained by solving five additional (linearized) IVP, the sensitivity equations (also known as variational equations), for $j \in \{0, \dots, M-1\}$. First, we differentiate (21a) where $u^j = u^j(s^j, q^j)$ w.r.t. s^j in direction δs and w.r.t. q^j in direction δq to obtain the equations

$$((\partial_t u_s^j, \varphi)) + a_u'(u^j)(u_s^j, \varphi) + (u_s^j(\tau_j) - \delta s^j, \varphi(\tau_j)) = 0, \quad (24a)$$

$$((\partial_t u_q^j, \varphi)) + a_u'(u^j)(u_q^j, \varphi) + b_q'(q^j)(\delta q^j, \varphi) + (u_q^j(\tau_j), \varphi(\tau_j)) = 0, \quad (24b)$$

which have to hold for all $\varphi \in X^j$. Having solved these problems (for given initial data $\delta s^{j,0}$ and $\delta q^{j,0}$), we end up with u_s^j, u_q^j which can now be inserted into the following three IVP that are obtained by differentiation of the adjoint equation (21b) w.r.t. all its arguments in corresponding directions and must hold for all $\psi \in X^j$:

$$J_{uu}^{j\prime\prime}(q^j, u^j)(u_s^j, \psi) - ((\partial_t z_s^j, \psi)) + a_{uu}''(u^j)(u_s^j, \psi, z^j)$$
$$+ a_u'(u^j)(\psi, z_s^j) + (z_s^j(\tau_{j+1}), \psi(\tau_{j+1})) = 0, \quad (25a)$$

$$J_{uu}^{j\prime\prime}(q^j, u^j)(u_q^j, \psi) - ((\partial_t z_q^j, \psi)) + a_{uu}''(u^j)(u_q^j, \psi, z^j)$$
$$+ a_u'(u^j)(\psi, z_q^j) + (z_q^j(\tau_{j+1}), \psi(\tau_{j+1})) = 0, \quad (25b)$$

$$-((\partial_t z_\lambda^j, \psi)) + a_u'(u^j)(\psi, z_\lambda^j) + (z_\lambda^j(\tau_{j+1}) - \delta\lambda^{j+1}, \psi(\tau_{j+1})) = 0. \quad (25c)$$

Solving these problems with initial data $\delta\lambda^{j+1,0}$ leaves us with a complete set of sensitivities, but only w.r.t. the chosen initial values $(\delta q^{j,0}, \delta s^{j,0}, \delta\lambda^{j+1,0})$. In order to assemble $\nabla_{(\mathbf{q},\mathbf{s},\boldsymbol{\lambda})}F$ explicitly, we have to solve the sensitivity equations for a whole basis of $\bigcup_{j=0}^{M-1}[Q^j \times H \times H]$, which is numerically expensive for fine temporal or spatial discretizations. Therefore, we choose an adjoint approach (matrix-free) where we handle Newton's system (23) with an iterative solver, for which we choose in our case, due to the asymmetric structure of the matrix, a GMRES method. We then have to solve the sensitivity equations only once per GMRES iteration. The adjoint approach thus avoids assembling the Jacobian and operates on the matrix-vector product $\nabla_{(\mathbf{q},\mathbf{s},\boldsymbol{\lambda})} F(\mathbf{q},\mathbf{s},\boldsymbol{\lambda}) \cdot (\delta\mathbf{q}, \delta\mathbf{s}, \delta\boldsymbol{\lambda})^\top$ instead.

This matrix-vector product, the left-hand side of (23), has the concrete form

$$\nabla_{(\mathbf{q},\mathbf{s},\boldsymbol{\lambda})} F(\mathbf{q},\mathbf{s},\boldsymbol{\lambda}) \cdot \begin{pmatrix} \delta\mathbf{q} \\ \delta\mathbf{s} \\ \delta\boldsymbol{\lambda} \end{pmatrix} = \left(\begin{array}{l} \qquad\qquad \delta s^0 \\ J^{0\prime\prime}_{qq}(q^0,u^0)(\delta q^0) + b''_{qq}(q^0)(\delta q^0, z^0) \\ \qquad + b'_q(q^0)[z^0_s(\delta s^0) + z^0_q(\delta q^0) + z^0_\lambda(\delta\lambda^0)] \\ \hline \delta\lambda^j - z^j_s(\tau_j;\delta s^j) - z^j_\lambda(\tau_j;\delta\lambda^{j+1}) \\ \delta s^{j+1} - u^j_s(\tau_{j+1};\delta s^j) - u^j_\lambda(\tau_{j+1};\delta\lambda^{j+1}) \\ J^{j\prime\prime}_{qq}(q^j,u^j)(\delta q^j) + b''_{qq}(q^j)(\delta q^j, z^j) \\ \qquad + b'_q(q^j)[z^j_s(\delta s^j) + z^j_q(\delta q^j) + z^j_\lambda(\delta\lambda^j)] \\ \hline \qquad\qquad \delta\lambda^M \end{array} \right)$$

where the middle part has to be interpreted for $j = 0, \ldots, M-1$; note the index shift we performed in the second component of the middle part to keep the presentation consistent. We can now formulate the following Algorithm 5 which yields the details of Step 6 of the above DMS algorithm:

Algorithm 5 Solution of the DMS shooting system (matrix-free approach)

Require: Shooting variables $(\mathbf{s}_k, \boldsymbol{\lambda}_k)$ and controls $\mathbf{q}_k$, intervalwise OCP solutions u^j, z^j

Build up residual $-F(\mathbf{q}_k, \mathbf{s}_k, \boldsymbol{\lambda}_k)$.

Set $i = 0$, prescribe tolerance TOL and choose $(\delta\mathbf{q}_k^{(0)}, \delta\mathbf{s}_k^{(0)}, \delta\boldsymbol{\lambda}_k^{(0)})$.

while $\|\nabla F(\mathbf{q}_k, \mathbf{s}_k, \boldsymbol{\lambda}_k)(\delta\mathbf{q}_k^{(i)}, \delta\mathbf{s}_k^{(i)}, \delta\boldsymbol{\lambda}_k^{(i)}) + F(\mathbf{q}_k, \mathbf{s}_k, \boldsymbol{\lambda}_k)\| > TOL$ **do**

  Compute matrix-vector product $\nabla F(\mathbf{q}_k, \mathbf{s}_k, \boldsymbol{\lambda}_k)(\delta\mathbf{q}_k^{(i)}, \delta\mathbf{s}_k^{(i)}, \delta\boldsymbol{\lambda}_k^{(i)})$ by solving the state and adjoint sensitivity equations (24) and (25).

  Solve system $\nabla F(\mathbf{q}_k, \mathbf{s}_k, \boldsymbol{\lambda}_k)(\delta\mathbf{q}_k^{(i)}, \delta\mathbf{s}_k^{(i)}, \delta\boldsymbol{\lambda}_k^{(i)}) = -F(\mathbf{q}_k, \mathbf{s}_k, \boldsymbol{\lambda}_k)$ by a Newton-GMRES type method (this requires the renewed solution of (24) and (25) in each iteration).

end while

Set $k \leftarrow k+1$ and $\mathbf{q}_{k+1} = \mathbf{q}_k + \delta\mathbf{q}_k^{end}$, $\mathbf{s}_{k+1} = \mathbf{s}_k + \delta\mathbf{s}_k^{end}$, $\boldsymbol{\lambda}_{k+1} = \boldsymbol{\lambda}_k + \delta\boldsymbol{\lambda}_k^{end}$.

3.4 Comparison of Both Approaches

From the last two Sects. 3.2 and 3.3 we see that IMS and DMS on a purely algebraic level, i.e. looking merely at the equations to be solved [which are in both cases the extended KKT conditions (14)], are in fact equivalent. The differences are due to the respective algorithmic realizations that result from the different splittings of system (16) leading to different internal dependencies of the arguments of the common starting point, the extended formulation of the optimal control problem (10)–(11).

Remark 7 Note that there are further possibilities how to split the set of arguments of the extended Lagrangian (13), although they might not work as well as the ones discussed so far. This might give reason to further research.

In the algorithmic description, we concentrated on the solution of the shooting systems (19) resp. (23) by Newton's method, which may lead to the supposition that DMS is more expensive (in this context, see also Remark 12 below). However, in the DMS context we only have to solve a (nonlinear) IVP (21) on each subinterval, and also the additional sensitivity equations are only intervalwise (linear) IVP, which can be solved in a rather straightforward manner. Despite the IMS shooting system (19) being much smaller than the DMS one, the solution of the (nonlinear) subinterval problems (17) and the corresponding (linearized) problems for the Newton-GMRES method necessitates large effort, because they constitute smaller versions of the original OCP and its linearization. It is thus not clear a priori which of the two methods (IMS or DMS) is to be preferred.

Both IMS and DMS still comprise a variety of algorithms, depending on how we solve the subinterval problems (reduced approach as above vs. all-at-once approach), on how we solve Newton's system (iterative matrix-free solver as above vs. direct solver, inclusion of globalization techniques or of an SQP-like inexact Newton method), on how we solve the sensitivity equations (simply by a fixed-point method or by more sophisticated approaches) etc.

We will now turn our attention to some of these differences which are responsible for the seemingly large differences between 'classical' DMS (introduced in Sect. 4.1 for parabolic OCP) and the DMS approach given in Algorithm 4.

4 Variants of DMS

The following considerations were initiated by the observation that the DMS method derived in Sect. 3.3, despite being called 'direct', seems to be rather an indirect method judged by the standard classification (see Introduction), because it is based on the optimality conditions (14), i.e., the optimization has already been done when shooting comes into play. According to our classification discussed in the Introduction, the DMS approach shown here is a direct approach since the control is included in the shooting system. Therefore, it benefits from the advantages of direct methods in terms of convergence as shown in our nonlinear example Sect. 6.2. Further study to compare IMS and DMS in the PDE context is needed, but is left for future work, especially considering state and control constraints.

Typical implementations of direct methods [1, 28], that use the 'first discretize, then optimize' approach, avoid the use of an adjoint equation either by using a sensitivity approach or by applying automatic differentiation techniques. We therefore present in Sect. 4.1 the 'classical' DMS approach but tailored to the parabolic situation, and show in Sect. 4.2 that this approach, relying on a reduced formulation of the extended optimal control problem (10)–(11), is equivalent to our variant of DMS. In the classical framework, the reduced problem is discretized, and the resulting nonlinear programming problem (NLP) is usually solved by an all-at-once approach, e.g. a sequential quadratic programming (SQP) method. In contrast, our DMS variant relies on a continuous unreduced formulation. The dichotomy of reduced versus unreduced approaches is discussed in [16] for the global OCP (2)–(3), and we use the notational framework established there. Note that the whole discussion takes place on the abstract function space level.

4.1 DMS Based on a Reduced Form of the Extended OCP Formulation

We will now embed a DMS method that was developed (mainly by Bock and his co-workers) in the 1980s (see, e.g. [3–5]) into the context of parabolic OCP.

DMS methods for problem (10)–(11) are usually based on a reformulation of this extended OCP formulation completely in terms of the primal shooting variables s^j and the intervalwise controls q^j, i.e. $u^j = u^j(q^j, s^j)$. Pursuing this strategy, we end up with the minimization problem

$$\min_{(\mathbf{q},\mathbf{s})} \bar{J}(\mathbf{q},\mathbf{s}) := \sum_{j=0}^{M-1} J^j(q^j, u^j(q^j, s^j)) \tag{26a}$$

$$\text{s. t.} \qquad s^0 - u_0 = 0, \tag{26b}$$

$$s^{j+1} - u^j(\tau_{j+1}; q^j, s^j) = 0, \tag{26c}$$

where (26c) comprises the continuity conditions for $j = 0, \dots, M-1$. We call (26) a reduced formulation of the extended OCP formulation (10)–(11). It is formulated in terms of the independent variables q^j and s^j and relies upon the solution of the IVP

$$e^j(q^j, s^j, u^j(q^j, s^j)) = \begin{pmatrix} \partial_t u^j(q^j, s^j) + \mathscr{A}(u^j(q^j, s^j)) + \mathscr{B}(q^j) - f|_{I_j} \\ u^j(\tau_j; q^j, s^j) - s^j \end{pmatrix} = \begin{pmatrix} 0 \\ 0 \end{pmatrix} \tag{27}$$

that has been solved on all subintervals I_j for $j = 0, \ldots, M-1$. We assume unique solvability of the subinterval problems which implies the existence of a solution operator mapping $Q^j \times H$ to X^j. In (27), $e^j(q^j, s^j, u^j(q^j, s^j))$ is an intervalwise counterpart of the abstract side condition in (1) which is, in contrast to the preceding sections, again strongly formulated. This abstract notation helps us to keep the proof of the equivalence result in the next section short.

Remark 8 In the reformulation (26)–(27) of the extended OCP formulation, the local control variable $q^j(x,t)$ is a function of both spatial variables x and time t. In DMS methods for ODE control problems which depend only on t, the control is usually parameterized as a piecewise polynomial of order $p \leq 3$ on the subintervals I_j, i.e. $q^j \equiv q^j(q_0^j, \cdots, q_p^j)$ (see, e.g., [20]). This parameterization of the control saves a large amount of computing time and storage (up to four control parameters per shooting interval I_j as opposed to a number of control values on each I_j determined by the control discretization, which is usually much finer than the mentioned parameterization). Furthermore, so-called condensing techniques (reducing the shooting system to the control variables) are frequently employed; they are not efficient if q is discretized on a similarly fine level as the state u. However, reducing the control to a much smaller space by parameterization leads to only suboptimal solutions of the given control problems. In the PDE case, a parameterization of the control $q(x,t)$ may lead to a loss of structural information in the spatial variables. Some ideas in this regard are briefly discussed in Sect. 5 (see Remark 11 below).

Starting from (26), we derive the corresponding Lagrange functional (introducing a Lagrange multiplier $\boldsymbol{\lambda} = (\lambda^j)_{j=0}^M \in H^{m+1}$):

$$\mathcal{L}(\mathbf{q}, \mathbf{s}, \boldsymbol{\lambda}) = \bar{J}(\mathbf{q}, \mathbf{s}) + (s^0 - u_0, \lambda^0) + \sum_{j=0}^{M-1} (s^{j+1} - u^j(\tau_{j+1}; q^j, s^j), \lambda^{j+1}). \tag{28}$$

We obtain the (reduced) optimality system as usual by differentiation w.r.t. the arguments $(\mathbf{q}, \mathbf{s}, \boldsymbol{\lambda})$. This yields the system [where $j \in \{0, \ldots, M-1\}$ in Eqs. (29b), (29c) and (29e)]

$$\mathcal{L}'_{\lambda^0}(\delta\lambda) = (s^0 - u_0, \delta\lambda), \tag{29a}$$

$$\mathcal{L}'_{\lambda^j}(\delta\lambda) = (s^{j+1} - u^j(\tau_{j+1}; q^j, s^j), \delta\lambda), \tag{29b}$$

$$\mathcal{L}'_{s^j}(\delta s) = \langle J_u^{j\prime}, u_s^{j\prime}(\delta s)\rangle_{X^{j*} \times X^j} + (\lambda^j, \delta s) - (\lambda^{j+1}, u_s^{j\prime}[\tau_{j+1}](\delta s)), \tag{29c}$$

$$\mathcal{L}'_{s^M}(\delta s) = (\lambda^M, \delta s), \tag{29d}$$

$$\mathcal{L}'_{q^j}(\delta q) = \langle J_q^{j\prime}, \delta q\rangle_{Q^{j*} \times Q^j} + \langle J_u^{j\prime}, u_q^{j\prime}(\delta q)\rangle_{X^{j*} \times X^j} - (\lambda^{j+1}, u_q^{j\prime}[\tau_{j+1}](\delta q)). \tag{29e}$$

Here, $u_q^{j'} : Q^j \to X^j$ and $u_s^{j'} : H \to X^j$ are operators mapping the controls and initial values to the respective (variational) states $u_{q/s}^{j'}$ that are denoted by the same symbols. Furthermore, the notation $u_{q/s}^{j'}[\tau_{j+1}]$ means that the respective variational state obtained by application of the operator $u_{q/s}^{j'}$ is evaluated at time-point τ_{j+1}. The classical DMS method consists in the solution of system (29). In the framework of ODE optimal control, [20] gives a detailed description of SQP methods that solve (29) without employing an adjoint equation. Therefore, either the Jacobian of (29) has to be assembled, or additional sophisticated algorithms are needed to circumvent this matrix assembly. An alternative matrix-free SQP approach has been proposed by Ulbrich [31]. This procedure corresponds to the sensitivity approach for generating derivative information that is needed during the solution process; this sensitivity approach is usually too expensive in PDE optimal control, because one has to solve an additional linearized problem for each basis vector δq of the (discrete) control space (see [16] or [24]).

4.2 *Equivalence of the Two DMS Approaches*

Comparing the two DMS variants presented in Sects. 3.3 and 4.1 yields, besides several minor differences, a central distinction that influences the structure of the solution process strongly. It consists of the absence of any adjoint problem in the DMS method from Sect. 4.1, whereas the variant discussed in Sect. 3.3 is based on the same full optimality system (14) as the IMS approach of Sect. 3.2, including the adjoint part.

Remark 9 Generally, modern implementations of DMS for ODE optimal control problems, which are capable of handling parabolic OCP by transforming the PDE side condition into a huge ODE system via the method of lines (MOL) approach, make use of adjoint methods for sensitivity generation, which constitute a suitable alternative to the above described approach. They often compute the adjoint equations (which may be nasty to derive by hand in case of large ODE systems with complicated nonlinear terms) by automatic differentiation (see [1]).

The following theorem, which is the main result of this section, shows the equivalence of the two DMS approaches. Moreover, the proof reveals that the seemingly so different DMS variant of Sect. 3.3 is merely a reformulation of 'classical' DMS by means of an adjoint approach for sensitivity generation. It is performed in an abstract function space setting, meaning that the argumentation is not affected by discretization.

Theorem 2 *The solution of the reduced formulation* (26) *of the modified OCP* (10)–(11) *by means of an adjoint approach leads to the DMS method described in Sect. 3.3.*

The following outline prepares the proof of Theorem 2. Classical DMS for problem (26) relies upon the solution of system (27) and necessitates the solution of (29). Analogously, the DMS approach from Sect. 3.3 for problem (10)–(11) relies upon (21) having been solved and necessitates the solution of (22). Comparing the two settings, the following correspondences are evident: (27) is the strong formulation of (21a), and (29a), (29b) and (29d) are identical to (22a), (22d) and (22f), respectively. It is thus our goal to derive the adjoint equation (21b), the continuity conditions (22c) and the control equations (22b) and (22e) from (29c) and (29e). To achieve this, we extend the ideas and techniques of Sect. 1.6 from [16] to the more complex multiple shooting situation. The following proof has already been outlined in [12].

Proof For the following discussion, we introduce adjoint operators $u_q^{j\prime *} : X^{j^*} \to Q^{j^*}$ and $u_s^{j\prime *} : X^{j^*} \to H^* \equiv H$ (as well as their restrictions to the final time-point, $u_q^{j\prime *}[\tau_{j+1}] : H \equiv H^* \to Q^{j^*}$ and $u_s^{j\prime *}[\tau_{j+1}] : H \equiv H^* \to H^* \equiv H$) corresponding to the differential operators $u_q^{j\prime} : Q^j \to X^j$ and $u_s^{j\prime} : H \to X^j$ and rewrite Eqs. (29c) and (29e) in an abstract adjoint form:

$$\mathscr{L}'_{s^j}(\delta s) = (u_s^{j\prime *}(J_u^{j\prime}), \delta s) + (\lambda^j, \delta s) - (u_s^{j\prime *}[\tau_{j+1}](\lambda^{j+1}), \delta s), \tag{29c*}$$

$$\mathscr{L}'_{q^j}(\delta q) = \langle J_q^{j\prime}, \delta q\rangle_{Q^{j*}\times Q^j} + \langle u_q^{j\prime *}(J_u^{j\prime}), \delta q\rangle_{Q^{j*}\times Q^j}$$
$$- \langle u_q^{j\prime *}[\tau_{j+1}](\lambda^{j+1}), \delta q\rangle_{Q^{j*}\times Q^j}. \tag{29e*}$$

We discuss the adjoint operators first on an abstract level which enables a clear presentation of the formal framework. By differentiating the interval state equations (27) w.r.t. q^j in direction δq and w.r.t. s^j in direction δs, we obtain (the arguments $(q^j, s^j, u^j(q^j, s^j))$ are omitted for brevity)

$$e_u^{j\prime}(\delta u_q) = -e_q^{j\prime}(\delta q), \quad e_u^{j\prime}(\delta u_s) = -e_s^{j\prime}(\delta s). \tag{30}$$

Here, δu_q and δu_s are abbreviations for $u_q^{j}(\delta q)$ and $u_s^{j}(\delta s)$, respectively. Assuming $e_u^{j\prime}$ to have a bounded inverse, we use the implicit function theorem to obtain

$$u_q^{j\prime} = -(e_u^{j\prime})^{-1} \circ e_q^{j\prime}, \quad u_s^{j\prime} = -(e_u^{j\prime})^{-1} \circ e_s^{j\prime}.$$

Now, the definition of adjoint operators gives us the following representation:

$$u_q^{j\prime *} = -e_q^{j\prime *} \circ (e_u^{j\prime})^{-*}, \quad u_s^{j\prime *} = -e_s^{j\prime *} \circ (e_u^{j\prime})^{-*}.$$

Inserting these expressions for $u_s^{j\prime *}$ and $u_q^{j\prime *}$ into the corresponding terms of (29c*) and (29e*), we get

$$(u_s^{j\prime *}(J_u^{j\prime}), \delta s) = -(e_s^{j\prime *}((e_u^{j\prime})^{-*}(J_u^{j\prime})), \delta s), \tag{31a}$$

$$(u_s^{j'^*}[\tau_{j+1}](\lambda^{j+1}), \delta s) = -(e_s^{j'^*}((e_u^{j'})^{-*}[\tau_{j+1}](\lambda^{j+1})), \delta s), \tag{31b}$$

$$\langle u_q^{j'^*}(J_u^{j'}), \delta q\rangle_{Q^{j*}\times Q^j} = -\langle e_q^{j'^*}((e_u^{j'})^{-*}(J_u^{j'})), \delta q\rangle_{Q^{j*}\times Q^j}, \tag{31c}$$

$$\langle u_q^{j'^*}[\tau_{j+1}](\lambda^{j+1}), \delta q\rangle_{Q^{j*}\times Q^j} = -\langle e_q^{j'^*}((e_u^{j'})^{-*}[\tau_{j+1}](\lambda^{j+1})), \delta q\rangle_{Q^{j*}\times Q^j}. \tag{31d}$$

We notice that, in (31a) and (31c), the same argument $(e_u^{j'})^{-*}(J_u^{j'})$ is inserted into both operators $e_s^{j'^*}$ and $e_q^{j'^*}$. The same holds for $(e_u^{j'})^{-*}[\tau_{j+1}](\lambda^{j+1})$ in (31b) and (31d). We define the variables $z_J^j := -(e_u^{j'})^{-*}(J_u^{j'})$ and $z_\lambda^j := (e_u^{j'})^{-*}[\tau_{j+1}](\lambda^{j+1})$, which then fulfil the following equations, respectively:

$$e_u^{j'^*}(z_J^j) = -J_u^{j'}, \qquad e_u^{j'^*}[\tau_{j+1}](z_\lambda^j) = \lambda^{j+1}. \tag{32}$$

These are the (formal) adjoint equations; we will see below that they can be merged into one equation, due to the linearity of the operator $e_u^{j'^*}$ and a superposition principle, which interprets z_J^j as a solution belonging to a homogeneous initial value and non-homogeneous right-hand side $-J_u^{j'}$, and z_λ^j as a solution belonging to a homogeneous right-hand side and non-homogeneous initial value λ^{j+1}.

In our concrete situation, we start from the weak formulation of (27) given by

$$((\partial_t u^j, \varphi)) + a(u^j)(\varphi) + b(q^j)(\varphi) - ((f|_{I_j}, \varphi)) + (u^j(\tau_j) - s^j, \varphi(\tau_j)) = 0. \tag{33}$$

The differential operator $u_q^{j'} : Q^j \to X^j$ mentioned above is the solution operator of the following linearized equation (which is the derivative of (33) w.r.t. q^j in direction δq):

$$((\partial_t \delta u_q, \varphi)) + a_u'(u^j)(\delta u_q, \varphi) + (\delta u_q(\tau_j), \varphi(\tau_j)) = -b_q'(q^j)(\delta q, \varphi). \tag{34}$$

Analogously, $u_s^{j'} : H \to X^j$ is the solution operator of the derivative of (33) w.r.t. s^j in direction δs:

$$((\partial_t \delta u_s, \varphi)) + a_u'(u^j)(\delta u_s, \varphi) + (\delta u_s(\tau_j), \varphi(\tau_j)) = (\delta s, \varphi(\tau_j)). \tag{35}$$

Here, δu_q and δu_s denote the respective solution variables. We call (34) and (35) the sensitivity equations belonging to (27). They correspond to the formal equations (30).

Now we are able to substantiate the adjoint equations (32). We have seen above that the application of the adjoint operator $u_q^{j'^*}$ resp. $u_s^{j'^*}$ to a functional $J_u^{j'} \in X^{j^*}$ (or of $u_q^{j'^*}[\tau_{j+1}]$ resp. $u_s^{j'^*}[\tau_{j+1}]$ to λ^{j+1}) corresponds to carrying out the following steps:

1. Solve the adjoint equation $e_u^{j'^*}(z_J^j) = -J_u^{j'}$ (or $e_u^{j'^*}[\tau_{j+1}](z_\lambda^j) = \lambda^{j+1}$).
2. Apply the adjoint operators $e_q^{j'^*}$ resp. $e_s^{j'^*}$ to the solution z_J^j (or z_λ^j).

Discussing the first step will lead us to the adjoint equation, while the second step yields the continuity conditions for the adjoint equation and the control equations. The general adjoint equation corresponding to both (34) and (35) is given by

$$-((\partial_t \delta u^*_{q/s}, \psi)) + a'_u(u^j)(\delta u^*_{q/s}, \psi) + (\delta u^*_{q/s}(\tau_{j+1}), \psi(\tau_{j+1})) = \mathrm{rhs}(\psi). \tag{36}$$

Here, the term $\mathrm{rhs}(\psi)$ which is fanning the dynamics represents either a distributed source term or an end-time initial condition. Our situation comprises the following two adjoint equations (where the abstract adjoint variable $\delta u^*_{q/s}$ has been suitably replaced):

$$-((\partial_t z^j_J, \psi)) + a'_u(u^j)(z^j_J, \psi) + (z^j_J(\tau_{j+1}), \psi(\tau_{j+1})) = -J^{j\prime}_u(q^j, u^j)(\psi), \tag{37a}$$

$$-((\partial_t z^j_\lambda, \psi)) + a'_u(u^j)(z^j_\lambda, \psi) + (z^j_\lambda(\tau_{j+1}), \psi(\tau_{j+1})) = (\lambda^{j+1}, \psi(\tau_{j+1})). \tag{37b}$$

Evidently, Eq. (37) are both fully linear, as the possibly nonlinear operators $a(\cdot)(\cdot)$ and $J^j(\cdot)$ enter only in linearized form. Therefore, we may write (37a) and (37b) as one single equation by defining $z^j := z^j_J - z^j_\lambda$. The resulting adjoint equation reads

$$J^{j\prime}_u(q^j, u^j)(\psi) - ((\partial_t z^j, \psi)) + a'_u(u^j)(z^j, \psi) \\ +(z^j(\tau_{j+1}) - \lambda^{j+1}, \psi(\tau_{j+1})) = 0. \tag{38}$$

A comparison of (21b) and (38) shows that, substituting the test function ψ by δu, our first objective, namely the introduction of the adjoint equation into the reduced DMS method by means of an adjoint approach to sensitivity generation, has been achieved.

We finally explain the second step of the above proceeding in detail. By means of the described superposition $z^j := z^j_J - z^j_\lambda$, system (31) diminishes to

$$(u^{j\prime *}_s(J^{j\prime}_u) - u^{j\prime *}_s[\tau_{j+1}](\lambda^{j+1}), \delta s) = -(e^{j\prime *}_s(z^j), \delta s), \tag{39a}$$

$$\langle u^{j\prime *}_q(J^{j\prime}_u) - u^{j\prime *}_q[\tau_{j+1}](\lambda^{j+1}), \delta q\rangle_{Q^{j*}\times Q^j} = -\langle e^{j\prime *}_q(z^j), \delta q\rangle_{Q^{j*}\times Q^j}, \tag{39b}$$

where the adjoint solution has been inserted into the right-hand side terms. Since the right-hand sides of the second equation of (30) and of (35) coincide, we get the following equalities (making use of the weak form $e^{j\prime}_s(\delta s)(\varphi) := (e^{j\prime}_s(\delta s), \varphi)$):

$$(\delta s, e^{j\prime *}_s(z^j)) = (e^{j\prime}_s(\delta s), z^j) = e^{j\prime}_s(\delta s)(z^j) = (\delta s, z^j(\tau_j)).$$

Thus, replacing (for all $j \in \{0, \dots, M-1\}$) the adjoint terms in (29c*) by the corresponding term in (39a) and using the last equality, we end up with

$$\mathcal{L}'_{s^j}(\delta s) = (\lambda^j, \delta s) - (z^j(\tau_j), \delta s), \tag{40}$$

which is exactly the adjoint continuity condition (22c). Analogously, we can exploit (39b). The right-hand sides of the first equation of (30) and of (34) coincide, which leaves us with

$$\langle e_q^{j\prime *}(z^j), \delta q\rangle_{Q^{j*}\times Q^j} = \langle e_q^{j\prime}(\delta q), z^j\rangle_{X^{j*}\times X^j} = e_q^{j\prime}(\delta q)(z^j) = -b_q'(q^j)(\delta q, z^j).$$

Here, we have made use of the definition $\langle e_q^{j\prime}(\delta q), \varphi\rangle_{X^{j*}\times X^j} := e_q^{j\prime}(\delta q)(\varphi)$. We can now replace the adjoint terms in (29e*) by the corresponding term in (39b), use the last equality and obtain for all $j \in \{0, \dots, M-1\}$:

$$\mathscr{L}'_{q^j}(\delta q) = \langle J_q^{j\prime}, \delta q\rangle_{Q^{j*}\times Q^j} + b_q'(q^j)(\delta q, z^j). \tag{41}$$

Since the last relationship is identical to (22e) [and, in the case $j = 0$, to (22b)], this completes the proof. □

5 Discretization

For the sake of completeness, we briefly present the discretization schemes that will be used for computing the examples in Sect. 6 below. We treat the discretization of the time variable and the spatial variables separately, but choose the same discretization schemes for IMS and DMS.

5.1 *Time Semi-Discretization*

The decomposition (9) of the solution interval I into subintervals $I_j = (\tau_j, \tau_{j+1}]$ is not to be seen as a discretization, but rather as a reformulation of the problem. As we have seen, it is compensated by introducing the additional equality constraints (11). Here, we describe a further decomposition of each shooting interval I_j into smaller time intervals $I_j^n = (t_j^n, t_j^{n+1}]$ of length $k_j^n := t_j^{n+1} - t_j^n$ with timepoints

$$\tau_j = t_j^0 < t_j^1 < \cdots < t_j^{N_j} = \tau_{j+1},$$

which is now an actual discretization of the time variable. On each shooting interval, we use the discontinuous Galerkin method of order r (dG(r) method). All semi-discrete variables are indexed with a symbol k denoting a piecewise constant function $k|_{I_j^n} := k_j^n$. We denote space-time scalar products and semilinear forms on I_j^n by $((\cdot,\cdot))_n, a_n(\cdot)(\cdot)$ and $b_n(\cdot)(\cdot)$, suppressing the shooting interval index j. The same holds for the functional $J_n^j(\cdot)$.

Starting point for the dG(r) method is the space of semi-discrete functions

$$X_k^r(I_j) := \{v_k \in L^2(I_j; V) \mid v_k(t_j^0) \in H, v_k|_{I_j^n} \in P_r(I_j^n; V), n = 1, \dots, N_j\}$$

where the space $P_r(I_j^n)$ contains all polynomials on I_j^n up to degree r. The time-discrete functions on I_j may be discontinuous at the timepoints t_n, hence we introduce the following notation for describing the jumps:

$$v_{k,n}^{j,+} := \lim_{t \searrow 0} v_k(t_j^n + t), \quad v_{k,n}^{j,-} := \lim_{t \nearrow 0} v_k(t_j^n + t), \quad [v_k^j]_n := v_{k,n}^{j,+} - v_{k,n}^{j,-}.$$

Now we are prepared to formulate the time semi-discrete state and adjoint equations: Find $u_k^j, z_k^j \in X_k^r(I_j)$, such that for all $\delta z_k, \delta u_k \in X_k^r(I_j)$, the following equations hold:

$$\sum_{n=0}^{N_j-1} \left[((\partial_t u_k^j, \delta z_k))_n + a_n(u_k^j)(\delta z_k) + b_n(q^j)(\delta z_k) - ((f|_{I_j^n}, \delta z_k))_n \right]$$
$$+ \sum_{n=1}^{N_j-1} ([u_k^j]_n, \delta z_{k,n}^+) + (u_k^j(\tau_j) - s^j, \delta z_k(\tau_j)) = 0, \quad \text{(42a)}$$

$$\sum_{n=0}^{N_j-1} \left[\overline{J}_{n,u}^{j\prime}(q^j, u_k^j)(\delta u_k) - ((\partial_t z_k^j, \delta u_k))_n + a_{n,u}^\prime(u_k^j)(\delta u_k, z_k^j) \right]$$
$$- \sum_{n=1}^{N_j-1} ([z_k^j]_n, \delta u_{k,n}^-) + (z_k^j(\tau_{j+1}) - \lambda^{j+1}, \delta u_k(\tau_{j+1})) = 0. \quad \text{(42b)}$$

Additional linearized equations needed to compute sensitivities (e.g., the tangent and additional adjoint equations in the IMS case) are discretized analogously. In the context of parabolic OCP without shooting, this is described in detail in [2]. Note that the control has not been discretized.

Remark 10 We will only consider the case $r = 0$, where the Galerkin method is equivalent to the backward Euler time-stepping scheme up to a quadrature error induced by the box rule. To keep the presentation short, we omit the discussion of alternatives like the continuous Galerkin method of order r (cG(r) method); details can be found, e.g., in [24].

5.2 *Space-Time Discretization*

For discretizing the spatial variables, we use a conforming finite element method on a shape-regular mesh $\mathcal{T}_h$ decomposing the domain $\overline{\Omega}$ into closed cells K (for definitions and details we refer to, e.g., the textbook [7]). Spatially discrete quantities are indexed by h, where $h|_K := h_K$ is the (cellwise constant) diameter of K. On the mesh $\mathscr{T}_h$, we define the space $V_h^s \subset V$ of finite element functions of polynomial degree s by

$$V_h^s := \left\{ v_h \,\middle|\, v_h|_K \in Q^s(K), K \in \mathscr{T}_h \right\}.$$

We denote by $Q^s(K)$ the space of functions that result from isoparametric transformations of polynomials defined on a reference unit cell $\hat{K}$. As we only consider the case $s = 1$, we exclusively deal with bilinear transformations. To formulate the fully discretized problem, we need the function space

$$X_{k,h}^{r,s}(I_j) := \{v_{kh} \in L^2(I_j; V_h^s) \mid v_{kh}(t_j^0) \in V_h^s, v_{kh}|_{I_j^n} \in P_r(I_j^n; V_h^s), n = 1, \ldots, N_j\}$$

which consists of all piecewise polynomials of degree r on the time intervals with values in the finite element space V_h^s. The space-time discrete problem now consists in finding $u_{kh}, z_{kh} \in X_{k,h}^{r,s}(I_j)$, such that for all $\delta u_{kh}, \delta z_{kh} \in X_{k,h}^{r,s}(I_j)$ the following equations hold:

$$\sum_{n=0}^{N_j-1} \left[((\partial_t u_{kh}^j, \delta z_{kh}))_n + a_n(u_{kh}^j)(\delta z_{kh}) + b_n(q_{kh}^j)(\delta z_{kh}) - ((f|_{I_j^n}, \delta z_{kh}))_n \right]$$
$$+ \sum_{n=1}^{N_j-1} ([u_{kh}^j]_{n-1}, \delta z_{kh,n-1}^+) + (u_{kh}^j(\tau_j) - s^j, \delta z_{kh}(\tau_j)) = 0, \quad \text{(43a)}$$

$$\sum_{n=0}^{N_j-1} \left[\bar{J}_{n,u}^{j'}(q_{kh}^j, u_{kh}^j)(\delta u_{kh}) - ((\partial_t z_{kh}^j, \delta u_{kh}))_n + a_{n,u}'(u_{kh}^j)(\delta u_{kh}, z_{kh}^j) \right]$$
$$- \sum_{n=1}^{N_j-1} ([z_{kh}^j]_n, \delta u_{kh,n}^-) + (z_{kh}^j(\tau_{j+1}) - \lambda^{j+1}, \delta u_{kh}(\tau_{j+1})) = 0. \quad \text{(43b)}$$

This system is, apart from the additional index h and the different function spaces, evidently equal to (42).

Remark 11 Note that for the fully discrete formulation, we discretized the control q. In doing so, we followed the approach suggested in [15] and let the control discretization be induced by the corresponding one for the states u and z, thus we did not describe it in detail. Alternatively, one might discretize q explicitly; this allows for a coarser resolution of the control in both space and time, either by choosing a coarser mesh (here, one could use hierarchically structured meshes for the state and control variables) or by employing time-stepping schemes or finite element methods of lower order than those chosen for state discretization.

Remark 12 We are now in a position to compare the dimension of systems (19) for IMS and (23) for DMS. Assume I is decomposed into 10 shooting intervals each of which is discretized by 50 time steps, and the spatial domain Ω is the unit square divided into 256 identical elements (for other configurations, see the examples in Sect. 6). In the IMS case, the solution vector $(\delta\mathbf{s}, \delta\boldsymbol{\lambda})$ of (19) comprises 10 initial values for the state and 10 for the adjoint solution, each of the size of the spatial discretization. This amounts to a system dimension of 5120. With DMS, the control enters into the system. As we resolve the control completely in the IMS case, we do not use any condensing techniques here, either, in order to keep the results comparable and the comparison fair. Thus the control is distributed in space and time and comprises a total of $10 \cdot 50 \cdot 256 = 128{,}000$ degrees of freedom. Therefore, the solution vector $(\delta\mathbf{q}, \delta\mathbf{s}, \delta\boldsymbol{\lambda})$ of (23) has dimension 133,120.

6 Numerical Results

In this section, we discuss the practical realization of the theoretical results from Sect. 3 by regarding two examples. In Sect. 6.1 we consider the case of a linear-quadratic optimal control problem. Afterwards, we introduce a nonlinear reaction term into the PDE side condition and discuss the semilinear case in Sect. 6.2. All computations have been done using the finite element software `deal.ii` and rely upon the discretization routines presented in the last section.

6.1 Linear Example

The following linear quadratic OCP is considered on the space-time domain $\Omega \times \bar{I} = ([-1;3] \times [-1,1]) \times [0,1]$ and aims at matching a given state profile $\hat{u}_T$ at the time interval endpoint $T = 1$:

$$\min_{(q,u)} \left[\frac{1}{2} \|u(x,1) - \hat{u}_T(x)\|^2_{L^2(\Omega)} + \frac{\alpha}{2} \int_0^1 \|q(x,t)\|^2_{L^2(\Omega)}\, dt \right],$$

subject to the parameterized nonstationary Helmholtz equation

$$\begin{aligned} \partial_t u(x,t) - \Delta u(x,t) - \omega u(x,t) &= q(x,t) && \text{in } \Omega \times (0,1], \\ u(x,t) &= 0 && \text{on } \partial\Omega \times [0,1], \\ u(x,0) &= \max\left\{0, \cos\left(\frac{\pi}{2}x_1\right)\cos\left(\frac{\pi}{2}x_2\right)\right\} && \text{on } \Omega \end{aligned}$$

The prescribed profile is chosen as $\hat{u}_T(x_1,x_2) = \min\left\{0, \cos\left(\frac{\pi}{2}x_1\right)\cos\left(\frac{\pi}{2}x_2\right)\right\}$. Thus, we expect the state solution to be a cosine bump moving from the left to the right half of the spatial domain over time, thereby changing its sign.

We compute solutions of this problem for different values of the parameter ω and $\alpha = 0.01$ by means of IMS and DMS (in the variant of Sect. 3.2) We use a four times globally refined spatial mesh (512 cells) and five equidistant shooting intervals discretized each by 100 time steps. The results are shown in Table 1: from left to right, we see that only one Newton step is needed, furthermore the number of GMRES iterations, the functional value, the residual of the respective shooting system (which also yields the stopping criterion) and the computing time measured in seconds.

Both methods have been implemented as described in Sect. 3, without any additional tuning (like condensing, reduction of control spaces etc.). Since we use the same implementation for both linear and nonlinear problems, we solve the shooting system by a Newton-type method, which requires only one iteration in the current example, as can be expected for a linear problem. For increasing ω, the number of inner GMRES iterations also increases in both cases, which reflects the worsening conditioning of the respective problems; indeed, there is a value of ω where five shooting intervals are not sufficient to solve the problem. With DMS, altogether more GMRES steps are needed than with IMS, which is due to the much larger linear system. The functional values $J(q,u)$ coincide for both methods, and also the shooting residual $\|F\|$ is of comparable size. However, the DMS algorithm takes longer (by a factor of 1.5 up to 2) than IMS to solve the problem with this same accuracy. Finally, in Fig. 1 we see that after convergence of the shooting methods (here: IMS) the expected wandering and inversion of the cosine bump is reproduced.

Table 1 Comparison of IMS (left) and DMS (right) for varying ω (required: $\|F\| < 5.0e^{-5}$) in a linear framework

ω	$\#_{New}$	$\#_{GMRES}$	$J(q,u)$	$\|F\|$	$t(s)$	$\#_{New}$	$\#_{GMRES}$	$J(q,u)$	$\|F\|$	$t(s)$
0	1	52	0.0446	$1.6e^{-11}$	1497	1	110	0.0446	$1.9e^{-10}$	2507
1	1	64	0.0367	$1.8e^{-11}$	1825	1	128	0.0367	$2.2e^{-10}$	2909
2	1	76	0.0290	$2.1e^{-11}$	2149	1	156	0.0290	$2.3e^{-10}$	3531
3	1	83	0.0218	$2.4e^{-11}$	2347	1	192	0.0218	$2.3e^{-10}$	4360
4	1	130	0.0163	$2.6e^{-11}$	3601	1	248	0.0163	$2.6e^{-10}$	5586
5	1	165	0.0148	$2.9e^{-11}$	4571	1	416	0.0148	$2.8e^{-10}$	9423

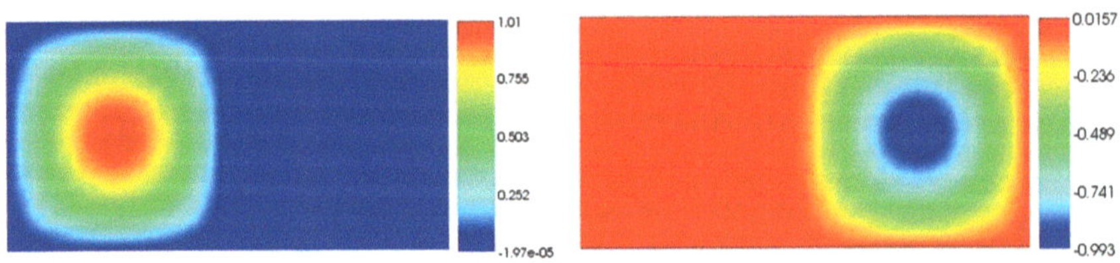

Fig. 1 Contour plot of the IMS solution on 5 shooting intervals after convergence: initial time $T = 0$ (*left*), final time $T = 1$ (*right*)

Table 2 Comparison of IMS (left) and DMS (right) for varying ω (required: $\|F\| < 1.0e^{-3}$) in a nonlinear framework

ω	$\#_{New}$	$\#_{GMRES}$	$J(q,u)$	$\|F\|$	$t(s)$	$\#_{New}$	$\#_{GMRES}$	$J(q,u)$	$\|F\|$	$t(s)$
0	4	24/51	0.1639	$3.1e^{-6}$	2530	4	28/53	0.1639	$3.1e^{-5}$	2088
1	4	26/52	0.1420	$6.4e^{-6}$	2795	4	38/62	0.1420	$1.5e^{-5}$	2427
2	4	28/56	0.1187	$2.5e^{-6}$	3118	4	43/74	0.1187	$9.0e^{-5}$	2926
3	4	28/75	0.0948	$3.9e^{-6}$	4201	4	51/84	0.0948	$1.4e^{-4}$	3280
4	4	28/79	0.0735	$5.6e^{-6}$	4713	4	68/108	0.0735	$2.1e^{-4}$	4201
5	4	28/94	0.0645	$1.2e^{-5}$	5658	4	80/139	0.0645	$2.6e^{-4}$	5376

6.2 *Nonlinear Example*

The second problem is a slight modification of the first one. We now choose the regularization parameter as $\alpha = 0.05$. Furthermore, we add a polynomial nonlinearity to the PDE side condition and consider the problem

$$\partial_t u(x,t) - \Delta u(x,t) - \omega u(x,t) + u(x,t)^3 = q(x,t) \quad \text{in } \Omega \times (0,1],$$

whereas the initial condition, the boundary values, the objective functional and the computational domain are chosen identical to the configuration in Sect. 6.1.

We need several Newton iterations before convergence, but at the beginning, the shooting variables are still far away from their true values. It is thus not necessary to carry out the first Newton steps on a highly refined spatial mesh. On the contrary, the shooting process becomes far more efficient if the first iterations are carried out on a coarse mesh until a good approximation has been obtained and if the spatial mesh is only then refined. In the above example, we therefore start on a mesh of only 8 cells and alternate between computing a Newton update for the shooting variables and refining the spatial mesh. This is repeated until we reach the finest mesh with 512 cells. Table 2 shows the results of this approach with global mesh refinement; in the second column of the respective method, we have given the minimum and maximum number of GMRES iterations needed within one Newton step. We emphasize that this process carries the potential of including adaptive mesh refinement into the shooting process, which is a starting point for further research. Furthermore, the IMS and DMS methods are comparable w.r.t. computing time, which is contrary to the linear example above. An extended comparison of IMS and DMS has to include additional control or state constraints, which is currently being examined.

7 Conclusion and Outlook

To underline the algebraic equivalence of DMS and IMS on an abstract function space level, we have rigorously derived the two approaches starting from a common formulation. Furthermore, we have given a detailed description of their algorithmic realization. We have shown that our DMS approach is equivalent to the 'classical' approach known in ODE context, which leads, in contrast to our approach, to a reduced formulation. The advantage of keeping the derivation of a DMS method at the continuous level, in contrast to the 'first discretize then optimize' classical approach, is the possibility to derive an a posteriori error estimation for the space and time discretization, including the continuity conditions in the shooting system. This important aspect is part of our current research.

References

1. Albersmeyer, J.: Adjoint-based algorithms and numerical methods for sensitivity generation and optimization of large scale dynamic systems. Ph.D. thesis, Heidelberg University (2010)
2. Becker, R., Meidner, D., Vexler, B.: Efficient numerical solution of parabolic optimization problems by finite element methods. Optim. Methods Softw. **22**, 813–833 (2007)
3. Bock, H.G.: Numerical treatment of inverse problems in chemical reaction kinetics. In: Modelling of Chemical Reaction Systems. Proceedings of an International Workshop in Heidelberg, pp. 102–125. Springer, Berlin (1981)
4. Bock, H.G.: Recent advances in parameter identification problems for ODE. In: Numerical Treatment of Inverse Problems in Differential and Integral Equations, pp. 95–121. Birkhäuser, Boston (1983)
5. Bock, H.G., Plitt, K.: A multiple shooting algorithm for direct solution of optimal control problems. In: Proceedings of the 9th IFAC World Congress Budapest. Pergamon Press (1984)
6. Bock, H.G., Kostina, E., Schlöder, J.P.: Direct multiple shooting and generalized Gauss-Newton method for parameter estimation problems in ODE models. In: Carraro, T., Geiger, M., Körkel, S., Rannacher, R. (eds.) Multiple Shooting and Time Domain Decomposition Methods. Contributions in Mathematical and Computational Sciences. Springer, Berlin/Heidelberg (2015)
7. Braess, D.: Finite Elements: Theory, Fast Solvers, and Applications in Solid Mechanics, 3rd edn. Cambridge University Press, Cambridge (2007)
8. Bulirsch, R., Stoer, J.: Introduction to Numerical Analysis. Texts in Applied Mathematics, vol. 12, 3rd edn. Springer, New York (2002)
9. Carraro, T., Geiger, M., Rannacher, R.: Indirect multiple shooting for nonlinear parabolic optimal control problems with control constraints. SIAM J. Sci. Comput. **36**, A452–A481 (2014)
10. Dautray, R., Lions, J.-L.: Mathematical Analysis and Numerical Methods for Science and Technology. Evolution Problems I, vol. 5. Springer, Berlin (1992)
11. Gander, M., Vandewalle, S.: Analysis of the parareal time-parallel time-integration method. SIAM J. Sci. Comput. **29**, 556–578 (2007)
12. Geiger, M.E.: Adaptive multiple shooting for boundary value problems and constrained parabolic optimization problems. Ph.D. thesis, Heidelberg University (2015)
13. Heinkenschloss, M.: A time-domain decomposition iterative method for the solution of distributed linear quadratic optimal control problems. J. Comput. Appl. Math. **173**, 169–198 (2005)
14. Hesse, H.K., Kanschat, G.: Mesh adaptive multiple shooting for partial differential equations. Part I: linear quadratic optimal control problems. J. Numer. Math. **17**, 195–217 (2009)
15. Hinze, M.: A variational discretization concept in control constrained optimization: the linear-quadratic case. Comput. Optim. Appl. **30**, 45–63 (2005)
16. Hinze, M., Pinnau, R., Ulbrich, M., Ulbrich, S.: Optimization with PDE Constraints. Mathematical Modelling, vol. 23. Springer, New York (2009)

17. Kirches, C., Bock, H.G., Schlöder, J.P., Sager, S.: Block-structured quadratic programming for the direct multiple shooting method for optimal control. Optim. Methods Softw. **26**(2), 239–257 (2011)
18. Kirches, C., Wirsching, L., Bock, H.G., Schlöder, J.P.: Efficient direct multiple shooting for nonlinear model predictive control on long horizons. J. Process Control **22**(3), 540–550 (2012)
19. Kirches, C., Bock, H.G., Schlöder, J.P., Sager, S.: Complementary condensing for the direct multiple shooting method. In: Modeling, Simulation and Optimization of Complex Processes. Springer, Berlin (2012)
20. Leineweber, D.B.: The theory of MUSCOD in a nutshell. Diploma thesis, Heidelberg University (1995)
21. Leineweber, D.B., Bauer, I., Bock, H.G., Schlöder, J.P.: An efficient multiple shooting based reduced {SQP} strategy for large-scale dynamic process optimization. Part 1: theoretical aspects. Comput. Chem. Eng. **27**(2), 157–166 (2003)
22. Lions, J.-L., Maday, Y., Turinici, G.: Resolution d'edp par un schema en temps 'parareel'. C. R. Acad. Sci. Paris Ser. I **332**, 661–668 (2001)
23. Lory, P.: Enlarging the domain of convergence for multiple shooting by the homotopy method. Numer. Math. **35**, 231–240 (1980)
24. Meidner, D.: Adaptive space-time finite element methods for optimization problems governed by nonlinear parabolic systems. Ph.D. thesis, Heidelberg University (2008)
25. Pontryagin, L.S., Boltyanski, V.G., Gamkrelidze, R.V., Miscenko, E.F.: The Mathematical Theory of Optimal Processes. Wiley, Chichester (1962)
26. Potschka, A.: A direct method for the numerical solution of optimization problems with time-periodic PDE constraints. Ph.D. thesis, Heidelberg University (2012)
27. Potschka, A.: Direct multiple shooting for parabolic PDE constrained optimization. In: Carraro, T., Geiger, M., Körkel, S., Rannacher, R. (eds.) Multiple Shooting and Time Domain Decomposition Methods. Contributions in Mathematical and Computational Sciences. Springer, Berlin/Heidelberg (2015)
28. Schäfer, A., Kuhl, P., Diehl, M., Schlöder, J., Bock, H.G.: Fast reduced multiple shooting methods for nonlinear model predictive control. Chem. Eng. Process. **46**(11), 1200 - 1214 (2007)
29. Serban, R., Li, S., Petzold, L.: Adaptive algorithms for optimal control of time-dependent partial differential-algebraic equation systems. Int. J. Numer. Methods Eng. **57**, 1457–1469 (2003)
30. Tröltzsch, F.: Optimal Control of Partial Differential Equations: Theory, Methods and Applications. American Mathematical Society, Providence (2010)
31. Ulbrich, S.: Generalized SQP methods with 'parareal' time-domain decomposition for time-dependent PDE-constrained optimization. In: Real-time PDE-Constrained Optimization. Computational Science and Engineering, vol. 3, pp. 145–168. SIAM, Philadelphia (2007)
32. Wloka, J.: Partial Differential Equations. Cambridge University Press, Cambridge (1987)

50 Years of Time Parallel Time Integration

Martin J. Gander

Abstract Time parallel time integration methods have received renewed interest over the last decade because of the advent of massively parallel computers, which is mainly due to the clock speed limit reached on today's processors. When solving time dependent partial differential equations, the time direction is usually not used for parallelization. But when parallelization in space saturates, the time direction offers itself as a further direction for parallelization. The time direction is however special, and for evolution problems there is a causality principle: the solution later in time is affected (it is even determined) by the solution earlier in time, but not the other way round. Algorithms trying to use the time direction for parallelization must therefore be special, and take this very different property of the time dimension into account.

We show in this chapter how time domain decomposition methods were invented, and give an overview of the existing techniques. Time parallel methods can be classified into four different groups: methods based on *multiple shooting*, methods based on *domain decomposition and waveform relaxation*, *space-time multigrid* methods and *direct time parallel methods*. We show for each of these techniques the main inventions over time by choosing specific publications and explaining the core ideas of the authors. This chapter is for people who want to quickly gain an overview of the exciting and rapidly developing area of research of time parallel methods.

1 Introduction

It has been precisely 50 years ago that the first visionary contribution to time parallel time integration methods was made by Nievergelt [64]. We show in Fig. 1 an overview of the many important contributions over the last 50 years to this field of

M.J. Gander (✉)
Section of Mathematics, University of Geneva, Rue du Lievre 2-4, CP 64, 1211 Geneva 4, Switzerland
e-mail: martin.gander@unige.ch

T. Carraro et al. (eds.), *Multiple Shooting and Time Domain Decomposition Methods*, Contributions in Mathematical and Computational Sciences 9,
DOI 10.1007/978-3-319-23321-5_3

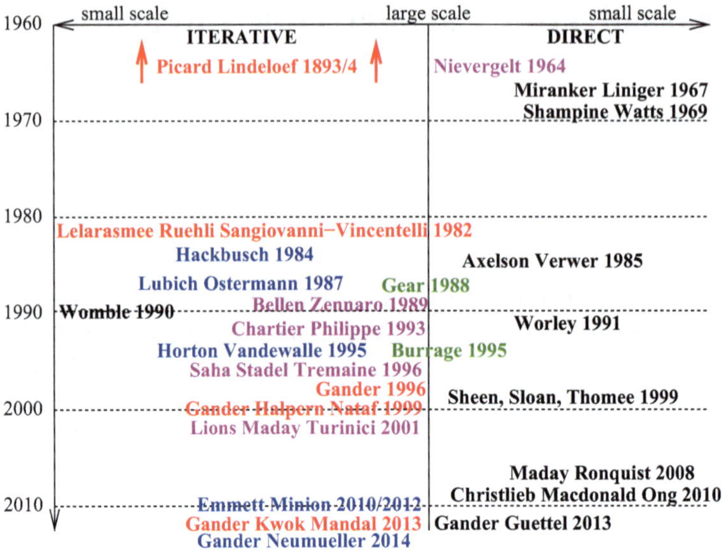

Fig. 1 An overview over important contributions to time parallel methods

research. The methods with iterative character are shown on the left, and the direct time parallel solvers on the right, and large scale parallel methods are more toward the center of the figure, whereas small scale parallel methods useful for multicore architectures are more towards the left and right borders of the plot.

We also identified the four main classes of space-time parallel methods in Fig. 1 using color:

1. methods based on *multiple shooting* are shown in magenta,
2. methods based on *domain decomposition and waveform relaxation* are shown in red,
3. methods based on *multigrid* are shown in blue,
4. and *direct time parallel methods* are shown in black.

There have also been already overview papers, shown in green in Fig. 1, namely the paper by Gear [40], and the book by Burrage [12].

The development of time parallel time integration methods spans now half a century, and various methods have been invented and reinvented over this period. We give a detailed account of the major contributions by presenting seminal papers and explaining the methods invented by their authors.

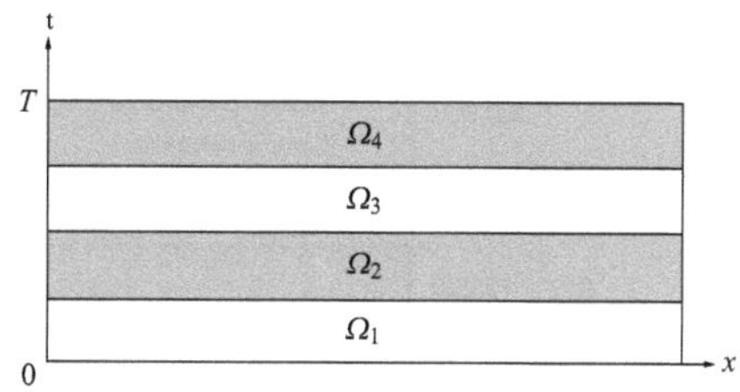

Fig. 2 Decomposition of the space-time domain in time for multiple shooting type methods

2 Shooting Type Time Parallel Methods

Time parallel methods based on shooting solve evolution problems in parallel using a decomposition of the space-time domain in time, as shown in Fig. 2. An iteration is then defined, or some other procedure, which only uses solutions in the time subdomains, to obtain an approximate solution over the entire time interval $(0, T)$.

2.1 Nievergelt 1964

Nievergelt was the first to consider a pure time decomposition for the parallel solution of evolution problems [64]. He stated precisely 50 years ago at the time of writing of this chapter, how important parallel computing was to become in the near future:

> For the last 20 years, one has tried to speed up numerical computation mainly by providing ever faster computers. **Today, as it appears that one is getting closer to the maximal speed of electronic components**, emphasis is put on allowing operations to be performed in parallel. In the near future, much of numerical analysis will have to be recast in a more 'parallel' form.

As we now know, the maximal speed of electronic components was only reached 40 years later, see Fig. 3.

Nievergelt presents a method for parallelizing the numerical integration of an ordinary differential equation, a process which "by all standard methods, is entirely serial". We consider in Nievergelt's notation the ordinary differential equation (ODE)

$$y' = f(x, y), \quad y(a) = y_0, \tag{1}$$

and we want to approximate its solution on the interval $[a, b]$. Such an approximation can be obtained using any numerical method for integrating ODEs, so-called time stepping methods, but the process is then entirely sequential. Nievergelt proposes instead to partition the interval $[a, b]$ into subintervals $x_0 = a < x_1 < \ldots < x_N = b$, as shown in his original drawing in Fig. 4, and then introduces the following direct

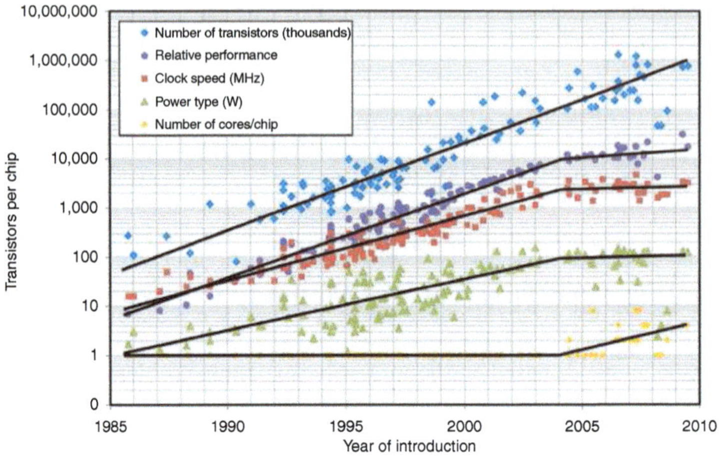

Fig. 3 Maximal speed of electronic components reached 40 years after the prediction of Nievergelt (taken from a talk of Bennie Mols at the VINT symposium 12.06.2013)

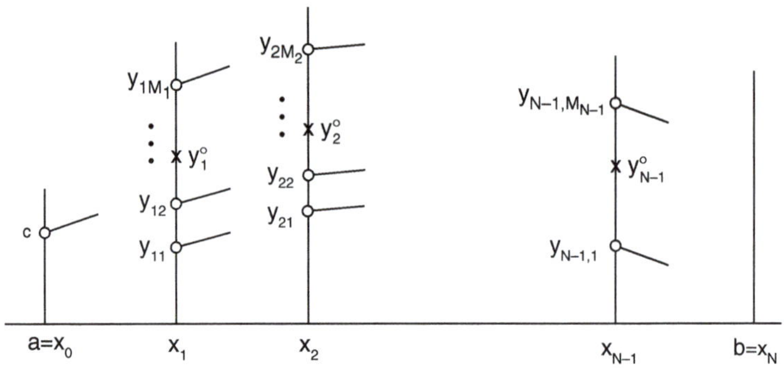

Fig. 4 First idea by Nievergelt to obtain a parallel algorithm for the integration of a first order ODE

time parallel solver:

1. Compute a rough prediction y_i^0 of the solution $y(x_i)$ at each interface (see Fig. 4), for example with one step of a numerical method with step size $H = (b - a)/N$.
2. For a certain number M_i of starting points $y_{i,1}, \dots, y_{i,M_i}$ at x_i in the neighborhood of the approximate solution y_i^0 (see Fig. 4), compute accurate (we assume here for simplicity exact) trajectories $y_{i,j}(x)$ in parallel on the corresponding interval $[x_i, x_{i+1}]$, and also $y_{0,1}(x)$ on the first interval $[x_0, x_1]$ starting at y_0.
3. Set $Y_1 := y_{0,1}(x_1)$ and compute sequentially for each $i = 1, \dots, N - 1$ the interpolated approximation by
   - finding the interval j such that $Y_i \in [y_{i,j}, y_{i,j+1}]$,
   - determining p such that $Y_i = py_{i,j} + (1 - p)y_{i,j+1}$, i.e. $p = \frac{Y_i - y_{i,j+1}}{y_{i,j} - y_{i,j+1}}$,

- setting the next interpolated value at x_{i+1} to $Y_{i+1} := py_{i,j}(x_{i+1}) + (1 - p)y_{i,j+1}(x_{i+1})$.

For linear ODEs, this procedure does actually produce the same result as the evaluation of the accurate trajectory on the grid, i.e. $Y_i = y(x_i)$ in our case of exact local solves, there is no interpolation error, and it would in fact suffice to have only two trajectories, $M_i = 2$ in each subinterval, since one can also extrapolate.

In the non-linear case, there is an additional error due to interpolation, and Nievergelt defines a class of ODEs for which this error remains under control if one uses Backward Euler for the initial guess with a coarse step H, and also Backward Euler for the accurate solver with a much finer step h, and he addresses the question on how to choose M_i and the location of the starting points $y_{i,j}$ in the neighborhood. He then concludes by saying

> The integration methods introduced in this paper are to be regarded as tentative examples of a much wider class of numerical procedures in which parallelism is introduced at the expense of redundancy of computation. As such, their merits lie not so much in their usefulness as numerical algorithms as in their potential as prototypes of better methods based on the same principle. It is believed that more general and improved versions of these methods will be of great importance when computers capable of executing many computations in parallel become available.

What a visionary statement again! The method proposed is inefficient compared to any standard serial integration method, but when many processors are available, one can compute the solution faster than with just one processor. This is the typical situation for time parallel time integration methods: the goal is not necessarily perfect scalability or efficiency, it is to obtain the solution faster than sequentially.

The method of Nievergelt is in fact a direct method, and we will see more such methods in Sect. 5, but it is the natural precursor of the methods based on multiple shooting we will see in this section.

2.2 Bellen and Zennaro 1989

The first to pick up the idea of Nievergelt again and to formally develop an iterative method to connect trajectories were Bellen and Zennaro in [6]:

> In addition to the two types of parallelism mentioned above, we wish to isolate a third which is analogous to what Gear has more recently called parallelism *across the time*. Here it is more appropriately called parallelism across the steps. In fact, the algorithm we propose is a realization of this kind of parallelism. Without discussing it in detail here, we want to point out that **the idea is indeed that of multiple shooting** and parallelism is introduced at the cost of redundancy of computation.

Bellen and Zennaro define their method directly at the discrete level, for a recurrence relation of the form

$$y_{n+1} = F_{n+1}(y_n), \quad y_0 \text{ known.} \tag{2}$$

This process looks entirely sequential, one needs to know y_n in order to be able to compute y_{n+1}. Defining the vector of unknowns $\mathbf{y} := (y_0, y_1, \ldots, y_n, \ldots)$ however, the recurrence relation (2) can be written simultaneously over many levels in the fixed point form

$$\mathbf{y} = \boldsymbol{\phi}(\mathbf{y}), \tag{3}$$

where $\boldsymbol{\phi}(\mathbf{y}) = (y_0, F_1(y_0), F_2(y_1), \ldots, F_n(y_{n-1}), \ldots)$. Bellen and Zennaro propose to apply a variant of Newton's method called Steffensen's method to solve the fixed point equation (3). Like when applying Newton's method and simplifying, as we will see in detail in the next subsection, this leads to an iteration of the form

$$\mathbf{y}^{k+1} = \boldsymbol{\phi}(\mathbf{y}^k) + \Delta\boldsymbol{\phi}(\mathbf{y}^k)(\mathbf{y}^{k+1} - \mathbf{y}^k), \tag{4}$$

where $\Delta\boldsymbol{\phi}$ is an approximation to the differential $D\boldsymbol{\phi}$, and they choose as initial guess $y_n^0 = y_0$. Steffensen's method for a nonlinear scalar equation of the form $f(x) = 0$ is

$$\begin{aligned} x_{k+1} &= x_k - g(x_k)^{-1} f(x_k) \\ g(x) &:= \frac{f(x + f(x)) - f(x)}{f(x)}, \end{aligned}$$

and one can see how the function $g(x)$ becomes a better and better approximation of the derivative $f'(x)$ as $f(x)$ goes to zero. As Newton's method, Steffensen's method converges quadratically once one is close to the solution.

Bellen and Zennaro show several results about Steffensen's method (4) applied to the fixed point problem (3):

1. They observe that each iteration gives one more exact value, i.e. after one iteration, the exact value $y_1^1 = y_1$ is obtained, and after two iterations, the exact value $y_2^2 = y_2$ is obtained, and so on. Hence convergence of the method is guaranteed if the vector $\mathbf{y}^k$ is of finite length.
2. They prove that convergence is locally quadratic, as it holds in general for Steffensen's method applied to non-linear problems.
3. The corrections at each step of the algorithm can be computed in parallel.
4. They also present numerically estimated speedups of 29–53 for a problem with 400 steps.

In contrast to the ad hoc interpolation approach of Nievergelt, the method of Bellen and Zennaro is a systematic parallel iterative method to solve recurrence relations.

2.3 Chartier and Philippe 1993

Chartier and Philippe return in [13] to the formalization of Bellen and Zennaro,[1] which was given at the discrete level, and formulate their parallel integration method at the continuous level for evolution problems:

> Parallel algorithms for solving initial value problems for differential equations **have received only marginal attention in the literature compared to the enormous work devoted to parallel algorithms for linear algebra**. It is indeed generally admitted that the integration of a system of ordinary differential equations in a step-by-step process is inherently sequential.

The underlying idea is to apply a shooting method, which was originally developed for boundary value problems, see [47] and references therein, to an initial value problem, see also [48]. For a boundary value problem of the form

$$u'' = f(u), \quad u(0) = a, \quad u(1) = b, \tag{5}$$

a shooting method also considers the same differential equation, but as an initial value problem,

$$u'' = f(u), \quad u(0) = a, \quad u'(0) = s, \tag{6}$$

and one then tries to determine the so-called shooting parameter s, the 'angle of the cannon to shoot with', such that the solution passes through the point $u(1) = b$, which explains the name of the method. To determine the shooting parameter s, one needs to solve in general a non-linear equation, which is preferably done by Newton's method, see for example [47].

If the original problem is however already an initial value problem,

$$u' = f(u), \quad u(0) = u^0, \qquad x \in [0, 1], \tag{7}$$

then there is in no target to hit at the other end, so at first sight it seems shooting is not possible. To introduce targets, one uses the idea of *multiple shooting*: one splits the time interval into subintervals, for example three, $[0, \frac{1}{3}]$, $[\frac{1}{3}, \frac{2}{3}]$, $[\frac{2}{3}, 1]$, and then solves on each subinterval the underlying initial value problem

$$\begin{array}{lll} u_0' = f(u_0), & u_1' = f(u_1), & u_2' = f(u_2), \\ u_0(0) = U_0, & u_1(\frac{1}{3}) = U_1, & u_2(\frac{2}{3}) = U_2, \end{array}$$

[1] "In diesem Artikel studieren wir verschiedene Versionen einer Klasse paralleler Algorithmen, die ursprünglich von A. Bellen und M. Zennaro für Differenzengleichungen konzipiert und von ihnen 'across the steps' Methode genannt worden ist."

together with the matching conditions

$$U_0 = u^0, \quad U_1 = u_0(\frac{1}{3}, U_0), \quad U_2 = u_1(\frac{2}{3}, U_1).$$

Since the shooting parameters U_n, $n = 0, 1, 2$ are not known (except for $U_0 = u^0$), this leads to a system of non-linear equations one has to solve,

$$F(\mathbf{U}) := \begin{pmatrix} U_0 - u^0 \\ U_1 - u_0(\frac{1}{3}, U_0) \\ U_2 - u_1(\frac{2}{3}, U_1) \end{pmatrix} = 0, \qquad \mathbf{U} = (U_0, U_1, U_2)^T.$$

If we apply Newton's method to this system, like in the classical shooting method, to determine the shooting parameters, we obtain for $k = 0, 1, 2, \ldots$ the iteration

$$\begin{pmatrix} U_0^{k+1} \\ U_1^{k+1} \\ U_2^{k+1} \end{pmatrix} = \begin{pmatrix} U_0^k \\ U_1^k \\ U_2^k \end{pmatrix} - \begin{bmatrix} 1 & & \\ -\frac{\partial u_0}{\partial U_0}(\frac{1}{3}, U_0^k) & 1 & \\ & -\frac{\partial u_1}{\partial U_1}(\frac{2}{3}, U_1^k) & 1 \end{bmatrix}^{-1} \begin{pmatrix} U_0^k - u^0 \\ U_1^k - u_1(\frac{1}{3}, U_0^k) \\ U_2^k - u_1(\frac{2}{3}, U_1^k) \end{pmatrix}.$$

Multiplying through by the Jacobian matrix, we find the recurrence relation

$$\begin{aligned} U_0^{k+1} &= u^0, \\ U_1^{k+1} &= u_0(\tfrac{1}{3}, U_0^k) + \tfrac{\partial u_0}{\partial U_0}(\tfrac{1}{3}, U_0^k)(U_0^{k+1} - U_0^k), \\ U_2^{k+1} &= u_1(\tfrac{2}{3}, U_1^k) + \tfrac{\partial u_1}{\partial U_1}(\tfrac{2}{3}, U_1^k)(U_1^{k+1} - U_1^k). \end{aligned}$$

In the general case with N shooting intervals, solving the multiple shooting equations using Newton's method gives thus a recurrence relation of the form

$$\begin{aligned} U_0^{k+1} &= u^0, \\ U_{n+1}^{k+1} &= u_n(t_{n+1}, U_n^k) + \tfrac{\partial u_n}{\partial U_n}(t_{n+1}, U_n^k)(U_n^{k+1} - U_n^k), \quad n = 0, 1, 2, \ldots N, \end{aligned} \tag{8}$$

and we recognize the form (4) of the method by Bellen and Zennaro. Chartier and Philippe prove that (8) converges locally quadratically. They then however already indicate that the method is not necessarily effective on general problems, and restrict their analysis to dissipative right hand sides, for which they prove a global convergence result. Finally, also discrete versions of the algorithm are considered.

2.4 Saha, Stadel and Tremaine 1996

Saha, Stadel and Tremaine cite the work of Bellen and Zennaro [6] and Nievergelt [64] as sources of inspiration, but mention already the relation of their algorithm to waveform relaxation [52] in their paper on the integration of the solar system over very long time [67]:

> We describe how long-term solar system orbit integration could be implemented on a parallel computer. The interesting feature of our algorithm is that **each processor is assigned not to a planet or a pair of planets but to a time-interval**. Thus, the 1st week, 2nd week, ..., 1000th week of an orbit are computed concurrently. The problem of matching the input to the $(n+1)$-st processor with the output of the n-th processor can be solved efficiently by an iterative procedure. Our work is related to the so-called **waveform relaxation methods**....

Consider the system of ordinary differential equations

$$\dot{y} = f(y), \quad y(0) = y_0,$$

or equivalently the integral formulation

$$y(t) = y(0) + \int_0^t f(y(s))ds.$$

Approximating the integral by a quadrature formula, for example the midpoint rule, we obtain for each time t_n for $y(t_n)$ the approximation

$$y_n = y_0 + h \sum_{m=0}^{n-1} f(\frac{1}{2}(y_{m+1} + y_m)), \quad n = 1, \ldots, N. \tag{9}$$

Collecting the approximations y_n in a vector $\mathbf{y} := (y_0, y_1, \ldots, y_N)$, the relation (9) can again be written simultaneously over many steps as a fixed point equation of the form

$$\mathbf{y} = F(\mathbf{y}), \tag{10}$$

which can be solved by an iterative process. Note that the quadrature formula (9) can also be written by reusing the sums already computed at earlier steps,

$$y_n = y_{n-1} + hf(\frac{1}{2}(y_n + y_{n-1})), \quad n = 1, \ldots, N, \tag{11}$$

so the important step here is not the use of the quadrature formula. The interesting step comes from the application of Saha, Stadel and Tremaine, namely a Hamiltonian problem with a small perturbation:

$$\dot{p} = -\partial_q H, \; \dot{q} = \partial_p H, \quad H(p, q, t) = H^0(p) + \epsilon H^1(p, q, t).$$

Denoting by $y := (p, q)$, and $f(y) := (-H_q(y), H_p(y))$, Saha, Stadel and Tremaine derive Newton's method for the associated fixed point problem (10), as Chartier and

Philippe derived (8). Rewriting (8) in their notation gives

$$Y_{n+1}^{k+1} = y_n^{\epsilon}(t_{n+1}, Y_n^k) + \frac{\partial y_n^{\epsilon}}{\partial Y_n}(t_{n+1}, Y_n^k)(Y_n^{k+1} - Y_n^k), \tag{12}$$

where the superscript ϵ denotes the solution of the perturbed Hamiltonian system.

The key new idea of Saha, Stadel and Tremaine is to propose an approximation of the derivative by a cheap difference for the unperturbed Hamiltonian problem,

$$Y_{n+1}^{k+1} = y_n^{\epsilon}(t_{n+1}, Y_n^k) + y_n^0(t_{n+1}, Y_n^{k+1}) - y_n^0(t_{n+1}, Y_n^k). \tag{13}$$

They argue that with the approximation for the Jacobian used in (13), each iteration now improves the error by a factor ϵ; instead of quadratic convergence for the Newton method (12), one obtains linear convergence.

They show numerical results for our solar system: using for H^0 Kepler's law, which leads to a cheap integrable system, and for ϵH^1 the planetary perturbations, they obtain the results shown in Fig. 5. They also carefully verify the possible speedup with this algorithm for planetary simulations over long time. Figure 6 shows the iterations needed to converge to a relative error of $1e-15$ in the planetary orbits.

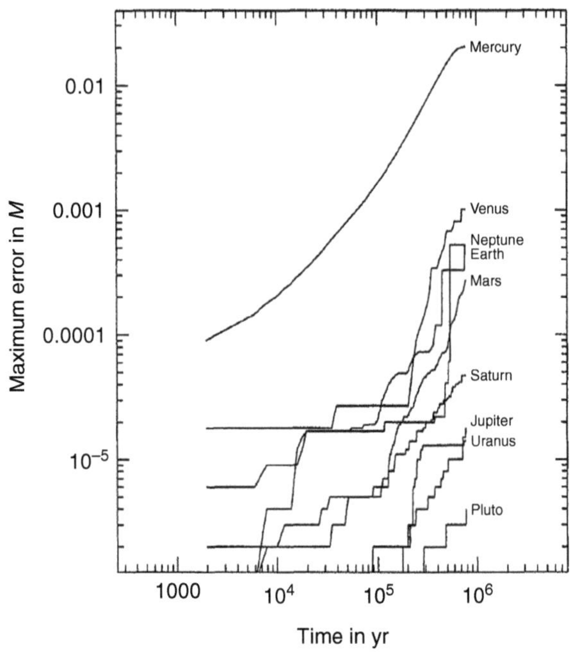

Fig. 5 Maximum error in mean anomaly M versus time, $h = 7\frac{1}{32}$ days, compared to results from the literature, from [67]

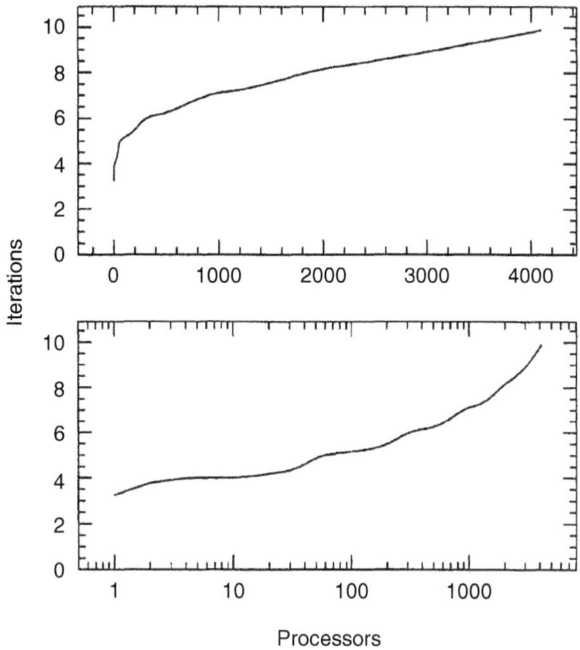

Fig. 6 *Top* linear scaling, and *bottom* logarithmic scaling of the number of iterations to reach a relative error of $1e - 15$ as a function of the number of processors (time intervals) used

2.5 Lions, Maday and Turinici 2001

Lions, Maday and Turinici invented the parareal algorithm in a short note [54], almost independently of earlier work; they only cite the paper by Chartier and Philippe [13]:

> On propose dans cette Note un schéma permettant de profiter d'une architecture parallèle pour la discrétisation en temps d'une équation d'évolution aux dérivées partielles. Cette méthode, basée sur un schéma d'Euler, **combine des résolutions grossières et des résolutions fines et indépendantes en temps en s'inspirant de ce qui est classique en espace. La parallélisation qui en résulte se fait dans la direction temporelle ce qui est en revanche non classique**. Elle a pour principale motivation les problèmes en temps réel, d'où la terminologie proposée de '**pararéel**'.

Lions, Maday and Turinici explain their algorithms on the simple scalar model problem[2]

$$\dot{y} = -ay, \quad \text{on } [0, T], \quad y(0) = y_0. \tag{14}$$

[2] "Pour commencer, on expose l'idée sur l'exemple simple."

The solution is first approximated using Backward Euler on the time grid T_n with coarse time step ΔT,

$$Y_{n+1}^1 - Y_n^1 + a\Delta T Y_{n+1}^1 = 0, \quad Y_0^1 = y_0. \tag{15}$$

The approximate solution values Y_n^1 are then used to compute on each time interval $[T_n, T_{n+1}]$ exactly and in parallel the solution of

$$\dot{y}_n^1 = -a y_n^1, \quad y_n^1(T_n) = Y_n^1. \tag{16}$$

One then performs for $k = 1, 2, \ldots$ the correction iteration

1. Compute the jumps $S_n^k := y_{n-1}^k(T_n) - Y_n^k$.
2. Propagate the jumps $\delta_{n+1}^k - \delta_n^k + a\Delta T \delta_{n+1}^k = S_n^k$, $\delta_0^k = 0$.
3. Set $Y_n^{k+1} := y_{n-1}^k(T_n) + \delta_n^k$ and solve in parallel

$$\dot{y}_n^{k+1} = -a y_n^{k+1}, \quad \text{on } [T_n, T_{n+1}], \quad y_n^{k+1}(T_n) = Y_n^{k+1}.$$

The authors prove the following error estimate for this algorithm [3]

Proposition 1 (Lions, Maday and Turinici 2001) *The parareal scheme is of order k, i.e. there exists c_k s.t.*

$$|Y_n^k - y(T_n)| + \max_{t \in [T_n, T_{n+1}]} |y_n^k(t) - y(t)| \leq c_k \Delta T^k.$$

This result implies that with each iteration of the parareal algorithm, one obtains a numerical time stepping scheme which has a truncation error that is one order higher than before. So for a fixed iteration number k, one can obtain high order time integration methods that are naturally parallel. The authors then note that the same proposition also holds for Forward Euler. In both discretization schemes however, the stability of the higher order methods obtained with the parareal correction scheme degrades with iterations, as shown in Fig. 7 taken from the original publication [54]. The authors finally show two numerical examples: one for a heat equation where they obtain a simulated speedup of a factor 8 with 500 processors, and one for a semi-linear advection diffusion problem, where a variant of the algorithm is proposed by linearization about the previous iterate, since the parareal algorithm was only defined for linear problems. Here, the speedup obtained is 18.

Let us write the parareal algorithm now in modern notation, directly for the non-linear problem

$$u' = f(u), \quad u(t_0) = u_0. \tag{17}$$

[3]"C'est alors un exercice que de montrer la:"

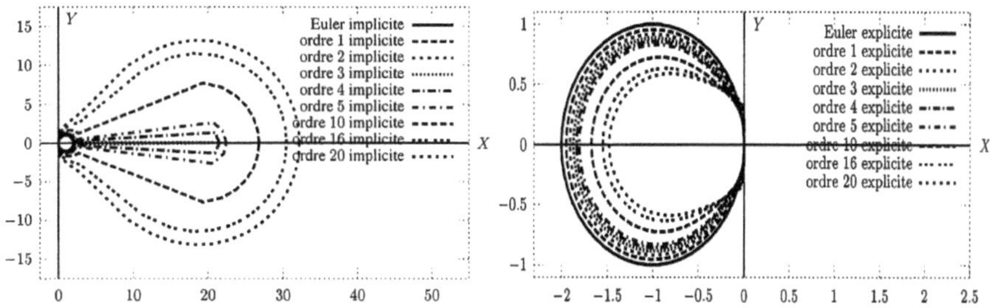

Fig. 7 Stability of the parareal algorithm as function of the iteration, on the *left* for Backward Euler, and on the *right* for Forward Euler

The algorithm is defined using two propagation operators:

1. $G(t_2, t_1, u_1)$ is a rough approximation to $u(t_2)$ with initial condition $u(t_1) = u_1$,
2. $F(t_2, t_1, u_1)$ is a more accurate approximation of the solution $u(t_2)$ with initial condition $u(t_1) = u_1$.

Starting with a coarse approximation U_n^0 at the time points $t_0, t_1, t_2, \ldots, t_N$, for example obtained using G, the parareal algorithm performs for $k = 0, 1, \ldots$ the correction iteration

$$U_{n+1}^{k+1} = F(t_{n+1}, t_n, U_n^k) + G(t_{n+1}, t_n, U_n^{k+1}) - G(t_{n+1}, t_n, U_n^k). \tag{18}$$

Theorem 1 (Parareal is a Multiple Shooting Method [35]) *The parareal algorithm is a multiple shooting method*

$$U_{n+1}^{k+1} = u_n(t_{n+1}, U_n^k) + \frac{\partial u_n}{\partial U_n}(t_{n+1}, U_n^k)(U_n^{k+1} - U_n^k), \tag{19}$$

where the Jacobian has been approximated in (18) by a difference on a coarse grid.

We thus have a very similar algorithm as the one proposed by Saha, Stadel and Tremaine [67], the only difference being that the Jacobian approximation does not come from a simpler model, but from a coarser discretization.

We now present a very general convergence result for the parareal algorithm applied to the non-linear initial value problem (17), which contains accurate estimates of the constants involved:

Theorem 2 (Convergence of Parareal [28]) *Let $F(t_{n+1}, t_n, U_n^k)$ denote the exact solution at t_{n+1} and $G(t_{n+1}, t_n, U_n^k)$ be a one step method with local truncation error bounded by $C_1 \Delta T^{p+1}$. If*

$$|G(t + \Delta T, t, x) - G(t + \Delta T, t, y)| \le (1 + C_2 \Delta T)|x - y|,$$

then the following error estimate holds for (18):

$$\max_{1\leq n\leq N} |u(t_n) - U_n^k| \leq \frac{C_1 \Delta T^{k(p+1)}}{k!}(1 + C_2\Delta T)^{N-1-k} \prod_{j=1}^{k}(N-j) \max_{1\leq n\leq N} |u(t_n) - U_n^0| \tag{20}$$

$$\leq \frac{(C_1 T)^k}{k!} e^{C_2(T-(k+1)\Delta T)} \Delta T^{pk} \max_{1\leq n\leq N} |u(t_n) - U_n^0|. \tag{21}$$

The proof uses generating functions and is just over a page long, see [28]. One can clearly see the precise convergence mechanisms of the parareal algorithm in this result: looking in (20) on the right, the product term is initially growing, for $k = 1, 2, 3$ we get the products $N-1$, $(N-1)(N-2)$, $(N-1)(N-2)(N-3)$ and so on, but as soon as $k = N$ the product contains the factor zero, and the method has converged. This is the property already pointed out by Bellen and Zennaro in [6]. Next looking in (21), we see that the method's order increases at each iteration k by p, the order of the coarse propagator, as already shown by Lions, Maday and Turinici in their proposition for the Euler method. We have however also a precise estimate of the constant in front in (21), and this constant contracts faster than linear, since it is an algebraic power of $C_1 T$ divided by $k!$ (the exponential term is not growing as the iteration k progresses). This division by $k!$ is the typical convergence behavior found in waveform relaxation algorithms, which we will see in more detail in the next section.

3 Domain Decomposition Methods in Space-Time

Time parallel methods based on domain decomposition solve evolution problems in quite a different way in parallel from multiple shooting based methods. The decomposition of the space-time domain for such methods is shown in Fig. 8. Again

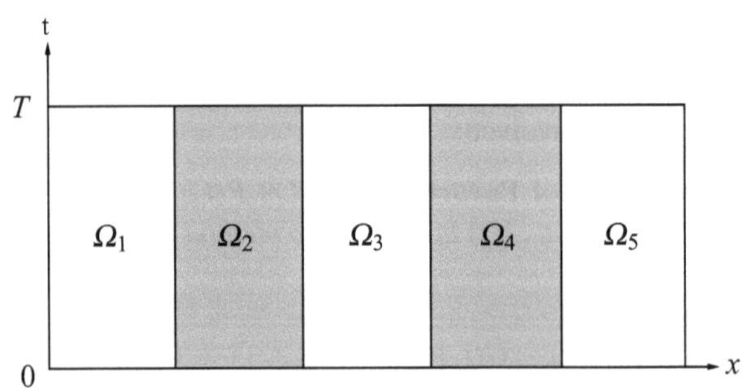

Fig. 8 Decomposition of the space-time domain for domain decomposition time parallel methods

an iteration is then used, which computes only solutions on the local space-time subdomains Ω_j. Since these solutions are obtained over the entire so-called time window $[0, T]$ before accurate interface values are available from the neighboring subdomains over the entire time window, these methods are also time parallel in this sense, and they are known under the name waveform relaxation.

3.1 Picard and Lindelöf 1893/1894

The roots of waveform relaxation type methods lie in the existence proofs of solutions for ordinary differential equations of Picard [65] and Lindelöf [53]. Like the alternating Schwarz method invented by Schwarz to prove the Dirichlet principle [70] and hence existence of solutions of Laplace's equation on general domains, Picard invented his method of successive approximations to prove the existence of solutions of the specific class of ordinary differential equations:

> Les méthodes d'approximation dont nous faisons usage sont théoriquement susceptibles de s'appliquer à toute équation, **mais elles ne deviennent vraiment intéressantes** pour l'étude des propriétés des fonctions définies par les équations différentielles que **si l'on ne reste pas dans les généralités et si l'on envisage certaines classes d'équations**.

Picard thus considers ordinary differential equations of the form

$$v'(t) = f(v(t)), \quad t \in [0, T], \tag{22}$$

with given initial condition $v(0)$. In order to analyze if a solution of such a non-linear problem exists, he proposed the nowadays called Picard iteration

$$v^n(t) = v(0) + \int_0^t f(v^{n-1}(\tau))d\tau, \quad n = 1, 2, \ldots, \tag{23}$$

where $v^0(t)$ is some initial guess. This transforms the problem of solving the ordinary differential equation (22) into a sequence of problems using only quadrature, which is much easier to handle. Picard proved convergence of this iteration in [65], which was sufficient to answer the existence question. It was Lindelöf a year later who gave the following convergence rate estimate in [53]:

Theorem 3 (Superlinear Convergence) *On bounded time intervals $t \in [0, T]$, the iterates (23) satisfy the superlinear error bound*

$$||v - v^n||_\infty \leq \frac{(CT)^n}{n!}||v - v^0||_\infty, \tag{24}$$

where C is the Lipschitz constant of the nonlinear right hand side f.

We see in the convergence estimate (24) the same term appear as in the parareal convergence estimate (21). This term is typical for the convergence of waveform relaxation methods we will see next, and thus the comment of Saha, Stadel and Tremaine in the quote at the beginning of Sect. 2.4 is justified.

3.2 *Lelarasmee, Ruehli and Sangiovanni-Vincentelli 1982*

The Picard iteration was not very successful as an iterative method for concrete computations,[4] but in the circuit community, an interesting variant was developed based on a decomposition of the circuit by Lelarasmee, Ruehli and Sangiovanni-Vincentelli [52]:

> The Waveform Relaxation (WR) method is an iterative method for analyzing nonlinear dynamical systems in the time domain. The method, at each iteration, **decomposes the system into several dynamical subsystems, each of which is analyzed for the entire given time interval.**

The motivation for this method was really the extremely rapid growth of integrated circuits, which made it difficult to simulate a new generation of circuits on the present generation computers.[5] Lelarasmee, Ruehli and Sangiovanni-Vincentelli explain the waveform relaxation algorithm on the concrete example of a MOS ring oscillator shown in Fig. 9. The reason why this circuit is oscillating can be seen as follows: suppose the voltage at node v_1 equals 5 V. Then this voltage is connected to the gate of the transistor to the right, which will thus open, and hence the voltage at node v_2 will be pulled down to ground, i.e. 0 V. This is however connected to the gate of the next transistor to the right of v_2, which will thus close, and v_3 will be pulled up to 5 V. These 5 V will now feedback to the gate of the transistor to the left of v_1, which will thus open, and thus v_1, which was by assumption at 5 V, will be pulled down to ground at 0 V, and we see how the oscillation happens.

Using the laws of Ohm and Kirchhoff, the equations for such a circuit can be written in form of a system of ordinary differential equations

$$\begin{aligned} \mathbf{v}'(t) &= \mathbf{f}(\mathbf{v}(t)),\ 0 < t < T, \\ \mathbf{v}(0) &= \mathbf{g}, \end{aligned}$$

where $\mathbf{v} = (v_1, v_2, v_3)$, and $\mathbf{g}$ is the initial state of the circuit.

If the circuit is extremely large, so that it does not fit any more on one single computer, the waveform relaxation algorithm is based on the idea of decomposing the circuit into subcircuits, as shown in Fig. 10. The idea is to cut the wires with

[4]"Actually this method of continuing the computation is highly inefficient and is not recommended", see [59].

[5]"The spectacular growth in the scale of integrated circuits being designed in the VLSI era has generated the need for new methods of circuit simulation. "Standard" circuit simulators, such as SPICE and ASTAP, simply take too much CPU time and too much storage to analyze a VLSI circuit", see [52].

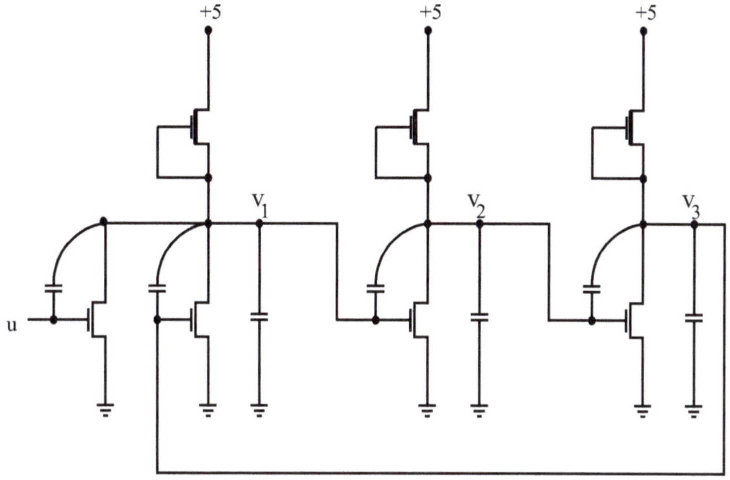

Fig. 9 Historical example of the MOS ring oscillator, for which the waveform relaxation algorithm was derived

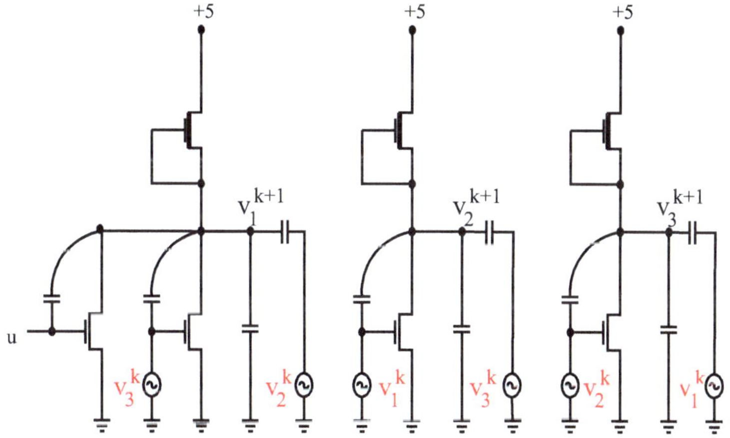

Fig. 10 Decomposition of the MOS ring oscillator circuit for the waveform relaxation algorithm

which the subcircuits are connected, and then to assume that there are small voltage sources on the wires that were cut, which feed in the voltage that was calculated at the previous iteration. This leads to the iterative method

$$\begin{aligned} \partial_t v_1^{k+1} &= f_1(v_1^{k+1}, v_2^k, v_3^k), \\ \partial_t v_2^{k+1} &= f_2(v_1^k, v_2^{k+1}, v_3^k), \\ \partial_t v_3^{k+1} &= f_3(v_1^k, v_2^k, v_3^{k+1}). \end{aligned} \tag{25}$$

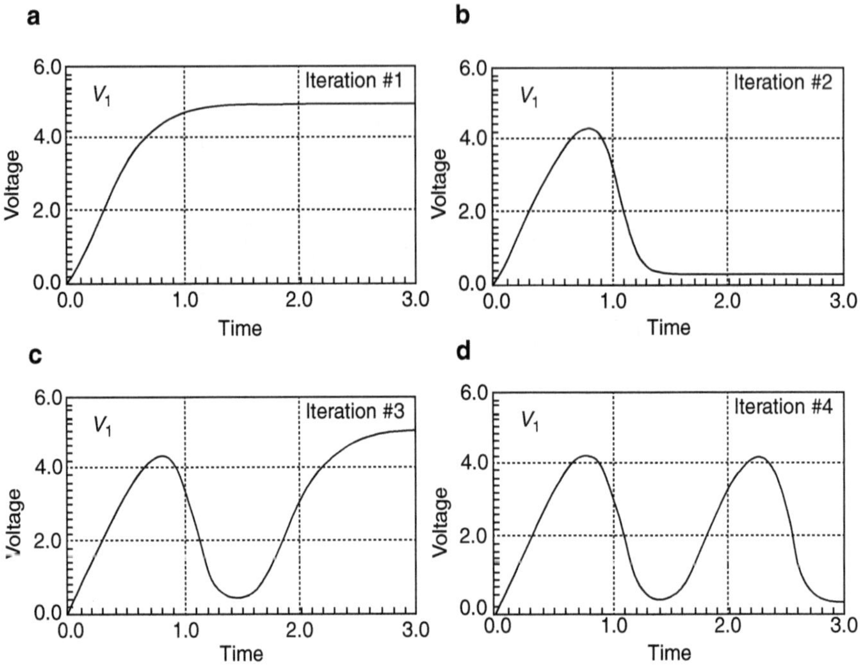

Fig. 11 Historical convergence result for the MOS ring oscillator from [52]. The x axis represents time here and the y axis voltage values

Since in the circuit simulation community signals along wires are called 'waveforms', this gave the algorithm the name *Waveform Relaxation*. We see in (25) that on the right all neighboring waveforms have been relaxed to the previous iteration, which results in a Jacobi type relaxation known in numerical linear algebra, which is entirely parallel. Naturally one could also use a Gauss-Seidel type relaxation which would then be sequential.

We show in Fig. 11 a historical numerical convergence study for the MOS ring oscillator taken from [52]. We can see that this circuit has the property that the waveform relaxation algorithm converges in a finite number of steps. This can be understood by the finite propagation speed of the information in this circuit,[6] and we will see this again when looking at hyperbolic equations in the following section. The convergence of waveform relaxation methods depends strongly on the type of equations that are being solved, and the general convergence estimate of Lindelöf (24), also valid for waveform relaxation, is not always sharp.

[6] "Note that since the oscillator is highly non unidirectional due to the feedback from v_3 to the NOR gate, the convergence of the iterated solutions is achieved with the number of iterations being proportional to the number of oscillating cycles of interest", see [52].

3.3 Gander 1996

The waveform relaxation algorithm from the previous subsection can be naturally extended to partial differential equations, as it was shown in [24]:

> Motivated by the work of Bjørhus [8], we show how one can use **overlapping domain decomposition to obtain a waveform relaxation algorithm** for the semi-discrete heat equation which converges at a rate independent of the mesh parameter.

The idea is best explained for the simple model problem of the one dimensional heat equation,

$$\partial_t u = \partial_{xx} u, \quad 0 < x < 1, \; t > 0 \tag{26}$$

with given initial condition $u(x, 0) = u_0(x)$ and homogeneous boundary conditions. Like in the waveform relaxation algorithm, where the circuit was partitioned into subcircuits, one partitions the domain $\Omega = (0, 1)$ into overlapping subdomains, say $\Omega_1 = (0, \beta)$ and $\Omega_1 = (\alpha, 1)$, $\alpha < \beta$, and then performs the iteration

$$\begin{aligned} \partial_t u_1^n &= \partial_{xx} u_1^n, & 0 < x < \beta, \; t > 0, \\ u_1^n(\beta, t) &= u_2^{n-1}(\beta, t), & \\ \partial_t u_2^n &= \partial_{xx} u_2^n, & \alpha < x < 1, \; t > 0, \\ u_2^n(\alpha, t) &= u_1^{n-1}(\alpha, t). & \end{aligned} \tag{27}$$

Since the decomposition is overlapping like in the classical overlapping Schwarz method for steady problems, and time dependent problems are solved in each iteration like in waveform relaxation, these algorithms are called *Schwarz Waveform Relaxation* algorithms. One can show that algorithm (27) converges linearly on unbounded time intervals, see [34], and superlinearly on bounded time intervals, see [41]. Both results can be found for nonlinear problems in [25]. The superlinear convergence rate in Schwarz waveform relaxation algorithms is faster than in classical waveform relaxation methods for circuits, since the heat kernel decay gives additional contraction. If the equation is a wave equation, then one obtains convergence in a finite number of steps, see for example [29]. Much better waveform relaxation methods can however be obtained using the new concept of optimized transmission conditions we will see next.

3.4 Gander, Halpern and Nataf 1999

It was shown in [36] that the Dirichlet transmission conditions used for the information exchange do not lead to good convergence behavior of the Schwarz waveform relaxation algorithm:

> We then show that the **Dirichlet conditions at the artificial interfaces inhibit the information exchange** between subdomains and therefore slow down the convergence of the algorithm.

This observation holds for all types of partial differential equations, also for steady state problems [62]. The key new idea is to introduce more effective transmission conditions, which leads for the model problem (26) to the new Schwarz waveform relaxation algorithm

$$\begin{aligned} \partial_t u_1^n &= \partial_{xx} u_1^n, && 0 < x < \beta,\ t > 0, \\ \mathscr{B}_1 u_1^n(\beta, t) &= \mathscr{B}_1 u_2^{n-1}(\beta, t), && \\ \partial_t u_2^n &= \partial_{xx} u_2^n, && \alpha < x < 1,\ t > 0, \\ \mathscr{B}_2 u_2^n(\alpha, t) &= \mathscr{B}_2 u_1^{n-1}(\alpha, t). && \end{aligned} \tag{28}$$

If one chooses $\mathscr{B}_1 = \partial_{n_1} + \text{DtN}_2$ and $\mathscr{B}_2 = \partial_{n_2} + \text{DtN}_1$, where ∂_{n_j} denotes the normal derivative, and DtN_j denotes the Dirichlet to Neumann operator of the subdomain j, then algorithm (28) converges in two iterations, independently of the overlap: it becomes a direct solver. This can be generalized to N iterations with N subdomains, or one iteration when using an alternating sweep, and is the underlying mechanism for the good convergence of the sweeping preconditioner recently presented in [20]. Since the DtN operators are in general expensive, so-called *optimized Schwarz waveform relaxation* methods use local approximations; for a complete treatment of advection reaction diffusion equations see [7, 30], and for the wave equation, see [29, 37]. An overview for steady problems and references can be found in [26]. We show in Fig. 12 as an illustration for an advection reaction diffusion equation and a decomposition into eight overlapping subdomains how much faster optimized Schwarz waveform relaxation methods converge compared to classical Schwarz waveform relaxation methods. While the Dirichlet transmission conditions in the left column greatly inhibit the information exchange, the absorbing conditions (here second order Taylor conditions) lead almost magically to a very good approximation already in the very first iteration. For more information, see [7, 30]. Waveform relaxation methods should thus never be used with classical transmission conditions, also when applied to circuits; optimized transmission conditions have also been proposed and analyzed for circuits, see for example [1, 2] and references therein.

3.5 Recent Developments

Other domain decomposition methods for steady problems have been recently proposed and analyzed for time dependent problems: for the convergence properties of the Dirichlet-Neumann Waveform Relaxation algorithm, see [39, 58], and for the Neumann-Neumann Waveform Relaxation algorithm, see [39, 50] for a convergence analysis, and [44] for well posedness of the algorithm.

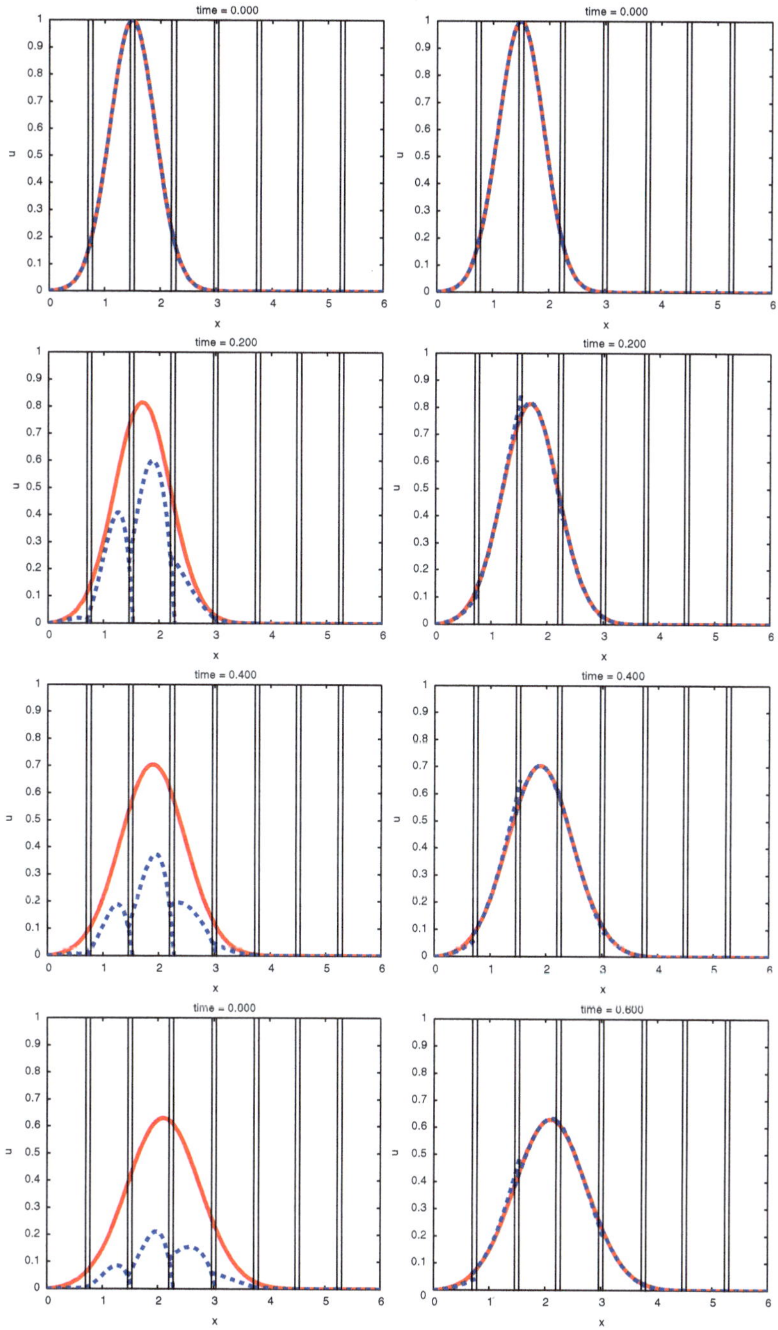

Fig. 12 Snapshots in time of the first classical Schwarz waveform relaxation iteration in the *left column*, and the first optimized Schwarz waveform relaxation iteration in the *right column*: the exact solution is shown in *solid red*, and the Schwarz waveform relaxation approximation in *dashed blue*

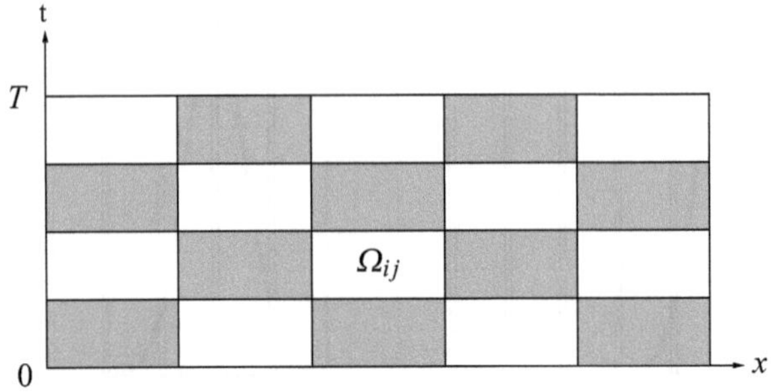

Fig. 13 Space-time decomposition for parareal Schwarz waveform relaxation

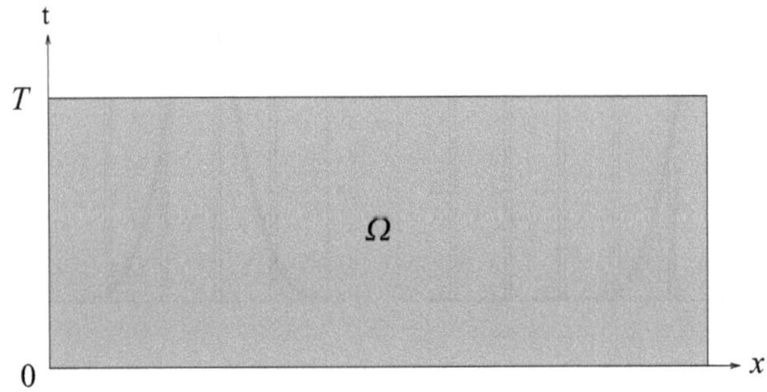

Fig. 14 Space-time multigrid methods work simultaneously on the entire space-time domain

It is also naturally possible to combine the multiple shooting method and the Schwarz waveform relaxation methods, which leads to a space-time decomposition of the form shown in Fig. 13. A parareal Schwarz waveform relaxation algorithm for such decompositions was proposed in [38], see also [57] for a method which uses waveform relaxation as a solver within parareal. These methods iterate simultaneously on sets of unknowns in the space-time domain, as the space-time multigrid methods we will see next.

4 Multigrid Methods in Space-Time

The methods we have seen so far were designed to be naturally parallel: the time decomposition methods based on shooting use many processors along the time axis, the waveform relaxation methods use many processors in the space dimensions. The multigrid methods we see in this section are not naturally parallel, but their components can be executed in parallel in space-time, since they work simultaneously on the entire space-time domain, as indicated in Fig. 14.

4.1 Hackbusch 1984

The first such method is the parabolic multigrid method developed by Hackbusch in [43]. Like other multigrid methods, the smoother can be naturally implemented in parallel in space, and in the parabolic multigrid method, the smoother operates over many time levels, so that interpolation and restriction can be performed in parallel in space-time.

In order to explain the method, we consider the parabolic PDE $u_t + \mathscr{L}u = f$ discretized by Backward Euler:

$$(I + \Delta t\mathscr{L})u_n = u_{n-1} + \Delta t f(t_n). \tag{29}$$

Hackbusch makes the following comment about this equation

> The conventional approach is to solve (29) time step by time step; u_n is computed from u_{n-1}, then u_{n+1} from u_n etc. The following process will be different. Assume that u_n is already computed or given as an initial state. **Simultaneously, we shall solve for u_{n+1}, $u_{n+2}, \ldots, u_{n+k}$ in one step of the algorithm.**

In the method of Hackbusch, one starts with a standard smoother for the problem at each time step. Let A be the iteration matrix, $A := I + \Delta t\mathscr{L}$; then one partitions the matrix into its lower triangular, diagonal and upper triangular part, $A = L + D + U$, and uses for example as a smoother the Gauss-Seidel iteration over many time levels:

for $n = 1 : N$
  for $j = 1 : \nu$
    $u_n^j = (L + D)^{-1}(-U{u_n}^{j-1} + u_{n-1}^{\nu} + \Delta t f(t_n));$
  end;
end;

We see that the smoothing process is sequential in time: one first has to finish the smoothing iteration at time step $n - 1$ in order to obtain u_{n-1}^{ν}, before one can start the smoothing iteration at time step n, since u_{n-1}^{ν} is needed on the right hand side.

After smoothing, one restricts the residual in space-time like in a classical multigrid method to a coarser grid, before applying the procedure recursively. Hackbusch first only considers coarsening in space, as shown in Fig. 15. In this case, one can prove that standard multigrid performance can be achieved for this

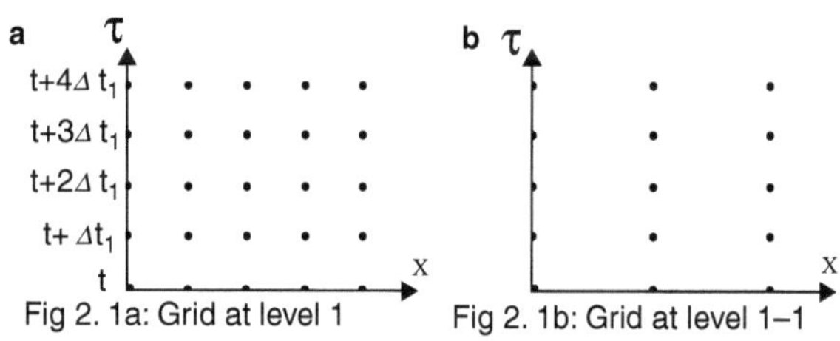

Fig. 15 Original figure of Hackbusch about the coarsening in the parabolic multigrid method. This figure was taken from [43]

method. If one however also coarsens in time, one does not obtain standard multigrid performance, and the method can even diverge. This is traced back by Hackbusch to errors which are smooth in space, but non smooth in time. Hackbusch illustrates the performance of the method by numerical experiments for buoyancy-driven flow with finite difference discretization.

4.2 Lubich and Ostermann 1987

Lubich and Ostermann [55] used the waveform relaxation context to define a space-time multigrid method:

> Multi-grid methods are known to be very efficient solvers for elliptic equations. Various approaches have also been given to extend multi-grid techniques to parabolic problems. A common feature of these approaches is that multi-grid methods are applied only *after* the equation has been discretized in time. In the present note we shall rather **apply multi-grid (in space) directly to the evolution equation**.

Their work led to the so-called *Multigrid Waveform Relaxation* algorithm. The easiest way to understand it is to first apply a Laplace transform to the evolution problem, assuming for simplicity a zero initial condition,

$$u_t + L_h u = f \quad \Longrightarrow \quad A(s)\hat{u} := s\hat{u} + L_h\hat{u} = \hat{f}.$$

One then applies a standard multigrid method to the Laplace transformed linear system $A(s)\hat{u} = \hat{f}$. Let $A(s) = L + D + sI + U$ be again the lower triangular, diagonal and upper triangular part of the matrix $A(s)$. A standard two grid algorithm would then start with the initial guess $\hat{u}_0^0(s)$, and perform for $n = 0, 1, 2, \ldots$ the steps

$$\begin{aligned}
&\text{for } j = 1 : \nu \\
&\qquad \hat{u}_n^j(s) = (L + D + sI)^{-1}(-U\hat{u}_n^{j-1}(s) + \hat{f}(s)); \\
&\text{end;} \\
&\hat{u}_{n+1}^0(s) = \hat{u}_n^\nu(s) + PA_c^{-1}R(\hat{f} - A\hat{u}_n^\nu(s)); \\
&\text{smooth again;}
\end{aligned}$$

where R and P are standard multigrid restriction and prolongation operators for steady problems, and the coarse matrix can be defined using a Galerkin approach, $A_c := RAP$.

Applying the inverse Laplace transform to this algorithm, we obtain the multigrid waveform relaxation algorithm: the smoothing step

$$(sI + L + D)\hat{u}_n^j(s) = -U\hat{u}_n^{j-1}(s) + \hat{f}(s)$$

becomes in the time domain

$$\partial_t u_n^j + (L + D)u_n^j + Uu_n^{j-1} = f,$$

which is a Gauss Seidel Waveform Relaxation iteration, see Sect. 3.2. The coarse correction

$$\hat{u}_{n+1}^{0}(s) := \hat{u}_{n}^{\nu}(s) + PA_{c}^{-1}R(\hat{f} - A\hat{u}_{n}^{\nu}(s))$$

becomes back in the time domain

$$\begin{aligned} &\text{solve } v_t + L_H v = R(f - \partial_t u_n^{\nu} - L_h u_n^{\nu}), \\ &u_{n+1}^{0} = u_n^{\nu} + Pv. \end{aligned}$$

This is a time continuous parabolic problem on a coarse spatial mesh.

Lubich and Ostermann prove for the heat equation and finite difference discretization that red-black Gauss Seidel smoothing is not as good as for the stationary problem, but still sufficient to give typical multigrid convergence, and that damped Jacobi smoothing is as good as for stationary problems. The authors show with numerical experiments that in the multigrid waveform relaxation algorithm one can use locally adaptive time steps.

4.3 Horton and Vandewalle 1995

Horton and Vandewalle are the first to try to address the problem of time coarsening in [45]:

> In standard time-stepping techniques multigrid can be used as an iterative solver for the elliptic equations arising at each discrete time step. By contrast, **the method presented in this paper treats the whole of the space-time problem simultaneously**.

They first show that time coarsening does not lead to multigrid performance, since the entire space-time problem is very anisotropic because of the time direction. To address this issue, they explain that one could either use line smoothers, which is related to the multigrid waveform relaxation algorithm we have seen in Sect. 4.2, or the following two remedies:

1. Adaptive semi-coarsening in space or time depending on the anisotropy,
2. Prolongation operators only forward in time.

For the heat equation with finite difference discretization and Backward Euler, BDF2 and Crank-Nicolson, Horton and Vandewalle perform a detailed local Fourier mode analysis, and show that good contraction rates can be obtain for space-time multigrid V-cycles, although not quite mesh independent. F-cycles are needed to get completely mesh independent convergence rates. These results are illustrated by numerical experiments for 1d and 2d heat equations.

4.4 Emmett and Minion 2012

There are several steps in the development of the solver PFASST, which stands for Parallel Full Approximation Scheme in Space-Time. The underlying iteration is a deferred correction method [60]:

> This paper investigates a variant of the parareal algorithm first outlined by Minion and Williams in 2008 that utilizes **a deferred correction strategy within the parareal iterations**.

We thus have to start by explaining the deferred correction method. Consider the initial value problem

$$u' = f(u), \quad u(0) = u_0. \tag{30}$$

We can rewrite this problem in integral form

$$u(t) = u(0) + \int_0^t f(u(\tau))d\tau. \tag{31}$$

Let $\tilde{u}(t)$ be an approximation with error $e(t) := u(t) - \tilde{u}(t)$. Inserting $u(t) = \tilde{u}(t) + e(t)$ into (31), we get

$$\tilde{u}(t) + e(t) = u(0) + \int_0^t f(\tilde{u}(\tau) + e(\tau))d\tau. \tag{32}$$

Defining the function $F(u) := u(0) + \int_0^t f(u(\tau))d\tau - u(t)$ from Eq. (31), the residual $r(t)$ of the approximate solution $\tilde{u}(t)$ is

$$r(t) := F(\tilde{u}) = \tilde{u}(0) + \int_0^t f(\tilde{u}(\tau))d\tau - \tilde{u}(t), \tag{33}$$

and thus from (32) the error satisfies the equation

$$\begin{aligned} e(t) &= u(0) + \int_0^t f(\tilde{u}(\tau) + e(\tau))d\tau - \tilde{u}(t) \\ &= r(t) + \int_0^t f(\tilde{u}(\tau) + e(\tau)) - f(\tilde{u}(\tau))d\tau, \end{aligned}$$

or written as a differential equation

$$e'(t) = r'(t) + f(\tilde{u}(t) + e(t)) - f(\tilde{u}(t)). \tag{34}$$

The idea of integral deferred correction is to choose a numerical method, for example Forward Euler, and to get a first approximation of the solution of (30) by computing

$$\tilde{u}_{m+1} = \tilde{u}_m + \Delta t f(\tilde{u}_m), \quad \text{for } m = 0, 1, \ldots, M-1.$$

With these values, one then computes the residual defined in (33) at the points t_m, $m = 0, 1, \ldots, M$ using a high order quadrature formula. One then solves the error equation (34) in differential form again with the same numerical method, here Forward Euler,

$$e_{m+1} = e_m + r_{m+1} - r_m + \Delta t(f(\tilde{u}_m + e_m) - f(\tilde{u}_m)). \tag{35}$$

Adding this correction, one obtains a new approximation $\tilde{u}_m + e_m$, for which one can show in our example of Forward Euler that the order has increased by one, i.e. it is now a second order approximation. One can continue this process and increase the order up to the order of the quadrature used.

This spectral deferred correction iteration can also be considered as an iterative method to compute the Runge-Kutta method corresponding to the quadrature rule used to approximate the integral: if we denote by $\mathbf{u}^0$ the initial approximation obtained by forward Euler, $\mathbf{u}^0 := (\tilde{u}_0, \tilde{u}_1, \ldots, \tilde{u}_M)^T$, each integral deferred correction corresponds to one step in the non-linear fixed point iteration

$$\mathbf{u}^k = F(\mathbf{u}^{k-1}, u_0), \tag{36}$$

where u_0 is the initial condition from (30). The classical application of integral deferred correction is to partition the time interval $[0, T]$ into subintervals $[T_{j-1}, T_j]$, $j = 1, 2, \ldots, J$, and then to start on the first subinterval $[T_0, T_1]$ to compute approximations $\mathbf{u}_1^k$ by performing K steps of (36) before passing to the next time interval $[T_1, T_2]$, see also Fig. 18 for an example with $M = 3$. The overall iteration is therefore

$u_{0,M}^K = u_0$;
for $j = 1 : J$
    compute $\mathbf{u}_j^0$ as Euler approximation on $[T_{j-1}, T_j]$;
    for $k = 1 : K$
        $\mathbf{u}_j^k = F(\mathbf{u}_j^{k-1}, u_{j-1,M}^K)$;
    end;
end;

We see that this is purely sequential, like a time stepping scheme: in each time subinterval, one first has to finish the K spectral deferred corrections, before one can pass to the next time subinterval. Minion proposed in [60] not to wait for each time subinterval to finish, and to replace the inner updating formula by

$$\mathbf{u}_j^k = F(\mathbf{u}_j^{k-1}, u_{j-1,M}^k), \quad \text{(note the lower case } k\text{ !)}, \tag{37}$$

which means that one can now perform the spectral deferred corrections on many time subintervals $[T_{j-1}, T_j]$ in parallel. This is very similar to the iteration of Womble we will see Sect. 5.3. In the approach of Minion, this is however combined with a coarse correction from the parareal algorithm, so it is like using a more and more accurate fine integrator, as the iteration progresses.

The PFASST algorithm proposed in [19] is based on using the parallel deferred correction iteration above as a smoother in a multigrid full approximation scheme in space-time for non-linear problems:

> The method is iterative with each iteration consisting of **deferred correction sweeps performed alternately on fine and coarse space-time discretizations**. The coarse grid problems are formulated using a space-time analog of the **full approximation scheme popular in multigrid methods for nonlinear equations**.

The method has successfully been applied to non-linear problems in [19, 75, 76], but there is so far no convergence analysis for PFASST.

4.5 *Neumüller 2014*

The new idea in this multigrid variant is to replace the classical point smoothers by block Jacobi smoothers. Suppose we discretize the heat equation

$$u_t = \Delta u + f$$

globally in space-time by an implicit method, for example Backward Euler. Then we obtain a block triangular linear system in space-time of the form

$$\begin{pmatrix} A_1 & & & & \\ B_2 & A_2 & & & \\ & B_3 & A_3 & & \\ & & \ddots & \ddots & \\ & & & B_n & A_n \end{pmatrix} \begin{pmatrix} \mathbf{u}_1 \\ \mathbf{u}_2 \\ \mathbf{u}_3 \\ \vdots \\ \mathbf{u}_n \end{pmatrix} = \begin{pmatrix} \mathbf{f}_1 \\ \mathbf{f}_2 \\ \mathbf{f}_3 \\ \vdots \\ \mathbf{f}_n \end{pmatrix}. \tag{38}$$

The space-time multigrid method consists now of applying a few damped block Jacobi smoothing steps, inverting the diagonal blocks A_n, before restricting by standard multigrid restriction operators in space-time to a coarser grid, and recursively continuing. One can show that for the heat equation, we have (see Martin Neumüller's PhD thesis [63]):

- The optimal damping parameter for the block Jacobi smoother is $\omega = \frac{1}{2}$.
- One always obtains good smoothing in time (semi-coarsening is always possible).
- For $\frac{\Delta t}{\Delta h^2} \geq C$, one also obtains good smoothing in space.

Table 1 Scaling results for the space-time multigrid method with block Jacobi smoother; all simulations performed by M. Neumüller

Weak scaling					Strong scaling			
Cores	$\frac{1}{\Delta T}$	dof	Iter	Time	$\frac{1}{\Delta T}$	dof	Iter	Time
1	4	59768	9	6.8	4096	61202432	9	6960.7
2	8	119536	9	8.1	4096	61202432	9	3964.8
4	16	239072	9	9.2	4096	61202432	9	2106.2
8	32	478144	9	9.2	4096	61202432	9	1056.0
16	64	956288	9	9.2	4096	61202432	9	530.4
32	128	1912576	9	9.3	4096	61202432	9	269.5
64	256	3825152	9	9.4	4096	61202432	9	135.2
128	512	7650304	9	9.4	4096	61202432	9	68.2
256	1024	15300608	9	9.4	4096	61202432	9	34.7
512	2048	30601216	9	9.4	4096	61202432	9	17.9
1024	4096	61202432	9	9.4	4096	61202432	9	9.4
2048	8192	122404864	9	9.5	4096	61202432	9	5.4

- One V-cycle in space suffices to invert the diagonal blocks A_n in the Jacobi smoother.

This multigrid method has excellent scaling properties for the heat equation, as it was shown in [63], from which the example in Table 1 is taken. The results are for the 3D heat equation, and computed on the Vienna Scientific Cluster VSC-2; see also [32].

5 Direct Solvers in Space-Time

The time parallel solvers we have seen so far were all iterative. There have been also attempts to construct direct time parallel solvers. There are both small scale parallel direct solvers and also large scale parallel direct solvers.

5.1 Miranker and Liniger 1967

The first small scale direct parallel solver was proposed by Miranker and Liniger [61], who also were very much aware of the naturally sequential nature of evolution problems:

> It appears at first sight that the sequential nature of the numerical methods do not permit a parallel computation on all of the processors to be performed. We say that **the front of computation** is too narrow to take advantage of more than one processor... Let us consider how we might widen the computation front.

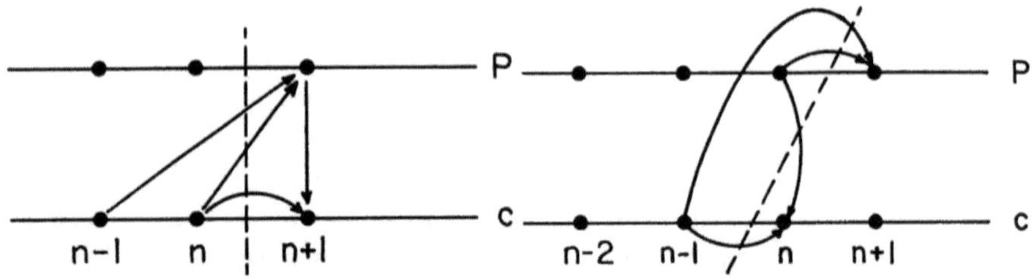

Fig. 16 Symbolic representation by Miranker and Liniger of an entirely sequential predictor corrector method on the *left*, and a parallel one on the *right*

For $y' = f(x, y)$, Miranker and Liniger consider the predictor corrector formulas

$$y_{n+1}^p = y_n^c + \frac{h}{2}(f(y_n^c) - f(y_{n-1}^c)), \quad y_{n+1}^c = y_n^c + \frac{h}{2}(f(y_{n+1}^p) + f(y_n^c)).$$

This process is entirely sequential as they illustrated with a figure, a copy of which is shown in Fig. 16 on the left. They then consider the modified predictor corrector formulas

$$y_{n+1}^p = y_{n-1}^c + 2hf(y_n^p), \quad y_n^c = y_{n-1}^c + \frac{h}{2}(f(y_n^p) + f(y_{n-1}^c)).$$

Those two formulas can now be evaluated in parallel by two processors, as illustrated in Fig. 16 on the right. Miranker and Liniger then show how one can systematically derive a general class of numerical integration methods which can be executed on $2s$ processors in parallel, and present a stability and convergence analysis for those methods.

Similar parallelism can also be obtained with the block implicit one-step methods developed by Shampine and Watts in [71]. These methods use different time stepping formulas to advance several time levels at once. For an early numerical comparison for parallel block methods and parallel predictor corrector methods, see Franklin [23]. These methods are ideally suited to be used on the few cores of a multicore processor, but they do not have the potential for large scale parallelism.

5.2 *Axelson and Verwer 1985*

Boundary value methods for initial value problems are a bit strange. A very good early introduction is the paper by Axelson and Verwer [4]:

> Hereby we concentrate on explaining the fundamentals of the method because for initial value problems the boundary value method seems to be fairly unknown [...] In the forward-step approach, the numerical solution is obtained by stepping through the grid [...] In this paper, we will tackle the numerical solution in a completely different way [...] **We will**

consider $\dot{y} = f(x, y)$ as a two point boundary value problem with a given value at the left endpoint and an implicitly defined value, by the equation $\dot{y} = f(x, y)$, at the right endpoint.

It is best to understand boundary value methods by looking at a simple example.[7] Suppose we discretize $\dot{y} = f(y)$ with the explicit midpoint rule

$$y_{n+1} - y_{n-1} - 2hf(y_n) = 0, \quad y_0 = y(0).$$

Since the explicit midpoint rule is a two step method, we also need an initial approximation for y_1. Usually, one defines y_1 from y_0 using a one step method, for example here by Backward Euler. In boundary value methods, one leaves y_1 as an unknown, and uses Backward Euler at the endpoint y_N to close the system, imposing

$$y_N - y_{N-1} - 2hf(y_N) = 0.$$

For a linear problem $\dot{y} = ay$, the midpoint rule and Backward Euler to define y_1 gives the triangular linear system

$$\begin{pmatrix} 1-ah & & & & \\ -2ah & 1 & & & \\ -1 & -2ah & 1 & & \\ & \ddots & \ddots & \ddots & \\ & & -1 & -2ah & 1 \end{pmatrix} \begin{pmatrix} y_1 \\ y_2 \\ y_3 \\ \vdots \\ y_N \end{pmatrix} = \begin{pmatrix} y_0 \\ y_0 \\ 0 \\ \vdots \\ 0 \end{pmatrix}. \tag{39}$$

For the boundary value method, leaving y_1 free and using Backward Euler on the right gives the tridiagonal system

$$\begin{pmatrix} -2ah & 1 & & & \\ -1 & -2ah & 1 & & \\ & \ddots & \ddots & \ddots & \\ & & -1 & -2ah & 1 \\ & & & -1 & 1-ah \end{pmatrix} \begin{pmatrix} y_1 \\ y_2 \\ y_3 \\ \vdots \\ y_N \end{pmatrix} = \begin{pmatrix} y_0 \\ 0 \\ 0 \\ \vdots \\ 0 \end{pmatrix}. \tag{40}$$

The tridiagonal system can now be solved either directly by factorization, or also by iteration, thus working on all time levels simultaneously.

It is very important however to realize that boundary value methods are completely different discretizations from initial value methods. The stability properties often are the contrary when one transforms an initial value method into a boundary value method. We show in Fig. 17 numerical examples for the initial value method (39) and boundary value method (40). We see that for a decaying solution,

[7] This example had already been proposed by Fox in 1954.

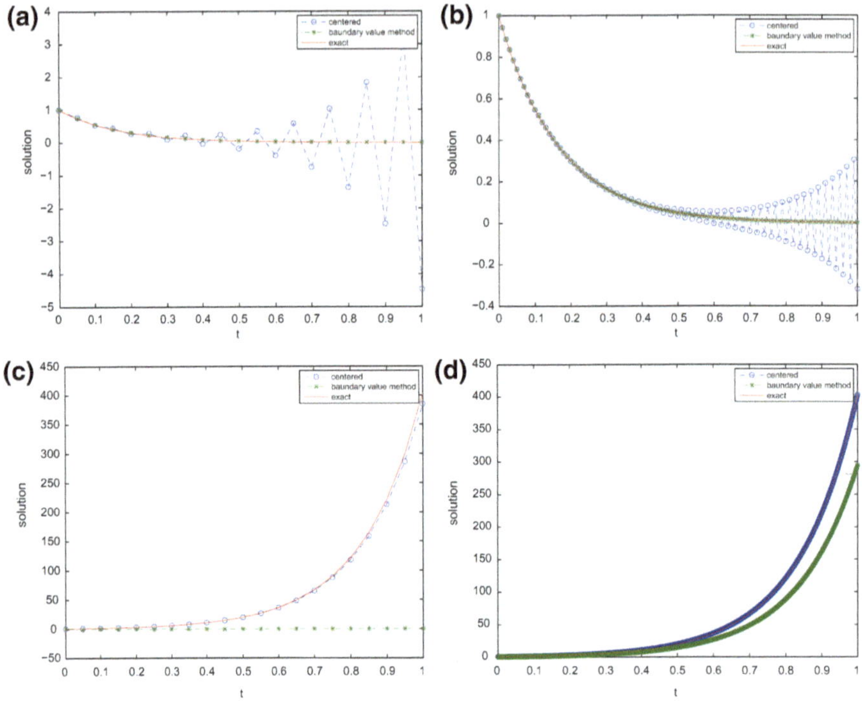

Fig. 17 Stability comparison of initial value and boundary value methods. (**a**) $a = -6$, $h = 1/20$. (**b**) $a = -6$, $h = 1/100$. (**c**) $a = 6$, $h = 1/20$. (**d**) $a = 6$, $h = 1/2000$

$a < 0$, the initial value method is exhibiting stability problems, while the boundary value method is perfectly stable (top row of Fig. 17). For growing solutions, $a > 0$ it is the opposite, the initial value method gives very good approximations, while the boundary value method needs extremely fine time steps to converge (bottom row of Fig. 17). One can therefore not just transform an initial value method into a boundary value method in order to obtain a parallel solver, one has to first carefully study the properties of the new method obtained, see [10, 11] and references therein. Now if the method has good numerical properties and the resulting system can well be solved in parallel, boundary value methods can be an attractive way of solving an evolution problem in parallel, see for example [9], where a Backward Euler method is proposed to precondition the boundary value method. This is still sequential, but if one only uses subblocks of the Backward Euler method as preconditioner, by dropping the connection after, say, every 10th time step, a parallel preconditioner is obtained. Such methods are called nowadays block boundary value methods, see for example [11]. If one introduces a coarse mesh with a coarse integrator, instead of the backward Euler preconditioner, and does not use as the underlying integrator a boundary value method any more, but just a normal time stepping scheme, the approach can be related to the parareal algorithm, see for example [3].

5.3 *Womble 1990*

The method presented by Womble in [79], see also the earlier work by Saltz and Naik [68], is not really a direct method, it is using iterations, but not in the same way of the iterative methods we have seen so far:

> Parabolic and hyperbolic differential equations are often solved numerically by time stepping algorithms. These algorithms have been regarded as sequential in time; that is, the solution on a time level must be known before the computation for the solution at subsequent time levels can start. While this remains true in principle, we demonstrate that **it is possible for processors to perform useful work on many time levels simultaneously**.

The relaxation idea is similar to the one later used by Minion in [60] as a smoother in the context of PFASST, see Sect. 4.4, but not for a deferred correction iteration. In order to explain the method, we discretize the parabolic problem

$$u_t = \mathscr{L}u + f$$

by an implicit time discretization and obtain at each time step a linear system of the form

$$A_n u_n = f_n + B_n u_{n-1}.$$

Such systems are often solved by iteration. If we want to use a stationary iterative method, for example Gauss-Seidel, we would partition the matrix $A_n = L_n + D_n + U_n$, its lower triangular, diagonal, and upper triangular parts. Then starting with an initial guess u_n^0, one solves for $k = 1, 2, \dots, K$

$$(L_n + D_n)u_n^k = -U_n u_n^{k-1} + f_n + B_n u_{n-1}^K. \tag{41}$$

The key idea to break the sequential nature is to modify this iteration slightly so that it can be performed in parallel over several time steps. It suffices to not wait for convergence of the previous time level, but to iterate like

$$(L_n + D_n)u_n^k = -U_n u_n^{k-1} + f_n + B_n u_{n-1}^{k-1}, \tag{42}$$

which is the same idea also used in (37). Womble obtained quite good results with this approach, and he was the first person to really demonstrate practical speedup results with this time parallel method on a 1024-processor machine. Even though it was later shown that only limited speedups are possible with this relaxation alone [16], the work of Womble made people realize that indeed time parallelization could become an important direction, and it drew a lot of attention toward time-parallel algorithms.

5.4 Worley 1991

Worley was already in his PhD thesis in 1988 very interested in theoretical limits on the best possible sequential and parallel complexity when solving PDEs. While the ultimate sequential algorithm for such a problem of size n is $O(n)$ on a sequential machine, it is $O(\log n)$ on a parallel machine. In [80], Worley presented an additional direct time parallelization method, which when combined with multigrid waveform relaxation leads to a nearly optimal time-parallel iterative method:

> The waveform relaxation multigrid algorithm is normally implemented in a fashion **that is still intrinsically sequential in the time direction**. But computation in the time direction only involves solving linear scalar ODEs. If the ODEs are solved using a linear multistep method with a statically determined time step, then each ODE solution corresponds to the solution of a banded lower triangular matrix equation, or, equivalently, a linear recurrence. Parallelizing linear recurrence equations has been studied extensively. In particular, **if a cyclic reduction approach is used to parallelize the linear recurrence, then parallelism is introduced without increasing the order of the serial complexity**.

The approach is based on earlier ideas for the parallel evaluation of recurrence relations [49] and the parallel inversion of triangular matrices [69]. Worley explains the fundamental idea as follows: suppose we want to solve the bidiagonal matrix equation

$$\begin{pmatrix} a_{11} & & & \\ a_{21} & a_{22} & & \\ & a_{32} & a_{33} & \\ & & a_{43} & a_{44} \end{pmatrix} \begin{pmatrix} x_1 \\ x_2 \\ x_3 \\ x_4 \end{pmatrix} = \begin{pmatrix} f_1 \\ f_2 \\ f_3 \\ f_4 \end{pmatrix}. \tag{43}$$

Then one step of the cyclic reduction algorithm leads to a new matrix equation of half the size,

$$\begin{pmatrix} a_{22} & \\ -\frac{a_{43}}{a_{33}} a_{32} & a_{44} \end{pmatrix} \begin{pmatrix} x_2 \\ x_4 \end{pmatrix} = \begin{pmatrix} f_2 - \frac{a_{21}}{a_{11}} f_1 \\ f_4 - \frac{a_{43}}{a_{33}} f_3 \end{pmatrix}, \tag{44}$$

i.e. we simply computed the Schur complement to eliminate variables with odd indices. For a bigger bidiagonal matrix, this process can be repeated, and we always obtain a bidiagonal matrix of half the size. Once a two by two system is obtained, one can solve directly, and then back-substitute the values obtained to recover the values of the eliminated variables. Each step of the cyclic reduction is parallel, since each combination of two equations is independent of the others. Similarly the back-substitution process is also parallel. Cyclic reduction is therefore a direct method to solve a linear forward recurrence in parallel, and the idea can be generalized to larger bandwidth using block elimination. The serial complexity of simple forward substitution in the above example is $3n$, whereas the cyclic reduction serial complexity is $7n$ (or $5n$ if certain quantities are precomputed), and thus the algorithm is not of interest for sequential computations. But performed in parallel, the complexity of cyclic reduction becomes a logarithm in n, and one can thus obtain the solution substantially faster in parallel than just by forward substitution. For further theoretical considerations and numerical results in combination with multigrid waveform relaxation, see [46]. A truly optimal time-parallel algorithm, based on a preconditioner in a waveform relaxation setting using a fast Fourier transform in space to decouple the unknowns, and cyclic reduction in time can be found in [74].

5.5 Sheen, Sloan and Thomée 1999

A new way to solve evolution problems with a direct method in parallel was proposed in [72]:

> **These problems are completely independent, and can therefore be computed on separate processors**, with no need for shared memory. In contrast, the normal step-by-step time-marching methods for parabolic problems are not easily parallelizable.

see also [15]. The idea is to Laplace transform the problem, and then to solve a sequence of steady problems at quadrature nodes used for the numerical evaluation of the inverse Laplace transform, and goes back to the solution in the frequency domain of hyperbolic problems, see for example [17]. Suppose we have the initial value problem

$$u_t + Au = 0, \quad u(0) = u_0,$$

where A represents a linear operator. Applying a Laplace transform to this problem in time with complex valued Laplace transform parameter s leads to the parametrized equation

$$s\hat{u} + A\hat{u} = u_0, \tag{45}$$

and to obtain the solution in the time domain, one has to perform the inverse Laplace transform

$$u(t) = \frac{1}{2\pi i}\int_{\Gamma} e^{st}\hat{u}(s)ds, \tag{46}$$

where Γ is a suitably chosen contour in the complex plane. If the integral in (46) is approximated by a quadrature rule with quadrature nodes s_j, one only needs to compute $u(s)$ from (45) at $s = s_j$, and these solutions are completely independent of one another, see the quote above, and can thus be performed in parallel. This direct time parallel solver is restricted to problems where Laplace transform can be applied, i.e. linear problems with constant coefficients in the time direction, and one needs a solver that works with complex numbers for (45). It is however a very successful and efficient time parallel solver when it can be used, see [18, 51, 73, 77].

5.6 Maday and Ronquist 2008

A new idea for a direct time parallel solver by diagonalization was proposed in [56]:

> Pour briser la nature intrinsèquement séquentielle de cette résolution, on utilise **l'algorithme de produit tensoriel rapide**.

To explain the idea, we discretize the linear evolution problem $u_t = Lu$ using Backward Euler,

$$\begin{pmatrix} \frac{1}{\Delta t_1} - L & & & \\ -\frac{1}{\Delta t_2} & \frac{1}{\Delta t_2} - L & & \\ & \ddots & \ddots & \\ & & -\frac{1}{\Delta t_N} & \frac{1}{\Delta t_N} - L \end{pmatrix} \begin{pmatrix} u_1 \\ u_2 \\ \vdots \\ u_N \end{pmatrix} = \begin{pmatrix} f_1 + \frac{1}{\Delta t_1} u_0 \\ f_2 \\ \vdots \\ f_N \end{pmatrix}.$$

Using the Kronecker symbol, this linear system can be written in compact form as

$$(B \otimes I_x - I_t \otimes L)\mathbf{u} = \mathbf{f}, \tag{47}$$

where I_x is the identity matrix of the size of the spatial unknowns, and I_t is the identity matrix of the size of the time direction unknowns, and the time stepping matrix is

$$B := \begin{pmatrix} \frac{1}{\Delta t_1} & & & \\ -\frac{1}{\Delta t_2} & \frac{1}{\Delta t_2} & & \\ & \ddots & \ddots & \\ & & -\frac{1}{\Delta t_N} & \frac{1}{\Delta t_N} \end{pmatrix}.$$

If B is diagonalizable, $B = SDS^{-1}$, one can rewrite the system (47) in factored form, namely

$$(S \otimes I_x)(\text{diag}(D - L))(S^{-1} \otimes I_x)\mathbf{u} = \mathbf{f}, \tag{48}$$

and we can hence solve it in 3 steps:

$$\begin{aligned} &(a) \quad (S \otimes I_x)\mathbf{g} = \mathbf{f}, \\ &(b) \; (\tfrac{1}{\Delta t_n} - L)\mathbf{w}^n = \mathbf{g}^n, \quad 1 \le n \le N, \\ &(c) \;\; (S^{-1} \otimes I_x)\mathbf{u} = \mathbf{w}. \end{aligned}$$

Note that the expensive step (b) requiring a solve with the system matrix L can now be done entirely in parallel for all time levels t_n. Maday and Ronquist obtain with this algorithm for the 1d heat equation close to perfect speedup. They recommend to use a geometric time mesh $\Delta t_k = \rho^{k-1}\Delta t_1$, with $\rho = 1.2$, since "choosing ρ much closer to 1 may lead to instabilities". This algorithm is not defined if identical time steps are used, since it is not possible to diagonalize a Jordan block ! For a precise truncation error analysis for a geometric time grid, a round-off error analysis due to the diagonalization, and error estimates based on the trade-off between the two, see [31].

5.7 Christlieb, Macdonald and Ong 2010

The integral deferred correction methods we have already seen in Sect. 4.4 in the context of PFASST can also be used to create naturally small scale parallelism [14]:

> ... we discuss a class of defect correction methods which is easily adapted to create **parallel time integrators for multicore architectures.**

As we see in the quote, the goal is small scale parallelism for multicore architectures, as in the early predictor corrector and block methods from Sect. 5.1. The new idea introduced by Christlieb, Macdonald and Ong is to modify integral deferred correction so that pipelining becomes possible, which leads to so called RIDC (Revisionist Integral Deferred Correction) methods. As we have seen already in Sect. 4.4, the classical integral deferred correction method is working sequentially on the time point groups $I_0, I_1, \ldots, I_{J-1}$ we show in Fig. 18 taken from [14], corresponding to the time intervals $[T_0, T_1], [T_1, T_2], \ldots, [T_{J-1}, T_J]$ in Sect. 4.4. For each time point group I_j, one has to evaluate in the step (35) of integral deferred correction the quadrature formula for (33) at time $t_{j,m+1}$, using quadrature points at time $t_{j,0}, t_{j,1}, \ldots, t_{j,M}$, $0 < m < M$, where $M = 3$ in the example shown in Fig. 18. Only once all deferred correction steps on the time point group I_j are finished, one can start with the next time point group I_{j+1}, the method is like a sequential time stepping method.

In order to obtain parallelism, the idea is to increase the size of the time point groups M to contain more points than the quadrature formula needs. One can then pipeline the computation, as shown in Fig. 19: the number of quadrature points is still four, but M is much larger, and thus the Euler prediction step and the correction steps of the integral deferred correction can be executed in parallel, since all the values represented by the black dots are available simultaneously to compute the next white ones, after an initial setup of this new 'computation front'.

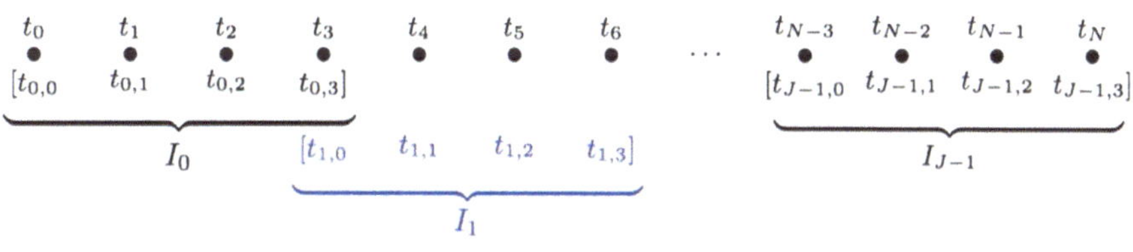

Fig. 18 Classical application of integral deferred correction, picture taken from [14]

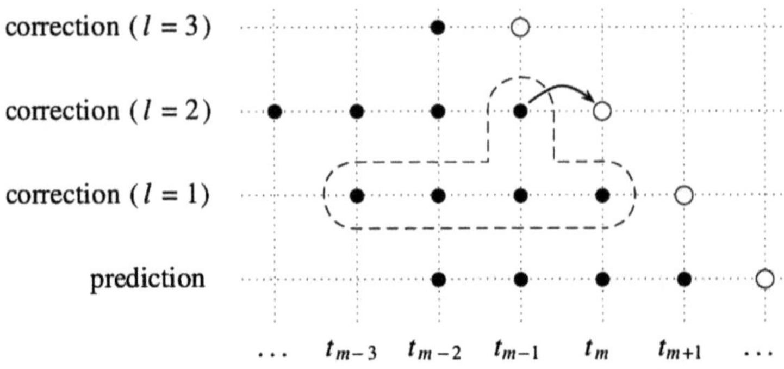

Fig. 19 RIDC way to compute integral deferred correction type methods in a pipelined way, figure taken from [14]

This leads to small scale parallel high order integrators which work very well on multicore architectures. When run in parallel, RIDC can give high order accuracy in a time comparable to the time of the low order integration method used, provided the startup costs are negligible.

5.8 Güttel 2012

A new direct time parallel method based on a completely overlapping decomposition of the time direction was proposed in [42]:

> We introduce an **overlapping time-domain decomposition method** for linear initial-value problems which gives rise to an efficient solution method for parallel computers without resorting to the frequency domain. This parallel method exploits the fact that **homogeneous initial-value problems can be integrated much faster than inhomogeneous problems** by using an efficient Arnoldi approximation for the matrix exponential function.

This method, called ParaExp [27], is only defined for linear problems, and especially suited for the parallel time integration of hyperbolic problems, where most large scale time parallel methods have severe difficulties (for a Krylov subspace remedy, see [22, 33, 66], but reorthogonalization costs might be high). ParaExp works very well also on diffusive problems [27], as we will also illustrate with a numerical experiment. To explain the method, we consider the linear system of evolution equations

$$\mathbf{u}'(t) = A\mathbf{u}(t) + \mathbf{g}(t), \quad t \in [0, T], \quad \mathbf{u}(0) = \mathbf{u}_0.$$

The ParaExp algorithm is based on a completely overlapping decomposition, as shown in Fig. 20: the time interval $[0, T]$ is decomposed into subintervals $[0, T_4 := T]$, $[T_1, T_4]$, $[T_2, T_4]$, and $[T_3, T_4]$. ParaExp is a direct solver, consisting of two steps: first one solves on the initial parts of the overlapping decomposition, $[0, T_1]$, $[T_1, T_2]$,

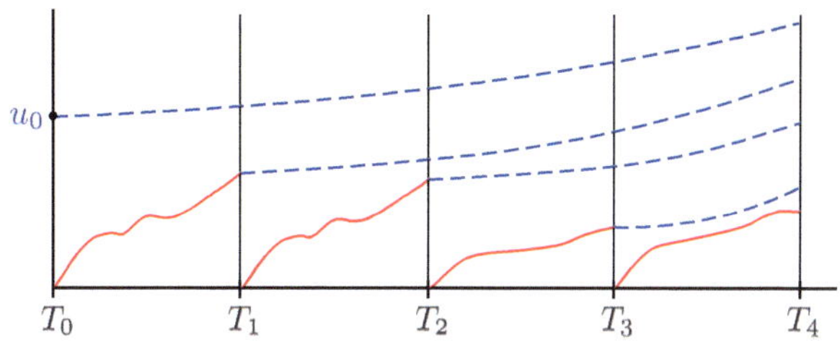

Fig. 20 Overlapping decomposition and solution strategy of ParaExp

$[T_2, T_3]$, and $[T_3, T_4]$ the non-overlapping inhomogeneous problems (shown in solid red in Fig. 20)

$$\mathbf{v}_j'(t) = A\mathbf{v}_j(t) + \mathbf{g}(t), \quad \mathbf{v}_j(T_{j-1}) = \mathbf{0}, \quad t \in [T_{j-1}, T_j],$$

and then one solves the overlapping homogeneous problems (shown in dashed blue in Fig. 20)

$$\mathbf{w}_j'(t) = A\mathbf{w}_j(t), \quad \mathbf{w}_j(T_{j-1}) = \mathbf{v}_{j-1}(T_{j-1}), \quad t \in [T_{j-1}, T]$$

By linearity, the solution is then simply obtained by summation,

$$\mathbf{u}(t) = \mathbf{v}_k(t) + \sum_{j=1}^{k} \mathbf{w}_j(t) \quad \text{with } k \text{ such that} \quad t \in [T_{k-1}, T_k].$$

Like in many time parallel methods, this seems to be absurd, since the overlapping propagation of the linear homogeneous problems is redundant, and the blue dashed solution needs to be integrated over the entire time interval $[0, T]$! The reason why substantial parallel speedup is possible in ParaExp is that near-optimal approximations of the matrix exponential are known, and so the homogeneous problems in dashed blue become very cheap. Two approaches work very well: *projection based methods*, where one approximates $\mathbf{a}_n(t) \approx exp(tA)\mathbf{v}$ from a Krylov space built with $S := (I - A/\sigma)^{-1}A$, and *expansion based methods*, which approximate $\exp(tA)\mathbf{v} \approx \sum_{j=0}^{n-1} \beta_j(t)p_j(A)\mathbf{v}$, where p_j are polynomials or rational functions. For more details, see [27].

We show in Table 2 the performance of the ParaExp algorithm applied to the wave equation from [27],

$$\begin{aligned} \partial_{tt}u(t,x) &= \alpha^2\partial_{xx}u(t,x) + \text{hat}(x)\sin(2\pi f t), \qquad x, t \in (0,1), \\ u(t,0) &= u(t,1) = u(0,x) = u'(0,x) = 0, \end{aligned}$$

where hat(x) is a hat function centered in the spatial domain. The problem is discretized with a second order centered finite difference scheme in space using $\Delta x = \frac{1}{101}$, and RK45 is used in time with $\Delta t_0 = \min\{5\cdot 10^{-4}/\alpha, 1.5\cdot 10^{-3}/f\}$ for the

Table 2 Performance of ParaExp on a wave equation

		Serial		Parallel			Efficiency (%)
α^2	f	τ_0	Error	$\max(\tau_1)$	$\max(\tau_2)$	Error	
0.1	1	2.54e−01	3.64e−04	4.04e−02	1.48e−02	2.64e−04	58
0.1	5	1.20e+00	1.31e−04	1.99e−01	1.39e−02	1.47e−04	71
0.1	25	6.03e+00	4.70e−05	9.83e−01	1.38e−02	7.61e−05	76
1	1	7.30e−01	1.56e−04	1.19e−01	2.70e−02	1.02e−04	63
1	5	1.21e+00	4.09e−04	1.97e−01	2.70e−02	3.33e−04	68
1	25	6.08e+00	1.76e−04	9.85e−01	2.68e−02	1.15e−04	75
10	1	2.34e+00	6.12e−05	3.75e−01	6.31e−02	2.57e−05	67
10	5	2.31e+00	4.27e−04	3.73e−01	6.29e−02	2.40e−04	66
10	25	6.09e+00	4.98e−04	9.82e−01	6.22e−02	3.01e−04	73

Table 3 Performance of ParaExp on the heat equation

		Serial		Parallel			Efficiency (%)
α	f	τ_0	Error	$\max(\tau_1)$	$\max(\tau_2)$	Error	
0.01	1	4.97e−02	3.01e−04	1.58e−02	9.30e−03	2.17e−04	50
0.01	10	2.43e−01	4.14e−04	7.27e−02	9.28e−03	1.94e−04	74
0.01	100	2.43e+00	1.73e−04	7.19e−01	9.26e−03	5.68e−05	83
0.1	1	4.85e−01	2.24e−05	1.45e−01	9.31e−03	5.34e−06	79
0.1	10	4.86e−01	1.03e−04	1.45e−01	9.32e−03	9.68e−05	79
0.1	100	2.42e+00	1.29e−04	7.21e−01	9.24e−03	7.66e−05	83
1	1	4.86e+00	7.65e−08	1.45e+00	9.34e−03	1.78e−08	83
1	10	4.85e+00	8.15e−06	1.45e+00	9.33e−03	5.40e−07	83
1	100	4.85e+00	3.26e−05	1.44e+00	9.34e−03	2.02e−05	84

inhomogeneous solid red problems. The homogeneous dashed blue problems were solved using a Chebyshev exponential integrator, and 8 processors were used in this set of experiments. We see that the parallel efficiency of ParaExp is excellent for this hyperbolic problem, and it would be difficult for other time parallel algorithms to achieve this.

ParaExp also works extremely well for parabolic problems. For the heat equation

$$\partial_t u(t,x) = \alpha \partial_{xx} u(t,x) + \text{hat}(x)\sin(2\pi f t), \qquad x,t \in (0,1),$$
$$u(t,0) = u(t,1) = 0, \quad u(0,x) = 4x(1-x),$$

we show numerical results from [27] in Table 3. The heat equation was discretized using centered finite differences in space with $\Delta x = \frac{1}{101}$, and again an RK45 method in time was used with $\Delta t_0 = \min\{5 \cdot 10^{-4}/\alpha, 1.5 \cdot 10^{-3}/f\}$ for the inhomogeneous problems, the homogeneous ones being solved also with a Chebyshev exponential integrator. For the heat equation, 4 processors were used. We see that again excellent parallel efficiency is obtained with the ParaExp algorithm. For more information and numerical results, see [27].

6 Conclusions

The efforts to integrate evolution problems in parallel span now five decades. We have seen that historically methods have grown into four different classes of time parallel methods: *shooting type methods* starting with Nievergelt, *domain decomposition and waveform relaxation methods* starting with Lelarasmee, Ruehli and Sangiovanni-Vincentelli, *space-time multigrid methods* starting with Hackbusch, and *direct time parallel solvers* starting with Miranker and Liniger. An important area which was not addressed in this review is a systematic comparison of the performance of these methods, and a demonstration that these algorithms are really faster than classical algorithms in a given parallel computational environment. Such a result on a 512 processor machine was shown for the space-time multigrid waveform relaxation algorithm in [78], compared to space parallel multigrid waveform relaxation and standard time stepping, see also [46] for results on an even larger machine, and [5]. A very recent comparison can be found in the short note [21], where the authors show that above a certain number of processors time-parallel algorithms indeed outperform classical ones. Time parallel methods are currently a very active field of research, and many new developments extending the latest directions we have seen, like Parareal, Schwarz-, Dirichlet Neumann and Neumann-Neumann waveform relaxation, PFASST and full space-time multigrid, and RIDC and ParaExp, are to be expected over the coming years. These will help leverage the power of new multicore and very large scale parallel computers in the near future.

Acknowledgements The author is very thankful for the comments of Stefan Vandewalle, which greatly improved this manuscript and also made the content more complete. We thank the Bibliotheque de Geneve for granting permission to reproduce pictures from the original sources.

References

1. Al-Khaleel, M., Ruehli, A.E., Gander, M.J.: Optimized waveform relaxation methods for longitudinal partitioning of transmission lines. IEEE Trans. Circuits Syst. Regul. Pap. **56**, 1732–1743 (2009)
2. Al-Khaleel, M.D., Gander, M.J., Ruehli, A.E.: Optimization of transmission conditions in waveform relaxation techniques for RC circuits. SIAM J. Numer. Anal. **52**, 1076–1101 (2014)
3. Amodio, P., Brugnano, L.: Parallel solution in time of ODEs: some achievements and perspectives. Appl. Numer. Math. **59**, 424–435 (2009)
4. Axelsson, A., Verwer, J.: Boundary value techniques for initial value problems in ordinary differential equations. Math. Comput. **45**, 153–171 (1985)

5. Bastian, P., Burmeister, J., Horton, G.: Implementation of a parallel multigrid method for parabolic differential equations. In: Parallel Algorithms for Partial Differential Equations. Proceedings of the 6th GAMM seminar Kiel, pp. 18–27 (1990)
6. Bellen, A., Zennaro, M.: Parallel algorithms for initial-value problems for difference and differential equations. J. Comput. Appl. Math. **25**, 341–350 (1989)
7. Bennequin, D., Gander, M.J., Halpern, L.: A homographic best approximation problem with application to optimized Schwarz waveform relaxation. Math. Comput. **78**, 185–223 (2009)
8. Bjørhus, M.: On domain decomposition, subdomain iteration and waveform relaxation. Ph.D. thesis, University of Trondheim, Norway (1995)
9. Brugnano, L., Trigiante, D.: A parallel preconditioning technique for boundary value methods. Appl. Numer. Math. **13**, 277–290 (1993)
10. Brugnano, L., Trigiante, D.: Convergence and stability of boundary value methods for ordinary differential equations. J. Comput. Appl. Math. **66**, 97–109 (1996)
11. Brugnano, L., Trigiante, D.: Solving Differential Equations by Multistep Initial and Boundary Value Methods. CRC Press, Boca Raton (1998)
12. Burrage, K.: Parallel and Sequential Methods for Ordinary Differential Equations. Clarendon Press, New York (1995)
13. Chartier, P., Philippe, B.: A parallel shooting technique for solving dissipative ODEs. Computing **51**, 209–236 (1993)
14. Christlieb, A.J., Macdonald, C.B., Ong, B.W.: Parallel high-order integrators. SIAM J. Sci. Comput. **32**, 818–835 (2010)
15. Crann, D., Davies, A.J., Lai, C.-H., Leong, S.H.: Time domain decomposition for European options in financial modelling. Contemp. Math. **218**, 486–491 (1998)
16. Deshpande, A., Malhotra, S., Schultz, M., Douglas, C.: A rigorous analysis of time domain parallelism. Parallel Algorithms Appl. **6**, 53–62 (1995)
17. Douglas, J.Jr., Santos, J.E., Sheen, D., Bennethum, L.S.: Frequency domain treatment of one-dimensional scalar waves. Math. Models Methods Appl. Sci. **3**, 171–194 (1993)
18. Douglas, C., Kim, I., Lee, H., Sheen, D.: Higher-order schemes for the Laplace transformation method for parabolic problems. Comput. Vis. Sci. **14**, 39–47 (2011)
19. Emmett, M., Minion, M.L.: Toward an efficient parallel in time method for partial differential equations. Commun. Appl. Math. Comput. Sci. **7**, 105–132 (2012)
20. Engquist, B., Ying, L.: Sweeping preconditioner for the Helmholtz equation: moving perfectly matched layers. Multiscale Model. Simul. **9**, 686–710 (2011)
21. Falgout, R., Friedhoff, S., Kolev, T.V., MacLachlan, S., Schroder, J.B.: Parallel time integration with multigrid. SIAM J. Sci. Comput. **36**, C635–C661 (2014)
22. Farhat, C., Cortial, J., Dastillung, C., Bavestrello, H.: Time-parallel implicit integrators for the near-real-time prediction of linear structural dynamic responses. Int. J. Numer. Methods Eng. **67**, 697–724 (2006)
23. Franklin, M.A.: Parallel solution of ordinary differential equations. IEEE Trans. Comput. **100**, 413–420 (1978)
24. Gander, M.J.: Overlapping Schwarz for linear and nonlinear parabolic problems. In: Proceedings of the 9th International Conference on Domain Decomposition, pp. 97–104 (1996). ddm.org
25. Gander, M.J.: A waveform relaxation algorithm with overlapping splitting for reaction diffusion equations. Numer. Linear Algebra Appl. **6**, 125–145 (1998)
26. Gander, M.J.: Optimized Schwarz methods. SIAM J. Numer. Anal. **44**, 699–731 (2006)
27. Gander, M.J., Güttel, S.: ParaExp: a parallel integrator for linear initial-value problems. SIAM J. Sci. Comput. **35**, C123–C142 (2013)
28. Gander, M.J., Hairer, E.: Nonlinear convergence analysis for the parareal algorithm. In: Widlund, O.B., Keyes, D.E. (eds.) Domain Decomposition Methods in Science and Engineering XVII. Lecture Notes in Computational Science and Engineering, vol. 60, pp. 45–56. Springer, Berlin (2008)
29. Gander, M.J., Halpern, L.: Absorbing boundary conditions for the wave equation and parallel computing. Math. Comput. **74**, 153–176 (2004)

30. Gander, M.J., Halpern, L.: Optimized Schwarz waveform relaxation methods for advection reaction diffusion problems. SIAM J. Numer. Anal. **45**, 666–697 (2007)
31. Gander, M.J., Halpern, L.: A direct solver for time parallelization. In: 22nd International Conference of Domain Decomposition Methods. Springer, Berlin (2014)
32. Gander, M.J., Neumüller, M.: Analysis of a new space-time parallel multigrid algorithm for parabolic problems (2015, submitted)
33. Gander, M.J., Petcu, M.: Analysis of a Krylov subspace enhanced parareal algorithm for linear problems. In: ESAIM: Proceedings. EDP Sciences, vol. 25, pp. 114–129 (2008)
34. Gander, M.J., Stuart, A.M.: Space-time continuous analysis of waveform relaxation for the heat equation. SIAM J. Sci. Comput. **19**, 2014–2031 (1998)
35. Gander, M.J., Vandewalle, S.: Analysis of the parareal time-parallel time-integration method. SIAM J. Sci. Comput. **29**, 556–578 (2007)
36. Gander, M.J., Halpern, L., Nataf, F.: Optimal convergence for overlapping and non-overlapping Schwarz waveform relaxation. In: Lai, C.-H., Bjørstad, P., Cross, M., Widlund, O. (eds.) Eleventh International Conference of Domain Decomposition Methods (1999). ddm.org
37. Gander, M.J., Halpern, L., Nataf, F.: Optimal Schwarz waveform relaxation for the one dimensional wave equation. SIAM J. Numer. Anal. **41**, 1643–1681 (2003)
38. Gander, M.J., Jiang, Y.-L., Li, R.-J.: Parareal Schwarz waveform relaxation methods. In: Widlund, O.B., Keyes, D.E. (eds.) Domain Decomposition Methods in Science and Engineering XX. Lecture Notes in Computational Science and Engineering, vol. 60, pp. 45–56. Springer, Berlin (2013)
39. Gander, M.J., Kwok, F., Mandal, B.: Dirichlet-Neumann and Neumann-Neumann waveform relaxation algorithms for parabolic problems (2015, submitted)
40. Gear, C.W.: Parallel methods for ordinary differential equations. Calcolo **25**, 1–20 (1988)
41. Giladi, E., Keller, H.B.: Space time domain decomposition for parabolic problems. Numer. Math. **93**, 279–313 (2002)
42. Güttel, S.: A parallel overlapping time-domain decomposition method for ODEs. In: Domain Decomposition Methods in Science and Engineering XX, pp. 459–466. Springer, Berlin (2013)
43. Hackbusch, W.: Parabolic multi-grid methods. In: Glowinski, R. Lions, J.-L. (eds.) Computing Methods in Applied Sciences and Engineering, VI. pp. 189–197. North-Holland, Amsterdam (1984)
44. Hoang, T.-P., Jaffré, J., Japhet, C., Kern, M., Roberts, J.E.: Space-time domain decomposition methods for diffusion problems in mixed formulations. SIAM J. Numer. Anal. **51**, 3532–3559 (2013)
45. Horton, G., Vandewalle, S.: A space-time multigrid method for parabolic partial differential equations. SIAM J. Sci. Comput. **16**, 848–864 (1995)
46. Horton, G., Vandewalle, S., Worley, P.: An algorithm with polylog parallel complexity for solving parabolic partial differential equations, SIAM J. Sci. Comput. **16**, 531–541 (1995)
47. Keller, H.B.: Numerical Solution for Two-Point Boundary-Value Problems. Dover Publications Inc, New York (1992)
48. Kiehl, M.: Parallel multiple shooting for the solution of initial value problems. Parallel Comput. **20**, 275–295 (1994)
49. Kogge, P.M., Stone, H.S.: A parallel algorithm for the efficient solution of a general class of recurrence equations. IEEE Trans. Comput. **100**, 786–793 (1973)
50. Kwok, F.: Neumann-Neumann waveform relaxation for the time-dependent heat equation. In: Domain Decomposition Methods in Science and Engineering, DD21. Springer, Berlin (2014)
51. Lai, C.-H.: On transformation methods and the induced parallel properties for the temporal domain. In: Magoulès, F. (ed.) Substructuring Techniques and Domain Decomposition Methods, pp. 45–70. Saxe-Coburg Publications, Scotland (2010)
52. Lelarasmee, E., Ruehli, A.E., Sangiovanni-Vincentelli, A. L.: The waveform relaxation method for time-domain analysis of large scale integrated circuits. IEEE Trans. CAD IC Syst. **1**, 131–145 (1982)
53. Lindelöf, E.: Sur l'application des méthodes d'approximations successives à l'étude des intégrales réelles des équations différentielles ordinaires. J. de Math. Pures et Appl. **10**, 117–128 (1894)

54. Lions, J.-L., Maday, Y., Turinici, G.: A parareal in time discretization of PDEs. C.R. Acad. Sci. Paris, Serie I **332**, 661–668 (2001)
55. Lubich, C., Ostermann, A.: Multi-grid dynamic iteration for parabolic equations. BIT **27**, 216–234 (1987)
56. Maday, Y., Rønquist, E.M.: Parallelization in time through tensor-product space–time solvers. C.R. Math. **346**, 113–118 (2008)
57. Maday, Y., Turinici, G.: The parareal in time iterative solver: a further direction to parallel implementation. In: Domain Decomposition Methods in Science and Engineering, pp. 441–448. Springer, Berlin (2005)
58. Mandal, B.: A time-dependent Dirichlet-Neumann method for the heat equation. In: Domain Decomposition Methods in Science and Engineering, DD21. Springer, Berlin (2014)
59. Milne, W.E., Wiley, J.: Numerical Solution of Differential Equations. vol. 19(3). Wiley, New York (1953)
60. Minion, M.L.: A hybrid parareal spectral deferred corrections method. Commun. Appl. Math. Comput. Sci. **5**, 265–301 (2010)
61. Miranker, W.L., Liniger, W.: Parallel methods for the numerical integration of ordinary differential equations. Math. Comput. **91**, 303–320 (1967)
62. Nataf, F., Rogier. F.: Factorization of the convection-diffusion operator and the Schwarz algorithm. M^3AS **5**, 67–93 (1995)
63. Neumüller, M.: Space-time methods: fast solvers and applications, Ph.D. thesis, University of Graz (2013)
64. Nievergelt, J.: Parallel methods for integrating ordinary differential equations. Commun. ACM **7**, 731–733 (1964)
65. Picard, E.: Sur l'application des méthodes d'approximations successives à l'étude de certaines équations différentielles ordinaires. J. de Math. Pures et Appl. **9**, 217–271 (1893)
66. Ruprecht, D., Krause, R.: Explicit parallel-in-time integration of a linear acoustic-advection system. Comput. Fluids **59**, 72–83 (2012)
67. Saha, P., Stadel, J., Tremaine, S.: A parallel integration method for solar system dynamics. Astron. J. **114**, 409–415 (1997)
68. Saltz, J.H., Naik, V.K.: Towards developing robust algorithms for solving partial differential equations on MIMD machines. Parallel Comput. **6**, 19–44 (1988)
69. Sameh, A.H., Brent, R.P.: Solving triangular systems on a parallel computer. SIAM J. Numer. Anal. **14**, 1101–1113 (1977)
70. Schwarz, H.A.: Über einen Grenzübergang durch alternierendes Verfahren. Vierteljahrsschrift der Naturforschenden Gesellschaft in Zürich **15**, 272–286 (1870)
71. Shampine, L., Watts, H.: Block implicit one-step methods. Math. Comput. **23**, 731–740 (1969)
72. Sheen, D., Sloan, I., Thomée, V.: A parallel method for time-discretization of parabolic problems based on contour integral representation and quadrature. Math. Comput. Am. Math. Soc. **69**, 177–195 (1999)
73. Sheen, D., Sloan, I.H., Thomée, V.: A parallel method for time discretization of parabolic equations based on Laplace transformation and quadrature. IMA J. Numer. Anal. **23**, 269–299 (2003)
74. Simoens, J., Vandewalle, S.: Waveform relaxation with fast direct methods as preconditioner. SIAM J. Sci. Comput. **21**, 1755–1773 (2000)
75. Speck, R., Ruprecht, D., Krause, R., Emmett, M., Minion, M., Winkel, M., Gibbon, P.: A massively space-time parallel n-body solver. In: Proceedings of the International Conference on High Performance Computing, Networking, Storage and Analysis, Salt Lake City, p. 92. IEEE Computer Society Press, Los Alamitos (2012)
76. Speck, R., Ruprecht, D., Emmett, M., Minion, M., Bolten, M., Krause, R.: A multi-level spectral deferred correction method. arXiv:1307.1312 (2013, arXiv preprint)
77. Thomée, V.: A high order parallel method for time discretization of parabolic type equations based on Laplace transformation and quadrature. Int. J. Numer. Anal. Model. **2**, 121–139 (2005)

78. Vandewalle, S., Van de Velde, E.: Space-time concurrent multigrid waveform relaxation. Ann. Numer. Math. **1**, 347–363 (1994)
79. Womble, D.E.: A time-stepping algorithm for parallel computers. SIAM J. Sci. Stat. Comput. **11**, 824–837 (1990)
80. Worley, P.: Parallelizing across time when solving time-dependent partial differential equations. In: Sorensen, D. (ed.) Proceedings of 5th SIAM Conference on Parallel Processing for Scientific Computing. SIAM, Houston (1991)

Direct Multiple Shooting for Nonlinear Optimum Experimental Design

Dennis Janka, Stefan Körkel, and Hans Georg Bock

Abstract Optimum experimental design (OED) for parameter identification has become a key technique in the model validation process for dynamical systems. This paper deals with optimum experimental design for systems modelled by differential-algebraic equations. We show how to formulate OED as a nonstandard nonlinear optimal control problem. The direct multiple shooting method is a state of the art method for the solution of standard optimal control problems that leads to structured nonlinear programs. We present two possibilities how to adapt direct multiple shooting to OED by introducing additional variables and constraints. We highlight special structures in the constraint and objective derivatives whose evaluation is usually the bottleneck when solving dynamic optimization problems by multiple shooting. We have implemented a structure exploiting algorithm that takes all these structures into account. Two benchmark examples show the efficiency of the new algorithm.

1 Introduction

Many processes in engineering, chemistry, or physics can be described by dynamical systems given by ordinary differential equations (ODEs) or differential-algebraic equations (DAEs). These equations usually depend on *model parameters*, for example material specific constants that are not directly accessible by measurements. However, as the models are often highly nonlinear, simulation results can vary strongly depending on the values of the model parameters. Thus it is desirable to estimate the model parameters with high precision. This process is called model *validation*. Only a validated model can be used to make meaningful predictions.

The first step in the model validation process is usually a parameter estimation. That means the model is fitted to given measurement data, yielding a first estimate

D. Janka (✉) • S. Körkel • H.G. Bock
Interdisciplinary Center for Scientific Computing, Heidelberg University, Im Neuenheimer Feld 368, 69120 Heidelberg, Germany
e-mail: dennis.janka@iwr.uni-heidelberg.de; stefan.koerkel@iwr.uni-heidelberg.de; SciCom@iwr.uni-heidelberg.de

T. Carraro et al. (eds.), *Multiple Shooting and Time Domain Decomposition Methods*, Contributions in Mathematical and Computational Sciences 9,
DOI 10.1007/978-3-319-23321-5_4

for the parameters. Then one can perform a sensitivity analysis to obtain estimates for the covariance matrix of the parameters. The covariance matrix can reveal large uncertainties in the parameter values or correlations between different parameters. In this context, it is also possible to quantify the uncertainty of arbitrary model quantities of interest.

An important observation is that the covariance matrix depends not only on the parameters but also on the experimental conditions. This leads to the task of *optimum experimental design (OED)*: Choose experimental conditions such that a subsequent parameter estimation yields parameters with minimum uncertainty. The uncertainty of the vector of parameters is characterized by a functional on the predicted covariance matrix. OED for general statistical models has been studied for several decades and it is a well-established field of research, see the textbooks [2, 10, 18]. Nonlinear OED for processes modeled by differential equations has been investigated by several authors, see, e.g., [4, 12, 15] for an overview.

In mathematical terms, optimum experimental design can be cast as a special (nonstandard) type of optimal control (OC) problem. As the objective, namely the functional on the covariance matrix, depends on first-order sensitivities of the states, *variational differential equations* or *sensitivity equations* must be explicitly included in the problem formulation, leading to large, but specially structured differential equation systems. We are interested in *direct* methods for OED problems, in particular direct multiple shooting, that transform the infinite dimensional optimal control problem into a finite dimensional nonlinear programming problem (NLP). The direct multiple shooting method for optimal control problems as described by Bock and Plitt [7] makes use of the partial separability of the objective function. This leads to a block-diagonal Hessian of the Lagrangian that can and should be exploited by Newton-type methods. However, a straightforward formulation of the OED objective function lacks the feature of partial separability, so special care must be taken to reformulate the OED problem as a standard optimal control problem. In [14, 16] direct multiple shooting has been applied to OED. In [13], a collocation discretization is applied to a similar problem. The main contribution of this paper consists in detailed descriptions of the structured NLPs that result from a multiple shooting discretization as well as numerical results that demonstrate their benefits and limitations.

The paper is organized as follows: In Sect. 2, we give an introduction to optimum experimental design and formulate it as a nonstandard optimal control problem. In Sect. 3, the direct multiple shooting method is described for standard optimal control problems. Afterwards, in Sect. 4 we propose two ways how to transform the OED problem to a specially structured standard OC problem. In Sect. 5 we show how to efficiently evaluate the constraint Jacobian as well as the gradient and Hessian of the objective. A numerical example from chemical engineering illustrates the efficacy of the approach in Sect. 6. Section 7 concludes.

2 Nonlinear Optimum Experimental Design for Parameter Estimation

Optimum experimental design aims at improving parameter estimation results in a statistical sense. We first introduce the type of parameter estimation problems whose results we seek to improve with OED along with some general notation that we use throughout the paper.

2.1 *Parameter Estimation for Dynamical Systems*

We consider a dynamical process on a fixed time horizon $[t_0, t_f]$ that is described by the following differential-algebraic equation (DAE) system with interior point and boundary conditions:

$$\dot{y}(t) = f(t, y(t), z(t), p, u(t)), \quad y(t_0) = y_0(p, u) \tag{1a}$$

$$0 = g(t, y(t), z(t), p, u(t)) \tag{1b}$$

$$0 = \sum_{i=1}^{N_r} r_i(t_i, y(t_i), z(t_i), p) \tag{1c}$$

where

$$\begin{aligned} y(t) &\in \mathbb{R}^{n_y} && \textit{(differential states)} \\ z(t) &\in \mathbb{R}^{n_z} && \textit{(algebraic states)} \\ p &\in \mathbb{R}^{n_p} && \textit{(parameters)} \\ u(t) &\in \mathbb{R}^{n_u} && \textit{(controls)} \\ r_i(t_i, y(t_i), z(t_i), p) &\in \mathbb{R}^{n_r} && \textit{(boundary conditions).} \end{aligned}$$

Throughout the text we will use the notation $x(t) = (y(t)^T, z(t)^T)^T$ to denote both differential and algebraic states.

Assume measurement data $\eta_1, \ldots, \eta_{N_M}$ are available at sampling times $t_1, \ldots, t_M$ such that

$$h_i(t_i, y(t_i), z(t_i), p^\star) = \eta_i + \varepsilon_i, \quad i = 1, \ldots, N_M,$$

where $p^\star$ are the true—but inaccessible—parameters, y and z the corresponding states, and ε_i are independently normally distributed with zero mean and standard deviation σ_i. This assumption states that the model is structurally correct and errors arise only due to inaccuracies in the measurement process. We call

$$h_i(t, y(t), z(t), p), \quad i = 1, \dots, N_M$$

the *model response* or *observable*.

The maximum likelihood parameter estimation problem can be stated as:

$$\min_{y_0, x, p} \sum_{i=1}^{N_m} \left(\frac{h_i(t_i, y(t_i), z(t_i), p) - \eta_i}{\sigma_i} \right)^2 \tag{2a}$$

$$\text{s.t. } \dot{y}(t) = f(t, y(t), z(t), p, u(t)), \quad y(t_0) = y_0(p, u) \tag{2b}$$

$$0 = g(t, y(t), z(t), p, u(t)) \tag{2c}$$

$$0 = \sum_{i=1}^{N_r} r_i(t_i, y(t_i), z(t_i), p) \tag{2d}$$

Parameter estimation problems constrained by differential equations can be solved by different approaches, e.g. by direct multiple shooting in combination with a generalized Gauss-Newton method, see [6].

2.2 Sensitivity Analysis

The solution $\hat{p}$ of the parameter estimation problem (2) is a random variable due to the fact that the measurements η_i are random. The variance-covariance matrix C of $\hat{p}$ is given by:

$$C = (I \; 0) \begin{pmatrix} \mathcal{J}_1^T \mathcal{J}_1 & \mathcal{J}_2^T \\ \mathcal{J}_2 & 0 \end{pmatrix}^{-1} \begin{pmatrix} I \\ 0 \end{pmatrix} \tag{3}$$

where $\mathcal{J}_1 \in \mathbb{R}^{N_m \times n_p}$ and $\mathcal{J}_2 \in \mathbb{R}^{n_r \times n_p}$ are the Jacobians of the residual vectors of the parameter estimation problem. We denote by $\mathcal{J}_{1,i}$ the rows of $\mathcal{J}_1$ and by $\mathcal{J}_{2,i}$ the summands that make up $\mathcal{J}_2$:

$$\mathcal{J}_{1,i} = \left(\frac{\sqrt{w_i}}{\sigma_i} \left(\frac{\partial h_i}{\partial x}(t_i, y(t_i), z(t_i), p) \frac{\partial x}{\partial p}(t_i) + \frac{\partial h_i}{\partial p}(t_i, y(t_i), z(t_i), p) \right) \right)_{i=1,\dots,N_m} \tag{4}$$

$$\mathcal{J}_2 = \sum_{i=1}^{N_r} \mathcal{J}_{2,i}, \quad \mathcal{J}_{2,i} = \frac{\partial r_i}{\partial x}(t_i, y(t_i), z(t_i), p) \frac{\partial x}{\partial p}(t_i) + \frac{\partial r_i}{\partial p}(t_i, y(t_i), z(t_i), p). \tag{5}$$

We assume that $\mathscr{J}_2$ has full rank and that $\mathscr{J}_1^T \mathscr{J}_1$ is positive definite on Ker $\mathscr{J}_2$ which implies existence of C.

In (4) we have also introduced *measurement weights* $w_i \in \{0, 1\}$, $i = 1, \ldots, N_m$ for each measurement time. They are fixed in the parameter estimation context but will be design variables in the experimental design where they allow us to select or de-select measurements.

The sensitivities of the states x with respect to the parameters p are subject to the following *variational differential-algebraic equations (VDAE)*, also called *sensitivity equations*:

$$\dot{y}_p(t) = \frac{\partial f}{\partial x}(t, y(t), z(t), p, u(t))x_p(t) + \frac{\partial f}{\partial p}(t, y(t), z(t), p, u(t)) \tag{6}$$

$$0 = \frac{\partial g}{\partial x}(t, y(t), z(t), p, u(t))x_p(t) + \frac{\partial g}{\partial p}(t, y(t), z(t), p, u(t)), \tag{7}$$

where

$$x_p(t) = \frac{\partial x}{\partial p}(t) = \left(\frac{\partial y}{\partial p}(t), \frac{\partial z}{\partial p}(t)\right).$$

Initial values for the VDAE are given by

$$y_p(t_0) = \frac{\partial y_0}{\partial p}$$

for the variational differential states and by (7) for the variational algebraic states. Note that (6) and (7) depend on $y(t)$ and $z(t)$ and therefore have to be solved together with (1a) and (1b).

2.3 *The Optimum Experimental Design Problem*

Based on the sensitivity analysis, we can predict the variance-covariance matrix for different experimental settings that are characterized by controls $u(t)$ as well as a choice of measurements. An experiment may also be constrained by external process constraints, e.g. safety or cost constraints.

The task of optimum experimental design is to choose experimental settings such that the predicted covariance matrix has the best properties in some statistical sense. The quality of the matrix is measured by a criterion ϕ from statistical experimental design:

- A-criterion: $\phi = \operatorname{tr} C$
- D-criterion: $\phi = \det C$
- E-criterion: $\phi = \max\{\lambda_i, i = 1, \ldots, n_p, \lambda_i \text{ eigenvalue of } C\} = ||C||_2$
- M-criterion: $\phi = \max\{C_{ii}, i = 1, \ldots, n_p\}$

The complete optimum experimental design problem is

$$\min_{y_0,x,x_p,u,w} \phi\left(\begin{pmatrix} I & 0 \end{pmatrix} \begin{pmatrix} \mathscr{J}_1^T \mathscr{J}_1 & \mathscr{J}_2^T \\ \mathscr{J}_2 & 0 \end{pmatrix}^{-1} \begin{pmatrix} I \\ 0 \end{pmatrix} \right) \tag{8a}$$

$$\text{s.t. } \dot{y}(t) = f(t, y(t), z(t), p, u(t)), \quad t \in [t_0, t_f] \tag{8b}$$

$$0 = g(t, y(t), z(t), p, u(t)), \quad t \in [t_0, t_f] \tag{8c}$$

$$y(t_0) = y_0(p, u) \tag{8d}$$

$$\dot{y}_p(t) = \frac{\partial f}{\partial x}(t, y(t), z(t), p, u(t))x_p(t) + \frac{\partial f}{\partial p}(t, y(t), z(t), p, u(t)), \; t \in [t_0, t_f] \tag{8e}$$

$$0 = \frac{\partial g}{\partial x}(t, y(t), z(t), p, u(t))x_p(t) + \frac{\partial g}{\partial p}(t, y(t), z(t), p, u(t)), \; t \in [t_0, t_f] \tag{8f}$$

$$y_p(t_0) = \frac{\partial y_0}{\partial p} \tag{8g}$$

$$0 = \sum_{i=1}^{N_r} r_i(t_i, y(t_i), z(t_i), p) \tag{8h}$$

$$0 \leq c\,(t, y(t), z(t), u(t), w)\,, \quad t \in [t_0, t_f] \tag{8i}$$

$$w_i \in \{0, 1\}, \quad i = 1, \ldots, N_m \tag{8j}$$

$$\mathscr{J}_{1,i} = \left(\frac{\sqrt{w_i}}{\sigma_i} \left(\frac{\partial h_i}{\partial x}(t_i, y(t_i), z(t_i), p)x_p(t_i) + \frac{\partial h_i}{\partial p}(t_i, y(t_i), z(t_i), p) \right) \right)_{i=1,\ldots,N_m} \tag{8k}$$

$$\mathscr{J}_2 = \sum_{i=1}^{N_r} \frac{\partial r_i}{\partial x}(t_i, y(t_i), z(t_i), p)x_p(t_i) + \frac{\partial r_i}{\partial p}(t_i, y(t_i), z(t_i), p) \tag{8l}$$

with the nominal DAE system (8b) and (8c) with initial values (8d), the variational DAE system (8e) and (8f) with initial values (8g), multipoint boundary constraints from the parameter estimation problem (8h), path and control constraints (8i), and integrality constraints for the measurement weights (8j). The Jacobians of the parameter estimation residuals (8k) and (8l) are given to define the covariance matrix on which a functional ϕ is minimized (8a). Note that while initial values for the nominal differential states (8b) may be degrees of freedom in the optimization, initial values for the variational differential states (8e) are explicitly defined by the relation (8g) and for the algebraic states $z(t)$ and $z_p(t)$ they are implicitly defined by the algebraic conditions (8c) and (8f).

3 The Direct Multiple Shooting Method for Optimal Control Problems

The direct multiple shooting method for optimal control problems has been first introduced in [7]. Let us first consider the standard optimal control problem

$$\min_{\tilde{y}_0,\tilde{x},u} \Phi(\tilde{x}(t_f)) \tag{9a}$$

$$\text{s.t. } \dot{\tilde{y}}(t) = \tilde{f}(t,\tilde{y}(t),\tilde{z}(t),\tilde{u}(t)), \quad \tilde{y}(t_0) = \tilde{y}_0 \tag{9b}$$

$$0 = \tilde{g}(t,\tilde{y}(t),\tilde{z}(t),\tilde{u}(t)) \tag{9c}$$

$$0 \le \tilde{c}(t,\tilde{y}(t),\tilde{z}(t),\tilde{u}(t)) \tag{9d}$$

$$0 \le \sum_{i=1}^{N_{\tilde{r}}} \tilde{r}_i(t_i,\tilde{y}(t_i),\tilde{z}(t_i)). \tag{9e}$$

In direct methods the infinite-dimensional optimal control problem (9) is approximated by a nonlinear programming problem (NLP) which is then solved by suitable numerical methods. The following infinite-dimensional objects of the optimal control problem must be treated adequately when setting up the finite-dimensional NLP:

- control functions $\tilde{u}$
- differential and algebraic states $\tilde{y}$ and $\tilde{z}$
- path constraints $0 \le \tilde{c}(t,\tilde{y}(t),\tilde{z}(t),\tilde{u}(t))$

3.1 Control Functions

We consider a time grid

$$t_0 = \tau_0^c < \tau_1^c < \cdots < \tau_{N_c}^c = t_f \tag{10}$$

on which the control function $\tilde{u}(\cdot)$ is parameterized by means of local basis functions:

$$\tilde{u}(t) = \varphi^j(t,q^j), \quad t \in [\tau_j^c, \tau_{j+1}^c],$$

where the $q^j \in \mathbb{R}^{n_u}$ are vectors of finitely many real optimization variables. We define $q := (q^0,\ldots,q^{N_c-1})^T$. The local functions φ^j are typically polynomials of low degree, e.g. linear or constant functions.

3.2 States

In shooting methods, initial value problem solvers are employed to obtain representations of the states x for given q and y_0. In the case of direct single shooting and pure ODEs, the states are regarded as dependent variables, and only q and y_0 are kept as variables in the optimization problem. Thus the tasks of simulation and optimization are kept separate.

The direct multiple shooting method for DAEs is a simultaneous strategy to resolve simulation and optimization in parallel. Again, we consider a discretization of the time horizon

$$t_0 = \tau_0^s < \tau_1^s < \cdots < \tau_{N_s}^s = t_f \tag{11}$$

where we assume without loss of generality the grid points to be a subset of the grid points of the control grid (10). On this shooting grid we consider the following set of initial value problems with initial values $s_x^j = (s_y^j, s_z^j)$ that become variables in the optimization problem:

$$\dot{\tilde{y}}(t) = \tilde{f}(t, \tilde{y}(t), \tilde{z}(t), p, \tilde{u}(t)) \qquad \tilde{y}(\tau_j^s) = s_y^j \tag{12a}$$

$$0 = \tilde{g}(t, \tilde{y}(t), \tilde{z}(t), p, \tilde{u}(t)) - \theta_j(t)\tilde{g}(\tau_j^s, s_y^j, s_z^j, p, \tilde{u}(t)) \qquad \tilde{z}(\tau_j^s) = s_z^j, \tag{12b}$$

where $\theta_j(\cdot)$ is a fast decreasing damping function with $\theta(\tau_j^s) = 1$. This relaxed formulation was proposed in [8] and means that the algebraic condition (12b) is automatically consistent for any initial values s_z^j. That means the DAE solver does not need to solve the (nonlinear) algebraic condition in every iteration of the optimization algorithm to find feasible initial values. Instead, the nonlinear algebraic consistency conditions

$$0 = \tilde{g}(\tau_j^s, s_y^j, s_z^j, \hat{q}^j), \quad j = 0, \ldots, N_s$$

are added to the optimization problem which ensures the solution of the original DAE at the solution of the optimization problem.

Note that the DAEs (12) are solved independently on the smaller time intervals $[\tau_j^s, \tau_{j+1}^s]$ as the initial values s_x^j are variables of the optimization problem. To ensure equivalence to the original system (1), continuity conditions are added to the optimization problem for every shooting interval. Let us denote by $\tilde{y}(\tau_{j+1}^s; s_y^j, s_z^j, \hat{q}^j)$ a representation of the solution to problem (12) on the intervals $[\tau_j^s, \tau_{j+1}^s]$, where $\hat{q}^j$ denotes the subvector of q that represents $\tilde{u}$ on the interval $[\tau_j^s, \tau_{j+1}^s]$. Then the continuity conditions read as:

$$\tilde{y}(\tau_{j+1}^s; s_y^j, s_z^j, \hat{q}^j) = s_y^{j+1}, \quad j = 0, \ldots, N_s - 1.$$

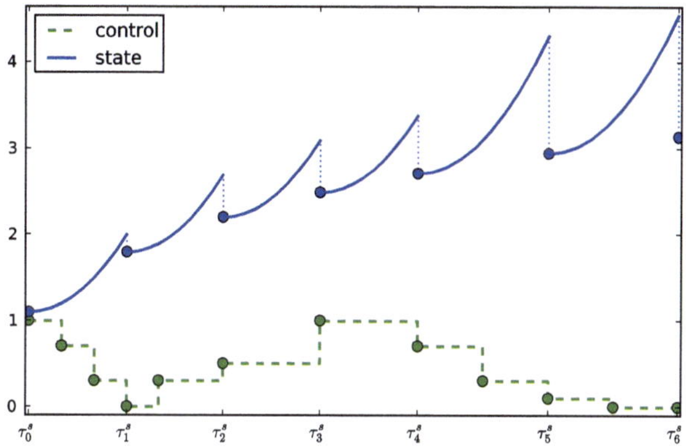

Fig. 1 Concept of direct multiple shooting for one state and one piecewise constant control. The continuity conditions are violated (*vertical dotted lines*). Note how the control is also allowed to switch within shooting intervals

Figure 1 illustrates the concept of direct multiple shooting. Note that we explicitly maintain separate grids for controls and states. A special case is of course to choose the same grid for both. However, in our experience, the decoupling of grids provides greater flexibility and a smaller number of shooting intervals can greatly accelerate convergence for problems where a relatively fine discretization of the controls is desirable.

3.3 Path Constraints

All path constraints such as (9d) that are required to hold at infinitely many points are evaluated on finitely many checkpoints only. Let us assume—without loss of generality—that the checkpoints are the grid points of the multiple shooting grid (10). Then the discretized path constraints read as

$$0 \leq \tilde{c}(\tau_j^s, s_y^j, s_z^j, \hat{q}^j), \quad j = 0, \ldots, N_s. \tag{13}$$

Depending on the choice of the time grid, the constraints might be violated in between grid points. There exist strategies how to adaptively add checkpoints, see, e.g., [17], but to keep the notation simple we assume for the scope of this paper that they match the grid points of the shooting grid.

3.4 Structured NLP

We have now addressed all constraints of the optimal control problem and can formulate the structured multiple shooting NLP as follows:

$$\min_{s_y, s_z, q} \phi(s_x^N) \tag{14a}$$

$$\text{s.t. } 0 = \tilde{y}(\tau_0^s) - s_y^0 \tag{14b}$$

$$0 = \tilde{y}(\tau_{j+1}^s; s_y^j, s_z^j, \hat{q}^j) - s_y^{j+1}, \qquad j = 0, \dots, N_s - 1 \tag{14c}$$

$$0 = \tilde{g}(\tau_j^s, s_y^j, s_z^j, \hat{q}^j), \qquad j = 0, \dots, N_s \tag{14d}$$

$$0 \leq c(\tau_j^c, \tilde{y}(\tau_j^c), \tilde{z}(\tau_j^c), \hat{q}^j), \qquad j = 0, \dots, N_s \tag{14e}$$

$$0 \leq \sum_{j=0}^{N_s} \sum_{\substack{i, \\ \tau_j \leq t_i < \tau_{j+1}}}^{N_{\tilde{r}}} \tilde{r}_i(t_i, \tilde{y}(t_i; s_y^j, s_z^j, \hat{q}^j), \tilde{z}(t_i; s_y^j, s_z^j, \hat{q}^j)), \tag{14f}$$

where (14c) and (14d) are the continuity and consistency conditions that guarantee the solution of the original DAE (1) at the solution of the optimization problem.

In Newton-type methods, the Jacobian of the constraints and the Hessian of the Lagrangian are of special importance. It is clear that the evaluation of the continuity and consistency constraints with index j only depend on variables s_x^j and $\hat{q}^j$ in a nonlinear way. This leads to a constraint Jacobian that has a banded structure and a Hessian of the Lagrangian with a block diagonal structure according to the shooting discretization. These structures can be seen in the structure of the KKT matrix as depicted in Fig. 2.

Depending on the shooting discretization, problems of type (14) can be very large, but sparse. Algorithmic techniques such as condensing (see [7]) exploit this sparsity and reduce the additional effort considerably that is caused by the larger matrices when using a fine multiple shooting discretization.

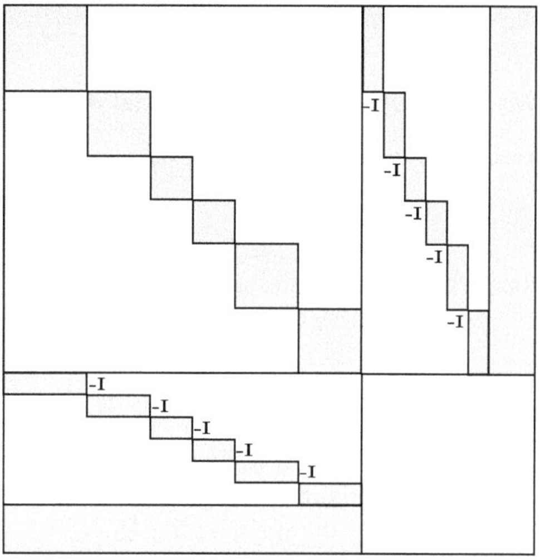

Fig. 2 Sparsity pattern of the KKT matrix of a multiple shooting discretized optimal control problem. The constraint Jacobian comprises continuity constraints that are responsible for the banded structure. Linearly coupled constraints give rise to a dense block. The Hessian of the Lagrangian (*upper left*) has block diagonal structure. Different block sizes may occur if the numbers of control variables on two intervals differ

4 Optimum Experimental Design as Separable NLP

We now want to apply the multiple shooting discretization as described in the previous section to the optimum experimental design problem (8). In particular we need to extend the problem formulation (14) to cope with the special kind of coupled objective that is characteristic for OED. Multiple Shooting for OED problems has first been applied in [16] and has been further investigated in [14].

4.1 Measurements

The grid of possible measurements depends on the process and should be independent of the shooting and control grid. In particular, more than one measurement could be taken at the same time, see [15].

In the original formulation, integrality of the measurement weights is required. In our formulation we employ a continuous relaxation:

$$0 \leq w_i \leq 1, \quad i = 1, \ldots, N_m$$

In practice, this often yields satisfactory results, as a bang-bang structure is observed for the measurements and so integrality is satisfied automatically. In fact there is also some theoretical evidence for this, see [20].

The measurement weights, along with the controls, are experimental design variables. All simple bounds and linear constraints on the measurement weights fit into the framework of general path constraints and linearly coupled interior point constraints (9d) and (9e).

4.2 Dynamical System

We combine nominal and variational states into one system:

$$\tilde{y}(t) = \begin{cases} y(t) \\ \downharpoonright y_p(t) \downharpoonleft \end{cases} \in \mathbb{R}^{n_y + n_y \cdot n_p}$$

for the differential states and

$$\tilde{z}(t) = \begin{cases} z(t) \\ \downharpoonright z_p(t) \downharpoonleft \end{cases} \in \mathbb{R}^{n_z + n_z \cdot n_p}$$

for the algebraic states, where we denote by $\downharpoonright \cdot \downharpoonleft$ the map that combines the columns of an $m \times n$ matrix into a single $m \cdot n$ column vector by stacking them one below the other.

That leaves us with the new DAE system

$$\dot{\tilde{y}}(t) = \tilde{f}(t, \tilde{y}, \tilde{z}, p, u) = \begin{cases} f(t, y(t), z(t), p, u(t)) \\ \downharpoonright \frac{\partial f}{\partial x}(t, y(t), z(t), p, u(t)) x_p(t) + \frac{\partial f}{\partial p}(t, y(t), z(t), p, u(t)) \downharpoonleft \end{cases} \tag{15}$$

$$0 = \tilde{g}(t, \tilde{y}, \tilde{z}, p, u) = \begin{cases} g(t, y(t), z(t), p, u(t)) \\ \downharpoonright \frac{\partial g}{\partial x}(t, y(t), z(t), p, u(t)) x_p(t) + \frac{\partial g}{\partial p}(t, y(t), z(t), p, u(t)) \downharpoonleft \end{cases} . \tag{16}$$

This system has of course a special structure that can and should be exploited in an efficient implementation. We will give details on this in Sect. 5.

4.3 Objective Function

An important difference between the problem of optimum experimental design (8) and the standard optimal control problem (9) is the nonlinear coupling in time in the objective that is due to the inversion when computing the covariance matrix, as it has been noted in [16]. In particular this violates the property of partial separation of the Lagrange function that is responsible for its sparse, block-diagonal Hessian. We present two approaches to resolve that nonlinear coupling.

4.3.1 Linearly Coupled Constraint for Information Matrix

Recall the definition of the covariance matrix:

$$C = (I\ 0)\begin{pmatrix} \mathscr{J}_1^T \mathscr{J}_1 & \mathscr{J}_2^T \\ \mathscr{J}_2 & 0 \end{pmatrix}^{-1} \begin{pmatrix} I \\ 0 \end{pmatrix}$$

with the rows of $\mathscr{J}_1$ and the summands that constitute $\mathscr{J}_2$ as defined by (4) and (5):

$$\mathscr{J}_{1,i} = \left(\frac{\sqrt{w_i}}{\sigma_i} \left(\frac{\partial h_i}{\partial x} x_p(t_i) + \frac{\partial h_i}{\partial p} \right) \right)_{i=1,\dots,N_m}$$

$$\mathscr{J}_{2,i} = \frac{\partial r_i}{\partial x}(t_i, y(t_i), z(t_i), p) x_p(t_i) + \frac{\partial r_i}{\partial p}(t_i, y(t_i), z(t_i), p).$$

In [16] it has been pointed out that

$$\begin{pmatrix} \mathscr{J}_1^T \mathscr{J}_1 & \mathscr{J}_2^T \\ \mathscr{J}_2 & 0 \end{pmatrix} = \begin{pmatrix} \sum_{i=0}^{N_m} \mathscr{J}_{1,i}^T \mathscr{J}_{1,i} & \sum_{i=0}^{N_r} \mathscr{J}_{2,i}^T \\ \sum_{i=0}^{N_r} \mathscr{J}_{2,i} & 0 \end{pmatrix}. \tag{17}$$

Note that in particular, $\mathscr{J}_{1,i}$ and $\mathscr{J}_{2,i}$ depend on evaluations of the nominal and variational states x and x_p at individual points t_i. Thus, (17) implies that the matrices $\mathscr{J}_1^T \mathscr{J}_1$ and $\mathscr{J}_2$ only exhibit a linear coupling in time. In a multiple shooting context, we assign the points t_i to the proper shooting intervals and plug in the representation of the solution $x(\tau_{j+1}^s; s_y^j, s_z^j, \hat{q}^j)$ and $x_p(\tau_{j+1}^s; s_{\tilde{y}}^j, s_{\tilde{z}}^j, \hat{q}^j)$, respectively. We write this as

$$\mathscr{J}_1^T \mathscr{J}_1 = \sum_{j=0}^{N_s-1} \sum_{\tau_j < t_i \le \tau_{j+1}}^{N_m} \mathscr{J}_{1,i}(s_{\tilde{y}}^j, s_{\tilde{z}}^j, \hat{q}^j, w_i)^T \mathscr{J}_{1,i}(s_{\tilde{y}}^j, s_{\tilde{z}}^j, \hat{q}^j, w_i)$$

$$\mathscr{J}_2 = \sum_{j=0}^{N_s-1} \sum_{\tau_j < t_i \le \tau_{j+1}}^{N_r} \mathscr{J}_{2,i}(s_{\tilde{y}}^j, s_{\tilde{z}}^j, \hat{q}^j).$$

We introduce additional variables H and J and linearly coupled constraints that fit into the framework of (14). The objective then only depends on the newly introduced variables H and J and we obtain the following structured NLP:

$$\min_{s_{\tilde{y}}, s_{\tilde{z}}, q, w, H, J} \phi\left((I\ 0) \begin{pmatrix} H & J^T \\ J & 0 \end{pmatrix}^{-1} \begin{pmatrix} I \\ 0 \end{pmatrix} \right) \tag{18a}$$

$$\text{s.t. } 0 = H - \sum_{j=0}^{N_s-1} \sum_{\tau_j < t_i \le \tau_{j+1}}^{N_m} \mathscr{J}_{1,i}(s_{\tilde{y}}^j, s_{\tilde{z}}^j, \hat{q}^j, w_i)^T \mathscr{J}_{1,i}(s_{\tilde{y}}^j, s_{\tilde{z}}^j, \hat{q}^j, w_i) \tag{18b}$$

$$0 = J - \sum_{j=0}^{N_s-1} \sum_{\tau_j < t_i \le \tau_{j+1}}^{N_r} \mathscr{J}_{2,i}(s_{\tilde{y}}^j, s_{\tilde{z}}^j, \hat{q}^j) \tag{18c}$$

$$0 = \tilde{y}(\tau_0^s; \hat{q}^0) - s_{\tilde{y}}^0 \tag{18d}$$

$$0 = \tilde{y}(\tau_{j+1}^s; s_{\tilde{y}}^j, s_{\tilde{z}}^j, \hat{q}^j) - s_{\tilde{y}}^{j+1}, \quad j = 0, \ldots, N_s - 1 \tag{18e}$$

$$0 = g(\tau_j^s, s_{\tilde{y}}^j, s_{\tilde{z}}^j, \hat{q}^j), \quad j = 0, \ldots, N_s \tag{18f}$$

$$0 \le c(\tau_j^s, s_y^j, s_z^j, \hat{q}^j, w^j), \quad j = 0, \ldots, N_s \tag{18g}$$

$$0 \le w_i \le 1, \quad i = 1, \ldots, N_m \tag{18h}$$

$$0 \le \sum_{j=0}^{N_s} \sum_{\tau_j \le t_i < \tau_{j+1}}^{N_r} r_i(t_i, y(t_i; s_y^j, s_z^j, \hat{q}^j), z(t_i; s_y^j, s_z^j, \hat{q}^j)). \tag{18i}$$

4.3.2 Pseudo States for Covariance Matrix

Another possibility to resolve the coupling in the objective is to move the computation of the covariance to the constraints by deriving a recursion formula.

Using this formula, we introduce matrix-valued variables

$$C^j = \begin{pmatrix} C_1^j & C_2^{jT} \\ C_2^j & C_3^j \end{pmatrix} \in \mathbb{R}^{(n_p+n_r)\times(n_p+n_r)}, j = 1, \ldots, N_s \tag{19}$$

and constraints for the recursion at the multiple shooting nodes and add them as additional variables to the NLP. This resembles the treatment of dynamical states in a multiple shooting method, hence we refer to C^j as pseudo states for the covariance matrix.

Let us first derive the formula in the unconstrained case: Let H^j denote the information matrix including all terms up to time τ_j. Then H^j is given as the sum of H^{j-1} and the information gain by measurements and constraint evaluations in the interval $(\tau_{j-1}, \tau_j]$:

$$H^{j+1} = H^j + \sum_{i:\tau_j < t_i \le \tau_{j+1}} \mathscr{J}_{1,i}(s_{\tilde{y}}^j, s_{\tilde{z}}^j, \hat{q}^j, w_i)^T \mathscr{J}_{1,i}(s_{\tilde{y}}^j, s_{\tilde{z}}^j, \hat{q}^j, w_i), \tag{20}$$

where we start with some initial information given as a positive definite matrix H^0, e.g., from previous experiments or from literature.

At every grid point τ_j the covariance matrix taking into account all measurements up to time τ_j is the inverse of H^j. From (20) we obtain as recursion formula for the covariance matrix:

$$C^0 = (H^0)^{-1} \tag{21}$$

$$C^{j+1} = \left((C^j)^{-1} + \sum_{i:\tau_j < t_i \leq \tau_{j+1}} \mathcal{J}_{1,i}(s^j_{\tilde{y}}, s^j_{\tilde{z}}, \hat{q}^j, w_i)^T \mathcal{J}_{1,i}(s^j_{\tilde{y}}, s^j_{\tilde{z}}, \hat{q}^j, w_i) \right)^{-1} . \tag{22}$$

Equation (22) can be simplified to

$$C^{j+1} = C^j \left(I + \sum_{i:\tau_j < t_i \leq \tau_{j+1}} \mathcal{J}_{1,i}(s^j_{\tilde{y}}, s^j_{\tilde{z}}, \hat{q}^j, w_i)^T \mathcal{J}_{1,i}(s^j_{\tilde{y}}, s^j_{\tilde{z}}, \hat{q}^j, w_i) C^j \right)^{-1} .$$

This can be easily generalized to the case of constrained parameter estimation problems and we obtain the complete NLP by replacing the coupled constraints (18b) and (18c) by the pseudo continuity constraints

$$0 = \left(\begin{pmatrix} C^j_1 & C^{jT}_2 \\ C^j_2 & C^j_3 \end{pmatrix}^{-1} + \sum_{i:\tau_j < t_i \leq \tau_{j+1}} \begin{pmatrix} \mathcal{J}^T_{1i} \mathcal{J}_{1i} & \mathcal{J}^T_{2i} \\ \mathcal{J}_{2i} & 0 \end{pmatrix} \right)^{-1} - \begin{pmatrix} C^{j+1}_1 & C^{j+1T}_2 \\ C^{j+1}_2 & C^{j+1}_3 \end{pmatrix}, \tag{23}$$

$$j = 0, \ldots, N-1.$$

As initial values for the pseudo states we take some a priori uncertainty given by a positive definite matrix C^0_1 and a full rank matrix C^0_2 such that $C^0_2 C^{0T}_2$ is positive definite to ensure invertibility of C^0. The a priori uncertainty should be chosen several orders of magnitude larger than the expected uncertainty after the OED to make sure that the choice of C^0_1 and C^0_2 does not interfere significantly with the solution. The objective for this formulation simplifies to

$$\phi \left((I \; 0) \begin{pmatrix} C^{N_s}_1 & C^{N_sT}_2 \\ C^{N_s}_2 & C^{N_s}_3 \end{pmatrix} \begin{pmatrix} I \\ 0 \end{pmatrix} \right) . \tag{24}$$

While this formulation seems to be very much in the spirit of multiple shooting—the covariance of the system is modelled as a kind of state on each interval which is coupled via continuity-type constraints, the nonlinearity of the objective is distributed over the shooting intervals—it is computationally much less appealing: The objective (24) is simpler because the matrix inversion is now contained in the constraints, however, the pseudo continuity constraints (23) are numerically delicate, each containing two matrix inversions (in fact, we can reduce this to one by appropriate reformulation) of potentially ill-conditioned matrices.

5 Evaluation of Problem Functions

After discretization we need to solve the large but structured NLP (18). In derivative based methods such as sequential quadratic programming often the number of variables and constraints corresponds to the overall runtime of the method. In shooting methods for optimal control and especially optimum experimental design, however, the runtime is often dominated by the evaluation of states and their derivatives within the continuity constraints because they comprise possibly expensive calls to external numerical integrators. Thus it is worthwhile to have a closer look at the structures of constraints and objective and derive efficient and accurate evaluation schemes. We concentrate on the formulation (18) of the problem—linearly coupled constraints for the information matrix—as it is numerically more promising.

5.1 *Constraint Derivatives*

The OED problem (18) has a special structure in the constraints due to the fact that the dynamic system consists of closely related states, namely nominal and corresponding variational states. Ideally, they are evaluated together using the principle of *internal numerical differentiation (IND)*, see [1, 5].

Let us now have a closer look at the derivatives of continuity and consistency constraints (18e) and (18f) for the system that consists of nominal and variational states.

We denote by

$$s^j_{\tilde{x}} = \left(s^j_x, s^j_{x,p_1}, \ldots, s^j_{x,p_{N_p}}\right)^T$$

the shooting variables for the nominal and variational states at one shooting node.

Observation 1. Variational states for different parameters are independent. This means

$$\frac{\partial x_{p_i}(\tau_{j+1})}{\partial s^j_{x,p_k}} = 0, \quad i \neq k.$$

Observation 2. By differentiating (6) and (7), we see that the derivative of a variational state with respect to its initial value satisfies the following $n_x \times n_x$ variational DAEs:

$$\dot{\left(\frac{\partial y_{p_i}(t)}{\partial s^j_{x,p_i}}\right)} = \frac{\partial}{\partial x_{p_i}}\left(\frac{\partial f}{\partial x} x_{p_i}(t) + \frac{\partial f}{\partial p}\right) \cdot \frac{\partial x_{p_i}(t)}{\partial s^j_{x,p_i}} = \frac{\partial f}{\partial x} \cdot \frac{\partial x_{p_i}(t)}{\partial s^j_{x,p_i}}$$

$$0 = \frac{\partial}{\partial x_{p_i}} \left(\frac{\partial g}{\partial x} x_{p_i}(t) + \frac{\partial g}{\partial p} \right) \cdot \frac{\partial x_{p_i}(t)}{\partial s^j_{x,p_i}} = \frac{\partial g}{\partial x} \cdot \frac{\partial x_{p_i}(t)}{\partial s^j_{x,p_i}}.$$

These are the same equations that describes the sensitivity of the nominal states x with respect to their initial values and hence

$$\frac{\partial x_{p_i}(\tau_{j+1})}{\partial s^j_{x,p_i}} = \frac{\partial x(\tau_{j+1})}{\partial s^j_x}.$$

In particular that means that we do not need to evaluate any additional state sensitivities when we explicitly discretize the variational equations by multiple shooting. Instead $\frac{\partial x(\tau)}{\partial s_x}$ needs to be computed only once for each shooting interval and then can be used multiple times in the constraint Jacobian.

The part of the constraint Jacobian that corresponds to the derivative of the continuity constraints (18e) coupling the states starting at τ_{j-1} and at τ_j has the following structure:

$$\begin{pmatrix} \frac{\partial y(\tau_j)}{\partial s_y^{j-1}} & & & \frac{\partial y(\tau_j)}{\partial s_z^{j-1}} & & & & \frac{\partial y(\tau_j)}{\partial q^{j-1}} & 0 \\ \frac{\partial y_{p_1}(\tau_j)}{\partial s_y^{j-1}} & \frac{\partial y(\tau_j)}{\partial s_y^{j-1}} & & \frac{\partial y_{p_1}(\tau_j)}{\partial s_z^{j-1}} & \frac{\partial y(\tau_j)}{\partial s_z^{j-1}} & & & \frac{\partial y_{p_1}(\tau_j)}{\partial q^{j-1}} & 0 \\ \vdots & & \ddots & \vdots & & \ddots & & \vdots & \vdots \\ \frac{\partial y_{pN_p}(\tau_j)}{\partial s_y^{j-1}} & & \frac{\partial y(\tau_j)}{\partial s_y^{j-1}} & \frac{\partial y_{pN_p}(\tau_j)}{\partial s_z^{j-1}} & & & \frac{\partial y(\tau_j)}{\partial s_z^{j-1}} & \frac{\partial y_{pN_p}(\tau_j)}{\partial q^{j-1}} & 0 \end{pmatrix},$$

where the variable vector is ordered as follows:

$$\left(s_y^{j-1}, s_{y,p_1}^{j-1}, \ldots, s_{y,pN_p}^{j-1}, s_z^{j-1}, s_{z,p_1}^{j-1}, \ldots, s_{z,pN_p}^{j-1}, q^{j-1}, w^{j-1} \right)^T.$$

A similar structure can be observed for the consistency conditions (18f).

Note that the derivatives $\frac{\partial y}{\partial(\cdot)}$ are first-order and $\frac{\partial y_p}{\partial(\cdot)}$ are second-order sensitivities of the states and must be supplied by the integrator.

5.2 *Objective Derivatives*

The objective (18a) also deserves special attention as it is a nontrivial—but more or less fixed for the whole problem class—function that in particular comprises the inversion of a symmetric matrix.

First of all we note that (18a) only depends on the newly introduced variables H and J. This allows us to derive explicit formulas for the first and second derivative of the objective. Furthermore H and J enter the constraints only *linearly* through the coupled constraints (18b) and (18c), so the second derivative of the objective contains the entire curvature of the problem with respect to H

and J. This can be used to cheaply compute part of the Hessian approximations in SQP methods without the need to evaluate state sensitivities of higher order.

We only discuss the case of unconstrained parameter estimation problems (i.e. no matrix J) and $\phi(\cdot) = \operatorname{tr}(\cdot)$ as optimization criterion.

5.2.1 Objective Gradient

We have an objective ϕ that maps a symmetric matrix to a scalar, however, we only include the lower triangular matrix as degrees of freedom in the nonlinear program. So whenever we need the gradient of the objective in a derivative based method we need derivatives of ϕ with respect to every entry H_{uv}, $1 \leq v \leq u \leq n_p$.

In the unconstrained case we have $H^{-1} = C$ and for every entry H_{uv}

$$\frac{\partial \phi}{\partial H_{uv}} = \frac{\partial \phi}{\partial C} \cdot \frac{\partial C}{\partial H_{uv}}.$$

Using the general formula

$$\frac{\partial H_{ij}^{-1}}{\partial H_{kl}} = -(H^{-1})_{ik} \cdot (H^{-1})_{lj} \tag{25}$$

we obtain for the derivative with respect to a fixed entry H_{uv}:

$$\frac{\partial \phi}{\partial H_{uv}} = \sum_{\substack{1 \leq i,j \leq n_p \\ 1 \leq k,l \leq n_p}} \frac{\partial \phi}{\partial C_{ij}} \cdot \frac{\partial C_{ij}}{\partial H_{kl}} \cdot \frac{\partial H_{kl}}{\partial H_{uv}} = -\sum_{\substack{1 \leq i,j \leq n_p \\ 1 \leq k,l \leq n_p}} \frac{\partial \phi}{\partial C_{ij}} \cdot C_{ik} \cdot C_{lj} \cdot \frac{\partial H_{kl}}{\partial H_{uv}}.$$

Taking symmetry into account, we have

$$\frac{\partial H_{kl}}{\partial H_{uv}} = \begin{cases} 1, & \text{if } (k,l) = (u,v) \text{ or } (l,k) = (u,v) \\ 0, & \text{else,} \end{cases}$$

and thus

$$\frac{\partial \phi}{\partial H_{uu}} = -\sum_{1 \leq i,j \leq n_p} \frac{\partial \phi}{\partial C_{ij}} \cdot C_{iu} \cdot C_{uj} \tag{26}$$

$$\frac{\partial \phi}{\partial H_{uv}} = -2 \sum_{1 \leq i,j \leq n_p} \frac{\partial \phi}{\partial C_{ij}} \cdot C_{iu} \cdot C_{vj}, \quad u > v. \tag{27}$$

For $\phi = \mathrm{tr}$ this simplifies to

$$\frac{\partial \phi}{\partial H_{uu}} = -\sum_{1 \le i \le n_p} C_{iu}^2 \tag{28}$$

$$\frac{\partial \phi}{\partial H_{uv}} = -2 \sum_{1 \le i \le n_p} C_{iu} \cdot C_{vi}, \quad u > v. \tag{29}$$

This can be calculated immediately once the covariance matrix C is computed.

5.2.2 Objective Hessian

We now use formulae (28) and (29) to derive an explicit formula for the Hessian of $\mathrm{tr}\,(H^{-1})$ with respect to the entries of H.

When computing $\frac{\partial^2 \,\mathrm{tr}\,(H^{-1})}{\partial H_{uv} \partial H_{rs}}$ we distinguish between three cases:

- H_{uv} and H_{rs} diagonal elements
- H_{uv} diagonal, H_{rs} off-diagonal element
- H_{uv} and H_{rs} off-diagonal elements.

For two diagonal elements we obtain:

$$\begin{aligned}
\frac{\partial^2 \phi(C)}{\partial H_{uu} \partial H_{rr}} &= \frac{\partial}{\partial H_{uu}} \left(\frac{\partial \phi}{\partial H_{rr}} \right) = \frac{\partial}{\partial H_{uu}} \left(-\sum_{1 \le i \le n_v} C_{ir}^2 \right) \\
&= -2 \sum_{1 \le i \le n_v} \frac{\partial C_{ir}}{\partial H_{uu}} \cdot C_{ir} = 2 \sum_{1 \le i \le n_v} C_{iu} \cdot C_{ur} \cdot C_{ir}.
\end{aligned}$$

If we have one diagonal and one off-diagonal element we compute:

$$\begin{aligned}
\frac{\partial^2 \phi(C)}{\partial H_{uu} \partial H_{rs}} &= \frac{\partial}{\partial H_{uu}} \left(\frac{\partial \phi}{\partial H_{rs}} \right) = \frac{\partial}{\partial H_{uu}} \left(-2 \sum_{1 \le i \le n_v} C_{ir} \cdot C_{si} \right) \\
&= -2 \sum_{1 \le i \le n_v} \frac{\partial C_{ir}}{\partial H_{uu}} \cdot C_{si} + C_{ir} \cdot \frac{\partial C_{si}}{\partial H_{uu}} \\
&= 2 \sum_{1 \le i \le n_v} C_{iu} \cdot C_{ur} \cdot C_{si} + C_{ir} \cdot C_{sp} \cdot C_{ui}.
\end{aligned}$$

Taking the second derivative with respect to two off-diagonal elements yields:

$$\begin{aligned}\frac{\partial^2 \phi(C)}{\partial H_{pq} \partial H_{rs}} &= \frac{\partial}{\partial H_{uv}} \left(\frac{\partial \phi}{\partial H_{rs}} \right) + \frac{\partial}{\partial H_{vu}} \left(\frac{\partial \phi}{\partial H_{rs}} \right) \\
&= \frac{\partial}{\partial H_{uv}} \left(-2 \sum_{1 \leq i \leq n_v} C_{ir} \cdot C_{si} \right) + \frac{\partial}{\partial H_{vu}} \left(-2 \sum_{1 \leq i \leq n_v} C_{ir} \cdot C_{si} \right) \\
&= -2 \sum_{1 \leq i \leq n_v} \frac{\partial C_{ir}}{\partial H_{uv}} \cdot C_{si} + C_{ir} \cdot \frac{\partial C_{si}}{\partial H_{uv}} + \frac{\partial C_{ir}}{\partial H_{vu}} \cdot C_{si} + C_{ir} \cdot \frac{\partial C_{si}}{\partial H_{vu}} \\
&= 2 \sum_{1 \leq i \leq n_v} C_{iu} \cdot C_{vr} \cdot C_{si} + C_{ir} \cdot C_{su} \cdot C_{vi} + C_{iv} \cdot C_{ur} \cdot C_{si} + C_{ir} \cdot C_{sv} \cdot C_{ui}.\end{aligned}$$

As the variables H enter the constraints only linearly we note that the objective Hessian with respect to H is in fact the same as the Hessian of the Lagrangian with respect to H.

6 Numerical Results

We test our methods on two examples: The first one is a predator-prey model adapted to OED [19] that serves as proof of concept for both formulations. The second one is a more involved example from chemical engineering: the urethane reaction [15]. We report SQP iterations which basically amount to derivative evaluations as well as CPU time and compare the results to a single shooting implementation.

6.1 *Implementation*

We implemented direct multiple shooting for OED within our software package VPLAN [15] that allows to formulate, simulate and optimize DAE models. From a user specified formulation of the nominal DAE system, parameters, controls, and process constraints it generates structured NLPs for OED as described in Sect. 4. A special focus is on the efficient sparse evaluation of the constraints and their derivatives as outlined in Sect. 5 and parallelization of state integration and derivative evaluation on multi experiment and shooting node level.

As integrator we use DAESOL [3], a variable order and stepsize BDF method that can efficiently compute sensitivities of first and second order. The absolute and relative tolerance for all computations was set to 10^{-9}.

For the solution of the structured nonlinear programs, we implemented a filter based line search SQP method as described in [21]. The block structured quadratic subproblems are solved by a modified version of the parametric active set solver qpOASES [11] that uses direct sparse linear algebra. The Hessian of the Lagrangian is approximated blockwise depending on the number of shooting intervals. For each block a positive definite damped BFGS update is employed which is scaled by the centered Oren-Luenberger sizing factor as described in [9]. Both full space and limited memory updates are available. In a multiple shooting context, the exact objective Hessian is computed cheaply as discussed in Sect. 5.2 and used for the lowermost diagonal block instead of a BFGS approximation. The optimality and nonlinear feasibility tolerance are set to 10^{-5}. Details of the SQP implementation will be discussed in an upcoming publication.

All results were obtained on a workstation with two Intel Xeon hexacore CPUs (2.4 GHz) allowing 24 parallel threads in total and 32 GB RAM running Ubuntu 12.04.

6.2 *Example 1: Predator-Prey Model*

We consider a predator-prey dynamics taken from [19] on the time horizon [0, 12] with fixed initial values and an additional fishing term $0 \leq u(t) \leq 1$ as control:

$$\begin{aligned}
\dot{y}_1(t) &= y_1(t) - p_1 \cdot y_1(t) \cdot y_2(t) - 0.4 \cdot u(t) \cdot y_1(t), \quad y_1(0) = 0.5 \\
\dot{y}_2(t) &= -y_2(t) + p_2 \cdot y_1(t) \cdot y_2(t) - 0.2 \cdot u(t) \cdot y_2(t), \quad y_2(0) = 0.7.
\end{aligned}$$

We assume $p_1 = p_2 = 1$ and that both states can be observed at most four times during the experiment with constant variances for the measurement errors, i.e., $\sigma_i = 1$. Both the control $u(t)$ and the grid of possible measurements are discretized on 50 equidistant intervals. The objective is to minimize the average variance of p_1 and p_2, i.e. $\frac{1}{2}\operatorname{tr}(C)$.

The initial guess for the controls is $u(t) \equiv 0.3$ and all 50 measurements selected, i.e. $w_i^1 = w_i^2 = 1$, $i = 1, \ldots, 50$. It yields an objective function value (average variance) of $\frac{1}{2}\operatorname{tr}(C) = 0.00683$ but note that this design is infeasible: all 50 possible measurements for both states are selected but only four per state are allowed.

6.3 *Example 2: Urethane Reaction*

The Urethane reaction is a well-known example from chemical reaction kinetics, see [15]. The reaction scheme is the following:

$$\begin{aligned}
A + B &\rightarrow C \\
A + C &\rightleftharpoons D \\
3A &\rightarrow E
\end{aligned}$$

Educts are phenylisocyanate A and butanol B in the solvent dimethylsulfoxide L. During the reaction the product urethane C, the byproduct allophanate D and the byproduct isocyanate E are formed.

The products C, D, and E are modelled as differential states while A, B, and L can be computed from C, D, and E using molar number balance. In total, six parameters have to be identified, namely the frequency factors and activation energies for the Arrhenius kinetics. The objective is to minimize $\frac{1}{6}\operatorname{tr}(C)$. To achieve this, one experiment is to be designed for the time horizon $[t_0, t_{end}] = [0h, 80h]$ and one out of three possible measurements can be taken at each of 11 equidistant points in time. The reactor is run in a stirrer tank with two feeds: Feed 1 contains phenylisocyanate and the solvent, feed 2 contains dimethylsulfoxide and the solvent. Both can be fed into the reactor during the process. Furthermore, the temperature can be controlled. For our numerical experiments, we parameterize the derivative of the control functions $\dot{T}(t), \dot{feed}_1(t)$, and $\dot{feed}_2(t)$ by piecewise constant functions on ten equidistant intervals. The actual process controls $T(t), feed_1(t)$, and $feed_2(t)$ are set up as additional differential states, so we end up with a total of six state variables. The constraints on the controls are formulated as path constraints. The full model including process constraints and measurement methods is summarized in Fig. 3. The variances of the measurement errors are assumed constant with $\sigma_i = 1$.

The initial guess for the controls is depicted in Fig. 4. It yields an objective function value (average variance) of $\frac{1}{6}\operatorname{tr}(C) = 1936.3$.

6.4 Results Predator-Prey

We solved the problem with both direct multiple shooting formulations introduced in Sect. 4. We keep the control discretization fixed but vary the number of shooting intervals giving rise to different, yet equivalent, NLPs. We discovered a number of different local minima that differ mainly with respect to the placing of the measurements. Table 1 shows the results for the first multiple shooting formulation—a coupled constraint for the information matrix—as well as the single shooting formulation while Table 2 shows results for the formulation where the covariance is distributed over the shooting intervals. Both algorithms terminated successfully when restarted in the minima found by the other one indicating the structural correctness of our approach.

Table 1 shows that the first formulation converges for all discretizations. However, we note how the number of SQP iterations increases when we increase the number of shooting intervals. Preliminary experiments show that this is probably due to the BFGS update that is unable to reflect negative curvature of the underlying Lagrangian. A block SQP method that can handle indefinite approximations which can reduce this effect is currently under development. When we look at the CPU time, however, we see that the benefits of the efficient evaluation scheme for the multiple shooting formulation outweighs the smaller number of SQP iterations for single shooting, making direct multiple shooting with a moderate number of

States

$$\dot{n}_C = V\cdot(r_1 - r_2 + r_3), \quad n_C(0) = 0$$
$$\dot{n}_D = V\cdot(r_2 - r_3), \quad n_D(0) = 0$$
$$\dot{n}_E = V\cdot r_4, \quad n_E(0) = 0$$
$$\dot{feed}_1 = u_1, \quad feed_1(0) = 0$$
$$\dot{feed}_2 = u_2, \quad feed_2(0) = 0$$
$$\dot{T} = u_3, \quad T(0) = 293.15$$

Educts

$$n_{A,e} = n_{A,e1,0}\cdot feed_1$$
$$n_{B,e} = n_{B,e2,0}\cdot feed_2$$
$$n_{L,e} = n_{L,e1,0}\cdot feed_1 + n_{L,e2,0}\cdot feed_2$$
$$n_A = n_{A,0} + n_{A,e} - n_C - 2\cdot n_D - 3\cdot n_E$$
$$n_B = n_{B,0} + n_{B,e} - n_C - n_D$$
$$n_L = n_{L,0} + n_{L,e}$$

Reaction Rates

$$r_1 = k_{ref1}\cdot\exp\left(-\frac{E_{a,1}}{R}\cdot\left(\frac{1}{T} - \frac{1}{363.16}\right)\right)\cdot\frac{n_A}{V}\cdot\frac{n_B}{V}$$
$$r_2 = k_{ref2}\cdot\exp\left(-\frac{E_{a,2}}{R}\cdot\left(\frac{1}{T} - \frac{1}{363.16}\right)\right)\cdot\frac{n_A}{V}\cdot\frac{n_C}{V}$$
$$r_3 = k_{ref2}\cdot\exp\left(-\frac{E_{a,2}}{R}\cdot\left(\frac{1}{T} - \frac{1}{363.16}\right)\right)\cdot\left(K_{C2}\cdot e^{-\frac{\Delta H_2}{R}\cdot\left(\frac{1}{T}-\frac{1}{T_{C2}}\right)}\right)^{-1}\cdot\frac{n_D}{V}$$
$$r_4 = k_{ref4}\cdot\exp\left(-\frac{E_{a,4}}{R}\cdot\left(\frac{1}{T} - \frac{1}{363.16}\right)\right)\cdot\left(\frac{n_A}{V}\right)^2$$
$$V = \frac{n_A\cdot M_A}{\rho_A} + \frac{n_B\cdot M_B}{\rho_B} + \frac{n_C\cdot M_C}{\rho_C} + \frac{n_D\cdot M_D}{\rho_D} + \frac{n_E\cdot M_E}{\rho_E} + \frac{n_L\cdot M_L}{\rho_L}$$

Time Dependent Controls

$$0 \le feed_1, feed_2 \le 1$$
$$0 \le u_1, u_2 \le 0.0125$$
$$273.15 \le T \le 473.15$$
$$-40 \le u_3 \le 40$$

Time Independent Controls

$$0.1 \le n_{A,0} \le 1, \qquad 0 \le n_{B,0}, n_{L,0} \le 1$$
$$\frac{n_{A,0}M_A + n_{B,0}M_B}{n_{A,0}M_A + n_{B,0}M_B + n_{L,0}M_L} \le 0.8$$
$$n_{A,0}M_A\rho_A^{-1} + n_{B,0}M_B\rho_B^{-1} + n_{L,0}M_L\rho_L^{-1} \le 7.5\cdot 10^{-4}$$

Measurements

$$h_1(t) = 100\cdot\frac{n_A\cdot M_A}{n_A\cdot M_A + n_B\cdot M_B + n_C\cdot M_C + n_D\cdot M_D + n_E\cdot M_E + n_L\cdot M_L}$$
$$h_2(t) = 100\cdot\frac{n_C\cdot M_C}{n_A\cdot M_A + n_B\cdot M_B + n_C\cdot M_C + n_D\cdot M_D + n_E\cdot M_E + n_L\cdot M_L}$$
$$h_3(t) = 100\cdot\frac{n_D\cdot M_D}{n_A\cdot M_A + n_B\cdot M_B + n_C\cdot M_C + n_D\cdot M_D + n_E\cdot M_E + n_L\cdot M_L}$$
$$h_4(t) = 100\cdot\frac{n_E\cdot M_E}{n_A\cdot M_A + n_B\cdot M_B + n_C\cdot M_C + n_D\cdot M_D + n_E\cdot M_E + n_L\cdot M_L}$$
$$w_i^2 = w_i^3, \quad w_i^1 + w_i^2 + w_i^4 \le 1 \quad i = 0,\dots,16$$

Parameters

$$k_{ref1} = 5.0\cdot 10^{-4}$$
$$E_{a,1} = 35240$$
$$k_{ref2} = 8.0\cdot 10^{-8}$$
$$E_{a,2} = 85000$$
$$k_{ref4} = 1.0\cdot 10^{-8}$$
$$E_{a,4} = 35000$$

Constants

$$M_A = 0.11911, \quad \rho_A = 1095$$
$$M_B = 0.07412, \quad \rho_B = 809$$
$$M_C = 0.19323, \quad \rho_C = 1415$$
$$M_D = 0.31234, \quad \rho_D = 1528$$
$$M_E = 0.35733, \quad \rho_E = 1451$$
$$M_L = 0.07806, \quad \rho_L = 1101$$
$$\Delta H_2 = -17031.0, \quad K_{C2} = 0.17$$

Fig. 3 Urethane reaction model

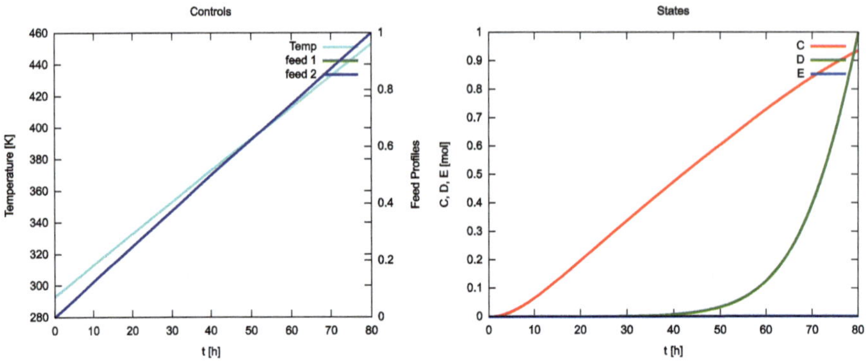

Fig. 4 Initial values for controls and corresponding states for the urethane example. The objective (average variance) is $\frac{1}{6}\,\mathrm{tr}(C) = 1936.3$

Table 1 Performance of multiple shooting for information matrix for predator-prey

Shooting intervals	Var.	Cons.	SQP iterations	Obj.	Time (s)
Single shooting	150	2	16	7.614527e-03	11.03
2	159	11	22	7.614526e-03	3.47
4	173	25	37	7.908877e-03	2.30
6	187	39	52	7.908865e-03	2.18
12	229	81	82	7.908856e-03	2.27
25	297	149	91	7.908878e-03	2.77
50	447	299	126	7.634122e-03	5.94

Table 2 Performance of multiple shooting for covariance matrix for predator-prey

Shooting intervals	Var.	Cons.	SQP iterations	Obj.	Time (s)
2	162	14	41	7.794461e-03	6.55
4	182	34	122	7.646316e-03	9.73
6	202	54	–	–	–
12	262	114	90	8.270562e-03	6.29
25	369	221	–	–	–
50	594	446	304	7.952309e-03	34.01

shooting intervals the best overall choice. The following aspects are responsible for this:

- derivatives with respect to controls are required only locally that means less directional derivatives of second order are needed
- derivative evaluation is easily parallelized on a multicore machine.

While the first formulation converges for all discretizations within a reasonable number of SQP iterations, the second formulation does not converge for every multiple shooting discretization. Furthermore, many of the SQP steps were reduced steps, which means many additional constraint and objective evaluations were necessary. Overall, the behaviour was competitive to the first formulation and single shooting only for certain problem instances.

6.5 *Results Urethane*

We solved the problem with the more promising variant of direct multiple shooting meaning we transform it to an NLP of the form (18) where we introduce a coupled constraint for the information matrix. Again, we use different numbers of shooting intervals with the same control discretization. The problem exhibits several structurally different local minima, however, all of them yield significantly better objective values than the initial guess. The states and controls for one of them are depicted in Fig. 5.

The SQP method was able to find a local minimum for every multiple shooting discretization. The results comprising the number of major iterations, the final objective value and the CPU time in seconds are summarized in Table 3.

We see that the method performs comparably well in terms of SQP iterations for single and multiple shooting. For CPU time, again the results shift strongly in favor of multiple shooting because the derivative evaluation can be done much cheaper.

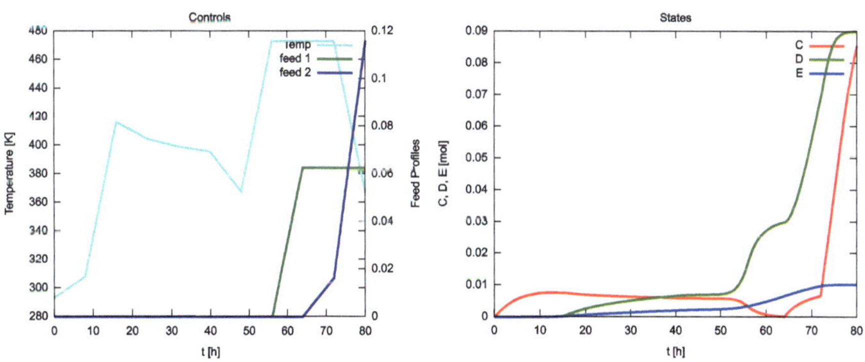

Fig. 5 Optimum experimental design for the urethane example. The objective (average variance) is $\frac{1}{6}\,\mathrm{tr}(C) = 0.0665$

Table 3 Performance of single and multiple shooting for urethane

Shooting intervals	Var.	Cons.	SQP iterations	Obj.	Time (s)
Single shooting	68	45	68	1.241664e-01	85.87
2	138	112	62	3.441588e-01	30.61
4	242	261	105	6.651601e-02	32.64
10	530	480	117	8.889849e-02	43.07

7 Conclusions

In this paper, we reviewed a nonstandard optimal control problem formulation of OED. For this formulation, we showed how to extend the classical direct multiple shooting method for optimal control problems for OED problems in two ways, leading to highly structured nonlinear programs. Special structures in the constraints and objective derivatives are highlighted that must be taken into account in an efficient implementation. The algorithms presented are implemented within the software package VPLAN. We presented two application examples, one of them a challenging example from chemical engineering that could be solved successfully with one of the new formulations. Our implementation outperforms an existing single shooting implementation in terms of CPU time.

We expect direct multiple shooting for OED to have more benefits for more challenging, large-scale real-life problems. Especially when nontrivial path constraints are present, as it is often the case for real-life systems, direct single shooting can run into problems finding feasible points. The direct multiple shooting method as introduced in this paper allows to choose fine shooting discretizations in critical regions and offers more flexibility for initialization. Another point is that OED, even for small nominal systems, basically requires the solution of an additional $n_x \times n_p$ variational system and even though this can be done efficiently using the principles of IND, its solution becomes very time consuming for large-scale systems. Here the lower sensitivity load as well as the excellent potential for parallelization provide great benefits for multiple shooting.

References

1. Albersmeyer, J.: Adjoint based algorithms and numerical methods for sensitivity generation and optimization of large scale dynamic systems. Ph.D. thesis, Ruprecht-Karls-Universität Heidelberg (2010)
2. Atkinson, A.C., Donev, A.: Optimum Experimental Designs. Oxford Statistical Sciences Series, vol. 8. Oxford University Press, Oxford (1992)
3. Bauer, I., Bock, H.G., Schlöder J.P.: DAESOL – a BDF-code for the numerical solution of differential algebraic equations. Internal Report, IWR, SFB 359, Universität Heidelberg (1999)
4. Bauer, I., Bock, H.G., Körkel, S., Schlöder, J.P.: Numerical methods for optimum experimental design in DAE systems. J. Comput. Appl. Math. **120**(1–2), 1–15 (2000)
5. Bock, H.G.: Numerical treatment of inverse problems in chemical reaction kinetics. In: Ebert, K.H., Deuflhard, P., Jäger, W. (eds.) Modelling of Chemical Reaction Systems. Springer Series in Chemical Physics, vol. 18, pp. 102–125. Springer, Heidelberg (1981)
6. Bock, H.G.: Randwertproblemmethoden zur parameteridentifizierung in systemen nichtlinearer differentialgleichungen. Bonner Mathematische Schriften, vol. 183. Universität Bonn, Bonn (1987)
7. Bock, H.G., Plitt, K.J.: A Multiple Shooting algorithm for direct solution of optimal control problems. In: Proceedings of the 9th IFAC World Congress, pp. 242–247. Pergamon Press, Budapest (1984). Available at http://www.iwr.uni-heidelberg.de/groups/agbock/FILES/Bock1984.pdf

8. Bock, H.G. Eich, E., Schlöder. J.P.: Numerical solution of constrained least squares boundary value problems in differential-algebraic equations. In: Strehmel, K. (ed.) Numerical Treatment of Differential Equations. Proceedings of the NUMDIFF-4 Conference, Halle-Wittenberg, 1987. Texte zur Mathematik, vol. 104, pp. 269–280. Teubner, Leipzig (1988)
9. Contreras, M., Tapia, R.A.: Sizing the BFGS and DFP updates: numerical study. J. Optim. Theory Appl. **78**(1), 93–108 (1993)
10. Fedorov, V.V.: Theory of Optimal Experiments. Elsevier, Amsterdam (1972)
11. Ferreau, H.J., Kirches, C., Potschka, A., Bock, H.G., Diehl, M.: qpOASES: A parametric active-set algorithm for quadratic programming. Math. Program. Comput. **6**(4), 327–363 (2014)
12. Franceschini, G., Macchietto, S.: Model-based design of experiments for parameter precision: state of the art. Chem. Eng. Sci. **63**, 4846–4872 (2008)
13. Hoang, M.D., Barz, T., Merchan, V.A., Biegler, L.T., Arellano-Garcia, H.: Simultaneous solution approach to model-based experimental design. AIChE J. **59**(11), 4169–4183 (2013)
14. Janka, D.: Optimum experimental design and multiple shooting. Master's thesis, Universität Heidelberg, Heidelberg (2010)
15. Körkel, S.: Numerische methoden für optimale versuchsplanungsprobleme bei nichtlinearen DAE-modellen. Ph.D. thesis, Universität Heidelberg, Heidelberg (2002)
16. Körkel, S., Potschka, A., Bock, H.G., Sager, S.: A multiple shooting formulation for optimum experimental design. Math. Program. (2012, submitted revisions)
17. Potschka, A., Bock, H.G., Schlöder, J.P.: A minima tracking variant of semi-infinite programming for the treatment of path constraints within direct solution of optimal control problems. Optim. Methods Softw. **24**(2), 237–252 (2009)
18. Pukelsheim, F.: Optimal design of experiments. In: Classics in Applied Mathematics, vol. 50. SIAM, Philadelphia (2006). ISBN:978-0-898716-04-7.
19. Sager, S.: MIOCP benchmark site. (2014) http://mintoc.de
20. Sager, S.: Sampling decisions in optimum experimental design in the light of Pontryagin's maximum principle. SIAM J. Control. Optim. **51**(4), 3181–3207 (2013)
21. Wächter, A., Biegler, L.T.: Line search filter methods for nonlinear programming: motivation and global convergence. SIAM J. Optim. **16**(1), 1–31 (2005)

Parameter Estimation for High-Dimensional PDE Models Using a Reduced Approach

Robert Kircheis and Stefan Körkel

Abstract Partial differential equations (PDE) are indispensable to describe complex processes. PDE constrained parameter estimation is still a prevailing topic of research. The increase in computation time with increasing complexity of the problem is one of the main problems.

With the application of multiple shooting, the number of required derivatives for the generalized Gauss–Newton method rises rapidly. We introduce a method to overcome this challenge. By using directional derivatives the computational effort can be reduced to the minimal number. We demonstrate our methods with help of the heat equation.

1 Introduction

Validated models are essential for process optimization and optimal control in chemistry, engineering etc. Usually these models depend on parameters that are not known initially but have to be identified from measurement data. Derivative based methods, such as the generalized Gauss–Newton method for direct multiple shooting by Bock [5], have shown good results for parameter estimation problems with ordinary differential equations (ODEs).

If spatially distributed processes are taken into account, we have to consider constraints given by partial differential equations (PDEs). If PDEs are discretized by means of a method of lines, we end up with a high-dimensional system of differential algebraic equations (DAEs). In general, we formulate the parameter estimation problem as a nonlinear least squares problem and apply the generalized Gauss–Newton method. In the context of multiple shooting, the effort for the computation of the Jacobians in each iteration of the generalized Gauss–Newton method is tremendous. We present a method which couples the evaluation of the Jacobians and the subsequent block-Gaussian elimination. Thus, the number of required derivatives is reduced to the minimal number. The *reduced approach* was

R. Kircheis (✉) • S. Körkel
IWR Heidelberg, Heidelberg University, Heidelberg, Germany
e-mail: robert.kircheis@iwr.uni-heidelberg.de; stefan.koerkel@iwr.uni-heidelberg.de

T. Carraro et al. (eds.), *Multiple Shooting and Time Domain Decomposition Methods*, Contributions in Mathematical and Computational Sciences 9,
DOI 10.1007/978-3-319-23321-5_5

introduced by Schlöder [13] for parameter estimation problems constrained by high-dimensional systems of ODEs in 1987. A first extension to DAE constrained problem was presented by Bauer [2]. In this paper, we develop a different formulation of the reduced approach for DAE constraints which we consider as an approximation of the solution of a partial differential equation. The first application of the reduced approach to PDE constrained parameter estimation problems was presented by Dieses [1]. Dieses considered only ODEs to approximate the solution of a PDE.

We first introduce the general formulation of a parameter estimation problem and the generalized Gauss–Newton method. Afterwards, we present the reduced approach. In the end, an application example is investigated to show the advantages of our method compared to the conventional approach. In the end, some conclusions are drawn.

2 Problem Formulation

We consider a dynamic system defined by partial differential equations,

$$0 = \mathscr{F}\left(t, x, u, \frac{\partial u_i}{\partial t}, \frac{\partial u_i}{\partial x_j}, \frac{\partial u_i}{\partial^2 x_j \partial x_k}, \dots, p\right), \tag{1}$$

on the bounded domain $\Omega \subset \mathbb{R}^d$. $\mathscr{F}$ is an arbitrary function with time $t = [t_0, t_{\text{end}}]$, independent variables $x \in \mathbb{R}^d$, dependent variables $u \in \mathbb{R}^{n_u}$ and parameters $p \in \mathbb{R}^{n_p}$. We examine transient problems. Thus, let initial values of the form

$$u(t_0) = u_0(p) \tag{2}$$

be given that may depend on the parameters. Additionally, let boundary conditions be defined by

$$au + b\frac{\partial u}{\partial n} = c \quad \text{on } \partial\Omega \tag{3}$$

for some constants a and b that can be zero but not at the same time and a given function c on the boundary of the domain.

Assume that n_{ex} experiments have been executed which have provided a set of measurements η_k^j, $k = 1, \dots, m^j$, $j = 1, \dots, n_{\text{ex}}$. By $h_k^j(t_k^j, u^{*j}(t_k^j), p^*)$, $k = 1, \dots, m^j$, $j = 1, \dots, n_{\text{ex}}$, we denote the corresponding model response evaluated for the true, but unknown parameters p^* and states $u^{*j}(t_k^j)$ computed by Eq. (1) for p^*. We assume the measurement errors

$$\varepsilon_k^j := \eta_k^j - h_k^j(t_k^j, u^{j,*}(t_k^j), p^*), \quad k = 1, \dots, m^j, j = 1, \dots, n_{\text{ex}},$$

to be normally distributed

$$\varepsilon_k^j \sim \mathcal{N}\left(0, \left(\sigma_k^i\right)^2\right), \quad k = 1, \ldots, m^j,\ j = 1, \ldots, n_{\text{ex}},$$

with mean 0 and standard deviation σ_k^j. The difference between measurement values and model response can be evaluated for other values of p and u, too, and we obtain the residuals

$$\eta_k^j - h_k^j(t_k^j, u^j(t_k^j), p), \quad k = 1, \ldots, m^j\ j = 1, \ldots, n_{\text{ex}}. \tag{4}$$

The parameter estimation problem is then to minimize the weighted sum of the residuals

$$\frac{1}{2}\sum_{j=1}^{n_{\text{ex}}}\sum_{k=1}^{m^j}\left(\frac{\eta_k^j - h_k^j(t_k^j, u^j(t_k), p)}{\sigma_k^j}\right)^2 \tag{5}$$

with the variances σ_k^j as weights. Equation (5) can also be interpreted as a log-likelihood estimator for the parameters, see Seber [14].

Often, we have to deal with additional interior point and boundary constraints as well:

$$0 = \sum_{l=1}^{n_r^j} r_l^j(u^j(t_l^j), p) \tag{6}$$

with $r_l : \mathbb{R}^{n_u} \times \mathbb{R}^{n_p} \to \mathbb{R}^r$.

Considering the model equations (1)–(3) and the interior point constraints (6) as additional constraints, we state the PDE constrained parameter estimation problem

$$\min_{y,p} \quad \frac{1}{2}\sum_{j=1}^{n_{\text{ex}}}\sum_{k=1}^{m^j}\left(\frac{\eta_k^j - h_k^j(t_k^j, u^j(t_k^j), p)}{\sigma_k^j}\right)^2 \tag{7a}$$

$$\text{s.t.} \quad 0 = \mathcal{F}^j\left(t, x, u^j, \frac{\partial u_i^j}{\partial t}, \frac{\partial u_i^j}{\partial x_j}, \frac{\partial u_i^j}{\partial^2 x_j \partial x_k}, \ldots, p\right) \quad j = 1, \ldots, n_{\text{ex}}. \tag{7b}$$

$$u^j(t_0) = u_0^j(p), \tag{7c}$$

$$c^j = a^j u^j + b^j \frac{\partial u^j}{\partial n} \quad \text{on } \partial\Omega, \tag{7d}$$

$$0 = \sum_{l=1}^{n_r^j} r_l^j(u^j(t_l^j), p). \tag{7e}$$

3 Discretization in Space and in Time

In this section, we present concepts for the discretization of the model equations (1) in space and the concept of multiple shooting. If not declared otherwise, the following methods are presented for the first experiment only. For the remaining experiments, the steps have to be repeated. We neglect the superscript 1.

The first step in the parametrization of the parameter estimation problem (7) consists of the discretization of the PDE constraints (1) by a method of lines, cf. Schiesser [12] , e.g., by a finite difference methods (FDMs) or a finite element methods (FEMs).

The approach leads to a high-dimensional system of differential algebraic equations

$$A(y(t),z(t),p)\dot{y} = f(t,y(t),z(t),p), \tag{8a}$$

$$0 = g(t,y(t),z(t)), \tag{8b}$$

$$y(t_0) = y_0(p), \tag{8c}$$

where $A(y(t),z(t),p)$ denotes the mass matrix that may depend on the spatially discretized states $y(t) \in \mathbb{R}^{n_y}$ and $z(t) \in \mathbb{R}^{n_z}$ and the parameters p. We consider only DAEs with differentiation index 1, i.e. the matrix $\frac{\partial g}{\partial z}$ is regular.

To solve System (8) in time, we apply direct multiple shooting, see Bock [6]. Thus, the parameter estimation problem is transformed into a problem with finite dimensional constraints.

We define the *shooting grid*, i.e., a partition of the time interval $[t_0, t_{\mathrm{end}}]$,

$$\tau_0 = t_0 < \tau_1 < \ldots < \tau_{n_{\mathrm{ms}}} < \tau_{n_{\mathrm{ms}}+1} = t_{end},$$

and the shooting intervals

$$\mathscr{I}_i = [\tau_i, \tau_{i+1}), \quad i = 0, \ldots, n_{\mathrm{ms}}.$$

We introduce artificial initial values $s_i = (s_i^{y^T}, s_i^{z^T})^T$, $i = 0, \ldots, n_{\mathrm{ms}}$, with

$$s_0^y = y_0(p) \tag{9}$$

for the differential and algebraic states $y(t)$ and $z(t)$, respectively, and examine the relaxed DAE system

$$A(y(t),z(t),p)\dot{y}(t) = f(t,y(t),z(t),p), \qquad t \in \mathscr{I}_i \tag{10a}$$

$$y(\tau_i) = s_i^y, \tag{10b}$$

$$0 = g(t,y(t),z(t),p) - \beta(t)g(\tau_i,s_i,p), \quad i = 0,\ldots,n_{\mathrm{ms}}, \tag{10c}$$

$$\beta(t) \in [0,1], \quad \beta(\tau_i) = 1, \tag{10d}$$

on each of the $n_{\mathrm{ms}} + 1$ subintervals. Here, $\beta(t)$ is a continuous, monotonically decreasing function with

$$\lim_{t \to \tau_{i+1}} \beta(t) = 0.$$

The evaluation of the DAE system leads to a step-by-step formulation of the trajectory

$$\begin{pmatrix} y(t) \\ z(t) \end{pmatrix} = \psi(t; s^i, p), \quad t \in \mathscr{I}_i, \; i = 0, \ldots, n_{\mathrm{ms}}. \tag{11}$$

We refer to $\psi_i(t; s^i, p)$, $t \in \mathscr{I}_i$, $i = 0, \ldots, n_{\mathrm{ms}}$ as the nominal trajectory

By Eq. (11), we obtain a piecewise continuous, finite dimensional parametrization of the nominal trajectories of (10). To assure continuity of the trajectory for the solution $\hat{p}$ of the parameter estimation problem for the whole time interval and consistency for the algebraic equations, we impose continuity constraints

$$c(\tau_i, s_i, s_{i-1}, p) := \psi^y(\tau_i; s_{i-1}^y, p) - s_i^y = 0, \quad i = 1, \ldots, n_{\mathrm{ms}} \tag{12}$$

and consistency constraints

$$g(\tau_i, s_i, p) = 0, \quad i = 0, \ldots, n_{\mathrm{ms}}. \tag{13}$$

The variables s_i, $i = 0, \ldots, n_{\mathrm{ms}}$ are additional degrees of freedom of the parameter estimation problem.

Before we formulate the finite dimensional constrained parameter estimation problem, we have to adjust the interior point constraints (6) to the shooting discretization. With the initial conditions (9) added to the set of constraints, we introduce a new vector of variables s^r that is locally uniquely determined by the interior point constraints, i.e. the matrix $\frac{\partial r}{\partial s^r}$, has full rank. Here, we use the definition

$$r := \sum_{l=1}^{n_r} r_l(\psi(t_l), p, s^r), \quad j = 1, \ldots, n_{\mathrm{ex}},$$

where we neglect the dependencies of the nominal trajectories ψ of s and p, respectively.

When we present the generalized Gauss–Newton method in Sect. 5, we will clarify the necessity of the variables s^r in more detail. For later considerations, we define the vector

$$s := (s_0, \ldots, s_{\mathrm{ms}}, s^r)$$

and the function

$$d(s,p) := \begin{pmatrix} y_0(s^r,p) - s_0^y, \\ r \end{pmatrix}. \tag{14}$$

Note, that the initial conditions may depend on the variables s^r, too.

Summarized, this results in the following finite dimensional nonlinear least squares problem for n_{ex} experiments:

$$\min_{s,p} \quad \frac{1}{2}\sum_{j=1}^{n_{\text{ex}}}\sum_{k=1}^{m^j}\left(\frac{\eta_k^j - h_k^j(t_k,\psi(t_k),p)}{\sigma_k^j}\right)^2 \tag{15a}$$

$$\text{s.t.} \quad 0 = c^j(\tau_i, s_i, s_{i-1}, p) - s_i^y, \qquad i = 1,\dots,n_{\text{ms}}^j,\ j = 1,\dots,n_{\text{ex}}, \tag{15b}$$

$$0 = g^j(\tau_i, s_i, p), \qquad i = 0,\dots,n_{\text{ms}}^j, \tag{15c}$$

$$0 = d^j(s,p). \tag{15d}$$

4 The Generalized Gauss–Newton Method

For readability, we introduce a shorter notation of Problem (15)

$$\min_{s,p} \quad \|F_1(s,p)\|_2^2 \tag{16a}$$

$$\text{s.t. } 0 = F_2(s,p). \tag{16b}$$

with $F_1(s,p) \in \mathbb{R}^{n_1}$, $F_2(s,p) \in \mathbb{R}^{n_2}$ and $(s,p) \in \mathbb{R}^n$. We use the definitions

$$F_1 := \left(\frac{\eta_k^j - h_k^j(t_k,\psi(t_k),p)}{\sigma_k^j}\right)_{\substack{k=1,\dots,m^j \\ j=1,\dots,n_{\text{ex}}}} \tag{17a}$$

$$F_2 := \begin{pmatrix} \left(c^j(\tau_i, s_i, s_{i-1}, p) - s_i^y\right)_{\substack{i=1,\dots,n_{ms}^j \\ j=1,\dots,n_{\text{ex}}}} \\ \left(g^j(\tau, s_i, p)\right)_{\substack{i=0,\dots,n_{ms}^j \\ j=1,\dots,n_{\text{ex}}}} \\ \left(d^j(s,p)\right)_{j=1,\dots,n_{\text{ex}}} \end{pmatrix} \tag{17b}$$

Bock [5] suggested to apply the generalized Gauss–Newton method to solve nonlinear least squares problems with ODE constraints. The first application to DAE constrained problems was presented by Bock et al. [7]. For a detailed description we refer to Körkel [10]. Problem (16) is solved iteratively by examining linearized equations

$$\min_{\Delta s, \Delta p} \quad \left\| F_1 + J_1 \begin{pmatrix} \Delta s \\ \Delta p \end{pmatrix} \right\|_2^2 \tag{18a}$$

$$\text{s.t. } 0 = F_2 + J_2 \begin{pmatrix} \Delta s \\ \Delta p \end{pmatrix}. \tag{18b}$$

Therefore, we need to compute the Jacobians

$$J_1 := \frac{dF_1}{d(s,p)}, \tag{19a}$$

$$J_2 := \frac{dF_2}{d(s,p)}. \tag{19b}$$

For problem (15) and $n_{\mathrm{ex}} = 1$, the Jacobian has the following structure:

$$J = \begin{pmatrix} J_1 \\ J_2 \end{pmatrix} = \left(\begin{array}{cccccc} D_0^1 & D_1^1 & \cdots & D_{n_{ms}}^1 & D_{s^r}^1 & D_p^1 \\ \hline G_0 & (-I,0) & & & G_0^{s^r} & G_0^p \\ & \ddots & \ddots & & \vdots & \vdots \\ & & G_{n_{\mathrm{ms}}-1} & (-I,0) & G_{n_{\mathrm{ms}}-1}^{s^r} & G_{n_{\mathrm{ms}}-1}^p \\ H_0 & & & & H_0^{s^r} & H_0^p \\ & & \ddots & & \vdots & \vdots \\ & & & H_{n_{ms}} & H_{n_{ms}}^{s^r} & H_{n_{ms}}^p \\ D_0^2 & D_1^2 & \cdots & D_{n_{ms}}^2 & D_{s^r}^2 & D_p^2 \end{array} \right), \tag{20}$$

with the derivatives of

- of the residual of the measurements

$$D_i^1 := \frac{\partial F_1}{\partial s_i},\ i = 0, \ldots, n_{\mathrm{ms}}, \quad D_{\hat{v}}^1 := \frac{\partial F_1}{\partial \hat{v}},$$

- of the continuity constraints

$$G_i := \frac{\partial \psi^y}{\partial s_i}(\tau_{i+1}; s_i, p),\ i = 0, \ldots, n_{\mathrm{ms}} - 1, \quad G_i^{\hat{v}} := \frac{\partial \psi^y}{\partial \hat{v}}(\tau_{i+1}, s_i, p),$$

- of the consistency conditions

$$H_i := \frac{\partial g}{\partial s_i}(\tau_i, s_i, p),\ i = 0, \ldots, n_{\mathrm{ms}}, \quad H_i^{\hat{v}} := \frac{\partial g}{\partial \hat{v}}(\tau_i, s_i, p),$$

- and the initial conditions and the interior point constraints

$$D_i^2 := \frac{\partial d}{\partial s_i}, \; i = 0, \ldots, n_{\mathrm{ms}}, \quad D_{\hat{v}}^2 := \frac{\partial d}{\partial \hat{v}}.$$

with $\hat{v} = (p, s^r)$.

To assure uniqueness of the solution of Problem (16), we assume that the following two conditions hold for all values of $(s, \; p)$, where we have to evaluate F and J :

- Constraint Qualification (CQ)

$$\operatorname{rank} J_2(s,p) = n_2, \tag{21}$$

- Positive Definiteness (PD)

$$\operatorname{rank} J(s,p) = n. \tag{22}$$

We recall from Sect. 3 the dimension of F_2. The continuity conditions (12) sum up to $n_y \cdot n_{\mathrm{ms}}$ constraints, the consistency constraints (13) provide $n_z \cdot (n_{\mathrm{ms}} + 1)$ additional constraints and the initial conditions and the interior point and boundary constraints (14) results in $n_y + n_r$ equations. We end up with

$$n_2 = (n_y + n_z) \cdot (n_{\mathrm{ms}} + 1) + n_r.$$

Since we have already defined

$$(n_y + n_z) \cdot (n_{\mathrm{ms}} + 1)$$

shooting variables, we need to introduce n_r additional variables s^r to guarantee that (CQ) is fulfilled.

The linearization of (16) leads to a comparatively large, but sparse Jacobian. Bock [5] introduced the condensing algorithm for ODE constrained parameter estimation problems that exploits the sparse structure of (20) and eliminates the shooting variables $s_i, \; i = 1, \ldots, n_{ms}$ by a block-Gaussian elimination. The condensed system depends only on $\Delta s_0^y, \; \Delta s^r$ and Δp

$$\min_{\Delta s_0, \Delta p} \quad \left\| u_1 + E_1 \Delta s_0^y + E_1^r \Delta s^r + E_1^p \Delta p \right\|_2^2 \tag{23a}$$

$$\text{s.t.} \quad 0 = u_2 + E_2 \Delta s_0^y + E_2^r \Delta s^r + E_2^p \Delta p \tag{23b}$$

By projecting on the algebraic variables $s_i^z, \; i = 0, \ldots, n_{\mathrm{ms}}$, the method can be applied to DAE constrained parameter estimation problems too, cf. Leineweber [11].

The procedure of evaluating the Jacobians first and applying the condensing method afterwards is referred to as the general approach. The general approach is implemented in the software package for parameter estimation *PARFIT* which is based on the methods presented in this section and in Bock [5] and Körkel [10].

5 The Reduced Approach

Especially for DAE constraints, that result from parametrized PDEs, the effort to compute (20) is tremendous due to the high dimension of s_i. The computation of submatrices G_i and H_i scales with the number of (discretized) states. For $(n_y + n_z) \gg n_p$, the effort to evaluate the blocks G_i and H_i dominates the evaluation of the Jacobian. The computation of these blocks requires the evaluation of $(n_y + n_z)$ variational differential equations. That is why the common approach is not suitable to solve PDE constrained parameter estimation problems in the context of multiple shooting.

Schlöder [13] developed an approach for high-dimensional ODE systems that couples the evaluation of the Jacobians and the subsequent condensing by using directional derivatives. Thereby, the effort of the computation of the Jacobian (20) reduces to the one of single shooting, i.e., the smallest possible number. Bauer [3] extended this method to parameter estimation problems with differential algebraic constraints.

We developed a different formulation of the reduced approach which fully eliminates the algebraic constraints and, thus, leads to a reduced condensed system of equal size as Problem (23). The approach of Bauer leads to redundant constraints which may cause numerical problems.

As in Sect. 3, we present the following method only for the first experiment and neglect the superscript 1. For the next steps, we assume that the interior point constraints and the residuals of the measurements can be written in the following form

$$0 = \sum_{i=0}^{n_{\text{ms}}} \sum_{t_k \in \mathscr{I}_i} \hat{h}_k(\psi(t_k), p) = \sum_{i=0}^{n_{\text{ms}}} R_i^1, \qquad \hat{h}_k = \frac{\eta_k - h_k}{\sigma_k}, \tag{24a}$$

$$0 = \sum_{i=0}^{n_{\text{ms}}} \sum_{t_l \in \mathscr{I}_i} r_l(\psi^j(t_l), p, s^r) = \sum_{i=0}^{n_{\text{ms}}} R_i^2. \tag{24b}$$

We refer to Eqs. (24) as separability conditions. We define the derivatives of (6) and (4) with respect to $\hat{v} = (p, s^r)$ according to

$$D_{\hat{v}}^1 = \frac{d}{d\hat{v}} \sum_{i=0}^{n_{\text{ms}}} \sum_{t_k \in \mathscr{I}_i} \hat{h}_k(\psi(t_k), p) = \sum_{i=0}^{n_{\text{ms}}} \frac{d}{d\hat{v}} \sum_{t_k \in \mathscr{I}_i} \hat{h}_k(\psi(t_k), p) = \sum_{i=0}^{n_{\text{ms}}} D_{\hat{v},i}^1.$$

$$D_{\hat{v}}^2 = \frac{d}{d\hat{v}} \sum_{i=0}^{n_{\text{ms}}} \sum_{t_l \in \mathscr{I}_i} r_l(\psi(t_l), p) = \sum_{i=0}^{n_{\text{ms}}} \frac{d}{d\hat{v}} \sum_{t_l \in \mathscr{I}_i} r_l(\psi(t_l), p) = \sum_{i=0}^{n_{\text{ms}}} D_{\hat{v},i}^2.$$

cf. Eq. (20). To eliminate the shooting variables at $t = \tau_0$, we define

$$d_0(s,p) := y_0(s^r,p) - s_0^y. \tag{25}$$

and we examine the rows of the Jacobian (20) that belong to the initial conditions (25) and to the consistency constraints at $t = \tau_0$

$$\begin{aligned} D_0^{s_0^y} \Delta s_0^y + D_0^{s_0^z} \Delta s_0^z + D_0^{s^r} \Delta s^r + D_0^p \Delta p + d_0(s,p) &= 0, \\ H_0^{s_0^y} \Delta s_0^y + H_0^{s_0^z} \Delta s_0^z + H_0^{s^r} \Delta s^r + H_0^p \Delta p + g(\tau_0, s_0, p) &= 0. \end{aligned} \tag{26}$$

Obviously, it holds

$$D_0^{s_0^y} = -I_{n_y}, \quad D_0^{s_0^z} = 0.$$

Since we consider only DAEs with differentiation index 1, the matrix

$$M := \begin{pmatrix} -I_{n_y} & 0 \\ H_0^{s_0^y} & H_0^{s_0^z} \end{pmatrix} \tag{27}$$

is regular. We eliminate $n_y + n_z$ variables Δs_0 formally from (26) and obtain

$$\begin{aligned} \begin{pmatrix} \Delta s_0^y \\ \Delta s_0^z \end{pmatrix} &= - \begin{pmatrix} -I_{n_y} & 0 \\ H_0^{s_0^y} & H_0^{s_0^z} \end{pmatrix}^{-1} \left[\begin{pmatrix} D_0^{s^r} \\ H_0^{s^r} \end{pmatrix} \Delta s^r + \begin{pmatrix} D_0^p \\ H_0^p \end{pmatrix} \Delta p + \begin{pmatrix} d_0 \\ g(\tau_0, s_0, p) \end{pmatrix} \right] \\ &= M_s^0 \Delta s^r + M_p^0 \Delta p + M_r^0 \end{aligned} \tag{28}$$

with

$$M_r^0 := - \begin{pmatrix} -I_{n_y} & 0 \\ H_0^{s_0^y} & H_0^{s_0^z} \end{pmatrix}^{-1} \begin{pmatrix} d_0 \\ g(\tau_0, s_0, p) \end{pmatrix}, \tag{29a}$$

$$M_s^0 := - \begin{pmatrix} -I_{n_y} & 0 \\ H_0^{s_0^y} & H_0^{s_0^z} \end{pmatrix}^{-1} \begin{pmatrix} D_0^{s^r} \\ H_0^{s^r} \end{pmatrix}, \tag{29b}$$

$$M_p^0 := - \begin{pmatrix} -I_{n_y} & 0 \\ H_0^{s_0^y} & H_0^{s_0^z} \end{pmatrix}^{-1} \begin{pmatrix} G_0^p \\ H_0^p \end{pmatrix}. \tag{29c}$$

We refer to (29) as *seed* matrices. The following steps are closely related to the ODE formulation of the reduced approach presented by Schlöder [13] with an extension to the consistency constraints which are solved locally for s_j^z, $j = 0, n_{\mathrm{ms}}$.

The idea is to apply the explicit representation for the increments Δs_0 given by Eq. (28) to the evaluation of the remaining constraints. We define

$$\hat{E}_l^{p,0} := D_0^l \cdot M_p^0 + D_{p,0}^l, \tag{30a}$$

$$\hat{E}_l^{s,0} := D_0^l \cdot M_s^0 + D_{s^r,0}^l, \quad l = 1, 2, \tag{30b}$$

$$\hat{u}_l^0 := D_0^l \cdot M_r^0 + R_0^l. \tag{30c}$$

Since we use the matrices $\hat{E}_l^{p,0}$, $\hat{E}_l^{s,0}$ and the vectors $\hat{u}_l^0$, $l = 1, 2$, for the computation of the reduced condensed system, the notations in Eqs. (30) correspond to (23). Then, we compute recursively

$$\hat{E}_l^{p,i} = \hat{E}_l^{p,i-1} + D_i^l \cdot M_p^i + D_{p,i}^l, \tag{31a}$$

$$\hat{E}_l^{s,i} = \hat{E}_l^{s,i-1} + D_i^l \cdot M_s^i + D_{s^r,i}^l, \quad i = 1, \ldots, n_{\text{ms}} \quad l = 1, 2, \tag{31b}$$

$$\hat{u}_l^i = \hat{u}_l^{i-1} + D_i^l \cdot M_r^i + R_i^l. \tag{31c}$$

We set

$$\tilde{E}_l^p := \hat{E}_l^{p,n_{\text{ms}}}, \tilde{E}_l^s := \hat{E}_l^{s,n_{\text{ms}}}, \tilde{u}_l := \hat{u}_i^{n_{\text{ms}}}, \quad l = 1, 2,$$

and obtain the reduced condensed system

$$\min_{\Delta p, \Delta s^r} \quad \frac{1}{2} \left\| \tilde{u}_1 + \tilde{E}_1^p \Delta p + \tilde{E}_1^s \Delta s^r \right\|_2^2 \tag{32a}$$

$$\text{s.t.} \quad 0 = \tilde{u}_2 + \tilde{E}_2^p \Delta p + \tilde{E}_2^s \Delta s^r. \tag{32b}$$

The seed matrices are updated iteratively by

$$M_p^i := \begin{pmatrix} G_i \\ H_{i+1} \end{pmatrix} \cdot M_p^{i-1} + \begin{pmatrix} G_i^p \\ H_{i+1}^p \end{pmatrix}, \tag{33a}$$

$$M_s^i := \begin{pmatrix} G_i \\ H_{i+1} \end{pmatrix} \cdot M_s^{i-1} + \begin{pmatrix} G_i^{s^r} \\ H_{i+1}^{s^r} \end{pmatrix}, \qquad k = 1, \ldots, n_{\text{ms}}, \tag{33b}$$

$$M_r^i := \begin{pmatrix} G_i \\ H_{i+1} \end{pmatrix} \cdot M_r^{i-1} + \begin{pmatrix} c(\tau_{i+1}, s_{i+1}, s_i, p) \\ g(\tau_{i+1}, s_{i+1}, p) \end{pmatrix}. \tag{33c}$$

In Eqs. (30), (31) and (33), the expressions given by " $\cdot$ " are evaluated by directional derivatives and do not denote matrix products. In difference to the common approach, described in Sect. 4, we have to evaluate $n_p + n_r + 1$ directions instead of $n_y + n_z + n_p$.

If the initial conditions (14) are given in the form

$$d_0(s, p) = y_0(p) - s_0^y,$$

i.e., Eq. (14) does not depend explicitly on the variables s^r, the number of required directions is independent of the number of states.

After we have solved Problem (32) for Δp and Δs^r, the increments for the shooting variables $(s_1, \ldots, s_{\mathrm{ms}})$ are determined by

$$\Delta s_i^j = M_r^{i,j} + M_s^{i,j} \Delta s^r + M_p^{i,j} \Delta p, \quad i = 0, \ldots, n_{\mathrm{ms}}^j, \; j = 1, \ldots, n_{\mathrm{ex}}. \tag{34}$$

The methods, which have been presented in this section, have been implemented in a software package for parameter estimation called *PAREMERA*. *PAREMERA* is a new implementation in Fortran90 that is suited for the treatment of multiple experiments.

6 Example

In the following, we examine the 1D heat equation, e.g., see Evans [9]. which describes the distribution of heat in a given region over time. The system is defined by

$$0 = \frac{\partial u}{\partial t} - p_1 \nabla^2 u \tag{35a}$$

$$0 = u(t, 0) = u(t, 1), \tag{35b}$$

$$u(0, x) = -4 \cdot x \cdot (x - 1) \tag{35c}$$

with homogeneous Dirichlet conditions on the domain $\Omega \times T = [0, 1] \times [0, 1]$ (Fig. 1).

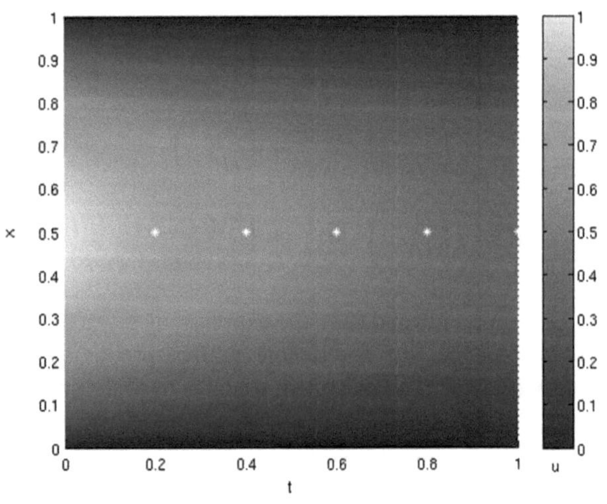

Fig. 1 Heat distribution over the domain $\Omega \times T$. The *white asterisks* mark the measurement points

Here, u is an arbitrary function, usually referred to as temperature. The parameter p_1 is the thermal diffusivity. We discretize (35) with second order central finite differences for three different mesh sizes $\Delta x = \{0.01, 0.002, 0.001\}$ to obtain an ODE system of 101, 501 and 1001 states, respectively.

To compare the results between the reduced approach and the common one, the time interval is decomposed into four subintervals $[\tau_i, \tau_{i+1})$ with $\tau_i = 0.25 \cdot i$, $i = 0, \ldots, 4$. We measure the peak at $x = 0.5$ at $t \in \{0.2, 0.4, 0.6, 0.8, 1.0\}$. The measurement data is determined by integrating the ODE system applying the software package *DAESOL* by Bauer [4] with known true parameter

$$p_1^* = 0.1$$

and adding Gaussian noise. For the calculations, p_1^* is scaled to one.

We compare the results of the two previously mentioned parameter estimation tools *PAREMERA* and *PARFIT*. Here, we use a version of *PARFIT* which is eligible for the exploitation of multiple experiment structures, see von Schwerin [15]. Both tools are embedded in software toolbox *VPLAN* by Körkel et al. [10].

We apply $p_1^0 = 2$ as starting parameter for all six settings (three mesh sizes and two algorithms). All computations are executed on a 64bit computer with 4 GB memory and an Intel® Core2Duo with 2×2.8 GHz. The results are listed in Table 1.

Both algorithms converge for all mesh sizes to approximately the same solution ($\hat{p}_1 \approx 1.00806$), but there is a significant difference in the time per iteration. For 101 states, PAREMERA is around four times faster then PARFIT. With increasing number of states, the difference in computational time increases. PAREMERA is 13 and 26 times faster for 501 and 1001 states, respectively.

Another fact that differs drastically, is the number of iterations. For all discretizations, PAREMERA achieves convergence after 5 iterations while PARFIT finds the solution after 10 iterations. This can be explained by the different types of globalization strategies. In PAREMERA, the restricted monotonicity test (RMT) is implemented as it is presented in Bock et al. [8]. In Parfit, only a first order Taylor series expansion is used to compute the curvature information.

Note, that the computed increments $(\Delta s, \Delta p)$ for the first iteration of both algorithms are exactly the same since both algorithms solve the same system.

Table 1 Survey of the results

# states		*PAREMERA*	*PARFIT*
101	# iterations	5	10
	Time per iteration	0.268 s	1.01 s
501	# iterations	5	10
	Time per iteration	3.44 s	45.62 s
1001	# iterations	5	10
	Time per iteration	15.06 s	392.5 s

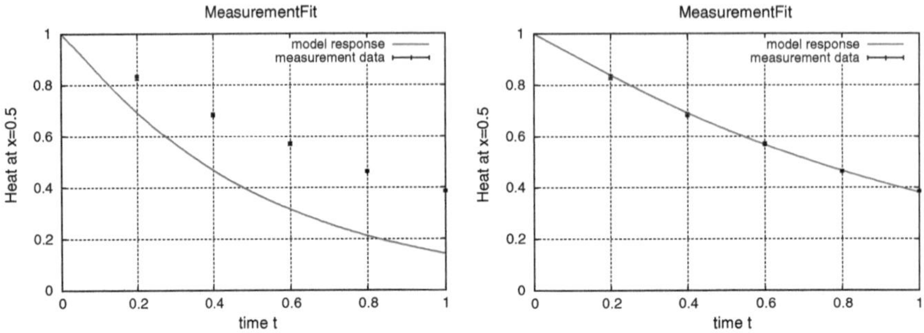

Fig. 2 Data fits for the starting parameter $p_1^0 = 2$ and the estimated parameter $\hat{p}_1 = 1.00806$

In Fig. 2, a comparison is shown between the two fitting curves with $p_1^0 = 2$ on the left hand side and the resulting parameter $\hat{p}_1 = 1.00806$ computed with *PAREMERA* on the right hand side, respectively.

Even for this rather small example we could show the advantages of the reduced approach. For more complex problems we expect even more significant savings in computation time. This is important to solve higher dimensional PDE problems or to do online parameter estimation. Thus, the reduced approach should be favored to solve this kind of problems.

Acknowledgements Financial support by BASF SE and HGS MathComp is gratefully acknowledged.

References

1. Altmann-Dieses, A., Schlöder, J.P., Bock, H.G., Richter, O.: Optimal experimental design for parameter estimation in column outflow experiments. Water Resour. Res. **38**, 1186ff (2002)
2. Bauer, I.: Numerische Verfahren zur Lösung von Anfangswertaufgaben und zur Generierung von ersten und zweiten Ableitungen mit Anwendungen bei Optimierungsaufgaben in Chemie und Verfahrenstechnik. Ph.D. thesis, Universität Heidelberg (1999)
3. Bauer, I.: Numerische Verfahren zur Lösung von Anfangswertaufgaben und zur Generierung von ersten und zweiten Ableitungen mit Anwendungen in Chemie und Verfahrenstechnik. Preprint, SFB 359, Universität Heidelberg (2001)
4. Bauer, I., Bock, H.G., Schlöder, J.P.: DAESOL—a BDF-code for the numerical solution of differential algebraic equations. Internal report, IWR, SFB 359, Universität Heidelberg (1999)
5. Bock, H.G.: Randwertproblemmethoden zur parameteridentifizierung in systemen nichtlinearer differentialgleichungen. Bonner Mathematische Schriften 183 (1987)
6. Bock, H.G., Plitt, K.J.: A Multiple Shooting algorithm for direct solution of optimal control problems. In: Proceedings of the 9th IFAC World Congress, pp. 242–247. Pergamon Press, Budapest (1984). Available at http://www.iwr.uni-heidelberg.de/groups/agbock/FILES/Bock1984.pdf
7. Bock, H.G., Schlöder, J.P., Schulz, V.H.: Numerik großer Differentiell-Algebraischer Gleichungen—Simulation und Optimierung. In: Schuler, H. (ed.), Prozeß-Simulation, Chap. 2, pp. 35–80. VCH, Germany (1995)

8. Bock, H.G., Kostina, E.A., Schlöder, J.P.: On the role of natural level functions to achieve global convergence for damped Newton methods. In: Powell, M.J.D., Scholtes, S. (eds.) System Modelling and Optimization: Methods, Theory and Applications, pp. 51–74. Kluwer, Dodrecht (2000)
9. Evans, L.C.: Partial Differential Equations. Graduate Studies in Mathematics, vol. 19, 2nd edn. American Mathematical Society, Providence, RI (2010)
10. Körkel, S.: Numerische Methoden für optimale Versuchsplanungsprobleme bei nichtlinearen DAE-Modellen. Ph.D. thesis, Universität Heidelberg, Heidelberg (2002)
11. Leineweber, D.B.: Efficient reduced SQP methods for the optimization of chemical processes described by large sparse DAE models. Fortschritt-Berichte VDI Reihe 3, Verfahrenstechnik, vol. 613. VDI Verlag, Düsseldorf (1999)
12. Schiesser, W.E.: The Numerical Method of Lines: Integration of Partial Differential Equations, vol. 17. Academic, San Diego (1991)
13. Schlöder, J.P.: Numerische Methoden zur Behandlung hochdimensionaler Aufgaben der Parameteridentifizierung. Dissertation, Hohe Mathematisch-Naturwissenschaftliche Fakultät der Rheinischen Friedrich-Wilhelms-Universität zu Bonn (1987)
14. Seber, G.A.F., Wild, C.J.: Nonlinear Regression. Wiley, New York (1989)
15. von Schwerin, R.: Numerical methods, algorithms, and software for higher index nonlinear differential-algebraic equations in multibody system simulation. Ph.D. thesis, Universität Heidelberg (1997)

Direct Multiple Shooting for Parabolic PDE Constrained Optimization

Andreas Potschka

Abstract Direct Multiple Shooting is a flexible and efficient method to solve difficult optimal control problems constrained by ordinary differential equations or differential-algebraic equations. The aim of this article is to concisely summarize the main conceptual and methodological approaches to solve also optimal control problems with parabolic partial differential equations constraints via a Direct Multiple Shooting method. The main obstacle is the sheer size of the discretized optimization problems. We explain a typical direct discretization approach and discuss an inexact SQP method based on two-grid Newton-Picard preconditioning. Special attention is given to a-posteriori κ-estimators that monitor contraction and to the structure-exploiting treatment of the resulting large-scale quadratic programming subproblems, including an extended condensing technique that exploits Multiple Shooting and two-grid Newton-Picard structures. Finally, we present numerical results for an advection-diffusion and a bacterial chemotaxis example.

1 Introduction

The early approaches for optimal control for ordinary differential equations (ODEs) were mostly based on Dynamic Programming [3] or Pontryagin's Maximum Principle [16]. Due to the curse of dimensionality, Dynamic Programming is not applicable to problems with more than a few state variables. The Maximum Principle is an indirect method in the sense that the controls are given as a closed form representation of adjoint states, the so-called co-states, which satisfy an adjoint differential equation and suitable boundary constraints. The resulting boundary value problems in the states and co-states can then be discretized by numerical methods for boundary value problems like Multiple Shooting [15] or Collocation [21]. The Maximum Principle is a typical example for an *optimize-then-discretize* approach. It was found out later that ideas from numerical methods for boundary value problems can also be used on the optimal control problem itself instead of

A. Potschka (✉)
Interdisciplinary Center for Scientific Computing, Heidelberg University, Heidelberg, Germany
e-mail: potschka@iwr.uni-heidelberg.de

T. Carraro et al. (eds.), *Multiple Shooting and Time Domain Decomposition Methods*, Contributions in Mathematical and Computational Sciences 9,
DOI 10.1007/978-3-319-23321-5_6

its optimality conditions, thus disposing of the need to formulate adjoint equations, which can be a time-consuming affair (*discretize-then-optimize*). Furthermore, this approach lends itself quite naturally to a direct formulation, in which the discretized controls are not eliminated in favor of adjoint variables. For Direct Collocation, we refer the reader to [4, 25] and for Direct Multiple Shooting to [6]. Direct formulations have been found to reflect the real conditioning of the optimal control problem better, in particular when the mapping of the adjoint to the control has a large Lipschitz constant. We want to remark that there are also direct optimize-then-discretize methods for which the control is kept in the system of optimality conditions and its discretization.

In optimal control for partial differential equations (PDEs), indirect optimize-then-discretize methods prevail (see, e.g., [24]). This is mostly due to the fact that indirect methods usually yield more accurate representations of the control (see also [11, Chap. 3]).

The contribution of this article is to summarize a Direct Multiple Shooting method for parabolic PDE constrained problems from Potschka [17, 18] in a concise form and to present numerical results for problems that could also be treated well with indirect methods. The main challenge in this approach is to tame the large-scale nature of the discretized Nonlinear Programming problems (NLPs), which can be mastered by the use of inexact Sequential Quadratic Programming (SQP) with two-grid Newton-Picard preconditioning.

The main strengths of the Direct Multiple Shooting approach can unfold in particular for practitioners who want to solve parabolic PDE constrained optimization problems and need an accurate representation of the system dynamics, but can live with lower resolution of the optimal control. This situation is not unusual due to physical restrictions of manipulator hardware. These practitioners can then benefit from time savings in the problem setup, because no adjoint equations need to be derived, but also in the problem solution, because the method can be easily parallelized on the basis of the Multiple Shooting structures. Furthermore, the solutions delivered by Direct Multiple Shooting can serve as initial guesses for other methods with higher accuracy.

2 Problem Formulation

Let $\Omega \subset \mathbb{R}^d$ be a bounded d-dimensional spatial domain with sufficiently regular boundary $\partial\Omega$. Regarding time, we only consider the fixed interval $(0, 1)$ here for simplicity. We assume that the control satisfies

$$q \in L^2((0,1); Q), \quad \text{where} \quad Q \subseteq L^2(\Omega)^{n_q^{\mathrm{d}}} \times L^2(\partial\Omega)^{n_q^{\mathrm{b}}},$$

with n_q^{d} distributed and n_q^{b} boundary controls. We would already like to point out that the pure boundary control case $n_q^{\mathrm{d}} = 0$ is especially advantageous in a Multiple Shooting approach, as we describe in Sect. 3.

We consider two kinds of coupled state variables: ODE states $v \in C^1([0,1];\mathbb{R}^{n_v})$, which are not spatially distributed, and PDE states $u \in W(0,1)^{n_u}$, for which we use the standard Hilbert space for parabolic PDEs

$$W(0,1) = \{u \in L^2((0,1);V) \mid \partial_t u \in L^2((0,1);V^*)\},$$

where $(V, L^2(\Omega), V^*)$ is a Gelfand triple and $\partial_t u$ denotes the time derivative of u in the sense of vectorial distributions (see, e.g., [29, Chap. IV]). We assume that the spatial regularity of u is $V \subseteq H^1(\Omega)$. According to [7, Chap. XVIII, Theorem 1], we can trade some spatial regularity for higher temporal regularity, because $W(0,1)$ is continuously embedded in $C^0([0,1];L^2(\Omega))$. Thus, we have existence of the trace $u(t) \in L^2(\Omega)^{n_u}$ for all $t \in [0,1]$, which is important for the formulation of boundary constraints in time.

We assume that the time derivative $\dot{v}$ equals a sufficiently regular nonlinear function $f^{\mathrm{ODE}} : Q \times L^2(\Omega)^{n_u} \times \mathbb{R}^{n_v} \to \mathbb{R}^{n_v}$ and $-\partial_t u$ equals a possibly nonlinear elliptic differential operator $A : Q \times V \times \mathbb{R}^{n_v} \to V^*$. Furthermore, we allow for generalized temporal boundary constraints, mixed control-ODE-state path constraints and final ODE state constraints via the functions

$$r^{\mathrm{b}} : L^2(\Omega)^{n_u} \times \mathbb{R}^{n_v} \to L^2(\Omega)^{n_u} \times \mathbb{R}^{n_v}, \quad r^{\mathrm{c}} : Q \times \mathbb{R}^{n_v} \to \mathbb{R}^{n_{\mathrm{r}}^{\mathrm{c}}}, \quad r^{\mathrm{e}} : \mathbb{R}^{n_v} \to \mathbb{R}^{n_{\mathrm{r}}^{\mathrm{e}}}.$$

The objective function $\Phi : L^2(\Omega)^{n_u} \times \mathbb{R}^{n_v} \to \mathbb{R}$ depends on the final values of the ODE and PDE states. Finally, we can state the problem of interest of this article as

$$\begin{aligned}
&\underset{\substack{q \in L^2((0,1);Q)\\ u \in W(0,1)^{n_u}\\ v \in C^1([0,1];\mathbb{R}^{n_v})}}{\text{minimize}} && \Phi(u(1), v(1)) && && \text{(1a)}\\
&\text{s.t.} && \partial_t u = -A(q(t), u(t), v(t)), && t \in (0,1), && \text{(1b)}\\
& && \dot{v} = f^{\mathrm{ODE}}(q(t), u(t), v(t)), && t \in (0,1), && \text{(1c)}\\
& && (u(0), v(0)) = r^{\mathrm{b}}(u(1), v(1)), && && \text{(1d)}\\
& && r^{\mathrm{c}}(q(t), v(t)) \geq 0, && t \in (0,1), && \text{(1e)}\\
& && r^{\mathrm{e}}(v(1)) \geq 0. && && \text{(1f)}
\end{aligned}$$

More general problems with free initial and final time or integral-type objectives can be equivalently reformulated using extra v variables to fit problem class (1). Further assumptions concerning regularity requirements on the occurring functions above is a delicate issue. From a practical point of view, we require that (1) is well-posed and that there exists an appropriate discretization for the involved variables and functions such that the solution of the resulting finite dimensional optimization problem is consistent with (1).

3 Discretization

In Potschka [17], problem (1) is discretized in space first and then in time. We explain here a different route, where time is discretized before space. In both cases, though, the resulting discretized problem (3) exhibits the same structure.

3.1 *Time Discretization*

We start out by discretizing the controls in time. To this end, let $0 = t^0 < \cdots < t^{n_{\mathrm{MS}}} = 1$ denote a partition of the interval $[0, 1]$. On the shooting intervals $I^i := (t^i, t^{i+1}), i = 0, \ldots, n_{\mathrm{MS}} - 1$, we perform a piecewise discretization of the controls in time. The piecewise nature is important for decoupling properties to be exploited in the numerics. We restrict ourselves here to the simplest case of a piecewise constant discretization in time such that

$$\tilde{q}(t) = \sum_{j=0}^{n_{\mathrm{MS}}-1} q^j \chi_{I^j}(t), \quad \text{with } q^i \in Q, i = 0, \ldots, n_{\mathrm{MS}} - 1,$$

where χ_{I^i} denotes the characteristic function of I^i.

In the next step, we parametrize the state variables by local initial values $u^i, \boldsymbol{v}^i, i = 0, \ldots, n_{\mathrm{MS}}$: We assume that the local initial value problems of the form (1b)–(1c) on I^i with initial values $u^i, \boldsymbol{v}^i$ admit a unique solution denoted by

$$\begin{pmatrix} \overline{u}^i(t; q^i, u^i, \boldsymbol{v}^i) \\ \overline{v}^i(t; q^i, u^i, \boldsymbol{v}^i) \end{pmatrix} \in L^2(\Omega)^{n_u} \times \mathbb{R}^{n_v}, \quad \text{for all } t \in I^i, i = 0, \ldots, n_{\mathrm{MS}} - 1.$$

Continuity of the states on the full time horizon $(0, 1)$ must then be enforced with matching conditions of the form

$$\overline{u}^i(t; q^i, u^i, \boldsymbol{v}^i) = u^{i+1}, \quad \overline{v}^i(t; q^i, u^i, \boldsymbol{v}^i) = \boldsymbol{v}^{i+1}, \quad i = 0, \ldots, n_{\mathrm{MS}} - 1. \tag{2}$$

At this stage of the discretization procedure, we have arrived at a problem depending only on variables that are either real vectors or functions distributed in space only, namely $q^i, u^i, \boldsymbol{v}^i, i = 0, \ldots, n_{\mathrm{MS}}$. Note, that we have en passant added an additional control variable $q^{n_{\mathrm{MS}}}$ to lend a common structure to the optimization variables corresponding to each shooting node t_i. We have to eliminate these additional degrees of freedom later via (3f).

3.2 Space Discretization

Now, we can discretize q^i and u^i in space, e.g., by Finite Differences, Finite Elements, or Finite Volumes. We denote the corresponding vectors by $\boldsymbol{q}^i$ and $\boldsymbol{u}^i$. The quantities $\overline{u}^i$ and $\overline{v}^i$ can then be discretized by an appropriate time stepping method. We denote the result by $\overline{\boldsymbol{u}}^i(t^{i+1}; \boldsymbol{q}^i, \boldsymbol{u}^i, \boldsymbol{v}^i)$ and $\overline{\boldsymbol{v}}^i(t^{i+1}; \boldsymbol{q}^i, \boldsymbol{u}^i, \boldsymbol{v}^i)$. It is not mandatory to use the same grids for the controls or states on all shooting intervals, or even within one shooting interval for the time steps of states in case of a Rothe-type time stepping scheme. According to [10], a good compromise for balancing good mesh adaptivity with few Finite Element matrix assembly calls is to fix the state mesh on each shooting interval and discretize the matching conditions (2) in a variational manner. In this case, a Method of Lines discretization in time seems to be a flexible and efficient approach, because existing ODE solvers can be used. For a more detailed discussion of discretization issues of parabolic PDEs, we refer the reader to [23].

For simplicity, we assume here that $\boldsymbol{q}^i$ and $\boldsymbol{u}^i$ are based on the same spatial discretization in each shooting node. However, we assume that there is a hierarchy of nested discretizations $V_h^1 \subset V_h^2 \subset \cdots \subset V$ for the states, which in turn yields on each level $\ell = 1, 2, \ldots$ discretized states and shooting solutions. The path constraint is discretized on the shooting grid only. For more sophisticated methods, see Potschka [19].

All remaining functions of problem (1) need to be appropriately discretized as well. Finally, we arrive at a highly structured, finite dimensional optimization problem on each spatial discretization level ℓ

$$\underset{(\boldsymbol{q}^i, \boldsymbol{u}^i, \boldsymbol{v}^i)_{i=0}^{n_{\mathrm{MS}}}}{\text{minimize}}\ \boldsymbol{\Phi}(\boldsymbol{u}^{n_{\mathrm{MS}}}, \boldsymbol{v}^{n_{\mathrm{MS}}}) \tag{3a}$$

$$\text{s.t.}\quad \boldsymbol{r}_u^{\mathrm{b}}(\boldsymbol{u}^{n_{\mathrm{MS}}}, \boldsymbol{v}^{n_{\mathrm{MS}}}) - \boldsymbol{u}^0 = 0, \tag{3b}$$

$$\overline{\boldsymbol{u}}^i(t^i; \boldsymbol{q}^{i-1}, \boldsymbol{u}^{i-1}, \boldsymbol{v}^{i-1}) - \boldsymbol{u}^i = 0, \quad i = 1, \ldots, n_{\mathrm{MS}}, \tag{3c}$$

$$\boldsymbol{r}_v^{\mathrm{b}}(\boldsymbol{u}^{n_{\mathrm{MS}}}, \boldsymbol{v}^{n_{\mathrm{MS}}}) - \boldsymbol{v}^0 = 0, \tag{3d}$$

$$\overline{\boldsymbol{v}}^i(t^i; \boldsymbol{q}^{i-1}, \boldsymbol{u}^{i-1}, \boldsymbol{v}^{i-1}) - \boldsymbol{v}^i = 0, \quad i = 1, \ldots, n_{\mathrm{MS}}, \tag{3e}$$

$$\boldsymbol{q}^{n_{\mathrm{MS}}} - \boldsymbol{q}^{n_{\mathrm{MS}}-1} = 0, \tag{3f}$$

$$\boldsymbol{r}^{\mathrm{i}}(\boldsymbol{q}^{i-1}, \boldsymbol{v}^{i-1}) \geq 0, \quad i = 1, \ldots, n_{\mathrm{MS}}, \tag{3g}$$

$$r^{\mathrm{e}}(\boldsymbol{v}^{n_{\mathrm{MS}}}) \geq 0. \tag{3h}$$

The numerical evaluation of (3c) and (3e) is expensive, because it comprises the solution of an initial value problem for the spatially discretized PDE. To this end, we use an adaptive backward differentiation formula with monitoring strategy [1, 2], which is implemented in the C++ code DAESOL-II. It also supplies efficient first and second order forward and backward directional derivatives of the solutions of the initial value problems on the basis of Internal Numerical Differentiation [5] and

Algorithmic Differentiation [9, 28], for which the backwards mode of Algorithmic Differentiation delivers an automated and efficient way to compute gradients in the style of adjoint equations.

We want to remark that special attention has to be paid to the question of consistency of the discretized problem (3) with the original problem (1) for fine discretizations. For instance, it is necessary to use correct weighting matrices for the discrete approximations of Hilbert space inner products and their corresponding norms. Their inverses must also be used for correct discrete Riesz representations of discrete variables that really live in the dual space, for example the Lagrange multipliers of (3b) and (3c). Then, it is typically possible to prove consistency of (3) and (1) for the particular problem at hand.

4 Newton-Picard Inexact SQP

When disregarding its special structure for a moment, we see that problem (3) is a Nonlinear Programming (NLP) problem of the form

$$\underset{x\in\mathbb{R}^n}{\text{minimize}} \quad f(x) \tag{4a}$$

$$\text{s.t.} \quad g_i(x) = 0, \quad i \in \mathcal{E}, \tag{4b}$$

$$g_i(x) \geq 0, \quad i \in \mathcal{I}, \tag{4c}$$

with $f : \mathbb{R}^n \to \mathbb{R}$ and $g : \mathbb{R}^n \to \mathbb{R}^m$. We assume throughout that f and g are twice continuously differentiable functions and that the sets $\mathcal{E}$ and $\mathcal{I}$ form a partition of $\{1, \dots, m\} = \mathcal{E} \dot{\cup} \mathcal{I}$. For $z = (x, y) \in \mathbb{R}^{n+m}$, we can then define the Lagrangian function

$$\mathcal{L}(z) = f(x) - \sum_{i=1}^{m} y_i g_i(x).$$

For the theory of NLP we refer the reader to [14].

4.1 The Equality Constrained Case

We first consider the equality constrained case $\mathcal{I} = \{\}$. If $x^* \in \mathbb{R}^n$ satisfies a constraint qualification and is a local minimizer of (3), then there exists a Lagrange multiplier $y^* \in \mathbb{R}^m$ such that with $z^* = (x^*, y^*)$

$$F(z^*) = \begin{pmatrix} F_1(z^*) \\ F_2(z^*) \end{pmatrix} = \begin{pmatrix} \nabla_x \mathcal{L}(z^*) \\ g(x^*) \end{pmatrix} = 0. \tag{5}$$

In practice, a Newton-type method is used to solve the necessary optimality condition (5) in order to obtain a critical point z^*. In general we must then check (second order) sufficient conditions to ensure that x^* is really a minimizer and not a maximizer or saddle-point.

Because of the large-scale nature of (3), it is a good idea to employ iterative methods for the solution of the linearized subproblems within the Newton-type method. We recede to a rather simple iterative method here, namely a Linear Iterative Splitting Approach (LISA), because we can generalize it from linear systems to Quadratic Programming (QP) problems in Sect. 4.4.

Let $N = n + m, J(z) = \frac{\mathrm{d}F}{\mathrm{d}z}(z)$, and $\hat{M} : \mathbb{R}^N \to \mathbb{R}^{N\times N}$ be given such that $\hat{M}(z)J(z) \approx \mathbb{I}_N$. Then, the LISA-Newton iterates are defined via

$$z^{k+1} = z^k + \alpha_k \Delta z^k, \quad z^0 \text{ given}, \tag{6}$$

where $\alpha_k \in (0, 1]$ is an appropriately chosen damping factor to ensure global convergence and Δz^k is computed via the inner LISA iteration

$$\Delta z^k_{i+1} = \Delta z^k_i - \hat{M}(z^k)\left[J(z^k)\Delta z^k_i + F(z^k)\right], \quad \Delta z^k_0 = 0. \tag{7}$$

The iteration (7) converges for all values of $F(x^k)$ and Δz^k_0 if and only if the spectral radius κ_{lin} of $A_k := \mathbb{I}_N - \hat{M}(z^k)J(z^k)$ is less than 1, see [22, Theorem 4.1].

The choice of the damping factors α_k exceeds the scope of this article. The interested reader is referred to the discussion in [8] and its specialization for the LISA-Newton method [17].

4.2 A-Posteriori κ-Estimators

Based on [17, Lemma 4.28], we observe that there is a representation of Δz^k_l for $l \geq 1$ in terms of a truncated Neumann series

$$\Delta z^k_l = -\left[\sum_{i=0}^{l-1} \left(\mathbb{I}_N - \hat{M}(z^k)J(z^k)\right)^i\right]\hat{M}(z^k)F(z^k) =: -M(z^k)F(z^k),$$

where $M(z^k)$ depends on l. On the basis of this M, we can observe the connection

$$\kappa_\varepsilon \in [(\kappa_{\text{lin}})^l, (\kappa_{\text{lin}})^l + \varepsilon] \quad \text{for all } \varepsilon > 0$$

between the linear convergence rate κ_{lin} of the LISA iteration (7) and the linear convergence rate κ_ε (which depends on an ε-dependent norm) of the nonlinear LISA-Newton iteration (6), if each Δz^k is computed from l LISA iterations [17, Theorem 4.29]. This implies that under the assumption $\kappa_{\text{lin}} < 1$, $l = 1$ is already enough for local convergence and, moreover, if we perform $l > 1$ LISA iterations per outer step, then this extra effort is compensated for by fewer outer LISA-Newton iterations, at least asymptotically for $k \to \infty$.

Furthermore, this result is the basis for a-posteriori κ-estimators [17]. To explain the different estimators, we fix k and define $\zeta_l := \Delta z_{l+1}^k - \Delta z_l^k$. We immediately observe that $\zeta_{l+1} = A_k \zeta_l$. The Rayleigh κ-estimator is based on the Rayleigh quotient

$$\kappa_l^{\text{Rayleigh}} = \frac{\zeta_l^{\mathrm{T}} A_k \zeta_l}{\zeta_l^{\mathrm{T}} \zeta_l} = \frac{\zeta_l^{\mathrm{T}} \zeta_{l+1}}{\zeta_l^{\mathrm{T}} \zeta_l}.$$

The sequence $(\kappa_l^{\text{Rayleigh}})$ converges to κ_{lin} for $l \to \infty$ if A^k is diagonalizable, has a single eigenvalue of largest modulus, and ζ_1 has a nonzero component in the corresponding eigenspace. For non-diagonalizable A_k, the Root κ-estimator

$$\kappa_l^{\text{Root}} = \left(\|\zeta_{l+1}\| \,/\, \|\zeta_l\| \right)^{1/l}$$

can be used. It converges if $\kappa_{\text{lin}} > 0$ and ζ_1 has a nonzero component in the dominant invariant subspace of A_k. However, convergence to κ_{lin} can be quite slow. In contrast, the Ritz κ-estimator converges in a finite number of iterations. It is based on the largest Ritz value κ_l^{Ritz} of A_k on the order-l Krylov subspace generated by A_k and ζ_1. The disadvantage is that an orthonormal basis of the Krylov subspace needs to be maintained, which can be prohibitive due to excessive memory consumption. In our numerical experience, this is not a problem because the Newton-Picard based preconditioners $\hat{M}$ described in Sect. 4.3 require only few LISA iterations.

The κ-estimators can also be used to asses the error of the LISA iteration (7) in the following sense: For all $\varepsilon \in (0, 1 - \kappa_{\text{lin}})$, there exists a vector-norm $\|.\|_{*,k}$ with corresponding matrix-norm $\|A_k\|_{*,k} \le \kappa_{\text{lin}} + \varepsilon < 1$ such that (compare [17, Lemma 4.30])

$$\left\| \Delta z_l^k - \Delta z^k \right\|_{*,k} \le \frac{(\kappa_{\text{lin}} + \varepsilon)^l}{1 - (\kappa_{\text{lin}} + \varepsilon)} \left\| \Delta z_1^k - \Delta z_0^k \right\|_{*,k}.$$

In our implementation, we have $l \ge 2$ and terminate the LISA iterations as soon as either the κ-estimate is lower than $\kappa_{\max} = \sqrt{1/2}$ or a maximum number of LISA iterations, in our case $l_{\max} = 7$, is reached. In the latter case, we ameliorate the preconditioner $\hat{M}$ in order to reduce the contraction rate κ_{lin} as discussed in the following section. In most outer iterations only $l = 2$ LISA iterations are required. We used the Ritz κ-estimator for the computations in Sect. 6.

4.3 *Two-Grid Newton-Picard Preconditioning*

The requirement $\kappa_{\text{lin}} < 1$ for convergence of the LISA iteration (7) is a strong requirement, which calls for specially tailored preconditioners. In our case, Newton-

Picard preconditioning is the method of choice for computing $\hat{M}$, see [20]. Although grid-independent convergence has so far only been proven for a linear quadratic model problem with single shooting, our numerical experience on nonlinear problems with Multiple Shooting is completely satisfying.

The general paradigm of Newton-Picard preconditioners is the following: Under reasonable assumptions, $\overline{u}^i$ has a compact Fréchet derivative with respect to q^i and u^i. Thus, even though the discretized Jacobian matrices $G_q^i = \partial \overline{\boldsymbol{u}}^i / \partial \boldsymbol{q}^i$ and $G_u^i = \partial \overline{\boldsymbol{u}}^i / \partial \boldsymbol{u}^i$ are in general large dense matrices, their eigenvalues and singular values cluster at 0 and they can thus be approximated well by low-rank matrices.

From a geometrical point of view, the compactness is a result of a smoothing property of $\overline{u}^i$, which can alternatively be exploited to form a low-rank approximation by a two-grid approach. To this end, we approximate the Jacobian matrices G_q^i and G_u^i on coarse grids with suitable projection and restriction matrices P and R. This is the preferred way because no large eigenvalue and singular value problems have to be solved. Furthermore, this approach can be extended in a straight-forward fashion to Multiple Shooting approaches.

4.4 The Inequality Constrained Case

We now consider the inequality constrained case when $\mathcal{I} \neq \{\}$. The Jacobian

$$J(z) = \begin{pmatrix} J_1(z) & -J_2(z)^{\mathrm{T}} \\ J_2(z) & 0 \end{pmatrix} = \begin{pmatrix} \nabla_{xx}^2 \mathcal{L}(z) & -\nabla g(x) \\ \nabla g(x)^{\mathrm{T}} & 0 \end{pmatrix}$$

is a saddle-point matrix in unsymmetric form. It is singular in general, e.g., due to rank deficiency of $J_2(z)$ if lower and upper variable bounds are present. Thus, we need to recede to a pseudo-inverse approach. Let us first consider the case of exact derivatives in the SQP method. In this case, we can define a suitable pseudo-inverse

$$\Delta z^k = J^{\oplus}(z^k, -F(z^k)) \quad \text{instead of } \Delta z^k = -M(z^k)F(z^k),$$

by computing the primal-dual solution $\tilde{z} = (\tilde{x}, \tilde{y}) \in \mathbb{R}^{n+m}$ of the QP

$$\underset{\tilde{x} \in \mathbb{R}^n}{\text{minimize}} \quad \frac{1}{2}\tilde{x}^{\mathrm{T}} J_1(z^k)\tilde{x} + \left(F_1(z^k) - J_1(z^k)x^k + J_2(z^k)^{\mathrm{T}} y^k\right)^{\mathrm{T}} \tilde{x} \tag{8a}$$

$$\text{s.t.} \quad \left(J_2(z^k)\tilde{x} + \left(F_2(z^k) - J_2(z^k)x^k\right)\right)_i = 0, \quad i \in \mathcal{E}, \tag{8b}$$

$$\left(J_2(z^k)\tilde{x} + \left(F_2(z^k) - J_2(z^k)x^k\right)\right)_i \geq 0, \quad i \in \mathcal{I}, \tag{8c}$$

and then the step $\Delta z^k = \tilde{z} - z^k$. One can show that if QP (8) has a unique solution at z^k and if some $\hat{z} = (\hat{x}, \hat{y}) \in \mathbb{R}^N$ satisfies $\hat{y}_i \geq -y_i$ for all $i \in \mathcal{I}$, then $J^{\oplus}$ acts linearly on the second argument like a pseudo-inverse in the sense that $J^{\oplus}(z^k, J(z^k)\hat{z}) = \hat{z}$ [17, Lemma 4.36].

Furthermore, if the active set of QP (8) does not change from iteration $k-1$ to k and the second order sufficient condition and strong complementarity hold, then we can construct a matrix M^k such that for a neighborhood U of $F(z^k)$ it holds that

$$-M^k\hat{F} = J^{\oplus}(z^k, -\hat{F}) \quad \text{for all } \hat{F} \in U.$$

In words, the pseudo-inverse $J^{\oplus}$ acts locally like a matrix [17, Theorem 4.37]. In addition, if the SQP method with $J^{\oplus}$ converges to z^*, then z^* is a critical point of NLP (4). Moreover, the second order sufficiency condition and strong complementarity transfer from QP (8) at z^* to NLP (4) at z^* [17, Theorem 4.41].

In the case of inexact derivatives on the basis of the Newton-Picard approximation from Sect. 4.3 we use a structured approximation of the Jacobian

$$J(z^k) = \begin{pmatrix} \nabla^2_{xx}\mathcal{L}(z^k) & -\nabla g(x^k) \\ \nabla g(x^k)^{\mathrm{T}} & 0 \end{pmatrix} \approx \begin{pmatrix} \tilde{B}^k & -(\tilde{C}^k)^{\mathrm{T}} \\ \tilde{C}^k & 0 \end{pmatrix} =: \hat{J}^k.$$

In analogy to the construction of $J^{\oplus}(z^k, .)$ from $J(z^k)$, we can construct a QP preconditioner $\hat{M}(z^k)$ for a QP-LISA iteration. We first compute the primal-dual solution $\tilde{z} = (\tilde{x}, \tilde{y})$ of the QP

$$\underset{\tilde{x}\in\mathbb{R}^n}{\text{minimize}} \quad \frac{1}{2}\tilde{x}^{\mathrm{T}}\tilde{B}^k\tilde{x} + \left(F_1(x^k) - \tilde{C}^k(x^k + \Delta x^k_{l-1}) + (\tilde{C}^k)^{\mathrm{T}}(y^k + \Delta y^k_{l-1})\right)^{\mathrm{T}}\tilde{x} \tag{9a}$$

$$\text{s.t.} \quad \left(\tilde{C}^k\tilde{x} + \left(F_2(x^k) - \tilde{C}^k(x^k + \Delta x^k_{l-1})\right)\right)_i = 0, \quad i \in \mathcal{E}, \tag{9b}$$

$$\left(\tilde{C}^k\tilde{x} + \left(F_2(x^k) - \tilde{C}^k(x^k + \Delta x^k_{l-1})\right)\right)_i \geq 0, \quad i \in \mathcal{I}, \tag{9c}$$

and then $\Delta z^k_l = \tilde{z} - z^k - \Delta z^k_{l-1}$.

In order to have valid κ-estimators, the working set of QP (9) should be fixed for $l > 1$, effectively deferring further changes of the active set to the next outer iteration.

5 Numerical Solution of the Large-Scale QPs

In order to solve the large-scale QP (9) efficiently, we need to exploit the specific block-sparse and low-rank structures generated in the Multiple Shooting and Newton-Picard approaches. To this end, we first regroup the n_1 PDE variables and n_2 non-PDE variables of NLP (3) in the order

$$(\boldsymbol{u}^0, \ldots, \boldsymbol{u}^{n_{\mathrm{MS}}} \mid \boldsymbol{v}^0, \ldots, \boldsymbol{v}^{n_{\mathrm{MS}}}, \boldsymbol{q}^0, \ldots, \boldsymbol{q}^{n_{\mathrm{MS}}}) = (x_1 \mid x_2) \in \mathbb{R}^{n_1+n_2}.$$

Then we observe that we can also divide the constraints of NLP (3) to obtain the structured NLP form

$$\underset{(x_1,x_2)\in\mathbb{R}^{n_1+n_2}}{\text{minimize}} \quad f(x_1,x_2) \tag{10a}$$

$$\text{s.t.} \quad g_i(x_1,x_2) = 0, \quad i \in \mathcal{E}_1, \tag{10b}$$

$$g_i(x_1,x_2) = 0, \quad i \in \mathcal{E}_2, \tag{10c}$$

$$g_i(x_1,x_2) \geq 0, \quad i \in \mathcal{I}, \tag{10d}$$

where $|\mathcal{E}_1| = n_1$. The constraints (10b) comprise the PDE boundary constraints (3b) and PDE matching conditions (3c).

5.1 Multiple Shooting Structure Exploitation

The QP constraint matrix C has the form

$$C = \begin{pmatrix} C_{11} & C_{12} \\ C_{21} & C_{22} \\ C_{31} & C_{32} \end{pmatrix}$$

$$= \left(\begin{array}{cccc|cccccccc}
-\mathbb{I}_{n_u} & & & R^{\text{b}}_{uu} & & & & R^{\text{b}}_{uv} & & & & \\
G^1_u & -\mathbb{I}_{n_u} & & & G^1_v & & & & G^1_q & & & \\
 & \ddots & \ddots & & & \ddots & & & & \ddots & & \\
 & & G^{n_{\text{MS}}}_u & -\mathbb{I}_{n_u} & & & G^{n_{\text{MS}}}_v & & & & G^{n_{\text{MS}}}_q & \\
\hline
 & & & R^{\text{b}}_{vu} & -\mathbb{I}_{n_v} & & & R^{\text{b}}_{vv} & & & & \\
H^1_u & & & & H^1_v & -\mathbb{I}_{n_v} & & & H^1_q & & & \\
 & \ddots & & & & \ddots & \ddots & & & \ddots & & \\
 & & H^{n_{\text{MS}}}_u & & & & H^{n_{\text{MS}}}_v & -\mathbb{I}_{n_v} & & & H^{n_{\text{MS}}}_q & \\
 & & & & & & & & & & \mathbb{I}_{n_q} & -\mathbb{I}_{n_q} \\
\hline
 & & & & R^{\text{i},1}_v & & & & R^{\text{i},1}_q & & & \\
 & & & & & \ddots & & & & \ddots & & \\
 & & & & & & R^{\text{i},n_{\text{MS}}}_v & & & & R^{\text{i},n_{\text{MS}}}_q & \\
 & & & & & & & R^{\text{e},n_{\text{MS}}}_v & & & &
\end{array}\right),$$

with the derivative abbreviations

$$
\begin{aligned}
R^{\mathrm{b}}_{uu} &= \frac{\partial \boldsymbol{r}^{\mathrm{b}}_{u}}{\partial \boldsymbol{u}^{n_{\mathrm{MS}}}}, & R^{\mathrm{b}}_{uv} &= \frac{\partial \boldsymbol{r}^{\mathrm{b}}_{u}}{\partial \boldsymbol{v}^{n_{\mathrm{MS}}}}, & R^{\mathrm{b}}_{vu} &= \frac{\partial \boldsymbol{r}^{\mathrm{b}}_{v}}{\partial \boldsymbol{u}^{n_{\mathrm{MS}}}}, & R^{\mathrm{b}}_{vv} &= \frac{\partial \boldsymbol{r}^{\mathrm{b}}_{v}}{\partial \boldsymbol{v}^{n_{\mathrm{MS}}}}, & R^{\mathrm{e}} &= \frac{r^{\mathrm{e}}}{\partial \boldsymbol{v}^{n_{\mathrm{MS}}}}, \\
G^{i}_{q} &= \frac{\partial \overline{\boldsymbol{u}}^{i}}{\partial \boldsymbol{q}^{i-1}}, & G^{i}_{u} &= \frac{\partial \overline{\boldsymbol{u}}^{i}}{\partial \boldsymbol{u}^{i-1}}, & G^{i}_{v} &= \frac{\partial \overline{\boldsymbol{u}}^{i}}{\partial \boldsymbol{v}^{i-1}}, & R^{\mathrm{i},i}_{q} &= \frac{\partial \boldsymbol{r}^{\mathrm{i}}}{\partial \boldsymbol{q}^{i-1}}, \\
H^{i}_{q} &= \frac{\partial \overline{\boldsymbol{v}}^{i}}{\partial \boldsymbol{q}^{i-1}}, & H^{i}_{u} &= \frac{\partial \overline{\boldsymbol{v}}^{i}}{\partial \boldsymbol{u}^{i-1}}, & H^{i}_{v} &= \frac{\partial \overline{\boldsymbol{v}}^{i}}{\partial \boldsymbol{v}^{i-1}}, & R^{\mathrm{i},i}_{v} &= \frac{\partial \boldsymbol{r}^{\mathrm{i}}}{\partial \boldsymbol{v}^{i-1}},
\end{aligned}
$$

at the current iterate z^k. We want to stress that contrary to the appearance the block C_{11} is several orders of magnitude larger than the blocks C_{22} and C_{32} on fine spatial discretization levels.

Let $M_B = \mathbb{I}_{n_u} - (\prod_{i=1}^{n_{\mathrm{MS}}} G^i_u) R^{\mathrm{b}}_{uu}$. It is easy to see that if M_B is invertible, then C_{11} is also invertible and its inverse is given by $C_{11}^{-1} =$

$$
\begin{pmatrix}
-\mathbb{I} & & & -(\prod_{i=1}^{0} G^i_u) R^{\mathrm{b}}_{uu} \\
& \ddots & & \vdots \\
& & -\mathbb{I} & -(\prod_{i=1}^{n_{\mathrm{MS}}-1} G^i_u) R^{\mathrm{b}}_{uu} \\
& & & -\mathbb{I}
\end{pmatrix}
\begin{pmatrix}
\mathbb{I} & & & \\
& \ddots & & \\
& & \mathbb{I} & \\
& & & M_B^{-1}
\end{pmatrix}
\begin{pmatrix}
\mathbb{I} & & & \\
\prod_{i=1}^{1} G^i_u & \mathbb{I} & & \\
\vdots & \ddots & \ddots & \\
\prod_{i=1}^{n_{\mathrm{MS}}} G^i_u & \cdots & \prod_{i=n_{\mathrm{MS}}}^{n_{\mathrm{MS}}} G^i_u & \mathbb{I}
\end{pmatrix}.
$$

This decomposition into block-triangular matrices yields a procedural recipe for evaluating matrix-vector products with C_{11}^{-1} that does not require to explicitly form either C_{11} or any part of C_{11}^{-1}, except for the M_B^{-1} block. In the following, we shall see that a two-grid Newton-Picard approximation of C_{11} also enables us to use a simple procedural form for matrix-vector products with the approximation of M_B^{-1}.

5.2 Two-Grid Newton-Picard Structure Exploitation

To this end, we need to establish grid transfer operators between a coarse and a fine grid. For two grid levels $\ell_{\mathrm{c}} \leq \ell_{\mathrm{f}}$, we denote the PDE-state degrees of freedom by $n_{\mathrm{c}} = n_u^{\ell_{\mathrm{c}}}$ and $n_{\mathrm{f}} = n_u^{\ell_{\mathrm{f}}}$. Typically, $n_{\mathrm{c}} \approx 100$, while $n_{\mathrm{f}} > 10{,}000$. We then assume that the prolongation operator $P \in \mathbb{R}^{n_{\mathrm{f}} \times n_{\mathrm{c}}}$ and the restriction operator $R \in \mathbb{R}^{n_{\mathrm{c}} \times n_{\mathrm{f}}}$ satisfy

$$RP = \mathbb{I}_{n_{\mathrm{c}}}. \tag{11}$$

On nested grids, it is rather simple to construct P via interpolation. The restriction of $v_{\mathrm{f}} \in \mathbb{R}^{n_{\mathrm{f}}}$ can then be computed as the unique coarse grid vector $v_{\mathrm{c}} \in \mathbb{R}^{n_{\mathrm{c}}}$ whose prolongation minimizes the L_2-distance to v_{f} on the fine grid. In case of a variational spatial discretization, we can use the coarse and fine grid mass matrices $M_{\mathrm{c}} \in \mathbb{R}^{n_{\mathrm{c}} \times n_{\mathrm{c}}}$ and $M_{\mathrm{f}} \in \mathbb{R}^{n_{\mathrm{f}} \times n_{\mathrm{f}}}$ in order to obtain the form

$$R = M_{\mathrm{c}}^{-1} P^{\mathrm{T}} M_{\mathrm{f}},$$

which satisfies (11) due to $P^{\mathrm{T}} M_{\mathrm{f}} P = M_{\mathrm{c}}$. Again, the matrix R should not be formed explicitly if M_{c}^{-1} is a dense matrix.

We then approximate the blocks of C via a two-grid approach. To this end, we denote all matrices on the coarse grid with hats ($\hat{\cdot}$) and approximate C by a matrix $\tilde{C}$ of the same block pattern, but consisting of the approximated blocks

$$\tilde{R}^{\mathrm{b}}_{uu} = P\hat{R}^{\mathrm{b}}_{uu}R, \qquad \tilde{R}^{\mathrm{b}}_{uv} = P\hat{R}^{\mathrm{b}}_{uv}, \qquad \tilde{R}^{\mathrm{b}}_{vu} = \hat{R}^{\mathrm{b}}_{vu}R, \qquad \tilde{R}^{\mathrm{b}}_{vv} = \hat{R}^{\mathrm{b}}_{vv}, \tag{12a}$$

$$\tilde{G}^{i}_{q} = P\hat{G}^{i}_{q}, \qquad \tilde{G}^{i}_{u} = P\hat{G}^{i}_{u}R, \qquad \tilde{G}^{i}_{v} = P\hat{G}^{i}_{v}, \tag{12b}$$

$$\tilde{H}^{i}_{q} = \hat{H}^{i}_{q}, \qquad \tilde{H}^{i}_{u} = \hat{H}^{i}_{u}R, \qquad \tilde{H}^{i}_{v} = \hat{H}^{i}_{v}, \tag{12c}$$

$$\tilde{R}^{\mathrm{i},i}_{q} = \hat{R}^{\mathrm{i},i}_{q}, \qquad \tilde{R}^{\mathrm{i},i}_{v} = \hat{R}^{\mathrm{i},i}_{v}, \qquad \tilde{R}^{\mathrm{e}} = \hat{R}^{\mathrm{e}}. \tag{12d}$$

Along the same lines, we use the approximations

$$\hat{G}_B := \left(\prod_{i=1}^{n_{\mathrm{MS}}} \hat{G}^{i}_{u}\right)\hat{R}^{\mathrm{b}}_{uu}, \qquad \hat{M}_B := \mathbb{I}_{n_{\mathrm{c}}} - \hat{G}_B,$$

$$\tilde{G}_B := \left(\prod_{i=1}^{n_{\mathrm{MS}}} \tilde{G}^{i}_{u}\right)\tilde{R}^{\mathrm{b}}_{uu}, \qquad \tilde{M}_B := \mathbb{I}_{n_{\mathrm{f}}} - \tilde{G}_B.$$

We can then observe that if $\hat{M}_B$ is invertible, then $\tilde{M}_B$ is also invertible and we obtain the procedural recipe [17, Lemma 7.2]

$$\tilde{M}_B^{-1} = \left(\mathbb{I}_{n_{\mathrm{f}}} - P\hat{G}_B R\right)^{-1} = \mathbb{I}_{n_{\mathrm{f}}} - P\left(\mathbb{I}_{n_{\mathrm{c}}} - \hat{M}_B^{-1}\right)R.$$

Thus, we do not need to explicitly form any matrix but $\hat{M}_B$ on the coarse grid in order to evaluate matrix-vector products with $\tilde{M}_B$ and thus also with $\tilde{C}_{11}^{-1}$. Furthermore, this form of $\tilde{M}_B$ allows for an even more concise representation of $\tilde{C}_{11}^{-1}$. To this end, we define the projectors

$$\Pi^{\mathrm{slow}} = \mathbb{I}_{n_{\mathrm{MS}}} \otimes (PR), \qquad \Pi^{\mathrm{fast}} = \mathbb{I}_{n_{\mathrm{MS}} n_{\mathrm{f}}} - \Pi^{\mathrm{slow}},$$

where $\otimes$ denotes the Kronecker product of matrices. Thus Π^{slow} is a block-diagonal matrix of n_{MS} blocks PR. Then, we can show [17, Theorem 7.3] that

$$\tilde{C}_{11}^{-1}\Pi^{\mathrm{slow}} = (\mathbb{I}_{n_{\mathrm{MS}}} \otimes P)\hat{C}_{11}^{-1}(\mathbb{I}_{n_{\mathrm{MS}}} \otimes R), \qquad \tilde{C}_{11}^{-1}\Pi^{\mathrm{fast}} = -\Pi^{\mathrm{fast}}.$$

This equation illustrates the name Newton-Picard: On the slowly converging modes, we approximate the inverse of C_{11} by its coarse grid counterpart and perform a Newton iteration on this subspace, while we only use a Picard (or fixed point) iteration on the anyway fast converging modes through approximating the Jacobian with a negative identity. Moreover, we can now easily derive [17, Corollary 7.4] that if it exists, the Newton-Picard approximation of block C_{11} has the inverse

$$\tilde{C}_{11}^{-1} = (\mathbb{I}_{n_{\mathrm{MS}}} \otimes P)\left(\hat{C}_{11}^{-1} + \mathbb{I}_{n_{\mathrm{MS}}n_{\mathrm{c}}}\right)(\mathbb{I}_{n_{\mathrm{MS}}} \otimes R) - \mathbb{I}_{n_{\mathrm{MS}}n_{\mathrm{f}}}.$$

5.3 QP Condensing

We now consider again QPs with a structure inherited from NLP (10)

$$\underset{(x_1,x_2)\in\mathbb{R}^{n_1+n_2}}{\text{minimize}} \quad \frac{1}{2}\begin{pmatrix}x_1\\x_2\end{pmatrix}^{\mathrm{T}}\begin{pmatrix}B_{11} & B_{12}\\B_{21} & B_{22}\end{pmatrix}\begin{pmatrix}x_1\\x_2\end{pmatrix} + \begin{pmatrix}b_1\\b_2\end{pmatrix}^{\mathrm{T}}\begin{pmatrix}x_1\\x_2\end{pmatrix} \tag{13a}$$

$$\text{s.t.} \quad C_{11}x_1 + C_{12}x_2 = c_1, \tag{13b}$$

$$C_{21}x_1 + C_{22}x_2 = c_2, \tag{13c}$$

$$C_{31}x_1 + C_{32}x_2 \geq c_3. \tag{13d}$$

We show that we can employ a partial null-space approach called *condensing* in order to efficiently solve the large-scale, structured QP (13b). Condensing is the key linear algebra ingredient for an efficient Direct Multiple Shooting method (see, e.g., [6]), but must be extended to also exploit the two-grid Newton-Picard structures. For this purpose, we prefer the following presentation of the concept of condensing.

Theorem 1 (Condensing) *Assume that C_{11} in QP (13) is invertible and define*

$$Z = \begin{pmatrix}-C_{11}^{-1}C_{12}\\ \mathbb{I}_{n_2}\end{pmatrix}, \qquad B' = Z^{\mathrm{T}}BZ,$$

$$c_1' = C_{11}^{-1}c_1, \qquad b' = B_{21}c_1' + b_2 - C_{12}^{\mathrm{T}}C_{11}^{-\mathrm{T}}(B_{11}c_1' + b_1),$$

$$c_2' = c_2 - C_{21}c_1', \qquad C_2' = C_{22} - C_{21}C_{11}^{-1}C_{12},$$

$$c_3' = c_3 - C_{31}c_1', \qquad C_3' = C_{32} - C_{31}C_{11}^{-1}C_{12}.$$

Let furthermore $(x_2^*, y_{\mathcal{E}_2}^*, y_{\mathcal{I}}^*) \in \mathbb{R}^{n_2+m_2+m_3}$ be a primal-dual solution of the QP

$$\underset{x_2\in\mathbb{R}^{n_2}}{\text{minimize}} \quad \frac{1}{2}x_2^{\mathrm{T}}B'x_2 + b'^{\mathrm{T}}x_2 \quad \text{s.t.} \quad C_2'x_2 = c_2', \quad C_3'x_2 \geq c_3'. \tag{14}$$

If we choose

$$x_1^* = C_{11}^{-1}(c_1 - C_{12}x_2^*), \tag{15a}$$

$$y_{\mathcal{E}_1}^* = C_{11}^{-\mathrm{T}}\left((B_{12} - B_{11}C_{11}^{-1}C_{12})x_2^* + B_{11}c_1' + b_1 - C_{21}^{\mathrm{T}}y_{\mathcal{E}_2}^* - C_{31}^{\mathrm{T}}y_{\mathcal{I}}^*\right) \tag{15b}$$

then $(x^*, y^*) := (x_1^*, x_2^*, y_{\mathcal{E}_1}^*, y_{\mathcal{E}_2}^*, y_{\mathcal{I}}^*)$ is a primal-dual solution of QP (13).

Proof See, e.g., [17, Theorem 7.6].

Theorem 1 shows how the solution of the structured, large-scale QP (13) can be boiled down to a medium-scale QP (14), which only contains non-PDE variables x_2. The pre- and post-processing steps only require matrix-vector products with the large-scale blocks, e.g., n_2 matrix-vector products for the computation of $C_{11}^{-1}C_{12}$.

5.4 Two-Grid Hessian Approximation

If we now use the approximated version of QP (13), we see that based on Sects. 5.1 and 5.2, the partial null-space basis can be evaluated purely on the coarse grid due to

$$\tilde{Z} = \begin{pmatrix} -\tilde{C}_{11}^{-1}\tilde{C}_{12} \\ \mathbb{I} \end{pmatrix} = \begin{pmatrix} -(\mathbb{I}\otimes P)\hat{C}_{11}^{-1}(\mathbb{I}\otimes R)(\mathbb{I}\otimes P)\hat{C}_{12} \\ \mathbb{I} \end{pmatrix} = \begin{pmatrix} -(\mathbb{I}\otimes P)\hat{C}_{11}^{-1}\hat{C}_{12} \\ \mathbb{I} \end{pmatrix}.$$

This observation suggests a projected Newton-Picard approximation of the Hessian matrix via

$$\begin{aligned}
\tilde{B}^{\text{fast}} &= \begin{pmatrix} (\Pi^{\text{fast}})^{\mathrm{T}} B_{11} \Pi^{\text{fast}} & 0 \\ 0 & 0 \end{pmatrix}, \\
\tilde{B}^{\text{slow}} &= \begin{pmatrix} (\mathbb{I}_{n_{\text{MS}}}\otimes R)^{\mathrm{T}}\hat{B}_{11}(\mathbb{I}_{n_{\text{MS}}}\otimes R) & (\mathbb{I}_{n_{\text{MS}}}\otimes R)^{\mathrm{T}}\hat{B}_{12} \\ \hat{B}_{21}(\mathbb{I}_{n_{\text{MS}}}\otimes R) & \hat{B}_{22} \end{pmatrix}, \\
\tilde{B} &= \tilde{B}^{\text{fast}} + \tilde{B}^{\text{slow}}.
\end{aligned}$$

Consequently, we have

$$\tilde{Z}^{\mathrm{T}}\tilde{B}^{\text{fast}}\tilde{Z} = 0$$

and thus we can also compute the condensed two-grid Newton-Picard Hessian matrix purely on the coarse grid according to

$$\tilde{B}' = \tilde{Z}^{\mathrm{T}}\tilde{B}\tilde{Z} = \hat{Z}^{\mathrm{T}}\hat{B}\hat{Z} \quad \text{with } \hat{Z} = \begin{pmatrix} -\hat{C}_{11}^{-1}\hat{C}_{12} \\ \mathbb{I}_{n_c} \end{pmatrix}.$$

Thus, it is possible to set up QP (14) efficiently with only a few fine-grid operations.

6 Numerical Examples

We now report on the numerical results for Direct Multiple Shooting applied to tracking problems for an advection-diffusion equation in 2D and a bacterial chemotaxis example in 1D. Further examples can be found in [17], including a real-world separation process from chemical engineering with economic objective and constraint functions. All computational results were computed on a Linux workstation with four 2.67 GHz Intel Core i7 cores and 24 GB of RAM.

6.1 Advection-Diffusion Equation

We consider an advection-diffusion equation on $\Omega \subset (-1,1)^2$, whose boundary $\partial\Omega$ is partitioned into disjoint sets $\Gamma_i, i = 0,1,2,3$. Let ν denote the outwards pointing normal on $\partial\Omega$ and let $\boldsymbol{U} \in L^2(\Omega)^2$ be a divergence-free velocity field. For $\gamma, \varepsilon, T > 0$, inflow profiles $u_i^{\text{in}} \in L^2(\Gamma_i), i = 1,2$, and a desired profile $\hat{u} \in L^2(\Omega)$, we apply the proposed method to the following periodic tracking-type boundary-control problem with control gradient constraints

$$\underset{\substack{u \in W(0,T)\\ q \in H^1(0,T)^2}}{\text{minimize}} \quad \frac{1}{2}\|u(T,.) - \hat{u}\|^2_{L^2(\Omega)} + \frac{\gamma}{2}\left(\|q_1\|^2_{H^1(0,T)} + \|q_2\|^2_{H^1(0,T)}\right) \tag{16a}$$

$$\text{s.t.} \quad \partial_t u = \varepsilon \nabla \cdot \nabla u - \boldsymbol{U} \cdot \nabla u, \qquad \text{in } \Omega, \tag{16b}$$

$$u(0,.) = u(T,.), \qquad \text{in } \Omega, \tag{16c}$$

$$\nu \cdot (\varepsilon \nabla u - \boldsymbol{U} u) = 0, \qquad \text{on } \Gamma_0, \tag{16d}$$

$$\nu \cdot (\varepsilon \nabla u - \boldsymbol{U} u) = u_i^{\text{in}} q_i - u, \qquad \text{on } \Gamma_i, i = 1,2, \tag{16e}$$

$$\nu \cdot (\varepsilon \nabla u - \boldsymbol{U} u) = -u, \qquad \text{on } \Gamma_3, \tag{16f}$$

$$q_i(0) = q_i(T), \qquad \text{for } i = 1,2, \tag{16g}$$

$$q_i(t) \geq 0, \qquad \text{for } t \in [0,T], i = 1,2, \tag{16h}$$

$$\frac{\mathrm{d}q_i}{\mathrm{d}t}(t) \in [-20, 20], \qquad \text{for a.a. } t \in [0,T], i = 1,2. \tag{16i}$$

The setting $q_i \in H^1(0,T)$ can be reformulated with $L^2(0,T)$ controls by introduction of ODE states v_i, whose time-derivative we control, and adequate adaption of all occurrences of q_i in (16).

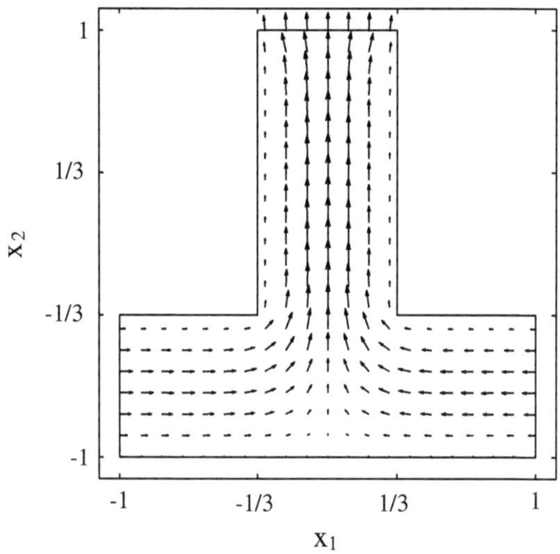

Fig. 1 Divergence-free velocity field obtained from the solution of a Stokes problem on an inverted-T-shaped domain with inflow from the *left* on Γ_1, inflow from the *right* on Γ_2 and outflow on *top* on Γ_3

We obtain a suitable velocity field $\boldsymbol{U}$ as the solution of a Stokes problem with parabolic inflow velocities on $\Gamma_1 = \{-1\} \times (-1, -1/3)$ and $\Gamma_2 = \{1\} \times (-1, -1/3)$ and free outflow velocity on $\Gamma_3 = (-1/3, 1/3) \times \{1\}$ (compare Fig. 1)

$$
\begin{aligned}
-\nabla \cdot \nabla \boldsymbol{U} + \nabla p &= 0, && \text{in } \Omega, \\
\nabla \cdot \boldsymbol{U} &= 0, && \text{in } \Omega, \\
\boldsymbol{U}(x) &= 0, && \text{for } x \in \Gamma_0, \\
\boldsymbol{U}(x) &= (9(x_2 + 1/3)(x_2 + 1), 0), && \text{for } x \in \Gamma_1, \\
\boldsymbol{U}(x) &= (-9(x_2 + 1/3)(x_2 + 1), 0), && \text{for } x \in \Gamma_2.
\end{aligned}
$$

We use Finite Differences on equidistant, staggered grids for the discretization of the Stokes problem. This discretization is compatible with an upwind-flux Finite Volume method for the advection-diffusion equation, because the velocity degrees of freedom align with the center of the volume interfaces.

We compute optimal controls for $T = 10$, $\hat{u}(x) = 1 + x_1$, $u_i^{\text{in}}(x) = \exp(-10(x_2 + 2/3)^2))$, $i = 1, 2$, and for varying values of $\varepsilon = 10^0, 10^{-1}, 10^{-2}, 5 \cdot 10^{-3}$ and $\gamma = 10^{-2}, 10^{-3}$ on $n_{\text{MS}} = 20$ shooting intervals. We use a 5-level hierarchy of grids with mesh sizes $h = 2/15, 1/15, 1/30, 1/60, 1/120$, giving rise to 125, 500, 2000, 8000, and 32,000 degrees of freedom on the respective level. As initial values, we chose $\boldsymbol{u}^i = 0, \boldsymbol{v}^i = (0.1, 0.1)^{\text{T}}, \boldsymbol{q}^i = (0.01, 0.01)^{\text{T}}$ for $i = 0, \ldots, 20$.

One purpose of treating an advection-diffusion problem here is to fathom the limits of the Newton-Picard approach, which exploits dispersion numerically. We can expect the problem to become more difficult to solve as we approach the hyperbolic limit $\varepsilon \to 0$. Because the upwind discretization achieves numerical stability by introduction of spurious numerical dispersion, the discretized problem

Table 1 The number of inexact SQP iterations and the computation times depend on the diffusion coefficient ε and the regularization parameter γ

	$\gamma = 10^{-1}$				$\gamma = 10^{-3}$			
$\varepsilon =$	10^0	10^{-1}	10^{-2}	$5 \cdot 10^{-3}$	10^0	10^{-1}	10^{-2}	$5 \cdot 10^{-3}$
Overall iterations	12	15	19	16	12	21	17	28
Iterations $l_f = 5$	5	7	7	7	5	6	6	8
Final coarse level	1	1	2	2	1	2	2	2
Time steps	657	760	884	903	655	741	859	935
Sim. time (min:s)	1:48	2:09	2:10	2:13	1:48	1:54	2:17	2:12
Time (h:min:s)	0:49:04	1:21:53	2:34:15	3:30:21	0:48:58	1:23:36	2:49:57	3:22:31

The overall computation time is given as the wall-clock time on four processors, while the simulation time is the CPU time on a single processor

exhibits more diffusion on coarser grid levels where the effect of numerical dispersion is more pronounced. Thus, the efficiency of the diffusion-exploiting Newton-Picard preconditioners is slightly better for coarser fine grids. We also want to remark that problem (16) is a linear-quadratic problem. We tackle it with a nonlinear solver and thus the reader should keep in mind that numerical linear algebra approaches (e.g., along the lines of Potschka et al. [20]) are more efficient, because they exploit linearity explicitly.

From the computational results in Table 1, we can observe the expected trend in the number of overall inexact SQP iterations that is growing for decreasing values of the diffusion coefficient ε. The increase is higher for lower regularization parameters γ. The number of inexact SQP iterations with the fine grid on the finest level grows only slightly. For high diffusion $\varepsilon = 10^0$, level $\ell_c = 1$ for the coarse grid is already enough for sufficient local contraction on all fine grid levels $\ell_f \leq 5$. For $\varepsilon = 10^{-1}$, $\ell_c = 1$ is still sufficiently fine for a high regularization parameter $\gamma = 10^{-1}$, but not for $\gamma = 10^{-3}$. For $\varepsilon = 10^{-2}, 5 \cdot 10^{-3}$, the κ-estimators trigger a refinement of the coarse level to $\ell_c = 2$ in order to obtain local contraction of the method on all fine grid levels $\ell_f \leq 5$. For $\varepsilon = 10^{-3}$, the coarse grid needs to be refined to $\ell_c = 3$. In this case, the memory consumption of DAESOL-II on the coarse and fine grids exceeds the available RAM space. This problem, however, is a momentary technical restriction of the software implementation and not a fundamental restriction of the method.

We can also observe in Table 1 that the number of adaptive time-steps for the initial value problem solver increases with decreasing diffusion coefficient γ. From the number of required time-steps we see that the time discretization has a higher resolution than the spatial discretization. The reason for this imbalance in accuracy are stability requirements for the solution of the initial value problems in (3c) and its first and second order derivatives as discussed in [17, Chap. 9, Sect. 3]. If we take into account the intermediate time steps, we obtain on the finest level between $2.1 \cdot 10^7$ and $3.0 \cdot 10^7$ state variables, although NLP (3) really only has $6.7 \cdot 10^5$ degrees of freedom in the PDE state variables $\boldsymbol{u}^i, i = 0, \ldots, n_{\mathrm{MS}}$, (independently of ε).

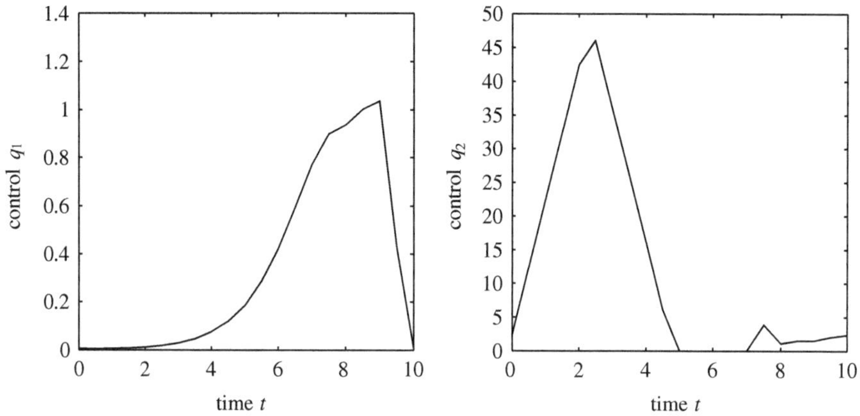

Fig. 2 The optimal controls for problem (16) with $\varepsilon = 10^{-2}, \gamma = 10^{-3}$ work with different magnitudes. The control gradient constraint is active for q_2 at the slopes of the initial peak

If we compare the serial simulation time for the numerical solution of the initial value problems for fixed initial values and fixed controls with the wall-clock time for the inexact SQP method parallelized on four cores, we obtain a ratio of 27–95. This is quite remarkable, because already for the forward problem of finding a time-periodic solution for fixed controls, several Newton-Picard iterations (in the sense of Lust et al. [13]) would be required.

For completeness, we depict the optimal control and the corresponding optimal final state for the case $\varepsilon = 10^{-2}$ and $\gamma = 10^{-3}$ in Figs. 2 and 3. We can observe that most of the control action happens at Γ_2, where high values of u are to be tracked. The control gradient constraint (16i) is active at the slopes of the initial peak of q_2. Furthermore, the optimal state at the end of the period matches the desired profile $\hat{u}$ well except for the area around the inflow boundary Γ_2.

6.2 Bacterial Chemotaxis

On the basis of a chemotaxis model by Tyson et al. [26, 27], we consider on $\Omega = (0, 1)$ with $\partial\Omega = \Gamma_1 \cup \Gamma_2 = \{0\} \cup \{1\}$ the nonlinear tracking-type boundary control problem (compare also [12])

$$\underset{\substack{z,c \in W(0,1) \\ q_1,q_2 \in L^2(0,1)}}{\text{minimize}} \quad \frac{1}{2}\int_\Omega (z(1,\cdot) - \hat{z})^2 + \frac{\gamma_c}{2}\int_\Omega (c(1,\cdot) - \hat{c})^2 + \frac{\gamma_q}{2}\int_0^1 (q_1^2 + q_2^2) \tag{17a}$$

$$\text{s.t.} \quad \partial_t z = \nabla \cdot \left(D_z \nabla z - \alpha \frac{z}{(1+c)^2} \nabla c \right), \qquad \text{in } (0,1) \times \Omega, \tag{17b}$$

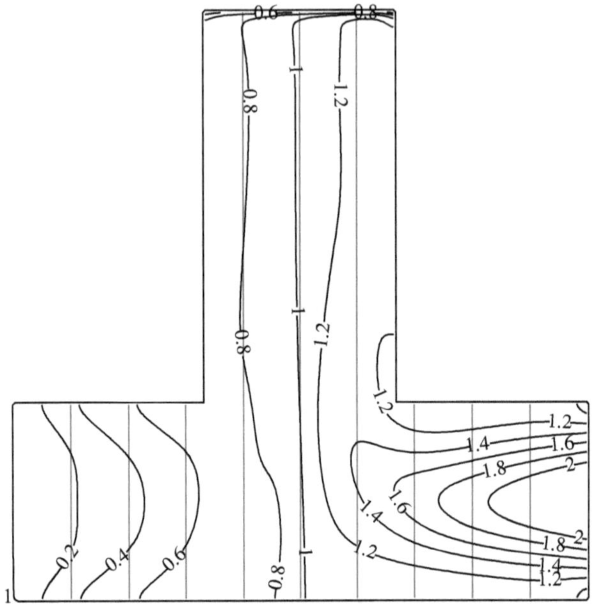

Fig. 3 For $\varepsilon = 10^{-2}, \gamma = 10^{-3}$, the *black isolines* of the optimal final state $u(T, .)$ match the *gray isolines* of the target well except for the south-east part of the domain. A boundary layer is visible at the northern outflow Γ_3

$$\partial_t c = \nabla \cdot \nabla c + w\frac{z^2}{(\mu + z^2)} - \rho c \qquad \text{in } (0,1) \times \Omega, \tag{17c}$$

$$\partial_\nu z = \frac{\alpha\beta(q_i - c)z}{D_z(1+c)^2}, \quad \partial_\nu c = \beta(q_i - c), \quad \text{in } (0,1) \times \Gamma_i, i = 1, 2, \tag{17d}$$

$$z(0, .) = z_0, \quad c(0, .) = c_0, \tag{17e}$$

$$q_i(t) \in [0, 1], \qquad \text{for a.a. } t \in (0, 1), \tag{17f}$$

where ∂_ν denotes the derivative in direction of the outwards pointing normal on Ω. The parameters considered here are $D_z = 0.3, \alpha = 4, \beta = 1, w = 0.1, \mu = 10, \rho = 0.1, \gamma_c = 10^{-2}, \gamma_q = 10^{-5}$. The targets to be tracked are $\hat{z}(x) = 2x, \hat{c}(x) = 0$ and the initial states are given as $z_0(x) = 1, c_0(x) = 0$.

On the basis of a Finite Difference discretization on a six-level hierarchy of nested grids with 322, 642, 1282, 2562, 5122, 10,242 degrees of freedom for the PDE states, we can use the proposed Direct Multiple Shooting method to obtain the optimal controls and states depicted in Fig. 4. Level $\ell_c = 1$ for the coarse level is sufficient to yield good local contraction for the inexact SQP method. In the solution, DAESOL-II uses 889 integration adaptive steps for the solution of the initial value

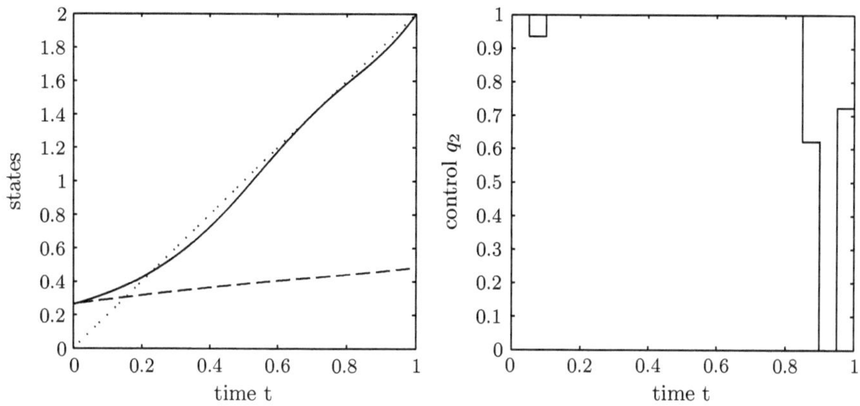

Fig. 4 *Left*: the desired cell distribution $\hat{z}$ (*dotted line*) can be reached well with the optimal final cell distribution $z(1,.)$ (*solid line*). The final chemoattractor concentration $c(1,.)$ is almost affine-linear (*dashed line*). *Right*: the control bound constraints are active most of the time for q_2. Not shown: the optimal control q_1 is always 0

Table 2 Cumulative CPU time in seconds for the chemotaxis example on each spatial discretization level for computations concerning the dynamic system, including system integration, forward and adjoint derivatives, and matrix-vector products with Hessian matrices

Level	Integration	Forward	Adjoint	Hessian
1	100.5	745.5	19.8	669.1
2	29.5	6.6	16.2	55.8
3	45.4	8.2	22.0	71.7
4	53.4	7.5	21.2	70.5
5	144.2	15.2	42.6	141.4
6	1314.9	73.5	190.3	597.9

problems on all 20 shooting intervals together. In this case, the spatial and temporal discretization errors are well balanced, because a high-order method is used in time.

In Table 2 we sum up the cumulative CPU time needed for the computation of values concerning the initial value problems on each spatial discretization level. This is the main part of the computational burden. The solution of all the condensed QPs (14) only takes 0.6 s, for instance. We observe that most of the effort is spent on levels 1 and 6. The effort on level 1 is due to the coarse grid computations in each iteration. Due to DAESOL-II's memory requirements, we could not refine the fine grid further. Were we able to do so, then the numerical effort spent on the finest grid level $\ell = 7$ would dominate the overall effort even more.

7 Conclusions

We elaborated on the challenges and a numerical approach for a Direct Multiple Shooting method for optimal control problems with coupled ODE and parabolic PDE constraints. Even difficult boundary conditions like time periodicity of the states can be treated efficiently in this framework. The large-scale nature of the resulting NLPs can be tamed by a structure exploiting two-grid Newton-Picard inexact SQP method. Encouraging numerical results indicate that challenging PDE constrained optimal control problems can be solved efficiently with a Direct Multiple Shooting method. The proposed method does not require the often time-consuming derivation of adjoint equations and can be easily parallelized. It is in particular suited for practitioners who need an accurate resolution of the system dynamics and can live with lower resolution of the optimal control.

We would like to thank the unknown reviewer for the constructive comments on the initial manuscript. Support by the EU through S. Engell's and H.G. Bock's ERC Advanced Investigator Grant MOBOCON (291 458) is gratefully acknowledged.

References

1. Albersmeyer, J.: Adjoint based algorithms and numerical methods for sensitivity generation and optimization of large scale dynamic systems. Ph.D. thesis, Ruprecht-Karls-Universität Heidelberg (2010)
2. Albersmeyer, J., Bock, H.G.: Sensitivity generation in an adaptive BDF-method. In: Bock, H.G., Kostina, E., Phu, X.H., Rannacher, R. (eds.) Modeling, Simulation and Optimization of Complex Processes. Proceedings of the International Conference on High Performance Scientific Computing, pp. 15–24, Hanoi, 6–10 March 2006. Springer, Berlin (2008)
3. Bellman, R.E.: Dynamic Programming, 6th edn. University Press, Princeton (1957)
4. Biegler, L.T.: Solution of dynamic optimization problems by successive quadratic programming and orthogonal collocation. Comput. Chem. Eng. **8**, 243–248 (1984)
5. Bock, H.G.: Recent advances in parameter identification techniques for ODE. In: Deuflhard, P., Hairer, E. (eds.) Numerical Treatment of Inverse Problems in Differential and Integral Equations, pp. 95–121. Birkhäuser, Boston (1983)
6. Bock, H.G., Plitt, K.J.: A multiple shooting algorithm for direct solution of optimal control problems. In: Proceedings of the 9th IFAC World Congress, pp. 242–247, Budapest. Pergamon Press (1984)
7. Dautray, R., Lions, J.-L.: Evolution problems I. In: Craig, A. (ed.) Mathematical Analysis and Numerical Methods for Science and Technology, vol. 5. Springer, Berlin (1992)
8. Deuflhard, P.: Newton Methods for Nonlinear Problems. Affine Invariance and Adaptive Algorithms. Springer Series in Computational Mathematics, vol. 35. Springer, Berlin (2006)
9. Griewank, A., Walther, A.: Evaluating Derivatives, 2nd edn. SIAM, Philadelphia (2008)
10. Hesse, H.K.: Multiple shooting and mesh adaptation for PDE constrained optimization problems. Ph.D. thesis, University of Heidelberg (2008)
11. Hinze, M., Pinnau, R., Ulbrich, M., Ulbrich, S.: Optimization with PDE Constraints. Springer, New York (2009)
12. Lebiedz, D., Brandt-Pollmann, U.: Manipulation of self-aggregation patterns and waves in a reaction-diffusion system by optimal boundary control strategies. Phys. Rev. Lett. **91**(20), 208301 (2003)

13. Lust, K., Roose, D., Spence, A., Champneys, A.R.: An adaptive Newton-Picard algorithm with subspace iteration for computing periodic solutions. SIAM J. Sci. Comput. **19**(4), 1188–1209 (1998)
14. Nocedal, J., Wright, S.J.: Numerical Optimization, 2nd edn. Springer, Berlin (2006)
15. Osborne, M.R.: On shooting methods for boundary value problems. J. Math. Anal. Appl. **27**, 417–433 (1969)
16. Pontryagin, L.S., Boltyanski, V.G., Gamkrelidze, R.V., Miscenko, E.F.: The Mathematical Theory of Optimal Processes. Wiley, Chichester (1962)
17. Potschka, A.: A direct method for the numerical solution of optimization problems with time-periodic PDE constraints. Ph.D. thesis, Universität Heidelberg (2011)
18. Potschka, A.: A Direct Method for Parabolic PDE Constrained Optimization Problems. Advances in Numerical Mathematics. Springer, Berlin (2013)
19. Potschka, A., Bock, H.G., Schlöder, J.P.: A minima tracking variant of semi-infinite programming for the treatment of path constraints within direct solution of optimal control problems. Optim. Methods Softw. **24**(2), 237–252 (2009)
20. Potschka, A., Mommer, M.S., Schlöder, J.P., Bock, H.G.: Newton-Picard-based preconditioning for linear-quadratic optimization problems with time-periodic parabolic PDE constraints. SIAM J. Sci. Comput. **34**(2), 1214–1239 (2012)
21. Russell, R.D., Shampine, L.F.: A collocation method for boundary value problems. Numer. Math. **19**, 1–28 (1972)
22. Saad, Y.: Iterative Methods for Sparse Linear Systems, 2nd edn. SIAM, Philadelpha (2003)
23. Thomée, V.: Galerkin Finite Element Methods for Parabolic Problems. Springer Series in Computational Mathematics, vol. 25, 2nd edn. Springer, Berlin (2006)
24. Tröltzsch, F.: Optimale Steuerung partieller Differentialgleichungen: Theorie, Verfahren und Anwendungen, 2nd edn. Vieweg+Teubner Verlag, Wiesbaden (2009)
25. Tsang, T.H., Himmelblau, D.M., Edgar, T.F.: Optimal control via collocation and non-linear programming. Int. J. Control. **21**, 763–768 (1975)
26. Tyson, R., Lubkin, S.R., Murray, J.D.: Model and analysis of chemotactic bacterial patterns in a liquid medium. J. Biol. **38**, 359–375 (1999). doi:10.1007/s002850050153
27. Tyson, R., Lubkin, S.R., Murray, J.D.: A minimal mechanism for bacterial pattern formation. Proc. R. Soc. B Biol. Sci. **266**, 299–304 (1999)
28. Walther, A., Kowarz, A., Griewank, A.: ADOL-C: a package for the automatic differentiation of algorithms written in C/C++. Technical report, Institute of Scientific Computing, Technical University Dresden (2005)
29. Wloka, J.: Partial Differential Equations. Cambridge University Press, Cambridge (1987). Translated from the German by C.B. Thomas and M.J. Thomas

Multiple Shooting in a Microsecond

Rien Quirynen, Milan Vukov, and Moritz Diehl

Abstract Nonlinear Model Predictive Control (NMPC) is a feedback control technique that uses the most current state estimate of a nonlinear system to compute an optimal plan for the future system behavior. This plan is recomputed at every sampling time, creating feedback. Thus, NMPC needs to repeatedly solve a nonlinear optimal control problem (OCP). Direct multiple shooting is since long known as a reliable approach for discretization of OCPs. This is mainly due to the fact that the approach shows good contraction properties within the NMPC framework. Moreover, the procedure is easy to initialize and parallelize. In the context of real-time NMPC, the multiple shooting method was tailored to the Real-Time Iteration (RTI) scheme. This scheme uses a strategy known as Initial Value Embedding to deal efficiently with the transition from one optimization problem to the next. It performs two algorithmic steps in each sampling time, a long preparation phase and a short feedback phase to minimize the feedback time to the system. The two phases respectively prepare and solve a convex Quadratic Program (QP) that depends parametrically on the estimated system state. The solution of this QP delivers quickly a generalized tangential predictor to the solution of the nonlinear problem. Recent algorithmic progress makes the solution of NMPC optimization problems possible at sampling times in the milli- or even microsecond range on modern computational hardware. An essential part is the simulation of the nonlinear model together with the propagation of its derivative information. This article describes the developments and their efficient software implementations that made it possible to solve a classical NMPC benchmark problem within 1 μs sampling time.

R. Quirynen (✉) • M. Vukov
KU Leuven, Kasteelpark Arenberg 10, 3001 Leuven, Belgium
e-mail: rien.quirynen@esat.kuleuven.be; milan.vukov@esat.kuleuven.be

M. Diehl
KU Leuven, Kasteelpark Arenberg 10, 3001 Leuven, Belgium

University of Freiburg, Georges-Koehler-Allee 102, 79110 Freiburg, Germany
e-mail: moritz.diehl@imtek.uni-freiburg.de

T. Carraro et al. (eds.), *Multiple Shooting and Time Domain Decomposition Methods*, Contributions in Mathematical and Computational Sciences 9,
DOI 10.1007/978-3-319-23321-5_7

1 Introduction

Model Predictive Control (MPC) needs to solve an Optimal Control Problem (OCP) at each sampling instant using the current system state $\bar{x}_0$ as initial value. This optimization task is almost exclusively executed using a *direct approach* which first discretizes the continuous time system to obtain a discrete time OCP formulation. Multiple Shooting (MS) will be motivated and shown to be such a time discretization with nice contraction properties and extra advantages, e.g. it is more flexible both to initialize and parallelize. One crucial dividing line in MPC is between convex and non-convex problems. In the case that the OCP is convex, e.g. for linear MPC, algorithms exist that find a global solution in a fast and reliable way. This paper will focus on the problems that use a nonlinear model i.e. the resulting OCP is non-convex and one generally has to be satisfied with approximations of locally optimal solutions. Nonlinear MPC (NMPC) has become a popular approach for real-time optimal control since it can explicitly handle constraints and nonlinear dynamics. Recent algorithmic progress [1, 2] allows to consider NMPC also for systems having rather fast dynamics. Among the available online algorithms, the Real-Time Iteration (RTI) scheme has been proposed as a highly competitive approach [3]. It is an SQP-type algorithm that uses a shooting discretization and a Gauss-Newton Hessian approximation.

It is important to use the right algorithmic tools to be able to meet the hard timing constraints of real-time applications. This paper focuses on the RTI scheme as an online algorithm to handle nonlinear OCPs using a multiple shooting discretization. It divides the computations at each sampling time into a preparation and a feedback phase [4]. The *preparation phase* takes care of the linearization and condensing resulting in a small scale Quadratic Program (QP). Reducing the computation time of this phase is crucial for achieving a certain sampling frequency since it will often dominate the total execution time. The prepared QP cannot be solved yet before the current state estimate is available. Once it becomes available, the *feedback phase* will quickly solve the subproblem to obtain an approximate solution to the original OCP. The faster this QP is solved, the faster the next control input can be fed back to the real process. Condensing is still a rather popular technique, because it leaves us with a dense but small scale QP to be solved in the feedback phase [5, 6]. This can be done by any embedded QP solver such as e.g. qpOASES [7] which uses an online active set strategy [8]. The alternative to this condensing approach would be to directly solve the multi-stage QP, using a structure exploiting convex solver such as FORCES [9], qpDUNES [10] or HPMPC [11]. An important disadvantage could be that the solution of the full QP then becomes part of the feedback phase in the context of the RTI framework.

Apart from using suitable algorithms, efficient implementations are needed to run them in real-time on embedded control hardware. One way to achieve this is by automatic code generation, i.e. by exporting a fully customized solver. Significant improvements in the computation time can be obtained by removing unnecessary computations, by optimizing the memory access and cache usage, and by exploiting the structure in the problem. This idea is already rather popular for convex optimization, examples of this are CVXGEN [12] and FORCES [9] which both generate tailored Interior Point (IP) convex solvers. In the context of NMPC, an important computational step is that of the integration and sensitivity generation for the nonlinear model. The export of tailored Explicit Runge-Kutta (ERK) integrators using the Variational Differential Equations (VDE) for sensitivity propagation has been presented and been experimentally validated in [13, 14]. This idea has been strongly extended in the work on automatic code generation for Implicit RK (IRK) methods with a tailored approach for computing their sensitivities [15, 16]. Embedded, implicit solvers allow their natural extension to Differential Algebraic Equations (DAE) and an efficient computation of continuous outputs [17]. The ACADO code generation tool pursues to export efficient C-code for the complete RTI scheme, assembled from the different necessary components [18]. It is part of the open-source ACADO Toolkit [19] with interfaces to multiple convex solvers [5, 10].

This paper is organized as follows. Section 2 presents the parametric optimization problem that needs to be solved at each sampling time together with its shooting discretization. This direct approach requires efficient integration and sensitivity generation for the nonlinear model, which is the topic of interest in Sect. 3. The optimization details are discussed in Sect. 4, focusing on Sequential Quadratic Programming (SQP) in a RTI framework. Finally, Sect. 5 shows us that NMPC can be done within 1 μs for a benchmark problem taken from the literature and this using the tools provided in this paper.

2 Nonlinear Model Predictive Control

NMPC is an approach of increasing popularity for real-time control due to the ability to explicitly handle constraints and nonlinear dynamics that characterize the system of interest. Section 2.1 presents the optimization problem that needs to be solved at each sampling time. Section 2.2 then describes multiple shooting as a reliable way of reformulating this as an approximate but tractable problem.

2.1 Parametric Optimization Problem

In what follows, the OCP that needs to be solved at each time point is assumed to be of the following form

$$\begin{aligned} \underset{x(\cdot),\, u(\cdot)}{\text{minimize}} \quad & \int_0^T \|F(t, x(t), u(t))\|_2^2 \, \mathrm{d}t + \|F_N(x(T))\|_2^2 && \text{(1a)} \\ \text{subject to} \quad & x(0) = \bar{x}_0, && \text{(1b)} \\ & \boldsymbol{\dot{x}(t) = f(t, x(t), u(t))}, \quad \forall t \in [0, T], && \text{(1c)} \\ & 0 \geq h(x(t), u(t)), \quad \forall t \in [0, T], && \text{(1d)} \\ & 0 \geq r(x(T)), && \text{(1e)} \end{aligned}$$

where $x(t) \in \mathbb{R}^{n_x}$ denotes the differential states at time t, $u(t) \in \mathbb{R}^{n_u}$ are the control inputs and Eq. (1a) defines the NMPC objective while Eqs. (1d) and (1e) are respectively the path and terminal constraints. The nonlinear dynamics in Eq. (1c) are described by an explicit system of Ordinary Differential Equations (ODE), although this could be generalized to e.g. an implicit DAE system of index 1. Note that $\bar{x}_0 \in \mathbb{R}^{n_x}$ is a parameter on which the OCP depends through the initial value constraint in Eq. (1b). What is mainly of interest is $u^*(\bar{x}_0)$, which denotes a locally optimal control trajectory to be applied as a function of the current system state $\bar{x}_0$.

2.2 Multiple Shooting Discretization

The continuous time OCP formulation from (1) leaves us with an infinite dimensional optimization problem which is impossible to solve in a general case. This problem is often discretized and subsequently optimized which is characteristic for any *direct approach* to optimal control. An important separator here is whether such a direct approach is either *sequential* or *simultaneous*. Variants of the latter approach are *direct discretization* [20] and *direct multiple shooting* [21], the focus of this paper. A shooting discretization of the problem in Eq. (1) results in the structured Nonlinear Program (NLP)

$$\begin{aligned} \underset{X,\, U}{\text{minimize}} \quad & \frac{1}{2} \sum_{i=0}^{N-1} \|F_i(x_i, u_i)\|_2^2 + \frac{1}{2} \, \|F_N(x_N)\|_2^2 && \text{(2a)} \\ \text{subject to} \quad & 0 = x_0 - \bar{x}_0, && \text{(2b)} \\ & \boldsymbol{0 = x_{i+1} - \Phi_i(x_i, u_i)}, \quad i = 0, \ldots, N-1, && \text{(2c)} \\ & 0 \geq h_i(x_i, u_i), \quad i = 0, \ldots, N-1, && \text{(2d)} \\ & 0 \geq r(x_N), && \text{(2e)} \end{aligned}$$

with state trajectory $X = [x_0^\top, \ldots, x_N^\top]^\top$ where $x_i \in \mathbb{R}^{n_x}$ and control trajectory $U = [u_0^\top, \ldots, u_{N-1}^\top]^\top$ where $u_i \in \mathbb{R}^{n_u}$. Note that the function $\Phi_i(x_i, u_i)$ here denotes the simulation of the nonlinear dynamics over one shooting interval, starting from the states x_i and using the control values u_i. This component is essential for any shooting method and will be the topic of interest in Sect. 3. When one addresses this optimization problem directly in a Newton-type framework, the variables in X generally represent a feasible state trajectory only at convergence. In direct multiple shooting, simulation and optimization are therefore performed simultaneously.

On the other hand, a sequential approach carries out the simulation task separately from solving the optimization problem. A reduced OCP formulation is obtained after replacing the variables x_i by the results $X_{\text{sim}}(\bar{x}_0, U)$ of a forward simulation starting from the initial states $\bar{x}_0$ using the control trajectory U. This technique is also known as *single shooting*. The equality constraints from Eq. (2c) are now automatically satisfied and can therefore be eliminated. Since the variable space of this problem is strongly reduced in dimension from $(N+1)n_x + Nn_u$ to only Nn_u, the task of solving this NLP appears to be simplified. But it has been shown that the cost per Newton iteration can be made equal for both approaches because of the sparsity structure in (2). Advantages of multiple shooting over single shooting are also the stronger flexibility in initializing the problem and parallelizing the algorithm, and the improved convergence properties especially in case of an unstable system [22].

3 Auto Generated Integrators

This section presents auto generated RK methods with efficient sensitivity generation. As one-step methods, they are particularly suited for simulation over relatively short shooting intervals such as needed in Eq. (2c). Their implementation is discussed in Sect. 3.1, for ODE as well as DAE systems of index 1. Extending these methods with an efficient computation of continuous outputs and first order sensitivity information is respectively described in Sects. 3.2 and 3.3. In Sect. 3.4, a common three stage model formulation is briefly introduced.

3.1 Runge-Kutta Methods

An integration method in general needs to solve the following Initial Value Problem (IVP) over a certain time interval $t \in [0, T_s]$:

$$\begin{aligned} 0 &= g(t, \dot{x}(t), x(t), z(t), u(t)), \\ x(0) &= x_0, \end{aligned} \tag{3}$$

with $x(t)$ a vector of n_x differential states, $\dot{x}(t)$ the corresponding time derivatives and $z(t)$ a vector of n_z algebraic states. This covers models ranging from explicit ODE to fully implicit DAE systems, which can be dealt with by IRK methods [15]. The only assumption is that the Jacobian matrix $\frac{\partial g(\cdot)}{\partial(z,\dot{x})}$ is invertible, i.e. the DAE system is of index 1. Mainly because of their good stability properties, the focus is often on A-stable schemes such as the Gauss methods [23].

With real-time applications in mind, a code generation tool exports a tailored integrator with a deterministic runtime. Applying an s-stage IRK method to the model in (3) results in a nonlinear system that can be solved using a Newton-type method. The step size, the order of the method and the number of Newton iterations have to be fixed such that there is no adaptivity. In the context of shooting methods, a good initialization of the variables using the previous solution is available so that a small amount of iterations is typically sufficient. Even a custom linear solver can be exported to perform the iterations, e.g. based on an LU decomposition. Note that an ERK method can be used in case of a model described by an explicit ODE system. Its code generation implementation is relatively trivial and these methods can also be more efficient when their use is restricted to non-stiff models [13].

3.2 Continuous Output

Some promising possibilities of auto generated integration methods with continuous output have been illustrated in [13, 17]. The general idea is to define some output function $y = \psi(t, \dot{x}(t), x(t), z(t))$ that can be evaluated efficiently on an arbitrarily fine grid, independent of the integration grid. These output functions can then be used to define the objective function or some constraint functions in the NMPC formulation. In case of collocation methods which are a specific family of IRK methods, this continuous extension comes rather naturally. But it is also possible to define continuous extensions for explicit or semi-implicit RK methods.

3.3 Sensitivity Generation

In the context of dynamic optimization, at least first order sensitivities with respect to the variables are needed in addition to the simulated values of states and outputs. A thorough discussion on techniques of forward sensitivity propagation for IRK methods can be found in [24]. The conclusion from that work is that the most efficient way is to apply the Implicit Function Theorem (IFT) to the nonlinear system of the IRK method. This direct approach also provides very accurate derivatives which is important for optimization. Note that for an explicit method, it is efficient to compute the first order derivatives by simulating the system extended with the Variational Differential Equations (VDE):

$$\begin{aligned}\dot{x}(t) &= f(t, x(t), u(t)),\\ \dot{S}_x(t) &= \frac{\partial f(t, x(t), u(t))}{\partial x} S_x(t),\\ \dot{S}_u(t) &= \frac{\partial f(t, x(t), u(t))}{\partial x} S_u(t) + \frac{\partial f(t, x(t), u(t))}{\partial u},\end{aligned} \tag{4}$$

with $x(0) = x_0$ and the sensitivity matrices are defined as $S_x(t) = \partial x(t)/\partial x_0$ and $S_u(t) = \partial x(t)/\partial u$ for which it holds that $[\, S_x(0) \,|\, S_u(0) \,] = [\, \mathbb{1} \,|\, \mathbb{0} \,]$. Since a fixed step size is assumed for the auto generated integrators, this approach is equivalent to performing algorithmic differentiation in forward mode [18].

3.4 Linear Subsystems

When modeling a system, the result is typically a set of nonlinear differential equations with possibly some algebraic states but one would often recognize one or more of the following three subsystems in this specific order:

$$C_1 \dot{x}_{[1]} = A_1 x_{[1]} + B_1 u, \tag{5a}$$

$$0 = f_2(\dot{x}_{[1]}, x_{[1]}, \dot{x}_{[2]}, x_{[2]}, z, u), \tag{5b}$$

$$C_3 \dot{x}_{[3]} = A_3 x_{[3]} + f_3(\dot{x}_{[1]}, x_{[1]}, \dot{x}_{[2]}, x_{[2]}, z, u), \tag{5c}$$

with matrices A_1, B_1, A_3 and invertible matrices C_1 and C_3 and the nonlinear functions f_2 and f_3. The main assumption is that the Jacobian matrix $\frac{\partial f_2(\cdot)}{\partial (z, \dot{x}_{[2]})}$ is invertible, i.e. the second subsystem represents a DAE of index 1. In the case that $A_3 = 0$ and C_3 is an identity matrix, Eq. (5c) reduces to

$$\dot{x}_{[3]} = f_3(\dot{x}_{[1]}, x_{[1]}, \dot{x}_{[2]}, x_{[2]}, z, u) \tag{6}$$

which are better known as quadrature states [25]. They are typically used to formulate objective and constraint functions, similar to how the more general linear input and output states can be used respectively from Eqs. (5a) and (5c). The exploitation of this three-stage model formulation in auto generated integration methods has been presented and illustrated in [16].

4 Sequential Quadratic Programming

A Newton-type algorithm to solve the NLP in (2) is dedicated to find a locally optimal solution by solving the nonlinear Karush-Kuhn-Tucker (KKT) conditions. There are different options to treat the inequality constraints in this system. One way which is done by nonlinear IP methods is to smoothen the corresponding KKT conditions, a popular software implementation of this is the code IPOPT [26]. The focus in this article is on Sequential Quadratic Programming (SQP), which defines another family of algorithms. The QP subproblem to be solved is described in Sect. 4.1 together with the popular Gauss-Newton Hessian approximation. Section 4.2 then presents two alternative ways to deal with this structured, multi-stage QP. Some important implementation details of the RTI scheme are discussed eventually in Sect. 4.3.

4.1 Generalized Gauss-Newton

After linearizing the nonlinear functions in the KKT system, it becomes equivalent to solving the following Quadratic Program (QP)

$$\underset{X,U}{\text{minimize}} \quad \Phi_{\text{quad}}(X,U;X^{[k]},U^{[k]},Y^{[k]},\lambda^{[k]}) \tag{7a}$$

$$\text{subject to} \quad G_{\text{eq,lin}}(\cdot) = \begin{bmatrix} x_0 - \bar{x}_0 \\ x_1 - \phi_0(x_0^{[k]},u_0^{[k]}) - \left[A_0^{[k]}, B_0^{[k]}\right]\begin{bmatrix} x_0 - x_0^{[k]} \\ u_0 - u_0^{[k]} \end{bmatrix} \\ \vdots \end{bmatrix} = 0, \tag{7b}$$

$$G_{\text{ineq,lin}}(\cdot) = \begin{bmatrix} h_0(x_0^{[k]},u_0^{[k]}) + \left[C_0^{[k]}, D_0^{[k]}\right]\begin{bmatrix} x_0 - x_0^{[k]} \\ u_0 - u_0^{[k]} \end{bmatrix} \\ \vdots \\ r(x_N^{[k]}) + C_N^{[k]}(x_N - x_N^{[k]}) \end{bmatrix} \leq 0, \tag{7c}$$

where $Y^{[k]}$ and $\lambda^{[k]}$ are the Lagrange multipliers for respectively the equality and inequality constraints and $A_i^{[k]}, B_i^{[k]}, C_i^{[k]}, D_i^{[k]}$ denote the Jacobian matrices of the corresponding functions evaluated at the current iterate k. Note that $\phi_i(\cdot)$, A_i and B_i for $i = 0, \ldots, N-1$ form the information that needs to be provided by the integration methods from Sect. 3. The result is a sequence of QPs of which the solution provides a sequence of iterations, converging to a local solution of the original, nonlinear optimal control problem under the assumptions as discussed in [27].

There are different variants of the SQP algorithm that use certain expressions for the QP objective in (7a). In case of an *exact Hessian* variant, the sparse Hessian of the Lagrangian $\nabla^2\mathscr{L}$ needs to be evaluated and used in the QP subproblem leading to a locally quadratic convergence rate in solving the NLP. The Hessian is often approximated resulting in a trade-off between computational complexity per step and convergence speed. The special case of using a least squares objective such as in Eq. (2a) allows a popular Hessian approximation that is known as the Generalized Gauss-Newton (GGN) method [28]. The objective function to be used in Eq. (7a) then reads as

$$\begin{aligned}\Phi_{\text{quad}} = \frac{1}{2}\sum_{i=0}^{N-1} \begin{bmatrix} x_i - x_i^{[k]} \\ u_i - u_i^{[k]} \end{bmatrix}^\top J_i^{[k]\top} J_i^{[k]} \begin{bmatrix} x_i - x_i^{[k]} \\ u_i - u_i^{[k]} \end{bmatrix} + \sum_{i=0}^{N-1} F_i^{[k]\top} J_i^{[k]} \begin{bmatrix} x_i - x_i^{[k]} \\ u_i - u_i^{[k]} \end{bmatrix} \\ + \frac{1}{2}\,(x_N - x_N^{[k]})^\top J_N^{[k]\top} J_N^{[k]}\,(x_N - x_N^{[k]}) + F_N^{[k]\top} J_N^{[k]}\,(x_N - x_N^{[k]})\end{aligned} \tag{8}$$

where $F_i^{[k]} = F_i(x_i^{[k]}, u_i^{[k]})$ and $F_N^{[k]} = F_N(x_N^{[k]})$ are evaluations of the residual functions from Eq. (2a) and $J_i^{[k]}$, $J_N^{[k]}$ are respectively their Jacobians $\frac{\partial F_i(x_i^{[k]}, u_i^{[k]})}{\partial(x_i, u_i)}$ and $\frac{\partial F_N(x_N^{[k]})}{\partial x_N}$. The GGN method is based on the observation that $J_i^{[k]\top} J_i^{[k]}$ for each $i = 0, \ldots, N$ forms a good approximation for the Hessian result $\nabla_{w_i}^2 \mathscr{L}$ where $w_i = (x_i, u_i)$, as long as the residual evaluations $F_i(\cdot)$ remain small [27]. The convergence rate of the GGN method is only linear but its implementation is rather simple and works very well in practice for such small residual problems. A special case of this is when the original objective function $\Phi(X, U)$ was already convex quadratic, meaning that it can be used directly in the QP. The method can also be further generalized to Sequential Convex Programming e.g. in case of NMPC problems with elliptic terminal regions [29].

4.2 *Sparsity Exploitation*

The QP subproblem presented in the previous Subsection shows a specific sparsity structure which should be exploited. One option is to reduce the variable space by a procedure called *condensing*, and then to solve the smaller, condensed QP using a suitable solver [5]. Another option is to directly use a tailored QP solver that can efficiently exploit this structure.

4.2.1 The Condensed Problem

To simplify notation, let us define the trajectories $\Delta X = X - X^{[k]}$ and $\Delta U = U - U^{[k]}$. The constraints in (7b) can be used to eliminate ΔX from the QP using

$$\Delta X = d + C\Delta\bar{x}_0 + E\Delta U \quad \text{with} \quad \Delta\bar{x}_0 = \bar{x}_0 - x_0^{[k]}, \tag{9}$$

and

$$d = \begin{bmatrix} \Delta\phi_0 \\ \Delta\phi_1 + A_1\Delta\phi_0 \\ \Delta\phi_2 + A_2\Delta\phi_1 + A_2A_1\Delta\phi_0 \\ \vdots \end{bmatrix}, \quad C = \begin{bmatrix} A_0 \\ A_1A_0 \\ A_2A_1A_0 \\ \vdots \end{bmatrix} \text{ and}$$

$$E = \begin{bmatrix} B_0 & & & \\ A_1B_0 & B_1 & & \\ A_2A_1B_0 & A_2B_1 & B_2 & \\ \vdots & & & \ddots \end{bmatrix}, \tag{10}$$

where $\Delta\phi_i = \phi_i(x_i^{[k]}, u_i^{[k]}) - x_{i+1}^{[k]}$. Note that a compact notation A_i, B_i has been used here for respectively the matrices $A_i^{[k]}, B_i^{[k]}$ at the current iteration. Insertion of the expression $\Delta X = d + C\Delta\bar{x}_0 + E\Delta U$ into (7c) and (8) yields an equivalent, but smaller scale QP of the following form:

$$\underset{\Delta U}{\text{minimize}} \quad \frac{1}{2}\Delta U^\top H_c\, \Delta U + \Delta U^\top g_c \tag{11a}$$

$$\text{subject to} \quad w + K\Delta U \leq 0. \tag{11b}$$

Let us omit the lengthy explicit expressions for the matrices H_c, K and the vectors g_c and w. For a simplified setting of a quadratic objective and simple bound constraints, these expressions can be found in [5]. Although the QP subproblem can now be solved in the reduced variable space $\Delta U \in \mathbb{R}^{Nn_u}$, the variables in X are still updated using an expansion step based on Eq. (9). The fact that the iterations are still performed in the full variable space, is the crucial difference with using a single shooting formulation. The bottleneck in an implementation of condensing is the computation of the condensed Hessian H_c, which has been shown to be of complexity $O(N^2)$ [6, 30]. It is hereby important to exploit the lower block triangular structure of matrix E from (10), the separability of the objective function in (8) and the symmetry of the Hessian matrix [5, 31]. Note that the small scale QP can be solved by an efficient, dense linear algebra solver such as qpOASES. This significantly reduces the feedback delay time between receiving the new state estimate $\bar{x}_0$ and applying the next control input $u_0^{[k+1]} = u_0^{[k]} + \Delta u_0^\star$.

4.2.2 Solving the Sparse Problem

Using a condensing approach, the corresponding cost per iteration is of order $O(N^2)$ including the factorization cost as discussed in [30, 32]. Alternatively, one would directly solve the sparse QP problem from (7) in the full variable space and exploiting the sparsity structure then becomes essential. Both for active set and Interior Point (IP) algorithms to solve this multi-stage QP, the cost per iteration of the solver can be made of order $O(N)$ [33]. Code generation tools exist that export tailored convex IP solvers, popular examples are CVXGEN [34] and FORCES [9]. An efficient implementation of a structure exploiting, primal-barrier IP method can be found in [35]. Employing the condensing technique is known to perform very well for relatively short horizon lengths N but can be outperformed by using a structure exploiting convex solver for longer horizons. Comparative simulation results can be found in [5]. Classical condensing is generally a good approach in case of many state variables $n_x > n_u$, while complementary condensing [36] was proposed as a competitive alternative in case of many controls $n_u > n_x$. A known issue with IP methods in a real-time framework is that it is difficult to warm-start them efficiently. There is ongoing research on combining the beneficial properties of both an active set method and a structure exploiting IP solver. A promising example based on a dual Newton strategy to solve structured multi-stage QPs is the open-source software qpDUNES, presented in [10].

4.3 *Real-Time Iterations*

The RTI scheme has already been mentioned multiple times in this paper, but it is important to elaborate on some of its properties since it is the key idea that allows nonlinear optimal control with microsecond execution times.

4.3.1 Initial Value Embedding

In NMPC, a sequence of optimal control problems with different initial values $\bar{x}_0^{[0]}, \bar{x}_0^{[1]}, \ldots$ needs to be solved. For the transition from one problem to the next, it is beneficial to take into account the fact that the optimal solution $U^*(\bar{x}_0)$ depends almost everywhere differentiably on x_0 which is the idea behind a continuation method. The solution manifold has smooth parts whenever the active set does not change, but non-differentiable points occur where the active set changes. After linearizing at such a point in the context of a nonlinear IP method, a simple *tangential predictor* would lead to a rather bad approximation. One remedy would be to increase the path parameter τ, which decreases the nonlinearity but it comes at the expense of generally less accurate solutions.

One can deal with active set changes naturally in an SQP type framework by the following procedure proposed and analysed in [4, 29, 37]: first of all, the parameter $\bar{x}_0$ needs to enter the NLP linearly, which is automatically the case for a simultaneous OCP formulation such as in Eq. (2b). The problem needs to be addressed using an exact Hessian SQP method. Finally, the solution trajectories $X^{[k]}$ and $U^{[k]}$ for the current problem in $\bar{x}_0^{[k]}$ are used as initial guess to solve the

OCP for the new parameter value $\bar{x}_0^{[k+1]}$. In the context of NMPC with a quadratic cost, this continuation technique can also be used with a Gauss-Newton Hessian approximation. This is done in the RTI algorithm and yields a multiplier free, *generalized* tangential predictor i.e. one that works across active set changes [2].

4.3.2 Reducing the Computational Delay

Ideally, the solution to a new optimal control problem is obtained instantly which is however impossible due to computational delays. Several ingredients of the RTI scheme can help us in dealing with this issue. A first and important one is to divide the computations at each sampling time into a preparation and a feedback phase [4]. The typically more CPU intensive *preparation phase* is performed with a predicted state, before the state estimate is even available. Once the new value $\bar{x}_0$ becomes available, the *feedback phase* quickly delivers an *approximate* solution to the original problem by solving the prepared, convex subproblem. The idea is to always work with the most current information in each iteration, i.e. not to iterate until convergence for an MPC problem that is only getting older. It can also be seen as a distributed-in-time optimization procedure.

Another important ingredient is to transfer solution information from one OCP to the next one, i.e. to efficiently warm-start each solution procedure. This can be done by using the previously optimal trajectories as an initial guess at the next time step and this either directly or in a shifted version. A last technique concerns code generation which has become a quite popular way to do code optimizations based on a high-level description of the problem to be solved. Multiple tools already exist to automatically generate custom solvers in a low-level language [34, 38]. Also for NMPC, the consecutive optimal control problems are similar and many computations can be done offline before the controller starts. The auto generated code then exploits problem dimensions and sparsity structures, it avoids dynamic memory allocation and has a nearly deterministic runtime. The latter is important to be able to satisfy the hard timing constraints in real-time applications. A tailored RTI algorithm for nonlinear optimal control can be generated as plain C-code by the open-source software ACADO Toolkit [18].

5 A Classical Benchmark Problem

This section presents numerical results that allow for interesting comparisons to be made between different NMPC formulations, algorithms and their implementation. The problem formulation is first presented in Sect. 5.1, followed by a description in Sect. 5.2 of the three test cases that are used in simulation. Some results of the corresponding numerical experiments are eventually shown and discussed in Sect. 5.3. The simulations presented in this section are performed using the ACADO code generation tool on a modern computer equipped with Intel i7-3720QM processor, running a 64-bit version of Ubuntu 12.04. All programs are compiled using the Clang 3.0 compiler.

5.1 Problem Formulation

Throughout this section, the simple NMPC problem from [39] will be used as a benchmark example. Its corresponding continuous time OCP reads

$$\min_{x(\cdot),u(\cdot)} \quad \int_t^{t+T} (\|x(\tau)\|_Q^2 + \|u(\tau)\|_R^2)\,\mathrm{d}\tau \; + \; \|x(t+T)\|_P^2 \tag{12a}$$

$$\text{s.t.} \quad x(t) = \bar{x}_t, \tag{12b}$$

$$\dot{x}_1(\tau) = x_2(\tau) + u(\tau)\,(\mu + (1-\mu)\,x_1(\tau)),$$

$$\dot{x}_2(\tau) = x_1(\tau) + u(\tau)\,(\mu - 4(1-\mu)\,x_2(\tau)), \tag{12c}$$

$$-2 \le u(\tau) \le 2, \qquad \forall \tau \in [t, t+T], \tag{12d}$$

$$\|x(t+T)\|_P^2 \le \alpha, \tag{12e}$$

where Eq. (12c) defines the simple but unstable ODE system with two differential states x_1 and x_2, a control input u and constant value $\mu = 0.5$. The parameters $P \succ 0$ and $\alpha \ge 0$ from Eqs. (12a) and (12e) define the terminal penalty and the terminal region Ω_α of which the latter is preferably as large as possible, while still leading to closed-loop stability for the resulting NMPC approach. The following parameter values will be used in simulation:

$$T_s = 0.1s \qquad N = 15 \qquad T = 1.5s$$

$$Q = \begin{pmatrix} 2.0 & 0.0 \\ 0.0 & 2.0 \end{pmatrix} \qquad R = 0.1$$

$$P = \begin{pmatrix} 10.605 & -9.395 \\ -9.395 & 10.605 \end{pmatrix} \qquad \alpha = 0.7$$

w here T_s and N respectively define the size and number of shooting intervals over the horizon $[0, T]$ i.e. they define the shooting discretization of the OCP.

5.2 Simulation Test Cases

5.2.1 Case A: Original Formulation and QP

Starting from the formulation in (12), a first NMPC scheme is to solve this OCP until convergence at every time point and this will be further called *case A*. Note that the resulting NMPC controller is the same as case A in the paper from [39].

5.2.2 Case B: Tuned Formulation and QP

It is interesting to have a look at the tuning possibilities to achieve faster sampling times when necessary. First of all, the RTI scheme can be used instead of iterating the procedure until convergence. Then it is important to use a suitable integration method with efficient sensitivity generation for the shooting discretization, as discussed in Sect. 3. The ODE system in (12c) is rather simple and non-stiff, thus an explicit Euler discretization with a step size of 0.1s already suffices in this case. To further improve the computation time of one RTI iteration, the number of optimization variables can be reduced. This means that the number of shooting intervals N will be reduced while keeping the horizon length long enough for a good NMPC performance. To achieve this, a non equidistant control grid is used as depicted in Fig. 1. Eventually, the quadratic terminal constraint in (12e) is also removed and the resulting scheme will be referred to as *case B*.

To obtain results that are comparable to the original scheme (case A), a few details must be addressed. One is the terminal cost matrix P that needs to be altered since the new horizon is shorter. By integrating the differential Riccati equation backwards over 0.5s, a new terminal cost matrix $P(1.0)$ can be found:

$$P(1.5) = \begin{pmatrix} 10.605 & -9.395 \\ -9.395 & 10.605 \end{pmatrix} \quad \Rightarrow \quad P(1.0) = \begin{pmatrix} 4.432 & -3.558 \\ -3.558 & 4.432 \end{pmatrix}.$$

Because of the varying interval size in the alternative control horizon, the weighting matrices Q and R are scaled using a factor relative to this interval size. Also standard shifting approaches are not applicable anymore and therefore abandoned.

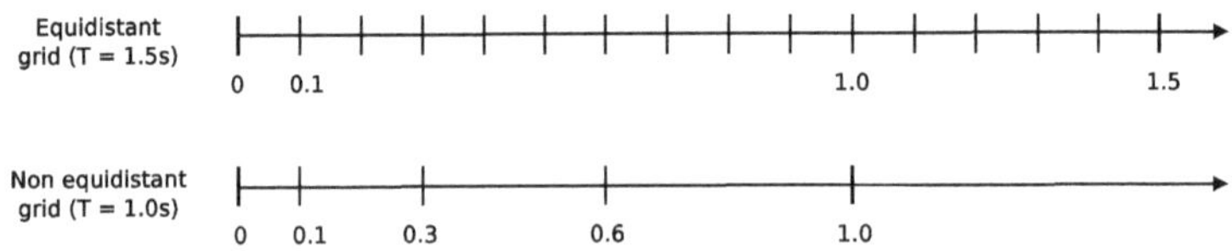

Fig. 1 Illustration of a non equidistant control grid of only four shooting intervals over ones

5.2.3 Case C: Original Formulation and QCQP

It is important to note that all presented techniques can be extended further towards Sequential Convex Programming (SCP) [29]. In Sect. 4.2.2, FORCES has been presented as a sparsity exploiting QP solver although the most general convex problem that it targets is a quadratically constrained QP (QCQP). For our benchmark example, the terminal inequality constraint from (12e) can be kept in the convex subproblem such that it becomes of this QCQP form. This will be referred to as *case C* and it uses the same, original OCP formulation as case A.

5.3 *Numerical Results*

5.3.1 Comparison of Single and Multiple Shooting

First of all, let us illustrate multiple shooting by comparing it with a single shooting discretization both on the OCP formulation of case A using initial value embedding. The performance of the NMPC controller will be measured using the Karush-Kuhn-Tucker (KKT) tolerance, computed as in [31]. Figure 2 illustrates the resulting convergence rate for both solvers in a simulation over a time period of 5s. It can be seen that the convergence is slightly better using the multiple shooting discretization. In both cases, the solver starts from the exact same initial guess which is the reference trajectory. Note that also the computational complexity is the same for both.

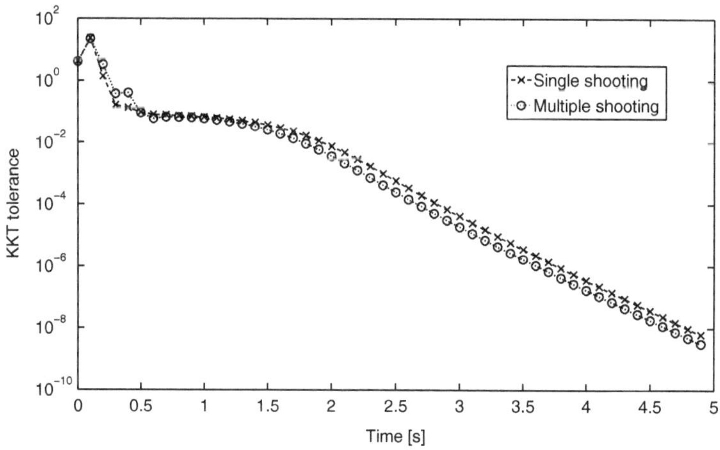

Fig. 2 Closed-loop NMPC performance using both single and multiple shooting

Table 1 Average computation times for NMPC using the RTI scheme

NMPC	Case A (μs)	Case B (μs)	Case C (μs)
Integration method	0.222	0.142	0.221
Condensing	3.370	0.088	–
QP solution	4.340	0.633	29.300
Remaining operations	1.608	0.087	0.679
One real-time iteration	**9.540**	**0.950**	**30.200**

5.3.2 Execution Times

The average computational times for the different components in one RTI iteration are shown in Table 1 and this for the three different cases. Using ACADO code generation, one iteration for case A, B and C on average takes respectively 9.54, 0.95 and 30.2 μs. Note that the formulation used in case B has been tuned precisely to result in a total execution time that is below 1 μs. The scheme which uses FORCES to solve a QCQP subproblem (case C) appears not to be competitive with the condensing based approach (case A) for this example. The reason is that the used horizon is relatively small as discussed more detailed in [5]. An important advantage of case C is that the terminal inequality becomes part of the subproblem to be solved and it is therefore guaranteed to be satisfied for a feasible trajectory. This is not necessarily true when linearizing that same constraint. Exploiting convexity as much as possible can therefore be a rather powerful tool.

5.3.3 Tracking Performance

Figure 3 compares the closed-loop tracking performance of the three NMPC schemes for five different initial values, also used in [39]. As a reference, the closed-loop behavior of the corresponding LQR scheme with control saturation is shown in the same figure. The latter controller appears to be unstable for one of these initial values while the NMPC schemes all exhibit a performance that is similar to one another. According to the RTI scheme, only one SQP iteration is performed per time step for cases B and C while the NMPC results for case A are iterated until convergence using a rather strict stopping criterion. Note that the feedback delay has not been taken into account in these closed-loop simulations.

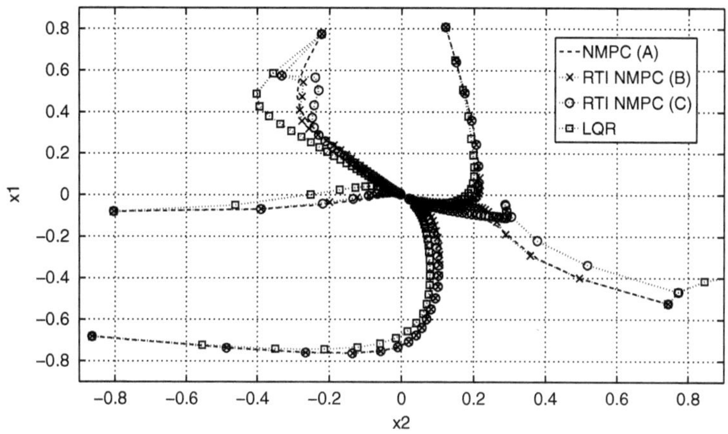

Fig. 3 Closed-loop performance of three different NMPC schemes and the LQR controller and this for five different initial values. This figure is comparable to one presented originally in [39]

6 Conclusions

This paper handled a general parametric optimization problem, which arises naturally in NMPC where one has to solve a nonlinear OCP at every sampling instant. It gave an outline of the different algorithmic developments that made these techniques real-time feasible today even for nonlinear systems with fast dynamics. Auto generated integration methods have been presented as an essential part to efficiently linearize the problem and compute derivative information. The RTI scheme has been described as an online algorithm that allows to perform real time NMPC while having a fast control feedback to the real process. A simple example taken from the literature, was used to illustrate the performance of the presented tools. The average execution time per time step for this nonlinear problem was eventually shown to be below 1 μs.

Acknowledgements The authors would like to thank all reviewers for their valuable comments, which helped to improve this article. This research was supported by Research Council KUL: PFV/10/002 Optimization in Engineering Center OPTEC, GOA/10/09 MaNet and GOA/10/11 Global real-time optimal control of autonomous robots and mechatronic systems, KUL-BOF OT/10/035. Flemish Government: FWO: PhD/postdoc grants, FWO KAN2013 1.5.189.13, FWO-G.0930.13; IWT: PhD Grants, projects: Eurostars SMART; Belgian Federal Science Policy Office: IUAP P7 (DYSCO, Dynamical systems, control and optimization, 2012–2017); EU: FP7-SADCO (MC ITN-264735), FP7-TEMPO (MC ITN-607957), H2020-AWESCO (MC ITN-642682), ERC HIGHWIND (259 166). R. Quirynen holds a PhD fellowship of the Research Foundation– Flanders (FWO).

References

1. Kirches, C., Wirsching, L., Sager, S., Bock, H.: Efficient numerics for nonlinear model predictive control. In: Diehl, M., Glineur, F.F., Michiels, E.J.W. (eds.) Recent Advances in Optimization and its Applications in Engineering, pp. 339–357. Springer, Berlin (2010)
2. Diehl, M., Ferreau, H.J., Haverbeke, N.: Efficient numerical methods for nonlinear MPC and moving horizon estimation. In: Nonlinear Model Predictive Control. Lecture Notes in Control and Information Sciences, vol. 384. pp. 391–417. Springer, Berlin (2009)
3. Diehl, M., Bock, H., Schlöder, J.: A real-time iteration scheme for nonlinear optimization in optimal feedback control. SIAM J. Control. Optim. **43**(5), 1714–1736 (2005)
4. Diehl, M., Bock, H., Schlöder, J., Findeisen, R. Nagy, Z., Allgöwer, F.: Real-time optimization and nonlinear model predictive control of processes governed by differential-algebraic equations. J. Process Control **12**(4), 577–585 (2002)
5. Vukov, M., Domahidi, A., Ferreau, H.J., Morari, M., Diehl, M.: Auto-generated algorithms for nonlinear model predicitive control on long and on short horizons. In: Proceedings of the 52nd Conference on Decision and Control (CDC) (2013)
6. Andersson, J.: A general-purpose software framework for dynamic optimization. Ph.D. thesis, Arenberg Doctoral School, KU Leuven, Department of Electrical Engineering (ESAT/SCD) and Optimization in Engineering Center, Kasteelpark Arenberg 10, 3001-Heverlee, Belgium (October 2013)
7. Ferreau, H., Kirches, C., Potschka, A., Bock, H., Diehl, M.: qpOASES: A parametric active-set algorithm for quadratic programming. Math. Program. Comput. **6**(4), 327–363 (2014)
8. Ferreau, H.J., Bock, H.G., Diehl, M.: An online active set strategy to overcome the limitations of explicit MPC. Int. J. Robust Nonlinear Control **18**(8), 816–830 (2008)
9. Domahidi, A., Zgraggen, A., Zeilinger, M., Morari, M., Jones, C.: Efficient Interior point methods for multistage problems arising in receding horizon control. In: IEEE Conference on Decision and Control (CDC) (Maui, HI, USA), pp. 668–674 (December 2012)
10. Frasch, J.V., Sager, S., Diehl, M.: A parallel quadratic programming method for dynamic optimization problems. Math. Program. Comput. **7**(3) 289–329 (September 2015)
11. Frison, G., Sorensen, H., Dammann, B., Jorgensen, J.: High-performance small-scale solvers for linear model predictive control. In: Proceedings of 2014 European Control Conference (ECC), pp. 128–133 (June 2014)
12. Mattingley, J., Boyd, S.: Automatic code generation for real-time convex optimization. In: Convex Optimization in Signal Processing and Communications. Cambridge University Press, Cambridge (2009)
13. Quirynen, R., Vukov, M., Zanon, M., Diehl, M.: Autogenerating microsecond solvers for nonlinear MPC: a tutorial using ACADO integrators. Optim. Control Appl. Methods (2014). http://onlinelibrary.wiley.com/doi/10.1002/oca.2152/abstract
14. Vukov, M., Loock, W.V., Houska, B., Ferreau, H., Swevers, J., Diehl, M.: Experimental Validation of Nonlinear MPC on an Overhead Crane using Automatic Code Generation. In: The 2012 American Control Conference, Montreal, Canada (2012)
15. Quirynen, R., Vukov, M., Diehl, M.: Auto generation of implicit integrators for embedded NMPC with microsecond sampling times. In: Lazar, M., Allgöwer, F. (eds.) Proceedings of the 4th IFAC Nonlinear Model Predictive Control Conference (2012)
16. Quirynen, R., Gros, S., Diehl, M.: Efficient NMPC for nonlinear models with linear subsystems. In: Proceedings of the 52nd IEEE Conference on Decision and Control (2013)
17. Quirynen, R., Gros, S., Diehl, M.: Fast auto generated ACADO integrators and application to MHE with multi-rate measurements. In: Proceedings of the European Control Conference (2013)
18. Houska, B., Ferreau, H., Diehl, M.: An auto-generated real-time iteration algorithm for nonlinear MPC in the microsecond range. Automatica **47**(10), 2279–2285 (2011)
19. Houska, B., Ferreau, H., Diehl, M.: ACADO toolkit – an open source framework for automatic control and dynamic optimization. Optim. Control Appl. Methods **32**(3), 298–312 (2011)

20. Biegler, L.: An overview of simultaneous strategies for dynamic optimization. Chem. Eng. Process. **46**, 1043–1053 (2007)
21. Bock, H., Plitt, K.: A multiple shooting algorithm for direct solution of optimal control problems. In: Proceedings 9th IFAC World Congress Budapest, pp. 242–247. Pergamon Press, New York (1984)
22. Albersmeyer, J., Diehl, M.: The lifted Newton method and its application in optimization. SIAM J. Optim. **20**(3), 1655–1684 (2010)
23. Hairer, E., Wanner, G.: Solving Ordinary Differential Equations II – Stiff and Differential-Algebraic Problems, 2nd edn. Springer, Berlin/Heidelberg (1991)
24. Quirynen, R.: Automatic code generation of Implicit Runge-Kutta integrators with continuous output for fast embedded optimization. Master's thesis, KU Leuven (2012)
25. Serban, R., Hindmarsh, A.: CVODES: The sensitivity-enabled ODE solver in SUNDIALS. In: Proceedings of IDETC/CIE (2005)
26. Wächter, A., Biegler, L.: On the Implementation of a primal-dual interior point filter line search algorithm for large-scale nonlinear programming. Math. Program. **106**(1), 25–57 (2006)
27. Nocedal, J., Wright, S.: Numerical Optimization. In: Springer Series in Operations Research and Financial Engineering, 2nd edn. Springer, Berlin (2006)
28. Bock, H. Recent advances in parameter identification techniques for ODE. In: Deuflhard, P., Hairer, E. (eds.) Numerical Treatment of Inverse Problems in Differential and Integral Equations. Birkhäuser, Boston (1983)
29. Tran-Dinh, Q., Savorgnan, C., Diehl, M.: Adjoint-based predictor-corrector sequential convex programming for parametric nonlinear optimization. SIAM J. Optim. **22**(4), 1258–1284 (2012)
30. Frison, G., Jorgensen, J.: A fast condensing method for solution of linear-quadratic control problems. In: Proceedings of the 52nd IEEE Conference on Decision and Control (2013)
31. Leineweber, D. Efficient reduced SQP methods for the optimization of chemical processes described by large sparse DAE models. Fortschritt-Berichte VDI Reihe 3, Verfahrenstechnik, vol. 613. VDI Verlag, Düsseldorf (1999)
32. Axehill, D., Morari, M.: An alternative use of the riccati recursion for efficient optimization. Syst. Control Lett. **61**(1), 37–40 (2012)
33. Rao, C., Wright, S., Rawlings, J.: Application of interior-point methods to model predictive control. J. Optim. Theory Appl. **99**, 723–757 (1998)
34. Mattingley, J., Wang, Y., Boyd, S.: Code generation for receding horizon control. In: Proceedings of the IEEE International Symposium on Computer-Aided Control System Design (Yokohama, Japan) (2010)
35. Wang, Y., Boyd, S.: Fast model predictive control using online optimization. IEEE Trans. Control Syst. Technol. **18**(2), 267–278 (2010)
36. Kirches, C., Bock, H., Schlöder, J., Sager, S.: Block structured quadratic programming for the direct multiple shooting method for optimal control. Optim. Methods Softw. **26**, 239–257 (2010)
37. Zavala, V., Anitescu, M.: Real-time nonlinear optimization as a generalized equation. SIAM J. Control Optim. **48**(8), 5444–5467 (2010)
38. Ohtsuka, T., Kodama, A.: Automatic code generation system for nonlinear receding horizon control. Trans. Soc. Instrum. Control Eng. **38**(7), 617–623 (2002)
39. Chen, H., Allgöwer, F.: A quasi-infinite horizon nonlinear model predictive control scheme with guaranteed stability. Automatica **34**(10), 1205–1218 (1998)

Preconditioners Based on "Parareal" Time-Domain Decomposition for Time-Dependent PDE-Constrained Optimization

Stefan Ulbrich

Abstract We consider optimization problems governed by time-dependent parabolic PDEs and discuss the construction of parallel preconditioners based on the parareal method for the solution of quadratic subproblems which arise within SQP methods. In the case without control constraints, the optimality system of the subproblem is directly reduced to a symmetric PDE system, for which we propose a preconditioner that decouples into a forward and backward PDE solve. In the case of control constraints we apply a semismooth Newton method and apply the preconditioner to the semismooth Newton system. We prove bounds on the condition number of the preconditioned system which shows no or only a weak dependence on the size of regularization parameters for the control. We propose to use the parareal time domain decomposition method for the forward and backward PDE solves within the PDE preconditioner to construct an efficient parallel preconditioner. Numerical results show the efficiency of the approach.

1 Introduction

We consider parallel preconditioners for time-dependent PDE-constrained optimization problems of the form

$$\min_{y \in Y, u \in U} f(y) + \frac{\alpha}{2}\|u\|_U^2 \quad \text{subject to } E(y,u) = 0, \quad u \in U_{ad}, \tag{1}$$

The author was supported by the DFG within the graduate school Computational Engineering, SFB 666 and SFB 805 and by the BMBF within SIMUROM.

S. Ulbrich (✉)
Department of Mathematics, Technische Universität Darmstadt, Dolivostraße 15, 64293 Darmstadt, Germany
e-mail: ulbrich@mathematik.tu-darmstadt.de

T. Carraro et al. (eds.), *Multiple Shooting and Time Domain Decomposition Methods*, Contributions in Mathematical and Computational Sciences 9,
DOI 10.1007/978-3-319-23321-5_8

where $u \in U$ is the control, $U_{ad} \subset U$ is closed and convex, $y \in Y \subset C([0,T];V)$ is a time-dependent state and $\alpha > 0$ is a regularization parameter. U is a Hilbert space, Y, V are Banach spaces, where $V \subset L^2(\Omega)$ with a domain $\Omega \subset \mathbb{R}^n$. The state equation $E(y,u) = 0$ represents an appropriate (usually weak) formulation of a time-dependent PDE (or system of PDEs)

$$\begin{aligned} &y_t + F(t,x,y,u) = 0, \quad (t,x) \in \Omega_T := (0,T) \times \Omega \\ &y(0,x) = y_0(x), \quad x \in \Omega \end{aligned} \tag{2}$$

with initial data $y_0 \in V$. For convenience we assume that boundary conditions are incorporated in the state space Y. For notational convenience, we use the abbreviations $W = Y \times U$ and $w = (y,u)$.

Throughout the paper we will work under the following assumptions.

Assumption 1 *With $W := Y \times U$ the following holds for a given open convex set $W_0 = Y_0 \times U_0 \subset W$ containing the feasible set of* (1).

- *The mappings*

$$y \in Y \mapsto f(y), \quad (y,u) \in W \mapsto E(y,u) \in Z^*$$

are continuously differentiable and the derivatives are uniformly bounded and Lipschitz on W_0.
- *For any $u \in U_0$ there exists a unique solution $y(u) \in Y_0$ of $E(y(u),u) = 0$.*
- *The derivative $E_y(y,u) \in \mathcal{L}(Y,Z^*)$ has an inverse that is uniformly bounded for all $(y,u) \in W_0$.*
- *$U_{ad} \subset U$ is closed and convex.*

In recent years, the design of efficient methods for the solution of PDE-constrained optimization problems (1) has received considerable attention, see for example [11, 13, 15–17, 23, 28–33].

Usually, iterative solvers are applied to solve the arising linear systems and the inexactness is controlled by the globalization mechanism of the optimization method [13, 23, 33]. Optimization methods such as SQP- or interior point methods usually lead to auxiliary problems with a saddle point structure, e.g. the optimality system for the SQP subproblem or the primal-dual Newton system of interior point methods. To exploit the sparsity of these systems it is therefore of importance to have fast iterative solvers for these systems available, which are usually ill conditioned. Therefore, preconditioners are required to achieve fast convergence of iterative, often Krylov-based, solvers. The development of preconditioning and multigrid techniques for optimality systems in PDE-constrained optimization is an active research topic. First approaches for preconditioners of optimality systems have been proposed in [3, 4]. Block preconditioners for such systems have been proposed e.g. in [7, 22, 25, 34]. Multigrid preconditioners have for example been considered in [5, 6]. While problems without inequality constraints are nowadays quite well understood, there are less results on the efficient preconditioning in the case of control and/or state constraints, see e.g. [14, 21, 24, 27]. While standard block preconditioners may depend strongly on critical parameters such as regularization or penalty parameter [14], the related preconditioners in [21, 24] show only a weak dependence and in [24] estimates for the condition number are given, which apply to problems for control constraints and regularized state constraints.

Time-domain decomposition methods based on a block Gauss-Seidel iteration for linear-quadratic optimal control problems have been proposed and analyzed in [12] and block parareal preconditioners for such problems in [20].

In this work we build on the class of preconditioners proposed in [24] and extend their analysis to parabolic problems. The preconditioner decouples into two PDE solves. To obtain a parallel preconditioner, we use the time-domain decomposition method parareal to approximate the PDE solves within the preconditioner.

The parareal method was proposed in [18] as a parallel numerical scheme to solve evolution problems. The method is a time domain decomposition method and consists of a parallel predictor step based on a sufficiently exact propagator on the time slabs and a sequential corrector step computed by a coarse propagator. The algorithm has been successfully applied, e.g., to the Navier-Stokes equations and fluid-structure interaction problems [8, 9]. Its stability and convergence properties have for example been studied in [1, 10, 18, 19, 26]. In [10] it was shown that the parareal algorithm can be considered as a multiple shooting method as well as a time-multigrid method.

In this paper, we combine the class of preconditioners in [24], see also [21], with the parareal method to approximate the PDE solves within the preconditioner. We focus on the construction of preconditioners for the fast solution of subproblems arising in optimization methods for (1). For example SQP-type methods solve, given a current iterate $w^k = (y^k, u^k)$, $u^k \in U_{ad}$ subproblems of the form

$$\min_{s=(s_y,s_u)\in W} q^k(s) := \langle f_y(y^k), s_y\rangle_{Y^*,Y} + \frac{1}{2}\langle s_y, H_k s_y\rangle_{Y,Y^*} + \frac{\alpha}{2}\|u^k + s_u\|_U^2$$

$$\text{subject to } E(w^k) + E_w(w^k)s = 0, \quad u^k + s_u \in U_{ad}, \tag{3}$$

where $H^k \in \mathcal{L}(Y, Y^*)$ is an approximation of $L_{yy}(w^k, \lambda^k)$ with the Lagrangian function

$$L(y, u, \lambda) = f(y) + \alpha\|u\|_U^2 + \langle\lambda, E(y, u)\rangle_{Z,Z^*}.$$

In many practical algorithms H_k is chosen in such a way that the quadratic problem (3) is strictly convex. This is usually achieved by using $H_k = M_k^* M_k$, where $M_k \in \mathcal{L}(Y, Q)$, which we assume from now on. Under Assumption 1, the unique solution s_k satisfies with a Lagrange multiplier (adjoint state) $\lambda^k \in Z$ the following optimality system

$$E(w^k) + E_w(w^k)s^k = 0, \tag{4}$$

$$q_y^k(s^k) + E_y(w^k)^* \lambda^k = 0, \tag{5}$$

$$s_u^k \in U_{ad} - u^k, \langle q_u^k(s^k) + E_u(w^k)^* \lambda^k, s_u - s_u^k \rangle_{U^*,U} \geq 0 \; \forall \, s_u \in U_{ad} - u^k. \tag{6}$$

Since U is a Hilbert space and $U_{ad} \subset U$ is convex and closed, it is well known that with the identification $U = U^*$ the variational inequality (6) can equivalently be replaced by

$$s_u^k = P_{U_{ad}-u^k}(s_u^k - \gamma(q_u^k(s^k) + E_u(w^k)^* \lambda^k)) \tag{7}$$

with any fixed $\gamma > 0$ and the projection $P_{U_{ad}-u^k}$ in U onto $U_{ad} - u^k$.

We consider now two cases, the unconstrained case $U_{ad} = U$ and the box constrained case $U = L^2(\omega)$, $U_{ad} = \{u \in U; \, a \leq u \leq b \text{ a.e.}\}$ with $a, b \in U$, $a \leq b$.

In the case $U_{ad} = U$ optimality system (4)–(6) simplifies to the linear system

$$\begin{aligned} E(w^k) + E_w(w^k)s^k &= 0, \\ f_y(y^k) + H_k s_y^k + E_y(w^k)^* \lambda^k &= 0, \\ \alpha(u^k + s_u^k) + E_u(w^k)^* \lambda^k &= 0. \end{aligned}$$

Solving the last equation for s_u^k and inserting in the first equation yields the reduced optimality system

$$\begin{pmatrix} H_k & E_y(w^k)^* \\ E_y(w^k) & -\alpha^{-1} E_u(w^k) E_u(w^k)^* \end{pmatrix} \begin{pmatrix} s_y^k \\ \lambda^k \end{pmatrix} = \begin{pmatrix} -f_y(y^k) \\ -E(w^k) + E_u(w^k)u^k \end{pmatrix}.$$

In the box constrained case $U = L^2(\omega)$, $U_{ad} = \{u \in U; \, a \leq u \leq b \text{ a.e.}\}$ with $a, b \in U$, $a \leq b$, we set $\gamma = \alpha^{-1}$ in (7) and obtain

$$s_u^k = -u^k + \max(a, \min(b, -\alpha^{-1} E_u(w^k)^* \lambda^k)).$$

Inserting this in (4), (5), we arrive at the reduced optimality system

$$\left. \begin{aligned} f_y(y^k) + H_k s_y^k + E_y(w^k)^* \lambda^k &= 0, \\ E(w^k) - E_u(w^k)u^k + E_y(w^k)s_y^k + & \\ + E_u(w^k) \max(a, \min(b, -\alpha^{-1} E_u(w^k)^* \lambda^k)) &= 0, \end{aligned} \right\} \quad G(s_y^k, \lambda^k) = 0.$$

It is well known that this is a semismooth system as long as $E_u(w^k)^* \in \mathcal{L}(Z, U)$ (note that we have identified $U^* = U$) satisfies in addition $E_u(w^k)^* \in \mathcal{L}(Z, L^p(\omega))$ for some $p > 2$, see [16, 30, 31]. If we now apply a semismooth Newton method then the iterates are given by $(s_y^{k,l+1}, \lambda^{k,l+1}) = (s_y^{k,l}, \lambda^{k,l}) + (\Delta y, \Delta \lambda)$, where $(\Delta y, \Delta \lambda)$

solves the linear system

$$\begin{pmatrix} H_k & E_y(w^k)^* \\ E_y(w^k) & -\alpha^{-1}E_u(w^k)D_{k,l}E_u(w^k)^* \end{pmatrix} \begin{pmatrix} \Delta y \\ \Delta\lambda \end{pmatrix} = -G(s_y^{k,l}, \lambda^{k,l}). \tag{8}$$

Here, $D_{k,l} = d_{k,l} \cdot \mathrm{id}$ is a multiplication operator with $d_{k,l} \in L^\infty(\omega)$ defined by $d_{k,l} = 1_{\{a \le -\alpha^{-1}E_u(w^k)^*\lambda^{k,l} \le b\}}$.

We conclude that in both cases we have for given $w = (y, u) \in W$, $\lambda \in Z$ and $H \in \mathcal{L}(Y, Y^*)$ to solve linear systems of the form

$$\begin{pmatrix} H & E_y(w)^* \\ E_y(w) & -\alpha^{-1}E_u(w)DE_u(w)^* \end{pmatrix} \begin{pmatrix} \Delta y \\ \Delta\lambda \end{pmatrix} = \begin{pmatrix} r_1 \\ r_2 \end{pmatrix},$$

where $r_1 \in Y^*$, $r_2 \in Z^*$ and $D = \mathrm{id}$ in the case $U_{ad} = U$ or $D = 1_{\{a \le -\alpha^{-1}E_u(w)^*\lambda \le b\}} \cdot \mathrm{id}$ in the case $U_{ad} = \{u \in U;\ a \le u \le b \text{ a.e.}\}$, $U = L^2(\omega)$.

Hence, introducing the operators

$$A := E_y(w), \quad CC^* := \alpha^{-1}E_u(w)DE_u(w)^*$$

and using that we consider Hessian approximations of the form $H = M^*M$, we arrive at saddle point systems of the form

$$\begin{pmatrix} M^*M & A^* \\ A & -CC^* \end{pmatrix} \begin{pmatrix} \Delta y \\ \Delta\lambda \end{pmatrix} = \begin{pmatrix} r_1 \\ r_2 \end{pmatrix}.$$

We note that also the application of interior point methods leads to a system of this structure.

Our aim is to develop a preconditioner for this type of systems. To this end, we use the preconditioner in [24] that requires essentially the solution of two linear systems with operators of the form $A + CIM$ and $(A + CIM)^*$, respectively. We then use the parareal time domain decomposition technique within the preconditioner to approximately solve these linear PDEs in parallel.

The paper is organized as follows. In Sect. 2 we introduce the general preconditioner of Schiela and Ulbrich [24] and extend its analysis to parabolic problems. We will derive estimates for the condition number which show only a weak dependence on critical parameters such as the regularization parameter α. We will treat the case with and without control constraints. In Sect. 3 we will recall the parareal method and its basic convergence properties. In Sect. 4 we propose a parareal based preconditioner by using the parareal algorithm as approximate PDE solver within the preconditioner of Sect. 2. Moreover, we present numerical results for parabolic control problems without and with control constraints. We end in Sect. 5 with some conclusions.

2 A Preconditioner for Optimality Systems

As we have seen, the solution of (3) leads—either directly or after applying a semismooth Newton or interior point method—to linear systems of the form

$$\begin{pmatrix} M^*M & A^* \\ A & -CC^* \end{pmatrix} \begin{pmatrix} \Delta y \\ \Delta\lambda \end{pmatrix} = \begin{pmatrix} r_1 \\ r_2 \end{pmatrix}. \tag{9}$$

Here, A corresponds to the linearized forward PDE operator $E_y(w)$ and A^* to its adjoint.

In order to apply the preconditioner and its analysis from [24], we make the following assumptions.

Assumption 2 (Basic Assumptions)

1. *Assume that the state space Y and the space of adjoints Z are reflexive Banach spaces and the control space U as well as the space Q are Hilbert spaces.*
2. *Let $A \in \mathcal{L}(Y, Z^*)$ be an isomorphism, which implies that its Banach-space adjoint $A^* : Z \to Y^*$ is an isomorphism as well.*
3. *Let $M \in \mathcal{L}(Y, Q)$ with dense range and let $C \in \mathcal{L}(U, Z^*)$.*
   We denote by the (Hilbert space) adjoint $C^* : Z \to U^* = U$ the mapping that satisfies

   $$(C^*p, u)_U = \langle \lambda, Cu \rangle_{Z,Z^*} \quad \forall u \in U$$

   It is continuous as well. Analogously, the (Hilbert space) adjoint $M^* : Q = Q^* \to Y^*$ is defined via

   $$\langle M^*q, y \rangle_{Y^*,Y} = (My, q)_Q \quad \forall y \in Y.$$

   It is continuous and injective since M has dense range.

Moreover, we need the following assumption which requires some compatibility between the control space and the objective function.

Assumption 3 *Let I be a non-zero continuous mapping*

$$I : Q \to U.$$

with $\|I\|_{Q,U} \leq 1$.

2.1 *Development of the Preconditioner*

To motivate the preconditioner, we start with some observations. As we have already noted, $M^* : Q \to Y^*$ is injective, and hence $M^* : Q \to \operatorname{ran} M^*$ is a bijective operator with inverse $M^{-*} : \operatorname{ran} M^* \to Q$. We define the new space $\hat{Z} \subset Z$

$$\hat{Z} := A^{-*}(\operatorname{ran} M^*) = \{z \in Z : A^* z \in \operatorname{ran} M^*\} \subset Z.$$

Then $M^{-*}A^*z$ is well defined for all $z \in \hat{Z}$ and $M^{-*}A^* : \hat{Z} \to Q$ is a bijective mapping. The following lemma shows that $\hat{Z}$ becomes a Hilbert space with the scalar product

$$(z_1, z_2) \in \hat{Z}^2 \mapsto (z_1, z_2)_{\hat{Z}} := (M^{-*}A^*z_1, M^{-*}A^*z_2)_Q. \tag{10}$$

Moreover, the bilinear form

$$(z_1, z_2) \in \hat{Z}^2 \mapsto (z_1, z_2)_K := (M^{-*}A^*z_1, M^{-*}A^*z_2)_Q + (C^*z_1, C^*z_2)_U \tag{11}$$

is a scalar product that defines an equivalent norm on $\hat{Z}$.

By using this we will in Lemma 2 reduce (9) to a system that will be used to construct our preconditioner.

Lemma 1 *$\hat{Z}$ is a Hilbert space with the scalar product $(\cdot,\cdot)_{\hat{Z}}$ in* (10). *Moreover, $(\hat{Z}, (\cdot,\cdot)_{\hat{Z}})$ is continuously embedded in Z and $(\cdot,\cdot)_K$ in* (11) *is a scalar product that defines an equivalent norm on $\hat{Z}$.*

Proof For $z_1, z_2 \in \hat{Z} \subset Z$ we have $A^*z_i \in \operatorname{ran} M^*$, $i = 1, 2$, and thus there are $q_1, q_2 \in Q$ with $A^*z_i = M^*q_i$. Hence, $(z_1, z_2)_{\hat{Z}} = (q_1, q_2)_Q < \infty$. Moreover, for $z \in \hat{Z}$ we have $(z, z)_{\hat{Z}} \geq 0$ and since $M^{-*}A^* : \hat{Z} \to Q$ is injective, $(z, z)_{\hat{Z}} = 0$ implies $z = 0$. Hence, $(\hat{Z}, (\cdot,\cdot)_{\hat{Z}})$ is a pre Hilbert space.

Let $z \in \hat{Z}$ be arbitrary. Then $A^*z = M^*q$ with some $q \in Q$ and

$$\|z\|_{\hat{Z}} = (z, z)_{\hat{Z}}^{\frac{1}{2}} = (q, q)_Q^{\frac{1}{2}} = \|q\|_Q. \tag{12}$$

Hence, $\|z\|_Z = \|A^{-*}M^*q\|_Z \leq \|A^{-*}M^*\|_{Q,Z}\|q\|_Q = \|A^{-*}M^*\|_{Q,Z}\|z\|_{\hat{Z}}$ and thus the embedding $(\hat{Z}, (\cdot,\cdot)_{\hat{Z}}) \hookrightarrow Z$ is continuous.

Finally, $(\hat{Z}, (\cdot,\cdot)_{\hat{Z}})$ is complete and thus a Hilbert space. In fact, any Cauchy sequence (z_k) in $(\hat{Z}, (\cdot,\cdot)_{\hat{Z}})$ satisfies $A^*z_k = M^*q_k$ with $q_k \in Q$ and (12) shows that (q_k) is a Cauchy sequence in Q and hence $q_k \to q$ in Q. This implies $A^*z_k = M^*q_k \to M^*q$ in Y^*. By the continuous embedding $(\hat{Z}, (\cdot,\cdot)_{\hat{Z}}) \hookrightarrow Z$, (z_k) is also a Cauchy sequence in Z and thus $z_k \to z$ in Z which implies $A^*z_k \to A^*z$ in Y^*. We conclude that $A^*z = M^*q \in \operatorname{ran} M^*$ and thus $z \in \hat{Z}$. This shows that $(\hat{Z}, (\cdot,\cdot)_{\hat{Z}})$ is complete.

Since $C^* \in \mathcal{L}(Z, U)$ and $(\hat{Z}, (\cdot,\cdot)_{\hat{Z}}) \hookrightarrow Z$, we have $C^* \in \mathcal{L}(\hat{Z}, U)$ and thus $(\cdot,\cdot)_K = (\cdot,\cdot)_{\hat{Z}} + (C^*\cdot, C^*\cdot)_U$ is a bounded bilinear form on $(\hat{Z}, (\cdot,\cdot)_{\hat{Z}})$. Moreover, for all $z \in \hat{Z}$ holds

$$(z, z)_{\hat{Z}} \le (z, z)_{\hat{Z}} + (C^*z, C^*z)_U = (z, z)_K \le (1 + \|C^*\|^2_{\hat{Z},U})(z, z)_{\hat{Z}}$$

and thus $(\cdot,\cdot)_K$ is a scalar product on $\hat{Z}$ that induces an equivalent norm. □

Lemma 2 *Let Assumption 2 hold. Then the system*

$$\begin{aligned}(M^{-*}A^*\hat{z}, M^{-*}A^*w)_Q + (C^*\hat{z}, C^*w)_U &= \\ = -\langle r_2, w\rangle_{Z^*,Z} - (C^*A^{-*}r_1, C^*w)_U \quad &\forall\, w \in \hat{Z}\end{aligned} \tag{13}$$

has a unique solution $\hat{z} \in \hat{Z}$ *and the solution of* (9) *can be obtained by*

$$\Delta\lambda = \hat{z} + A^{-*}r_1, \tag{14}$$

$$\Delta y = A^{-1}(r_2 + CC^*\Delta\lambda). \tag{15}$$

Proof A proof can be found in [24, Lemma 2.4]. We give here a more constructive proof. Inserting $\Delta\lambda = \hat{z} + A^{-*}r_1$ in (9) yields the following system for $\hat{z}$.

$$\begin{pmatrix} M^*M & A^* \\ A & -CC^* \end{pmatrix} \begin{pmatrix} \Delta y \\ \hat{z} \end{pmatrix} = \begin{pmatrix} 0 \\ r_2 + CC^*A^{-*}r_1 \end{pmatrix}. \tag{16}$$

This shows that $A^*\hat{z} \in \operatorname{ran} M^*$ and therefore $\hat{z} \in \hat{Z}$. To derive a reduced system for $\hat{z}$, we perform block elimination with the second equation. This yields the equation

$$(A^* + M^*MA^{-1}CC^*)\hat{z} = -M^*MA^{-1}(r_2 + CC^*A^{-*}r_1)$$

and we observe that all terms are in $\operatorname{ran} M^*$. Applying the bijective operator $M^{-*} : \operatorname{ran} M^* \to Q$ yields the equivalent system

$$(M^{-*}A^* + MA^{-1}CC^*)\hat{z} = -MA^{-1}(r_2 + CC^*A^{-*}r_1).$$

Since $M^{-*}A^* : \hat{Z} \to Q$ is bijective, we obtain an equivalent variational equation if we use $M^{-*}A^*w$ with $w \in \hat{Z}$ as test functions. This leads to

$$\begin{aligned}(M^{-*}A^*w, M^{-*}A^*\hat{z})_Q + (M^{-*}A^*w, MA^{-1}CC^*\hat{z})_Q &= \\ = -(M^{-*}A^*w, MA^{-1}(r_2 + CC^*A^{-*}r_1))_Q \quad &\forall\, w \in \hat{Z},\end{aligned}$$

and since $A^{-*}M^*M^{-*}A^*w = w$ for all $w \in \hat{Z}$, we obtain (13). Hence, we have shown that (9) is with (14) equivalent to (16). Moreover, (16) is equivalent to (13) and the second equation in (16), while the latter is by (14) equivalent to (15). We conclude that (13), (14) and (15) are equivalent to (9).

The unique solvability of (13) follows from Lemma 1 and the Riesz representation theorem, since the left hand side can be written as $(\hat{z}, w)_K$, and $(\hat{Z}, (\cdot,\cdot)_K)$ is a Hilbert space by Lemma 1. □

Hence, we have seen that (9) can by (14), (15) be reduced to (13), which can by (11) be written as

$$(\hat{z}, w)_K = \langle r, w\rangle_{Z^*,Z} \quad \forall\, w \in \hat{Z}. \tag{17}$$

Clearly, this system can be solved by a preconditioned conjugate gradient method.

Following [24] we construct a preconditioner by approximately decoupling the operator K induced by $(\cdot,\cdot)_K$ into the product of two PDE operators.

By Assumptions 2, 3 we have $CIM \in \mathcal{L}(Y, Z^*)$ and therefore $(CIM)^* = M^*I^*C^* \in \mathcal{L}(Y^*, Z)$. We consider the preconditioner $\hat{K} : \hat{Z} \to \hat{Z}^*$ defined by

$$(z, w)_{\hat{K}} := (M^{-*}(A + CIM)^*z, M^{-*}(A + CIM)^*w)_Q \quad \forall\, w \in \hat{Z}. \tag{18}$$

The application of the preconditioner is described in Algorithm 2.1.

Algorithm 2.1 ***Application of the Preconditioner*** $\hat{\mathbf{K}}$
Input: $\ell \in \hat{Z}^*$
Output: Solution $z \in \hat{Z}$ *of* $(z, w)_{\hat{K}} = \langle \ell, w\rangle_{\hat{Z}^*,\hat{Z}} \forall\, w \in \hat{Z}$

1. *Let* $q \in Q$ *be the solution of*

$$(q, M^{-*}(A + CIM)^*w)_Q = \langle \ell, w\rangle_{\hat{Z}^*,\hat{Z}} \quad \forall\, w \in \hat{Z}.$$

2. *Let* $z \in Z$ *be the solution of*

$$\langle (A + CIM)^*z, w\rangle_{Y^*,Y} = \langle M^*q, w\rangle_{Y^*,Y} \quad \forall\, w \in Y. \tag{19}$$

Note that the solution z satisfies $A^*z = M^*q - M^*I^*C^*z \in \operatorname{ran}(M^*)$ and therefore $z \in \hat{Z}$.

The following Lemma estimates the condition number of K relative to the preconditioner $\hat{K}$.

Lemma 3 *Consider the preconditioner* $\hat{K} : \hat{Z} \to \hat{Z}^*$ *defined in* (18). *Assume that the quantity*

$$\gamma_{\hat{K}} := \sup_{0 \neq z \in \hat{Z}} \frac{(C^*z, C^*z)_U}{(z, z)_{\hat{K}}}$$

is finite. Then we have the estimate

$$\frac{1}{2}(z,z)_{\hat{K}} \le (z,z)_K \le (2+3\gamma_{\hat{K}})(z,z)_{\hat{K}} \quad \forall \hat{z} \in \hat{Z}. \tag{20}$$

Hence, the condition number of K relative to $\hat{K}$ is bounded by $\kappa_{\hat{K}} \le 4 + 6\gamma_{\hat{K}}$. Moreover, if $((M^{-*}A^* + \frac{1}{2}I^*C^*)z, I^*C^*z)_Q \ge 0$ for all $z \in \hat{Z}$ then we obtain the improved estimate

$$\frac{1}{2}(z,z)_{\hat{K}} \le (z,z)_K \le (1+\gamma_{\hat{K}})(z,z)_{\hat{K}}.$$

Proof The first part was already shown in [24]. For convenience, we present a complete proof. We use the inequality $2(q_1,q_2)_Q \le \|q_1\|_Q^2 + \|q_2\|_Q^2$. This yields

$$\begin{aligned}(z,z)_{\hat{K}} &= \|M^{-*}A^*z + I^*C^*z\|_Q^2 \le 2(\|M^{-*}A^*z\|_Q^2 + \|I^*C^*z\|_Q^2)\\ &\le 2(\|M^{-*}A^*z\|_Q^2 + \|C^*z\|_Q^2) = 2(z,z)_K.\end{aligned}$$

This shows the first inequality in (20) and the second follows from

$$\begin{aligned}(z,z)_K &= \|M^{-*}A^*z + I^*C^*z - I^*C^*z\|_Q^2 + \|C^*z\|_U^2\\ &= (z,z)_{\hat{K}} + \|I^*C^*z\|_Q^2 + \|C^*z\|_U^2 - 2(M^{-*}A^*z + I^*C^*z, I^*C^*z)_Q\\ &\le 2(z,z)_{\hat{K}} + 3\|C^*z\|_U^2 = 2(z,z)_{\hat{K}} + 3\|C^*z\|_U^2 \le (2+3\gamma_{\hat{K}})(z,z)_{\hat{K}}.\end{aligned}$$

If $((M^{-*}A^* + \frac{1}{2}I^*C^*)z, I^*C^*z)_Q \ge 0$ then we can improve the first estimate in the last line to

$$\begin{aligned}(z,z)_K &= (z,z)_{\hat{K}} + \|I^*C^*z\|_Q^2 + \|C^*z\|_U^2 - 2(M^{-*}A^*z + I^*C^*z, I^*C^*z)_Q\\ &\le (z,z)_{\hat{K}} + \|C^*z\|_U^2 \le (1+\gamma_{\hat{K}})(z,z)_{\hat{K}}.\end{aligned}$$

□

As we will see, for parabolic operators there exists an appropriate imbedding operator $J \in \mathcal{L}(\hat{Z}, Y)$ such that

$$\langle (A + CIM)^*z, Jz\rangle_{Y^*,Y} > 0 \quad \forall\, 0 \ne z \in \hat{Z}. \tag{21}$$

Then we have the following result that will be helpful to estimate $\gamma_{\hat{K}}$ for practical applications.

Lemma 4 *Let Assumptions 2 and 3 hold and assume that* (21) *is satisfied with an imbedding operator* $J \in \mathcal{L}(\hat{Z}, Y)$. *Then* $\gamma_{\hat{K}}$ *in Lemma 3 can be estimated by*

$$\gamma_{\hat{K}} \leq \sup_{0 \neq z \in \hat{Z}} \frac{\|C^* z\|_U^2 \|MJz\|_Q^2}{(\langle Jz, A^* z\rangle_{Y,Y^*} + (C^* z, IMJz)_U)^2}. \tag{22}$$

Proof Let $0 \neq z \in \hat{Z}$ be arbitrary. Then $(A^* + M^* I^* C^*)z \in \operatorname{ran}(M^*)$ and thus

$$\begin{aligned} 0 &< \langle (A^* + M^* I^* C^*)z, Jz\rangle_{Y^*,Y} = (M^{-*}(A^* + M^* I^* C^*)z, MJz)_Q \\ &\leq \|M^{-*}(A^* + M^* I^* C^*)z\|_Q \|MJz\|_Q = (z,z)_{\hat{K}}^{\frac{1}{2}} \|MJz\|_Q. \end{aligned}$$

Hence, we obtain

$$\gamma_{\hat{K}} = \sup_{0 \neq z \in \hat{Z}} \frac{\|C^* z\|_U^2 \|MJz\|_Q^2}{(z,z)_{\hat{K}} \|MJz\|_Q^2} \leq \sup_{0 \neq z \in \hat{Z}} \frac{\|C^* z\|_U^2 \|MJz\|_Q^2}{(\langle A^* z, Jz\rangle_{Y^*,Y} + (C^* z, IMJz)_U)^2}.$$

□

2.2 Application to Parabolic Control Problems

We consider a parabolic state equation of the form

$$\begin{aligned} y_t - \nabla \cdot (\sigma \nabla y) + a_1 \cdot \nabla y + a_0 y &= b + Bu \quad && \text{on } \Omega_T := (0,T) \times \Omega, \\ y(0,\cdot) &= y_0 && \text{on } \Omega, \\ y &= 0 && \text{on } (0,T) \times \partial\Omega, \end{aligned} \tag{23}$$

where $\sigma \in L^\infty(\Omega)$, $a_0 \in L^\infty(\Omega_T)$, $a_1 \in (H^1 \cap L^\infty)(\Omega_T)^n$, $\sigma \geq \sigma_0 > 0$ and $y_0 \in L^2(\Omega)$.

We set $V = H_0^1(\Omega)$, $W(0,T) := \{v \in L^2(0,T;V) : v_t \in L^2(0,T;V^*)\}$ and $Y := W(0,T)$. Let $Z = Z_1 \times Z_2 := L^2(0,T;V) \times L^2(\Omega)$ and assume that $b \in Z_1^*$ and $B \in \mathcal{L}(U, Z_1^*)$. We work with the usual weak solutions and define the operator $A \in \mathcal{L}(Y, Z^*)$ by

$$\begin{aligned} \langle Ay, (\lambda,\mu)\rangle_{Z^*,Z} = (y(0),\mu)_{L^2(\Omega)} + \int_0^T \big(&\langle y_t(t), \lambda(t)\rangle_{V^*,V} \\ &+ (\sigma \nabla y(t), \nabla\lambda(t))_{L^2(\Omega)} + (a_1 \cdot \nabla y(t) + a_0 y(t), \lambda(t))_{L^2(\Omega)} \big)\, dt. \end{aligned}$$

Then the state equation is given by

$$\langle Ay, (\lambda, \mu)\rangle_{Z^*,Z} = \int_0^T \langle b(t), \lambda(t)\rangle_{V^*,V}\, dt + (y_0, \mu)_{L^2(\Omega)} \quad \forall\, (\lambda, \mu) \in Z.$$

It is well known that $A \in \mathcal{L}(Y, Z^*)$ has a bounded inverse $A^{-1} \in \mathcal{L}(Z^*, Y)$. For $z = (\lambda, \mu) \in W(0, T) \times L^2(\Omega_T)$ we have

$$\begin{aligned} \langle y, A^*(\lambda, \mu)\rangle_{Y,Y^*} &= (y(T), \lambda(T))_{L^2(\Omega)} + (y(0), \mu - \lambda(0))_{L^2(\Omega)} \\ &\quad + \int_0^T \Big(-\langle y(t), \lambda_t(t)\rangle_{V,V^*} + (\sigma \nabla y(t), \nabla\lambda(t))_{L^2(\Omega)} \\ &\qquad\qquad + (a_1 \cdot \nabla y(t) + a_0 y(t), \lambda(t))_{L^2(\Omega)}\Big)\, dt. \end{aligned} \tag{24}$$

2.2.1 Tracking Type Functional and Distributed Control

Now consider for example the case

$$f(y) = \tfrac{1}{2}\|y - y_d\|^2_{L^2(\Omega_T)}, \quad U = L^2(\Omega_T), \quad B = I_{L^2(\Omega_T),L^2(0,T;V^*)} \tag{25}$$

with the natural imbedding $I_{L^2(\Omega_T),L^2(0,T;V^*)}$. Then $Q = U = L^2(\Omega_T)$, $I = \mathrm{id}_{L^2(\Omega_T)}$, $M = I_{Y,L^2(\Omega_T)}$, $M^* = I_{L^2(\Omega_T),Y^*}$, $\operatorname{ran} M^* = L^2(\Omega_T) \subset Y^*$ and $M^{-*} = \mathrm{id}_{L^2(\Omega_T)}$. Using (24) we see that

$$\hat{Z} = \{z \in Z : A^* z \in L^2(\Omega_T)\} \subset \{(\lambda, \lambda(0)) : \lambda \in W(0, T), \lambda(T) = 0\}\,. \tag{26}$$

In particular

$$\bar{Z} := \big\{(\lambda, \lambda(0)) : \lambda \in L^2(0, T; V \cap H^2(\Omega)), \lambda_t \in L^2(0, T; L^2(\Omega)), \lambda(T) = 0\big\}$$

is a dense subset of $\hat{Z}$ (and under additional regularity assumptions on Ω and the initial data it coincides with $\hat{Z}$).

In the case without control constraints we have

$$C = \alpha^{-\frac{1}{2}} I_{L^2(\Omega_T),L^2(0,T;V^*)}, \quad C^* = \alpha^{-\frac{1}{2}} I_{L^2(0,T;V),L^2(\Omega_T)}$$

and in the case with control constraints

$$C = \alpha^{-\frac{1}{2}} I_{L^2(\Omega_T),L^2(0,T;V^*)} D, \quad C^* = \alpha^{-\frac{1}{2}} D I_{L^2(0,T;V),L^2(\Omega_T)}, \tag{27}$$

with the multiplication operator $Dv = 1_{\mathcal{I}} v$ and the current estimate $\mathcal{I}$ of the inactive set. For a unified notation we set $\mathcal{I} = \Omega$ in the unconstrained case.

The application of the preconditioner in Algorithm 2.1 consists now of the following steps.

Algorithm 2.2 ***Preconditioner $\hat{\mathbf{K}}$ for Parabolic Problems***
Input: $\ell \in \hat{Z}^*$
Output: Solution $z \in \hat{Z}$ *of* $(z, w)_{\hat{K}} = \langle \ell, w \rangle_{\hat{Z}^*, \hat{Z}} \forall\, w \in \hat{Z}$

1. *Compute the solution* $q \in Q = L^2(\Omega_T)$ *of*

$$(q, -w_t - \nabla \cdot (\sigma \nabla w + a_1 w) + (a_0 + \alpha^{-\frac{1}{2}} 1_{\mathcal{I}})w)_{L^2(\Omega_T)} = \langle \ell, (w, w(0)) \rangle_{\hat{Z}^*, \hat{Z}} \quad \forall\, (w, w(0)) \in \bar{Z}.$$

   Note that we can use the dense subset $\bar{Z}$ *of* $\hat{Z}$ *as test space.*
2. *Compute a solution* $z = (\lambda, \lambda(0))$, $\lambda \in W(0, T)$, $\lambda(T) = 0$ *of*

$$\begin{aligned} &- \langle w, \lambda_t \rangle_{L^2(0,T;V), L^2(0,T;V^*)} + (\sigma \nabla w, \nabla \lambda)_{L^2(\Omega_T)} \\ &+ (a_1 \cdot \nabla w + (a_0 + \alpha^{-\frac{1}{2}} 1_{\mathcal{I}})w, \lambda)_{L^2(\Omega_T)} = (q, w)_{L^2(\Omega_T)} \;\forall\, w \in Y. \end{aligned} \tag{28}$$

   Here we have already used the knowledge that the result $z \in Z$ *lives actually in* $\hat{Z}$.

We note that the application of the preconditioner decouples into the solution of two parabolic problems. Later we will apply the parareal time domain decomposition method to perform these two solves in parallel.

We obtain the following estimate for the condition number of the system (13) with preconditioner (18).

Theorem 1 *Consider the system* (8) *written in the form* (9) *for the linearized state equation* (23) *and the objective function* (25). *Then the condition number* $\kappa_{\hat{K}}$ *of the operator* K *in the reduced system* (13) *relative to the preconditioner* $\hat{K}$ *in* (18) *is bounded by*

$$\kappa_{\hat{K}} \le 4 + 6\gamma_{\hat{K}}, \quad \gamma_{\hat{K}} \le \begin{cases} e^{cT} & \text{in the case without control constraints,} \\ \frac{e^{cT}}{2 c_P \sigma_0 \alpha^{\frac{1}{2}}} & \text{in the case with control constraints,} \end{cases}$$

where c_P *is a Poincaré constant on* Ω *and*

$$c = \max \left\{ 0, \frac{1}{\sigma_0} \|a_1\|^2_{L^\infty(\Omega_T)} - 2 \operatorname{essinf}_{\Omega_T} a_0 \right\} \tag{29}$$

In particular in the case $a_0 \ge 0$, $a_1 = 0$ *we have* $c = 0$.

The estimate for the control constrained case applies also to the case $U = L^2((0, T) \times \Omega_c)$ *with a smaller control domain* $\Omega_c \subsetneq \Omega$.

Proof We apply Lemmas 3 and 4, where we choose the operator $J \in \mathcal{L}(\hat{Z}, Y)$ in Lemma 4 by

$$J : (\lambda(t), \mu) \mapsto e^{ct}\lambda(t)$$

with c as in (29). Then we have with $z = (\lambda, \lambda(0)) \in \hat{Z}$

$$\langle A^* z, Jz\rangle_{Y^*,Y} = \frac{1}{2}\left(\langle A^* z, Jz\rangle_{Y^*,Y} + \langle z, AJz\rangle_{Z,Z^*}\right) = \frac{\|\lambda(0)\|_{L^2}^2 + e^{cT}\|\lambda(T)\|_{L^2}^2}{2}$$

$$+ \int_0^T e^{ct}\left((\sigma\nabla\lambda(t), \nabla\lambda(t))_{L^2} + (a_1 \cdot \nabla\lambda(t) + (\tfrac{1}{2}c + a_0)\lambda(t), \lambda(t))_{L^2}\right) dt$$

$$\geq \int_0^T e^{ct}\left(\tfrac{\sigma_0}{2}\|\nabla\lambda(t)\|_{L^2}^2 + (\tfrac{1}{2}c + a_0 - \tfrac{\|a_1\|_{L^\infty}^2}{2\sigma_0})\|\lambda(t)\|_{L^2}^2\right) dt$$

$$\geq \frac{\sigma_0}{2}\|e^{ct/2}\nabla\lambda(t)\|_{L^2(0,T;L^2(\Omega))}^2 \geq \frac{c_P\sigma_0}{2}\|MJ^{\frac{1}{2}}z\|_Q^2$$

with a Poincaré constant c_P. Using the concrete definition of M and C (22) yields for the unconstrained case

$$\gamma_{\hat{K}} \leq \sup_{0\neq z\in\hat{Z}} \frac{\|C^* z\|_U^2\|MJz\|_Q^2}{(\langle Jz, A^* z\rangle_{Y,Y^*} + (C^* z, IMJz)_U)^2}$$

$$\leq \sup_{0\neq(\lambda,\lambda(0))\in\hat{Z}} \frac{\alpha^{-1}\|\lambda\|_{L^2(\Omega_T)}^2\|e^{ct}\lambda\|_{L^2(\Omega_T)}^2}{(\frac{c_P\sigma_0}{2}\|e^{ct/2}\lambda\|_{L^2(\Omega_T)}^2 + \alpha^{-\frac{1}{2}}\|e^{ct/2}\lambda\|_{L^2(\Omega_T)}^2)^2}$$

$$= \sup_{0\neq(\lambda,\lambda(0))\in\hat{Z}} \frac{\alpha^{-1}\|e^{-ct/2}\lambda\|_{L^2(\Omega_T)}^2\|e^{ct/2}\lambda\|_{L^2(\Omega_T)}^2}{(\frac{c_P\sigma_0}{2}\|\lambda\|_{L^2(\Omega_T)}^2 + \alpha^{-\frac{1}{2}}\|\lambda\|_{L^2(\Omega_T)}^2)^2} \leq e^{cT}.$$

In the constrained case we obtain with the inactive set $\mathcal{I} \subset \Omega_T$

$$\gamma_{\hat{K}} \leq \sup_{0\neq z\in\hat{Z}} \frac{\|C^* z\|_U^2\|MJz\|_Q^2}{(\langle Jz, A^* z\rangle_{Y,Y^*} + (C^* z, IMJz)_U)^2}$$

$$\leq \sup_{0\neq(\lambda,\lambda(0))\in\hat{Z}} \frac{\alpha^{-1}\|\lambda\|_{L^2(\mathcal{I})}^2\|e^{ct}\lambda\|_{L^2(\Omega_T)}^2}{(\frac{c_P\sigma_0}{2}\|e^{ct/2}\lambda\|_{L^2(\Omega_T)}^2 + \alpha^{-\frac{1}{2}}\|e^{ct/2}\lambda\|_{L^2(\mathcal{I})}^2)^2}$$

$$= \sup_{0\neq(\lambda,\lambda(0))\in\hat{Z}} \frac{\alpha^{-1}\|e^{-ct/2}\lambda\|_{L^2(\mathcal{I})}^2\|e^{ct/2}\lambda\|_{L^2(\Omega_T)}^2}{(\frac{c_P\sigma_0}{2}\|\lambda\|_{L^2(\Omega_T)}^2 + \alpha^{-\frac{1}{2}}\|\lambda\|_{L^2(\mathcal{I})}^2)^2} \leq \frac{e^{cT}}{2c_P\sigma_0\alpha^{\frac{1}{2}}}.$$

This estimate case applies also to the case $U = L^2((0,T) \times \Omega_c)$ with a smaller control domain $\Omega_c \subsetneq \Omega$, since we can choose $\mathcal{I} = \Omega_c$ in the unconstrained case and $\mathcal{I} \subset \Omega_c$ in the control constrained case. □

Now consider the case

$$f(y) = \tfrac{1}{2}\|y - y_d\|^2_{L^2(\Omega_T)} + \tfrac{1}{2}\|y(T) - y_{d,T}\|^2_{L^2(\Omega)}, \quad U = L^2(\Omega_T). \tag{30}$$

Then $Q = L^2(\Omega_T) \times L^2(\Omega)$, $I = \mathrm{id}_{L^2(\Omega_T)}$, $M : y \in Y \mapsto (y, y(T)) \in Q$, $M^*q = (q_1, \cdot)_{L^2(\Omega_T)} + (q_2, \cdot(T))_{L^2(\Omega)}$. Using (24) we see that

$$\hat{Z} = \{z \in Z : A^*z \in \mathrm{ran}(M^*)\} \subset \{(\lambda, \lambda(0)) : \lambda \in W(0,T)\}.$$

In particular

$$\bar{Z} := \left\{(\lambda, \lambda(0)) : \lambda \in L^2(0,T; V \cap H^2(\Omega)), \lambda_t \in L^2(0,T; L^2(\Omega))\right\}$$

is a dense subset of $\hat{Z}$.

Theorem 2 *Under the assumptions of Theorem 1 but with the objective function* (30) *the condition number* $\kappa_{\hat{K}}$ *of the operator* K *in the reduced system* (13) *relative to the preconditioner* $\hat{K}$ *in* (18) *is in the case with and without control constraints bounded by*

$$\kappa_{\hat{K}} \le 4 + 6\gamma_{\hat{K}}, \quad \gamma_{\hat{K}} \le \frac{e^{cT}}{2\min\{1, c_P\sigma_0\}\alpha^{\frac{1}{2}}}$$

where c_P *is a Poincaré constant on* Ω *and* c *is given by* (29). *In particular in the case* $a_0 \ge 0$, $a_1 = 0$ *we have* $c = 0$.

The estimate applies also to the case $U = L^2((0,T) \times \Omega_c)$ *with a smaller control domain* $\Omega_c \subsetneq \Omega$.

Proof We obtain exactly as in the proof of Theorem 1 for $z = (\lambda, \lambda(0)) \in \hat{Z}$

$$\langle A^*z, Jz\rangle_{Y^*,Y} = \frac{1}{2}\left(\langle A^*z, Jz\rangle_{Y^*,Y} + \langle z, AJz\rangle_{Z,Z^*}\right) = \frac{\|\lambda(0)\|^2_{L^2} + e^{cT}\|\lambda(T)\|^2_{L^2}}{2}$$

$$+ \int_0^T e^{ct}\left((\sigma\nabla\lambda(t), \nabla\lambda(t))_{L^2} + (a_1 \cdot \nabla\lambda(t) + (\tfrac{1}{2}c + a_0)\lambda(t), \lambda(t))_{L^2}\right) dt$$

$$\ge \frac{\|\lambda(0)\|^2_{L^2} + e^{cT}\|\lambda(T)\|^2_{L^2}}{2} + \frac{c_P\sigma_0}{2}\|e^{ct/2}\lambda\|^2_{L^2}.$$

with a Poincaré constant c_P. Using the concrete definition of M and C (22) yields for $\mathcal{I} = \Omega_T$ in the unconstrained case and $\mathcal{I} \subset \Omega_T$ in the control constrained case

$$\begin{aligned}
\gamma_{\hat{K}} &\le \sup_{0 \neq z \in \hat{Z}} \frac{\|C^* z\|_U^2 \|MJz\|_Q^2}{(\langle Jz, A^* z\rangle_{Y,Y^*} + (C^* z, IMJz)_U)^2} \\
&\le \sup_{0 \neq (\lambda, \lambda(0)) \in \hat{Z}} \frac{\alpha^{-1} \|\lambda\|_{L^2(\mathcal{I})}^2 (\|e^{ct}\lambda\|_{L^2(\Omega_T)}^2 + \|e^{cT}\lambda(T)\|_{L^2(\Omega)}^2)}{(\frac{e^{cT}}{2}\|\lambda(T)\|_{L^2(\Omega)}^2 + \frac{c_P\sigma_0}{2}\|e^{ct/2}\lambda\|_{L^2(\Omega_T)}^2 + \alpha^{-\frac{1}{2}}\|e^{ct/2}\lambda\|_{L^2(\mathcal{I})}^2)^2} \\
&= \sup_{0 \neq (\lambda, \lambda(0)) \in \hat{Z}} \frac{\alpha^{-1} \|e^{-ct/2}\lambda\|_{L^2(\mathcal{I})}^2 (\|e^{ct/2}\lambda\|_{L^2(\Omega_T)}^2 + \|e^{cT/2}\lambda(T)\|_{L^2(\Omega)}^2)}{(\frac{1}{2}\|\lambda(T)\|_{L^2(\Omega)}^2 + \frac{c_P\sigma_0}{2}\|\lambda\|_{L^2(\Omega_T)}^2 + \alpha^{-\frac{1}{2}}\|\lambda\|_{L^2(\mathcal{I})}^2)^2} \\
&\le \frac{e^{cT}}{2\min\{1, c_P\sigma_0\}\alpha^{\frac{1}{2}}}.
\end{aligned}$$

This estimate case applies also to the case $U = L^2((0,T) \times \Omega_c)$ with a smaller control domain $\Omega_c \subsetneq \Omega$, since we can choose $\mathcal{I} = \Omega_c$ in the unconstrained case and $\mathcal{I} \subset \Omega_c$ in the control constrained case. □

2.2.2 An Improved Estimate of the Condition Number

We will now use regularity theory to improve the condition number estimate for the preconditioned system further. To this end, we extend the result of Schiela and Ulbrich [24] for elliptic problems to parabolic problems.

We focus on the state equation (23) with objective function and control operator according to (25) and the case (27) of control constraints with current inactive set $\mathcal{I}$ or partial control domain $\mathcal{I} = (0,T) \times \Omega_c$ with $\Omega_c \subsetneq \Omega$. In this case we have as above

$$\gamma_{\hat{K}} = \sup_{0 \neq z \in \hat{Z}} \frac{\|C^* z\|_{L^2(\Omega_T)}^2}{(z,z)_{\hat{K}}} = \sup_{0 \neq (\lambda, \lambda(0)) \in \hat{Z}} \frac{\|\alpha^{-\frac{1}{2}} 1_{\mathcal{I}} \lambda\|_{L^2(\Omega_T)}^2}{\|A^*(\lambda, \lambda(0)) + \alpha^{-\frac{1}{2}} 1_{\mathcal{I}}\lambda\|_{L^2(\Omega_T)}^2}. \tag{31}$$

We now derive an improved lower bound for $\|A^*(\lambda, \lambda(0)) + \alpha^{-\frac{1}{2}} 1_{\mathcal{I}}\lambda\|_{L^2(\Omega_T)}$. We need the following regularity assumption.

Assumption 4 *The operator CIM is a multiplication operator of the form*

$$(CIM)v(t,x) = \phi(t,x)v(t,x), \quad \phi(t,x) = \phi_1 1_{\mathcal{I}}(t,x), \quad (t,x) \in \Omega_T,$$

where $\phi_1 > 0$ is a constant. Assume that $\mathcal{I} \subset \Omega$ and $\mathcal{A} := \Omega \setminus \mathcal{I}$ have Lipschitz boundary.

Moreover, we assume that $\hat{A}$ *is cylinder-like, i.e., there exists a bi-Lipschitzian map* $(t,x) \in \Omega_T \mapsto (t, \tau(t,x)) \in \Omega_T$ *with* $\tau(t, \mathcal{A}(t)) = \hat{A}$ *for all* $t \in (0,T)$, *where* $\mathcal{A}(t) = \{x : (t,x) \in \mathcal{A}\}$, *and denote by* $\partial\mathcal{A}(t)$ *the boundary of* $\mathcal{A}(t)$ *relative to* Ω. *Finally, we assume that for any* $q \in L^2(\mathcal{A})$ *the solution* $(\lambda_{\mathcal{A}}, \mu_{\mathcal{A}}) \in L^2(0,T; H_0^1(\mathcal{A}(\cdot))) \times L^2(\mathcal{A}(0))$ *of the problem*

$$\langle w, A^*(\lambda_{\mathcal{A}}, \mu_{\mathcal{A}})\rangle_{Y,Y^*} = (w,q)_{L^2(\mathcal{A})} \;\forall\, w \in W_{\mathcal{A}}(0,T), \tag{32}$$

where $W_{\mathcal{A}}(0,T) = \{w \in L^2(0,T; H_0^1(\mathcal{A}(\cdot))) : w_t \in L^2(0,T; H_0^1(\mathcal{A}(\cdot)))^*\}$, *satisfies* $\lambda_{\mathcal{A}} \in W_{\mathcal{A}}(0,T) \cap L^2(0,T; H^{3/2+\varepsilon}(\mathcal{A}(\cdot))))$ *and its normal trace the estimate*

$$\|\partial_\nu \lambda_{\mathcal{A}}(t)\|_{L^2(\partial\mathcal{A}(t))} \le c_{tr,1}\|q(t)\|_{L^2(\mathcal{A}(t))}. \tag{33}$$

with a constant $c_{tr,1}$ *independent of* q *and* $t \in (0,T)$.

We note that for y_0, σ and $\mathcal{A}$ sufficiently regular this follows from parabolic regularity theory. We believe that also sets $\mathcal{A}$ that are not cylinder-like could be handled but leave this to future work.

Lemma 5 *Let* $\mathcal{J}$ *be an open domain with Lipschitz boundary. Then the following trace estimate holds for all* $v \in H^1(\mathcal{J})$ *with a constant* $c_{tr,2}$

$$\|v\|_{L^2(\partial\mathcal{J})} \le c_{tr,2}\sqrt{\|v\|_{H^1(\mathcal{J})}\|v\|_{L^2(\mathcal{J})}}.$$

Proof See [24, Lemma 5]. □

In the following lemma we will for $q \in L^2(\Omega_T)$ consider the problem to find $(\lambda, \lambda(0)) \in \hat{Z}$ with

$$\langle w, A^*(\lambda, \lambda(0))\rangle_{Y,Y^*} + (w, \phi_1\lambda)_{L^2(\mathcal{I})} = (w,q)_{L^2(\Omega_T)} \;\forall\, w \in W(0,T), \tag{34}$$

which corresponds with $\phi_1 = \alpha^{-\frac{1}{2}}$ to (19) and its concrete form (28). As observed in (26), (24) yields $\lambda(T) = 0$.

Lemma 6 *Consider problem* (34) *where* $\phi_1 > 0$ *and* $\mathcal{I}$, $\mathcal{A} = \Omega_T \setminus \mathcal{I}$ *have the properties defined in Assumption 4. Assume that* $q \in L^2(\Omega_T)$. *Then the solution* λ *of* (34) *satisfies*

$$\phi_1\|e^{ct/2}\lambda\|_{L^2(\mathcal{I})} \le e^{cT/2}\left(\|q\|_{L^2(\mathcal{I})} + c_{\mathcal{I}}\phi_1^{\frac{1}{4}}\|q\|_{L^2(\mathcal{A})}\right). \tag{35}$$

Here, c is defined in (29) *and* $c_{\mathcal{I}}$ *depends on* $c_{\mathrm{tr},1}$, $c_{\mathrm{tr},2}$, σ_0, *and the Poincaré constant* c_P *of* Ω. *Note that* $c = 0$ *if* $a_0 \ge 0$ *and* $a_1 = 0$.

Proof As in the proof for the elliptic case [24] we split λ into two parts $\lambda = \lambda_0 + \lambda_1$. Here, λ_0 is the extension by zero of the solution $\lambda_{\mathcal{A}}$ of the problem (32). Then

$\lambda_0|_{\mathcal{A}} = \lambda_{\mathcal{A}}$ and $\lambda_0|_{\mathcal{I}} = 0$. By using (24), we observe similarly as above that (24) yields $\lambda_{\mathcal{A}}(T) = 0$ and $\mu_{\mathcal{A}} = \lambda_{\mathcal{A}}(0)$. Under the regularity asserted by Assumption 4, (32) reads

$$\begin{aligned} &-\langle w, (\lambda_{\mathcal{A}})_t\rangle_{L^2(0,T;H_0^1(\mathcal{A}(\cdot))),L^2(0,T;H_0^1(\mathcal{A}(\cdot))^*)} + (\sigma\nabla w, \nabla\lambda_{\mathcal{A}})_{L^2(\mathcal{A})} \\ &+ (a_1\cdot\nabla w + a_0 w, \lambda_{\mathcal{A}})_{L^2(\mathcal{A})} = (q, w)_{L^2(\mathcal{A})}\ \forall\, w \in W_{\mathcal{A}}(0,T), \end{aligned} \tag{36}$$

where $\lambda_{\mathcal{A}}(T) = 0$.

By our trace estimate (33) we conclude

$$\|\partial_\nu \lambda_{\mathcal{A}}(t)\|_{L^2(\partial\mathcal{A}(t))} \le c_{tr,1}\|q(t)\|_{L^2(\mathcal{A}(t))}. \tag{37}$$

Since λ_0 is an extension by zero of $\lambda_{\mathcal{A}}$ and (36) is only valid for testfunctions $w \in W_{\mathcal{A}}(0,T)$, integration by parts on $\mathcal{A}$ shows that λ_0 satisfies

$$\begin{aligned} \langle w, A^*(\lambda_0, \lambda_0(0))\rangle_{Y,Y^*} - \int_0^T (w(t), \sigma\partial_\nu\lambda_{\mathcal{A}}(t))_{L^2(\partial\mathcal{A}(t))}\,dt = \ &(w,q)_{L^2(\mathcal{A})} \\ &\forall\, w \in W(0,T). \end{aligned} \tag{38}$$

Hence, if we define $\lambda_1 \in W(0,T)$ as the solution of the problem

$$\begin{aligned} \langle w, A^*(\lambda_1, \lambda_1(0))\rangle_{Y,Y^*} + (w, \phi_1\lambda_1)_{L^2(\mathcal{I})} + \int_0^T (w(t), \sigma\partial_\nu\lambda_{\mathcal{A}}(t))_{L^2(\partial\mathcal{A}(t))}\,dt \\ = (w,q)_{L^2(\mathcal{I})}\ \forall\, w \in W(0,T) \end{aligned} \tag{39}$$

we see by adding (38) and (39) that $\lambda = \lambda_0 + \lambda_1$ solves the original problem (34) (note that $\lambda_0|_{\mathcal{I}} = 0$).

To obtain an estimate for λ_1 we test (39) with $e^{ct}\lambda_1(t)$, where c is defined in (29). Then we get as in the proof of Theorem 1

$$\begin{aligned} &\|q\|_{L^2(\mathcal{I})}\|e^{ct}\lambda_1\|_{L^2(\mathcal{I})} + \int_0^T \|\sigma\partial_\nu\lambda_{\mathcal{A}}(t)\|_{L^2(\partial\mathcal{A}(t))}\|e^{ct}\lambda_1(t)\|_{L^2(\partial\mathcal{A}(t))}\,dt \\ &\ge \langle e^{ct}\lambda_1, A^*(\lambda_1, \lambda_1(0))\rangle_{+} + \phi_1\|e^{ct/2}\lambda_1\|^2_{L^2(\mathcal{I})} \\ &\ge \frac{\sigma_0}{2}\|e^{ct/2}\nabla\lambda(t)\|^2_{L^2(0,T;L^2(\Omega))} + \phi_1\|e^{ct/2}\lambda_1\|^2_{L^2(\mathcal{I})} \\ &\ge \frac{c_P\sigma_0}{2}\|e^{ct/2}\lambda(t)\|^2_{L^2(0,T;H^1(\Omega))} + \phi_1\|e^{ct/2}\lambda_1\|^2_{L^2(\mathcal{I})}. \end{aligned} \tag{40}$$

By Lemma 5 we obtain

$$\|e^{ct}\lambda_1(t)\|_{L^2(\partial\mathcal{A}(t))} \le c_{tr,2}\sqrt{\|e^{ct}\lambda_1(t)\|_{H^1(\mathcal{I}(t))}\|e^{ct}\lambda_1(t)\|_{L^2(\mathcal{I}(t))}}.$$

Division of (40) by the square-root of its right-hand side yields with (37)

$$\phi_1^{\frac{1}{2}} \|e^{ct/2}\lambda_1\|_{L^2(\mathcal{I})} \le \Bigg(\|q\|_{L^2(\mathcal{I})}\|e^{ct}\lambda_1\|_{L^2(\mathcal{I})}$$

$$+ \int_0^T \|\sigma\partial_\nu\lambda_{\mathcal{A}}(t)\|_{L^2(\partial\mathcal{A}(t))} c_{tr,2}\sqrt{\|e^{ct}\lambda_1(t)\|_{L^2(\mathcal{I}(t))}\|e^{ct}\lambda_1(t)\|_{L^2(\mathcal{I}(t))}}\,dt\Bigg)\cdot$$

$$\cdot\left(\frac{c_P\sigma_0}{2}\|e^{ct/2}\lambda(t)\|^2_{L^2(0,T;H^1(\Omega))} + \phi_1\|e^{ct/2}\lambda_1\|^2_{L^2(\mathcal{I})}\right)^{-\frac{1}{2}}$$

$$\le e^{cT/2}\left(\|q\|_{L^2(\mathcal{I})}\phi_1^{-\frac{1}{2}} + c_{\mathcal{I}}\|q\|_{L^2(\mathcal{A})}\phi_1^{-\frac{1}{4}}\right).$$

Since $\lambda = \lambda_1$ on $\mathcal{I}$ we obtain from this the estimate (35). Tracing back the constant $c_{\mathcal{I}}$, we notice that it depends only on $c_{\text{tr},1}$, $c_{\text{tr},2}$, σ_0, and c_P. □

We obtain the following improved estimate for the condition number of the system (13) with preconditioner (18).

Theorem 3 *Consider the system* (8) *written in the form* (9) *for the linearized state equation* (23) *and the objective function* (25)*. Assume that* $\mathcal{I} \subset \Omega_T$ *satisfies Assumption 4, where* $\mathcal{I}$ *is either the current estimate of the inactive set for control constraints or* $\mathcal{I} = (0,T)\times\Omega_c$ *in the case* $U = L^2((0,T)\times\Omega_c)$ *with a smaller control domain* $\Omega_c \subsetneq \Omega$*. Then the condition number* $\kappa_{\hat{K}}$ *of the operator K in the reduced system* (13) *relative to the preconditioner* $\hat{K}$ *in* (18) *is bounded by*

$$\kappa_{\hat{K}} \le 4 + 6\gamma_{\hat{K}}, \quad \gamma_{\hat{K}} \le e^{cT/2}(1 + c_{\mathcal{I}}\alpha^{-\frac{1}{8}}),$$

where $c_{\mathcal{I}}$ *is the constant in Lemma 6 and c is defined in* (29)*. In the case* $a_0 \ge 0$, $a_1 = 0$ *we have* $c = 0$.

Remark 1 For the control constrained case or the case of a local control domain the condition number estimate improves from $O(\alpha^{-\frac{1}{2}})$ in Theorem 1 to $O(\alpha^{-\frac{1}{4}})$. Hence, the dependence on the regularization parameter is quite week, if the active set or control set is regular enough.

Proof We apply Lemma 3. For our problem we have by (31)

$$\gamma_{\hat{K}} = \sup_{0\neq(\lambda,\lambda(0))\in\hat{Z}} \frac{\|\alpha^{-\frac{1}{2}}1_{\mathcal{I}}\lambda\|^2_{L^2(\Omega_T)}}{\|A^*(\lambda,\lambda(0)) + \alpha^{-\frac{1}{2}}1_{\mathcal{I}}\lambda\|^2_{L^2(\Omega_T)}}.$$

Applying Lemma 6 with $\phi_1 = \alpha^{-\frac{1}{2}}$ and $q = A^*(\lambda,\lambda(0)) + \alpha^{-\frac{1}{2}}1_{\mathcal{I}}\lambda$ we have

$$\|\alpha^{-\frac{1}{2}}1_{\mathcal{I}}\lambda\|_{L^2(\Omega_T)} \le \alpha^{-\frac{1}{2}}\|e^{ct/2}\lambda\|_{L^2(\mathcal{I})} \le e^{cT/2}(1 + c_{\mathcal{I}}\alpha^{-\frac{1}{8}})\|q\|_{L^2(\Omega_T)}$$

and thus

$$\gamma_{\hat{K}} \le e^{cT}(1 + c_{\mathcal{I}}\alpha^{-\frac{1}{8}})^2.$$

□

3 Parareal Time-Domain Decomposition

For parabolic problems, the application of the preconditioner $\hat{K}$ in Algorithm 2.2 decouples into two parabolic PDE solves. Therefore, time domain decomposition techniques are directly applicable within the preconditioner. In the following, we will use the parareal time-domain decomposition method as approximate solvers within the preconditioner.

The parareal algorithm was proposed by Lions et al. [18] to speed up the numerical solution of time dependent partial differential equations by using parallel computers with a sufficiently large number of processors.

3.1 *Description of the Parareal Method*

Although the preconditioner requires only the solution of linearized PDEs we describe the parareal method more generally for nonlinear PDEs. We consider a time-dependent PDE (or a system) of the general form

$$\begin{aligned} &y_t + F(t,x,y) = 0, \quad (t,x) \in \Omega_T := (0,T) \times \Omega, \\ &y(0,x) = y_0(x), \quad x \in \Omega, \end{aligned} \tag{41}$$

where $y : [0,T] \to V$ maps time to a Banach space $V \subset L^2(\Omega)$, $y_0 \in V$ are initial data and $F(t,x,y)$ is a possibly time dependent partial differential operator in the variables $x \in \Omega$. The parareal technique, which was originally proposed in [18] and slightly modified in [1, 2], uses a time-domain decomposition

$$0 = t_0 < t_1 < \ldots < t_N = T, \tag{42}$$

which is uniform in the sense that

$$\eta_0 \Delta T \le t_{n+1} - t_n \le \Delta T, \quad \text{where } \Delta T := \max_{0 \le n < N} t_{n+1} - t_n.$$

We assume that on each time domain $[t_n, t_{n+1}]$, $0 \le n < N$, there exists a unique solution propagator

$$g(t_n, \cdot): \quad v \in V \mapsto g(t_n, v) := y(t_{n+1}) \in V,$$

where y is the solution of

$$\begin{aligned} y_t + F(t, x, y) &= 0, \quad (t, x) \in (t_n, t_{n+1}) \times \Omega, \\ y(t_n, x) &= v(x), \quad x \in \Omega. \end{aligned} \tag{43}$$

Moreover, we assume that we have a coarse grid approximation $g_c(t_n, v)$ of the exact propagator $g(t_n, v)$ available.

Example 1 We will later use a backward Euler step

$$\frac{g_c(t_n, v) - v}{t_{n+1} - t_n} + F(t_{n+1}, x, g_c(t_n, v)) = 0$$

as coarse propagator. As we will see the dissipativity of the backward Euler discretization is useful to stabilize the parareal scheme.

The parareal method uses a multiple shooting reformulation of the initial value problem (41) corresponding to the time domain decomposition (42). It combines parallel fine grid propagators with a coarse propagator for the iterative solution of the multiple shooting reformulation of (41).

More precisely, the parareal algorithm is defined as follows.

Algorithm 3.1 ***"Parareal" Time Integration Method:***
Compute approximations y_n^k of $y_n = y(t_n)$, $1 \le n \le N$, for the solution y of (41) as follows.

1. Initialize by coarse scheme: $y_0^1 = y_0$,

$$y_{n+1}^1 = g_c(t_n, y_n^1), \quad 0 \le n < N.$$

2. For $k = 1, \dots, K-1$: $y_0^{k+1} = y_0$

$$y_{n+1}^{k+1} = \underbrace{g_c(t_n, y_n^{k+1})}_{\text{predictor}} + \underbrace{\left(g(t_n, y_n^k) - g_c(t_n, y_n^k)\right)}_{\substack{\text{corrector, computable} \\ \text{in parallel on time slabs}}}, \quad 0 \le n < N. \tag{44}$$

The predictor step has to be computed sequentially to propagate the new approximation y_n^{k+1} by the coarse propagator. The error is corrected by using the approximation y_n^k of the previous iteration and can thus be computed in parallel.

It is obvious that the exact solution $y_n = y(t_n)$ is a fixed point of the parareal scheme, since $y_{n+1} = g(t_n, y_n)$. Moreover, after N iterations the exact solution is obtained. However, to obtain an efficient scheme, we want to apply only a few parareal iterations and are interested in the convergence speed of the parareal method. In practice the exact propagator $g(t_n, y_n^k)$ is replaced by an approximation $g_f(t_n, y_n^k)$ obtained by a sufficiently accurate fine grid scheme.

The parareal scheme can directly be applied to nonlinear equations and is then a nonlinear iteration. Thus, user-provided nonlinear fine grid solvers can be used directly. Besides the fact that nonlinear solvers can be more robust and efficient for some problems the nonlinear parareal algorithm has the advantage that the state in the time slabs has not to be stored in contrast to, e.g., a Newton iteration with inner parareal solver for the linearized equation.

3.2 Convergence Properties of the Parareal Algorithm

In the following, we collect several convergence results for the parareal scheme in the context of time-dependent PDEs. The first result shows that the parareal after k iterations can be considered as a scheme of order km provided the solution is smooth enough and the coarse propagator has order m and is Lipschitz.

Theorem 4 ([1]) *Let $V_k \subset V_{k-1} \subset \ldots \subset V_0 = V$ be a scale of Banach spaces. Assume that*

1. (41) *is stable in all spaces V_j, $0 \le j \le k$, i.e.*

$$\|y(t)\|_{V_j} \le C\|v_0\|_{V_j} \quad \forall\, v_0 \in V_j,\ \ t \in [0, T].$$

2. *The coarse propagator g_c is Lipschitz in the sense that for $0 \le j < k$*

$$\max_{0 \le n < N} \|g_c(t_n, v) - g_c(t_n, w)\|_{V_j} \le (1 + C\Delta T)\|v - w\|_{V_j} \quad \forall\, v, w \in V_j.$$

3. *The coarse propagator g_c has order m in the sense that for $0 \le j < k$ with $\delta g(t_n, v) := g(t_n, v) - g_c(t_n, v)$ the estimate holds*

$$\max_{0 \le n < N} \|\delta g(t_n, v) - \delta g(t_n, w)\|_{V_j} \le C(\Delta T)^{m+1}\|v - w\|_{V_{j+1}} \quad \forall\, v, w \in V_{j+1}.$$

Then for initial data $y_0 \in V_k$ the parareal scheme after k iterations has an accuracy of order mk, i.e.,

$$\max_{1 \le n \le N} \|y(t_n) - y_n^k\|_{V_0} \le C(\Delta T)^{mk}\|y_0\|_{V_k}$$

with C independent of ΔT and y_0.

Proof See [1, Theorem 1]. Our Assumption 3 is a modification of assumption (H3) in [1] and is adjusted directly to the proof of Bal [1, Theorem 1]. □

The above result requires additional regularity of the solution in order to achieve order *km*. By considering dissipative coarse propagators, Bal [1] was able to obtain an improved result for certain linear partial differential operators by using Fourier analysis.

Theorem 5 ([1]) *Consider a linear Mth order spatial operator F in 1D with constant coefficients and symbol* $P(\xi) = \alpha_0 + \sum_{j=1}^{M-1} \alpha_j \xi^j + |\xi|^M$, *where* $\alpha_j \in \mathbb{C}$ *and* $\alpha_0 \geq 0$ *such that* $P(\xi) \geq 0$ *or* $\Re(P(\xi)) > 0$ *(e.g., heat equation, advection-diffusion equation). If for* $t_n = n\Delta T$, $\Delta T = T/N$, *the coarse propagator is given by the* θ*-scheme*

$$\frac{g_c(t_n, v) - v}{\Delta T} + A(\theta g_c(t_n, v) + (1-\theta)v) = 0$$

then for $\theta \in (1/2, 1]$ *and* $k = O(1)$ *the parareal scheme is stable and*

$$\max_{1 \leq n \leq N} \|y(t_n) - y_n^k\|_{H^s(\mathbb{R})} \leq C(\Delta T)^k \|v_0\|_{H^s(\mathbb{R})}.$$

for all $s \in \mathbb{R}$. *Here,* $H^s(\mathbb{R})$ *denotes the usual Sobolev space of order s.*

Proof As shown in [1], the result follows from Theorems 2 and 3 in [1]. □

The convergence behavior with respect to the iteration index k was analyzed in [10], where in particular the following result was shown.

Theorem 6 *Consider the heat equation in 1D, i.e., the spatial operator has the symbol* $P(\xi) = \xi^2$. *If for* $t_n = n\Delta T$, $\Delta T = T/N$, *the coarse propagator is a one-step method that has a stability function* $R(z)$ *such that*

$$\gamma_l := \sup_{z \leq 0} \frac{|e^z - R(z)|}{1 - |R(z)|} \in (0, 1),$$

(for example the backward Euler scheme) then the parareal scheme is stable and

$$\max_{1 \leq n \leq N} \|y(t_n) - y_n^k\|_{H^s(\mathbb{R})} \leq \gamma_l^k \max_{1 \leq n \leq N} \|y(t_n) - y_n^0\|_{H^s(\mathbb{R})}$$

for all $s \in \mathbb{R}$.

Proof In [10, Theorem 4.9,Theorem 5.1] the estimate is shown pointwise for the Fourier transforms. Integration with the appropriate weights yields the estimate in the Sobolev norms.

The stability function for the backward Euler scheme is $R(z) = \frac{1}{1-z}$ and yields $\gamma_l = 0.2984256075$, see [10]. □

3.3 Parareal as Preconditioned Iteration

For our purposes it is useful to note that the parareal scheme can one hand be seen as a scheme of higher order on the coarse time grid but can on the other hand also be interpreted as a preconditioned iteration. Consider for simplicity the case that (41) is linear. Then

$$g(t_n, y_n) = G_n y_n + c_n, \quad g_c(t_n, y_n) = G_{c,n} y_n + c_{c,n}.$$

and the relation $y_{n+1} = g(t_n, y_n)$, $0 \le n < N$ can be written as

$$My := \begin{pmatrix} I & & & & \\ -G_1 & I & & & \\ & -G_2 & I & & \\ & & \ddots & \ddots & \\ & & & -G_{N-1} & I \end{pmatrix} \begin{pmatrix} y_1 \\ \vdots \\ y_N \end{pmatrix} = \begin{pmatrix} c_0 + G_0 y_0 \\ c_1 \\ \vdots \\ c_{N-1} \end{pmatrix} =: c.$$

If we rewrite the parareal scheme (44) as

$$y_{n+1}^{k+1} - g_c(t_n, y_n^{k+1}) = y_{n+1}^k - g_c(t_n, y_n^k) + g(t_n, y_n^k) - y_{n+1}^k, \quad 0 \le n < N,$$

we see that with the coarse grid approximation M_c of M

$$M_c = \begin{pmatrix} I & & & & \\ -G_{c,1} & I & & & \\ & -G_{c,2} & I & & \\ & & \ddots & \ddots & \\ & & & -G_{c,N-1} & I \end{pmatrix}$$

the parareal scheme (44) can be written as

$$y^{k+1} = y^k + M_c^{-1}(c - My^k). \tag{45}$$

Theorems 4 and 5 show that under their assumptions for fixed k the operator $I - M_c^{-1}M$ has for sufficiently large N (i.e., ΔT small enough) a condition number $\ll 1$. On the other hand, Theorem 6 shows that for the heat equation we have

$$\|I - M_c^{-1}M\|_{\max_n \|(\cdot)_n\|_{H^s(\mathbb{R})}, \max_n \|(\cdot)_n\|_{H^s(\mathbb{R})}} \le \gamma_l,$$

where for example $\gamma_l = 0.2984256075$, if the backward Euler scheme is selected as coarse propagator.

Hence, we see that under appropriate assumptions M_c^{-1} is a good preconditioner for the operator M.

In the nonlinear case we introduce the operators

$$\mathcal{G}(y) = \begin{pmatrix} y_1 - g(t_0, y_0) \\ \vdots \\ y_N - g(t_{N-1}, y_{N-1}) \end{pmatrix}, \quad \mathcal{G}_c(y) = \begin{pmatrix} y_1 - g_c(t_0, y_0) \\ \vdots \\ y_N - g_c(t_{N-1}, y_{N-1}) \end{pmatrix}.$$

Then the parareal scheme can be written as the preconditioned iteration

$$\mathcal{G}_c(y^{k+1}) = \mathcal{G}_c(y^k) - \mathcal{G}(y^k)$$

which is (45) in the linear case. Hence, as already observed in [10] the parareal scheme can also be seen as a deferred correction method. In [10] it is also shown that the parareal method can be interpreted as a time-multigrid method.

4 A Parareal-Based Preconditioner for Optimality Systems and Numerical Results

We use now the parareal method as an inexact implementation of the two PDE solves in the preconditioner (18), see Algorithms 2.1 and 2.2. As in Sect. 2.2 we focus on the case of parabolic state equations. In the following we always assume that the exact propagator $g(t_n, y_n)$ is replaced by a fine grid propagator consisting of a time stepping scheme on a finer time grid within the time slab $[t_n, t_{n+1}]$.

4.1 *Strategies to Combine the Parareal Scheme and the Preconditioner*

There are essentially two different approaches to apply the parareal scheme within the proposed preconditioner (18):

In the first approach the parareal scheme with a fixed iteration number K is used as time discretization scheme for the state equation in (1) and also for the linearized state equation in (3). Then using the same parareal scheme and its adjoint scheme within the preconditioner (18) in Algorithm 2.2 is just the application of the preconditioner (18) to the discretized problem.

In the second approach the state equation in (1) and also the linearized state equation in (3) is discretized by the fine grid propagator and the parareal scheme with a fixed iteration number K is used within the preconditioner to solve the corresponding optimality system. The resulting preconditioner is then an inexact version of the preconditioner (18) in Algorithm 2.2, since the two PDE solves are approximated by K parareal iterations.

In the following we present numerical results for the second approach.

4.2 *Numerical Results for a Parabolic Optimal Control Problem*

As an example for (3) we consider the problem

$$\min_{\substack{y\in W(0,T)\\ u\in L^2((0,T)\times\Omega)}} \quad \frac{\beta}{2}\int_\Omega (y(T)-y_T)^2\,dx + \frac{1}{2}\int_0^T\int_\Omega (y-y_d)^2\,dx\,dt + \frac{\alpha}{2}\int_0^T\int_\Omega u^2\,dx\,dt$$

$$\text{s.t.} \qquad \begin{aligned} y_t - \Delta y &= u \quad \text{on } (0,T)\times\Omega,\\ y|_{(0,T)\times\partial\Omega} &= 0,\\ y(0,\cdot) &= y_0 \quad \text{on } \Omega, \end{aligned} \tag{46}$$

where $\beta \geq 0$ is a constant. The state and control space are given by

$$U = L^2((0,T)\times\Omega),$$
$$Y = W(0,T) = \left\{y : y \in L^2(0,T;H_0^1(\Omega)),\ \ y_t \in L^2(0,T;H^{-1}(\Omega))\right\}.$$

4.2.1 Propagators in the Parareal Scheme

For the implementation of the parareal method we use the equidistant time decomposition

$$T_n = n\Delta T, \quad \Delta T = \frac{T}{N}.$$

For the coarse propagator $g_\Delta(t_n, v; u)$ we apply one backward Euler step in time and use a standard 5-point stencil for the Laplacian, i.e., $y_{n+1} = g_c(t_n, v; u)$ is given by

$$\frac{y_{n+1} - v}{\Delta T} - Ay_{n+1} = u_n.$$

To approximate the exact propagator $g(t_n, v; u)$ we apply N_f steps of a Crank-Nicholson scheme with time step $\Delta t = \Delta T/N_f$ and use a standard 5-point stencil for the Laplacian. Hence, $y^f_{n+1} = g_f(t_n, v; u)$ is given by

$$\begin{aligned} v_0 &= v,\\ \frac{v_{j+1}-v_j}{\Delta t} - A\frac{v_{j+1}+v_j}{2} &= u_{n,j}, \quad j = 0,\ldots,N_f-1,\\ y^f_{n+1} &= v_{N_f}. \end{aligned}$$

Remark 2 For the solution of the time step equations in the propagators, standard multigrid solvers can directly be used. It would be possible to control the inexactness of the multigrid solvers by the generalized SQP-methods.

In addition, a coarser space grid could be used for the coarse propagator.

4.2.2 Numerical Results

We consider the specific problem

$$T = 1, \quad \Omega = (0,1)^2, \quad y_0(x) = 0,$$
$$y_d(t,x) = \sin(t)\, x_1 x_2 (1 - x_1)(1 - x_2), \quad y_T(x) = y_d(T,x).$$

For the discretization we use $N = 20$ time domains, coarse time step $\Delta T = \frac{1}{N}$, fine time step $\Delta t = \frac{\Delta T}{40}$ and a 100×100 equidistant space grid.

Unconstrained Problem

We apply a preconditioned cg method to the discretized version of (13) and use the preconditioner (18), see Algorithm 2.2, where we replace the exact discrete PDE-solves on the fine grid by K parareal iterations according to Algorithm 3.1, where we add a parallel fine grid solve on the time slabs for the final initial data y_n^K obtained from the parareal iterations. Then the costs of the K parareal iterations are K sequential coarse solves and K parallel fine grid solves on the N time slabs.

The iteration history for different choices of α and β is shown in Table 1 for $K = 2$ and $K = 3$. In both cases the preconditioned cg method is stopped if the relative residual is $\leq 10^{-2}$.

Table 1 Preconditioned conjugate gradient iterations for the solution of (13) with preconditioner (18) for a relative stopping tolerance 10^{-2}

$K = 2$ parareal its.			$K = 3$ parareal its.		
α	β	pcgits	α	β	pcgits
1e−3	1	15	1e−3	1	8
1e−4	1	7	1e−4	1	6
1e−5	1	4	1e−5	1	4
1e−6	1	5	1e−6	1	5
1e−3	1e−3	3	1e−3	1e−3	3
1e−4	1e−3	3	1e−4	1e−3	3
1e−5	1e−3	3	1e−5	1e−3	3
1e−6	1e−3	2	1e−6	1e−3	2

The two PDE solves within the preconditioner are approximated by K parareal iterations

Table 2 Semismooth Newton method, where the semismooth Newton system is solved by a preconditioned conjugate gradient method with preconditioner (18)

$\alpha = 1e{-}3$		$\alpha = 1e{-}4$	
Residual	pcgits	Residual	pcgits
1.3e+1	5	1.3e+2	5
3.6e−1	6	7.5e+1	3
1.3e−2	4	1.6e+1	6
3.9e−4	5	1.1e−1	8
2.3e−5	4	5.4e−3	5
		2.3e−4	4

The two PDE solves within the preconditioner are approximated by three parareal iterations

In the case $K = 2$, $\beta = 1e - 3$ and $\alpha \in [1e - 6, 1e - 3]$ we obtain with $N = 20$ processors we obtain the following time ratio between the time for the parallel solution of the SQP subproblem and the serial solve of the state equation.

$$\frac{\text{time(parallel SQP solve)}}{\text{time(serial state solve)}} = \frac{3 \times 4 \times (20 \text{ coarse} + 40 \text{ fine steps})}{800 \text{ fine steps}} = 0.9.$$

Control Constrained Problem

We add the control constraint

$$u \geq \frac{1}{10}$$

and apply a semismooth Newton method. The semismooth Newton system is reduced to the system (13). As above, we apply a preconditioned cg method to the discretized version of (13) and use the preconditioner (18), see Algorithm 2.2, where we replace the exact discrete PDE-solves on the fine grid by K parareal iterations according to Algorithm 3.1. We set $\beta = 1$ and use $K = 3$ parareal iterations within the preconditioner. The preconditioned cg method is terminated if the relative residual is $\leq 10^{-1}$.

The iteration history for different choices of α is shown in Table 2.

5 Conclusions

We have proposed a parallel preconditioner for optimality systems or semismooth Newton systems arising in parabolic optimal control problems without or with control constraints. The preconditioner is based on [24], see also [21]. We extend the estimates for the condition number to the case of parabolic optimal control problems. The estimates show no or only a weak dependence on the regularization parameter α. Since the preconditioner decouples into two PDE solves, we propose to obtain a parallel preconditioner by using the parareal method as approximate PDE solver within the preconditioner. The efficiency and very weak dependence on the regularization parameter α is confirmed by numerical results. Since the preconditioner requires only fine and coarse propagators for a slightly modified forward and backward PDE, it is quite easy to implement. The numerical tests show that already for 2–3 parareal iterations within the preconditioner the number of preconditioned cg iterations is very moderate. Hence, the preconditioner has potential for the parallel solution of large scale optimal control problems.

References

1. Bal, G.: On the convergence and the stability of the parareal algorithm to solve partial differential equations. In: Domain Decomposition Methods in Science and Engineering. Lecture Notes in Computational Science and Engineering, vol. 40, pp. 425–432. Springer, Berlin (2005)
2. Bal, G., Maday, Y.: A "parareal" time discretization for non-linear PDE's with application to the pricing of an American put. In: Recent Developments in Domain Decomposition Methods (Zürich, 2001). Lecture Notes in Computational Science and Engineering, vol. 23, pp. 189–202. Springer, Berlin (2002)
3. Battermann, A., Heinkenschloss, M.: Preconditioners for Karush-Kuhn-Tucker matrices arising in the optimal control of distributed systems. In: Control and Estimation of Distributed Parameter Systems (Vorau, 1996). International Series of Numerical Mathematics, vol. 126, pp. 15–32. Birkhäuser, Basel (1998)
4. Battermann, A., Sachs, E.W.: Block preconditioners for KKT systems in PDE-governed optimal control problems. In: Fast Solution of Discretized Optimization Problems (Berlin, 2000). International Series of Numerical Mathematics, vol. 138, pp. 1–18. Birkhäuser, Basel (2001)
5. Borzì, A.: Smoothers for control- and state-constrained optimal control problems. Comput. Vis. Sci. **11**(1), 59–66 (2008)
6. Borzi, A., Schulz, V.: Multigrid methods for PDE optimization. SIAM Rev. **51**(2), 361–395 (2009)
7. Dollar, H.S., Gould, N.I.M., Stoll, M., Wathen, A.J.: Preconditioning saddle-point systems with applications in optimization. SIAM J. Sci. Comput. **32**(1), 249–270 (2010)
8. Farhat, C., Chandesris, M.: Time-decomposed parallel time-integrators: theory and feasibility studies for fluid, structure, and fluid-structure applications. Int. J. Numer. Methods Eng. **58**(9), 1397–1434 (2003)
9. Fischer, P.F., Hecht, F., Maday, Y.: A parareal in time semi-implicit approximation of the Navier-Stokes equations. In: Domain Decomposition Methods in Science and Engineering. Lecture Notes in Computational Science and Engineering, vol. 40, pp. 433–440. Springer, Berlin (2005)
10. Gander, M.J., Vandewalle, S.: Analysis of the parareal time-parallel time-integration method. SIAM J. Sci. Comput. **29**(2), 556–578 (2007)
11. Gong, W., Hinze, M., Zhou, Z.J.: Space-time finite element approximation of parabolic optimal control problems. J. Numer. Math. **20**(2), 111–145 (2012)
12. Heinkenschloss, M.: A time-domain decomposition iterative method for the solution of distributed linear quadratic optimal control problems. J. Comput. Appl. Math. **173**(1), 169–198 (2005)

13. Heinkenschloss, M., Vicente, L.N.: Analysis of inexact trust-region SQP algorithms. SIAM J. Optim. **12**(2), 283–302 (2001/2002)
14. Herzog, R., Sachs, E.: Preconditioned conjugate gradient method for optimal control problems with control and state constraints. SIAM J. Matrix Anal. Appl. **31**(5), 2291–2317 (2010). doi:10.1137/090779127. http://dx.doi.org/10.1137/090779127
15. Hintermüller, M., Hinze, M.: A SQP-semismooth Newton-type algorithm applied to control of the instationary Navier-Stokes system subject to control constraints. SIAM J. Optim. **16**(4), 1177–1200 (2006)
16. Hintermüller, M., Ito, K., Kunisch, K.: The primal-dual active set strategy as a semismooth Newton method. SIAM J. Optim. **13**(3), 865–888 (2003) (2002)
17. Hinze, M., Pinnau, R., Ulbrich, M., Ulbrich, S.: Optimization with PDE Constraints. Mathematical Modelling: Theory and Applications, vol. 23. Springer, New York (2009)
18. Lions, J.L., Maday, Y., Turinici, G.: Résolution d'EDP par un schéma en temps "pararéel". C. R. Acad. Sci. Paris Sér. I Math. **332**(7), 661–668 (2001)
19. Maday, Y., Turinici, G.: A parareal in time procedure for the control of partial differential equations. C. R. Math. Acad. Sci. Paris **335**(4), 387–392 (2002)
20. Mathew, T.P., Sarkis, M., Schaerer, C.E.: Analysis of block parareal preconditioners for parabolic optimal control problems. SIAM J. Sci. Comput. **32**(3), 1180–1200 (2010)
21. Pearson, J.W., Stoll, M., Wathen, A.J.: Preconditioners for state-constrained optimal control problems with Moreau-Yosida penalty function. Numer. Linear Algebra Appl. **21**(1), 81–97 (2014)
22. Rees, T., Stoll, M., Wathen, A.: All-at-once preconditioning in PDE-constrained optimization. Kybernetika (Prague) **46**(2), 341–360 (2010)
23. Schiela, A., Günther, A.: An interior point algorithm with inexact step computation in function space for state constrained optimal control. Numer. Math. **119**(2), 373–407 (2011)
24. Schiela, A., Ulbrich, S.: Operator preconditioning for a class of inequality constrained optimal control problems. SIAM J. Optim. **24**(1), 435–466 (2014)
25. Schöberl, J., Zulehner, W.: Symmetric indefinite preconditioners for saddle point problems with applications to PDE-constrained optimization problems. SIAM J. Matrix Anal. Appl. **29**(3), 752–773 (2007) (electronic)
26. Staff, G.A., Rønquist, E.M.: Stability of the parareal algorithm. In: Domain Decomposition Methods in Science and Engineering, Lecture Notes in Computational Science and Engineering, vol. 40, pp. 449–456. Springer, Berlin (2005)
27. Stoll, M., Wathen, A.: Preconditioning for partial differential equation constrained optimization with control constraints. Numer. Linear Algebra Appl. **19**(1), 53–71 (2012)
28. Tröltzsch, F.: On the Lagrange-Newton-SQP method for the optimal control of semilinear parabolic equations. SIAM J. Control Optim. **38**(1), 294–312 (1999) (electronic)
29. Tröltzsch, F.: Optimal Control of Partial Differential Equations. Theory, Methods and Applications. Graduate Studies in Mathematics, vol. 112. American Mathematical Society, Providence, RI (2010). Translated from the 2005 German original by Jürgen Sprekels
30. Ulbrich, M.: Semismooth Newton methods for operator equations in function spaces. SIAM J. Optim. **13**(3), 805–842 (2003) (2002)
31. Ulbrich, M.: Semismooth Newton Methods for Variational Inequalities and Constrained Optimization Problems in Function Spaces. MOS-SIAM Series on Optimization, vol. 11. Society for Industrial and Applied Mathematics (SIAM)/Mathematical Optimization Society, Philadelphia (2011)
32. Ulbrich, M., Ulbrich, S.: Primal-dual interior-point methods for PDE-constrained optimization. Math. Program. **117**(1–2, Ser. B), 435–485 (2009)
33. Ziems, J.C., Ulbrich, S.: Adaptive multilevel inexact SQP methods for PDE-constrained optimization. SIAM J. Optim. **21**(1), 1–40 (2011)
34. Zulehner, W.: Analysis of iterative methods for saddle point problems: a unified approach. Math. Comput. **71**(238), 479–505 (2002) (electronic)

Space-Time Discontinuous Galerkin Methods for Optimal Control Problems Governed by Time Dependent Diffusion-Convection-Reaction Equations

Tuğba Akman and Bülent Karasözen

Abstract In this paper, space-time discontinuous Galerkin finite element method for distributed optimal control problems governed by unsteady diffusion-convection-reaction equation without control constraints is studied. Time discretization is performed by discontinuous Galerkin method with piecewise constant and linear polynomials, while symmetric interior penalty Galerkin with upwinding is used for space discretization. We present some numerical results in order to evaluate the performance of the method.

1 Introduction

Optimal control problems (OCPs) governed by diffusion-convection-reaction equations arise in environmental control problems, optimal control of fluid flow and in many other applications. It is well known that the standard Galerkin finite element discretization causes non-physical oscillating solutions when convection dominates. Stable and accurate numerical solutions can be achieved by various effective stabilization techniques such as the streamline upwind/Petrov Galerkin (SUPG) finite element method [10], the local projection stabilization [4], the edge stabilization [19] for steady state OCPs. Recently, discontinuous Galerkin (dG) methods gain importance due to their better convergence behaviour, local mass conservation, flexibility in approximating rough solutions on complicated meshes, mesh adaptation and weak imposition of the boundary conditions in OCPs (see, e.g., [21, 22, 36, 37]).

In the recent years, much effort has been spent on parabolic OCPs (see, e.g., [2, 24]). However, there are few publications dealing with OCPs governed by nonstationary diffusion-convection-reaction equation. In many publications, for

T. Akman (✉) • B. Karasözen
Department of Mathematics and Institute of Applied Mathematics, Middle East Technical University, 06800 Ankara, Turkey
e-mail: tr.tugba.akman@gmail.com; bulent@metu.edu.tr

T. Carraro et al. (eds.), *Multiple Shooting and Time Domain Decomposition Methods*, Contributions in Mathematical and Computational Sciences 9,
DOI 10.1007/978-3-319-23321-5_9

space discretization, conforming finite elements are used. In [16, 17], a priori error estimates are derived for the characteristic finite element method, whereas for time discretization, backward Euler method is used. Crank-Nicolson time discretization is applied to OCP of diffusion-convection equation in [5]. In the study of Chrysafinos [7], a priori error estimates for unconstrained parabolic OCP, where conforming finite elements are combined with dG time discretization, are presented by decoupling the optimality system. In [17], error analysis concerning the characteristic finite element solution of the OCP with control constraints is discussed. The local dG approximation of the OCP which is discretized by backward Euler in time is derived in [38] and a priori error estimates for semi-discrete OCP are provided in [30]. We present a priori error analysis for SIPG discretization combined with backward Euler method in [1].

In this paper, we solve the OCP governed by diffusion-convection-reaction equation without control constraints by applying symmetric interior penalty Galerkin (SIPG) method in space and dG discretization in time [14]. We adapt the error analysis [7] to space-time dG discretization. For this purpose, we divide the error analysis into three parts as in [17] and use the error estimates for dG bilinear forms.

Discontinuous Galerkin discretization schemes have the pleasant property that discretization and dualization interchange, i.e. discretization and optimization commute. There are two different approaches for solving OCPs: *optimize-then-discretize (OD)* and *discretize-then-optimize (DO)*. In *OD* approach, first, the infinite dimensional optimality system is derived containing the state and the adjoint equation and the variational inequality. Then, the optimality system is discretized by using a suitable discretization method in space and time. In *DO* approach, the infinite dimensional OCP is discretized and then the finite-dimensional optimality system is derived. *OD* and *DO* approaches do not commute in general for OCPs governed by diffusion-convection-reaction equation [10]. However, commutativity is achieved in the case of SIPG discretization for steady state problems [21, 36]. Recently, dG time discretization has been applied to PDE constrained optimization problems, which is solved by the indirect multiple shooting method [18]. The multiple shooting method was formulated in function spaces and discretized afterwards, where dG time discretization commutes for both approaches. The spatial mesh was adapted at each constant time step using a dual weighted residual error estimate.

The rest of the paper is organized as follows. In Sect. 2, we define the model problem and then derive the optimality system. In Sect. 3, dG discretization and the semi-discrete optimality system follow. In Sect. 4, the fully discrete optimality system, which is discretized by space-time dG method and θ-scheme, are presented. Under dG time discretization, we show that *OD* and *DO* approaches commute for time-dependent problems, too. In Sect. 5, some auxiliary results accompanied with a priori error estimates for the fully discrete optimality system follow. In Sect. 6, numerical results are shown in order to evaluate the performance of the suggested method. Additionally, we give some numerical results for θ-method and compare them with the dG time discretization. The paper ends with some conclusions.

2 The Optimal Control Problem

We consider the following distributed optimal control problem governed by the unsteady diffusion-convection-reaction equation

$$\begin{aligned}
&\underset{u\in L^2(0,T;L^2(\Omega))}{\text{minimize}} \quad J(y,u) := \frac{1}{2}\int_0^T \left(\|y-y_d\|^2_{L^2(\Omega)} + \alpha \|u\|^2_{L^2(\Omega)} \right) dt, \\
&\text{subject to } \partial_t y - \epsilon \Delta y + \beta \cdot \nabla y + ry = f + u \quad (x,t) \in \Omega \times (0,T], \\
&\qquad\qquad y(x,t) = 0 \qquad (x,t) \in \partial\Omega \times [0,T], \\
&\qquad\qquad y(x,0) = y_0(x) \qquad x \in \Omega.
\end{aligned} \tag{1}$$

We adopt the standard notations for Sobolev spaces on computational domains and their norms. Let Ω be a bounded convex polygonal domain in $\mathbb{R}^2$ with Lipschitz boundary $\partial\Omega$. The inner product in $L^2(\Omega)$ is denoted by $(\cdot,\cdot)$. The source function and the desired state are denoted by $f \in L^2(0,T;L^2(\Omega))$ and $y_d \in L^2(0,T;L^2(\Omega))$, respectively. The initial condition is also defined as $y_0(x) \in H_0^1(\Omega)$. The diffusion and the reaction coefficients are $\epsilon > 0$ and $r \in L^\infty(\Omega)$, respectively. The velocity field $\boldsymbol{\beta} \in (W^{1,\infty}(\Omega))^2$ satisfies the incompressibility condition, i.e. $\nabla \cdot \boldsymbol{\beta} = 0$. Furthermore, we assume the existence of the constant C_0 such that $r \geq C_0$ a.e. in Ω so that the well-posedness of the OCP (1) is guaranteed. The trial and the test spaces are $Y = V = H_0^1(\Omega)$, $\forall t \in (0,T]$.

It is well known that the functions $(y,u) \in H^1(0,T;L^2(\Omega)) \cap L^2(0,T;Y) \times L^2(0,T;L^2(\Omega))$ solve (1) if and only if there is an adjoint $p \in H^1(0,T;L^2(\Omega)) \cap L^2(0,T;Y)$ such that (y,u,p) is the unique solution of the following optimality system [23, 32],

$$\begin{aligned}
(\partial_t y, v) + a(y,v) &= (f+u, v), \quad \forall v \in V, \quad y(x,0) = y_0, \\
-(\partial_t p, \psi) + a(\psi, p) &= -(y - y_d, \psi), \quad \forall \psi \in V, \quad p(x,T) = 0, \\
\int_0^T (\alpha u - p, w - u)\, dt &= 0, \quad \forall w \in L^2(0,T;L^2(\Omega))
\end{aligned} \tag{2}$$

with the bilinear form

$$a(y,v) = \int_\Omega (\epsilon \nabla y \cdot \nabla v + \beta \cdot \nabla y v + ryv)\, dx,$$

where $(\cdot,\cdot)$ is the inner product in $L^2(\Omega)$.

3 Discontinuous Galerkin Semidiscretization

Let $\{\mathcal{T}_h\}_h$ be a family of shape regular meshes such that $\overline{\Omega} = \cup_{K\in\mathcal{T}_h}\overline{K}$, $K_i \cap K_j = \emptyset$ for $K_i, K_j \in \mathcal{T}_h$, $i \neq j$. The diameters of elements K are denoted by h_K. The maximum diameter is $h = \max\limits_{K\in\mathcal{T}_h} h_K$. In addition, the length of an edge E is denoted by h_E.

In this paper, we consider discontinuous piecewise finite element spaces to define the discrete test, state and control spaces

$$V_{h,p} = Y_{h,p} = U_{h,p} = \left\{y \in L^2(\Omega) \,:\, y\,|_K \in \mathbb{P}^p(K) \quad \forall K \in \mathcal{T}_h\right\}. \tag{3}$$

Here, $\mathbb{P}^p(K)$ denotes the set of all polynomials on $K \in \mathcal{T}_h$ of degree p.

We split the set of all edges $\mathcal{E}_h$ into the set $\mathcal{E}_h^0$ of interior edges and the set $\mathcal{E}_h^\partial$ of boundary edges so that $\mathcal{E}_h = \mathcal{E}_h^\partial \cup \mathcal{E}_h^0$. Let $\mathbf{n}$ denote the unit outward normal to $\partial\Omega$. We define the inflow boundary

$$\Gamma^- = \{x \in \partial\Omega \,:\, \beta\cdot\mathbf{n}(x) < 0\}$$

and the outflow boundary $\Gamma^+ = \partial\Omega \setminus \Gamma^-$. The boundary edges are decomposed into edges $\mathcal{E}_h^- = \left\{E \in \mathcal{E}_h^\partial \,:\, E \subset \Gamma^-\right\}$ that correspond to inflow boundary and edges $\mathcal{E}_h^+ = \mathcal{E}_h^\partial \setminus \mathcal{E}_h^-$ that correspond to outflow boundary. The inflow and outflow boundaries of an element $K \in \mathcal{T}_h$ are defined by

$$\partial K^- = \{x \in \partial K \,:\, \beta\cdot\mathbf{n}_K(x) < 0\}, \quad \partial K^+ = \partial K \setminus \partial K^-,$$

where $\mathbf{n}_K$ is the unit normal vector on the boundary ∂K of an element K.

Let the edge E be a common edge for two elements K and K^e. For a piecewise continuous scalar function y, there are two traces of y along E, denoted by $y|_E$ from interior of K and $y^e|_E$ from interior of K^e. Then, the jump and average of y across the edge E are defined by:

$$[\![y]\!] = y|_E\mathbf{n}_K + y^e|_E\mathbf{n}_{K^e}, \quad \{\!\{y\}\!\} = \frac{1}{2}\big(y|_E + y^e|_E\big). \tag{4}$$

Similarly, for a piecewise continuous vector field ∇y, the jump and average across an edge E are given by

$$[\![\nabla y]\!] = \nabla y|_E\cdot\mathbf{n}_K + \nabla y^e|_E\cdot\mathbf{n}_{K^e}, \quad \{\!\{\nabla y\}\!\} = \frac{1}{2}\big(\nabla y|_E + \nabla y^e|_E\big). \tag{5}$$

For a boundary edge $E \in K \cap \Gamma$, we set $\{\!\{\nabla y\}\!\} = \nabla y$ and $[\![y]\!] = y\mathbf{n}$ where $\mathbf{n}$ is the outward normal unit vector on Γ.

The state equation (1) in space for fixed control u is discretized by the symmetric interior penalty method (SIPG). The convective term is discretized by the upwind method [3]. This leads to the following semi-discrete state equation

$$(\partial_t y_h, v_h) + a_h^s(y_h, v_h) + b_h(u_h, v_h) = (f, v_h) \quad \forall v_h \in V_{h,p}, \quad t \in (0, T], \tag{6}$$

with the (bi-)linear forms

$$\begin{aligned} a^d(y, v) = & \sum_{K \in \mathcal{T}_h} \int_K \epsilon \nabla y \cdot \nabla v \, dx \\ & - \sum_{E \in \mathcal{E}_h} \int_E \Big(\{\!\{\epsilon \nabla y\}\!\} \cdot [\![v]\!] + \{\!\{\epsilon \nabla v\}\!\} \cdot [\![y]\!] - \overbrace{\epsilon \frac{\sigma}{h_E} [\![y]\!] \cdot [\![v]\!]}^{J_\sigma(y,v)} \Big) \, ds \end{aligned} \tag{7}$$

and

$$\begin{aligned} a_h^s(y, v) = & \, a^d(y, v) + \sum_{K \in \mathcal{T}_h} \int_K \big(\beta \cdot \nabla y v + r y v \big) \, dx \\ & + \sum_{K \in \mathcal{T}_h} \int_{\partial K^- \setminus \Gamma^-} \beta \cdot \mathbf{n}(y^e - y) v \, ds - \sum_{K \in \mathcal{T}_h} \int_{\partial K^- \cap \Gamma^-} \beta \cdot \mathbf{n} y v \, ds, \end{aligned} \tag{8}$$

$$b_h(u, v) = - \sum_{K \in \mathcal{T}_h} \int_K u v \, dx. \tag{9}$$

The penalty parameter $\sigma > 0$ should be sufficiently large to ensure the stability of the dG discretization [26, Sect. 2.7.1] with a lower bound depending only on the polynomial degree.

Let f_h, y_h^d and y_h^0 be approximations of the source function f, the desired state function y_d and the initial condition y_0, respectively. Then, the semi-discrete approximation of the OCP (2) can be defined as follows:

$$\begin{aligned} & \underset{u_h \in L^2(0,T;U_{h,p})}{\text{minimize}} \int_0^T \Big(\frac{1}{2} \sum_{K \in \mathcal{T}_h} \|y_h - y_h^d\|_{L^2(K)}^2 + \frac{\alpha}{2} \sum_{K \in \mathcal{T}_h} \|u_h\|_{L^2(K)}^2 \Big) \, dt, \\ & \text{subject to } (\partial_t y_h, v_h) + a_h^s(y_h, v_h) + b_h(u_h, v_h) = (f_h, v_h), \; t \in (0, T], \, v_h \in V_{h,p} \\ & \qquad y_h(x, 0) = y_h^0. \end{aligned} \tag{10}$$

The semi-discrete optimality system is written as follows:

$$\begin{aligned} & (\partial_t y_h, v_h) + a_h^s(y_h, v_h) + b(u_h, v_h) = (f_h, v_h), \; \forall v_h \in V_{h,p} \quad y_h(x, 0) = y_h^0, \\ & -(\partial_t p_h, \psi_h) + a_h^a(p_h, \psi_h) = -(y_h - y_h^d, \psi_h), \; \forall \psi_h \in V_{h,p}, \quad p_h(x, T) = 0, \\ & \textstyle\int_0^T (\alpha u_h - p_h, w_h - u_h) \, dt = 0, \quad \forall w_h \in L^2(0, T; U_{h,p}), \end{aligned} \tag{11}$$

where

$$
\begin{aligned}
a_h^a(p,\psi) &= \sum_{K\in\mathscr{T}_h}\int_K \epsilon\nabla p\cdot\nabla\psi\, dx \\
&- \sum_{E\in\mathscr{E}_h}\int_E \left(\{\epsilon\nabla p\}\cdot[\![\psi]\!] + \{\epsilon\nabla\psi\}\cdot[\![p]\!] - \frac{\sigma\epsilon}{h_E}[\![p]\!]\cdot[\![\psi]\!]\right) ds \\
&+ \sum_{K\in\mathscr{T}_h}\int_K \left(-\beta\cdot\nabla p\psi + rp\psi\right) dx \\
&- \sum_{K\in\mathscr{T}_h}\int_{\partial K+\setminus\Gamma+} \beta\cdot\mathbf{n}(p^e - p)\psi\, ds + \sum_{K\in\mathscr{T}_h}\int_{\partial K+\cap\Gamma+} \beta\cdot\mathbf{n}p\psi\, ds.
\end{aligned}
$$

4 Time Discretization of the Optimal Control Problem

In this section, we derive the fully-discrete optimality system using θ-method and dG time stepping method [14]. The fully discrete optimality systems are compared for *optimize-then-discretize (OD)* and *discretize-then-optimize (DO)* approaches.

4.1 Time Discretization Using θ-Method

Let $0 = t_0 < t_1 < \cdots < t_{N_T} = T$ be a subdivision of $I = (0, T)$ with time intervals $I_m = (t_{m-1}, t_m]$ and time steps $k_m = t_m - t_{m-1}$ for $m = 1, \ldots, N_T$ and $k = \max_{1\le m\le N_T} k_m$.

We start with *OD* approach, so we discretize the semi-discrete optimality system (11) using θ-method as follows:

$$
\begin{aligned}
&(y_{h,m+1} - y_{h,m}, v) + ka_h^s((1-\theta)y_{h,m} + \theta y_{h,m+1}, v) = \\
&k((1-\theta)f_{h,m} + \theta f_{h,m+1}) + k((1-\theta)u_{h,m} + \theta u_{h,m+1}, v), \quad m = 0,\cdots,N-1, \\
&y_{h,0}(x,0) = y_0 \\
&(p_{h,m} - p_{h,m+1}, q) + ka_h^a(\theta p_{h,m} + (1-\theta)p_{h,m+1}, q) = \\
&-k\left(\theta(y_{h,m} - y_{h,m}^d, q) + (1-\theta)(y_{h,m+1} - y_{h,m+1}^d, q)\right), \quad m = N-1,\cdots,0, \\
&p_{h,N} = 0, \\
&(\alpha u_{h,m} - p_{h,m}, w - u_{h,m}) = 0, \quad m = 0,1,\ldots,N.
\end{aligned} \tag{12}
$$

In *DO* approach, the first and the second parts of the cost functional are discretized by the rectangle rule and the trapezoidal rule, respectively, so that the value of the adjoint at the final time becomes zero as in [29]. Then, we have the following fully-discrete OCP:

$$\text{minimize } \frac{k}{2}\sum_{m=0}^{N-1}(y_{h,m}-y^d_{h,m})^T M (y_{h,m}-y^d_{h,m})$$

$$+\alpha\frac{k}{2}\left(\frac{1}{2}u^T_{h,0}Mu_{h,0}+\sum_{m=1}^{N-1}u^T_{h,m}Mu_{h,m}+\frac{1}{2}u^T_{h,N}Mu_{h,N}\right)$$

subject to

$$(y_{h,m+1}-y_{h,m},v)+ka^s_h((1-\theta)y_{h,m}+\theta y_{h,m+1},v)=$$

$$k((1-\theta)f_{h,m}+\theta f_{h,m+1})+k((1-\theta)u_{h,m}+\theta u_{h,m+1},v),\quad m=0,\cdots,N-1,$$

$$(y_{h,0},v)=(y_0,v),$$

where M is the mass matrix.

Now, we construct the discrete Lagrangian

$$\mathscr{L}\,(y_{h,1},\ldots,y_{h,N},p_{h,0},\ldots,p_{h,N},u_{h,0},\ldots,u_{h,N})$$

$$=\frac{k}{2}\sum_{m=0}^{N-1}(y_{h,m}-y^d_{h,m})^T M (y_{h,m}-y^d_{h,m})$$

$$+\ \alpha\frac{k}{2}\left(\frac{1}{2}u^T_{h,0}Mu_{h,0}+\sum_{m=1}^{N-1}u^T_{h,m}Mu_{h,m}+\frac{1}{2}u^T_{h,N}Mu_{h,N}\right)+(y_{h,0}-y_0,p_{h,0})$$

$$+\ \sum_{m=0}^{N-1}((y_{h,m+1}-y_{h,m},p_{h,m+1})+ka^s_h((1-\theta)y_{h,m}+\theta y_{h,m+1},p_{h,m+1})$$

$$-\ k((1-\theta)f_{h,m}+\theta f_{h,m+1})+k((1-\theta)u_{h,m}+\theta u_{h,m+1},p_{h,m+1})). \tag{13}$$

By differentiating the Lagrangian (13), we derive the fully-discrete optimality system

$$(y_{h,m+1}-y_{h,m},v)+ka^s_h((1-\theta)y_{h,m}+\theta y_{h,m+1},v)=$$

$$k((1-\theta)f_{h,m}+\theta f_{h,m+1})+k((1-\theta)u_{h,m}+\theta u_{h,m+1},v),\quad m=0,\cdots,N-1$$

$$y_{h,0}(x,0)=y_0$$

$$(q,p_{h,N})+ka^s_h(q,\theta p_{h,N})=0,$$

$$(p_{h,m}-p_{h,m+1},q)+ka^s_h(q,\theta p_{h,m}+(1-\theta)p_{h,m+1})= \tag{14}$$

$$
\begin{aligned}
&-k(y_{h,m} - y^d_{h,m}, q), \quad m = N-1, \ldots, 1, \\
&(q, p_{h,0} - p_{h,1}) + k a^s_h(q, (1-\theta)p_{h,1}) = -k(y_{h,0} - y^d_{h,0}, q), \\
&(\frac{\alpha}{2} u_{h,0} - (1-\theta)p_{h,1}, w - u_{h,0}) = 0, \\
&(\alpha u_{h,m} - (\theta p_{h,m} + (1-\theta)p_{h,m+1}), w - u_{h,m}) = 0, \quad m = 1, \ldots, N-1, \\
&(\frac{\alpha}{2} u_{h,N} - \theta p_{h,N}, w - u_{h,N}) = 0.
\end{aligned}
$$

In the case of backward Euler method ($\theta = 1$), the value $u_{h,0}$ is not needed, as we observe from (14). As we mentioned before, approximating the first integral in the cost functional by using the rectangle rule leads to $p_{h,N} = 0$, $u_{h,N} = 0$, as we see from (14). Due to SIPG, we obtain $a^s_h(\psi_\delta, p_\delta) = a^a_h(p_\delta, \psi_\delta)$ [36]. Therefore, variational formulations (12) and (14) are the same.

In the case of Crank-Nicolson method ($\theta = 1/2$), we observe that some differences occur in the adjoint equation. In (12), the right-hand side of the adjoint equation is evaluated at two successive points, while it is evaluated at just one point in (14). Additional differences are seen in the variational inequalities (12) and (14), too. Thus, *OD* and *DO* approaches lead to different weak forms. Several variants of Crank-Nicolson method are used for optimal control of heat equation in [2]. For *DO* approach, the cost functional is discretized by using the midpoint rule. On the other hand, for *OD* approach, the semi-discrete state equation is discretized by using the midpoint rule and a variant of the trapezoidal rule is applied to the semi-discrete adjoint equation to obtain the fully-discrete optimality system. Then, the resulting optimality systems commute.

4.2 Discontinuous Galerkin Time Discretization

We define the space-time finite element space of piecewise discontinuous functions for test function, state and control as

$$
V^{k,q}_{h,p} = Y^{k,q}_{h,p} = U^{k,q}_{h,p} = \Big\{ v \in L^2(0,T; L^2(\Omega)) \, : \, v|_{I_m} = \sum_{s=0}^{q} t^s \phi_s, t \in I_m, \phi_s \in V_{h,p}, m = 1, \ldots, N \Big\} .
$$

We define the temporal jump of $v \in V^{k,q}_{h,p}$ as $[v]_m = v^m_+ - v^m_-$, where $w^m_\pm = \lim_{\varepsilon \to 0\pm} v(t_m + \varepsilon)$.

Let f_δ and y_δ^d be approximations of the source function f and the desired state function y^d on each interval I_m. Then, the fully-discrete OCP is written as

$$\underset{u_\delta \in U_{h,p}^{k,q}}{\text{minimize}} \int_0^T \Big(\frac{1}{2} \sum_{K\in\mathcal{T}_h} \|y_\delta - y_\delta^d\|_{L^2(K)}^2 + \frac{\alpha}{2} \sum_{K\in\mathcal{T}_h} \|u_\delta\|_{L^2(K)}^2\Big)\, dt,$$

$$\text{subject to} \sum_{m=1}^{N_T} \int_{I_m} (\partial_t y_\delta, v_\delta)\, dt + \int_0^T a_h^s(y_\delta, v_\delta)\, dt + \sum_{m=1}^{N_T} ([y_\delta]_{m-1}, v_{\delta,+}^{m-1}), \tag{15}$$

$$= \int_0^T (f_\delta + u_\delta, v_\delta)\, dt, \ \forall v_\delta \in V_{h,p}^{k,q}, \quad y_{\delta,-}^0 = (y_0)_\delta.$$

The OCP (15) has a unique solution (y_δ, u_δ) and that pair $(y_\delta, u_\delta) \in V_{h,p}^{k,q} \times U_{h,p}^{k,q}$ is the solution of (15) if and only if there is an adjoint $p_\delta \in V_{h,p}^{k,q}$ such that $(y_\delta, u_\delta, p_\delta) \in V_{h,p}^{k,q} \times U_{h,p}^{k,q} \times V_{h,p}^{k,q}$ is the unique solution of the fully-discrete optimality system

$$\begin{aligned}
\sum_{m=1}^{N_T} \int_{I_m} (\partial_t y_\delta, v_\delta)\, dt + \int_0^T a_h^s(y_\delta, v_\delta)\, dt + \sum_{m=1}^{N_T} ([y_\delta]_{m-1}, v_{\delta,+}^{m-1})& \\
= \int_0^T (f_\delta + u_\delta, v_\delta)\, dt, \forall v_\delta \in V_{h,p}^{k,q},& \\
y_{\delta,-}^0 = (y_0)_\delta,& \\
\sum_{m=1}^{N_T} \int_{I_m} (-\partial_t p_\delta, \psi_\delta)\, dt + \int_0^T a_h^u(p_\delta, \psi_\delta)\, dt - \sum_{m=1}^{N_T} ([p_\delta]_m, \psi_{\delta,-}^m)& \\
= -\int_0^T (y_\delta - y_\delta^d, \psi_\delta)\, dt, \forall \psi_\delta \in V_{h,p}^{k,q},& \\
p_{\delta,+}^N = 0,& \qquad (16) \\
\int_0^T (\alpha u_\delta - p_\delta, w_\delta - u_\delta)\, dt = 0 \quad \forall w_\delta \in U_{h,p}^{k,q}.&
\end{aligned}$$

In *DO* approach, firstly, we construct the discrete Lagrangian

$$\begin{aligned}
\mathcal{L}(y_\delta, u_\delta, p_\delta) = &\frac{1}{2} \int_0^T \left(\Big(\sum_{K\in\mathcal{T}_h} \|y_\delta - y_\delta^d\|_{L^2(K)}^2 + \alpha \sum_{K\in\mathcal{T}_h} \|u_\delta\|_{L^2(K)}^2 \Big) \right) dt \\
&+ \sum_{m=1}^{N_T} \Big(\int_{I_m} \big((\partial_t y_\delta, p_\delta) + a_h^s(y_\delta, p_\delta)\big)\, dt + ([y_\delta]_{m-1}, p_{\delta,+}^{m-1})\Big) \\
&- \sum_{m=1}^{N_T} \int_{I_m} (f_\delta + u_\delta, p_\delta)\, dt\Big) + ((y_0)_\delta - y_{\delta,-}^0, p_{\delta,-}^0).
\end{aligned}$$

Differentiating $\mathscr{L}$ with respect to y_δ and applying integration by parts, we obtain

$$\sum_{m=1}^{N_T} \int_{I_m} \left((\psi_\delta, -\partial_t p_\delta) + a_h^s(\psi_\delta, p_\delta) \right) dt + \sum_{m=1}^{N_T-1} (\psi_{\delta,-}^m, -[p_\delta]_m) + (q_{\delta,-}^{N_T}, p_{\delta,-}^{N_T})$$
$$= - \sum_{m=1}^{N_T} \int_{I_m} (y_\delta - y_\delta^d, \psi_\delta) \, dt, \quad \forall \psi_\delta \in V_{h,p}^{k,q}. \tag{17}$$

Now, we add and subtract $(\psi_{\delta,N_T}^-, p_{\delta,N_T}^+)$ to (17) and obtain

$$\sum_{m=1}^{N_T} \int_{I_m} \left(-(\partial_t p_\delta, \psi_\delta) + a_h^s(\psi_\delta, p_\delta) \right) dt - \sum_{m=1}^{N_T} ([p_\delta]_m, \psi_{\delta,-}^m) + (\psi_{\delta,-}^{N_T}, p_{\delta,+}^{N_T})$$
$$= - \sum_{m=1}^{N_T} \int_{I_m} (y_\delta - y_\delta^d, \psi_\delta) \, dt, \quad \forall \psi_\delta \in V_{h,p}^{k,q}. \tag{18}$$

On each subinterval I_m, the adjoint equation reads as

$$\int_{I_m} \left(-(\partial_t p_\delta, \psi_\delta) + a_h^s(\psi_\delta, p_\delta) \right) \, dt - ([p_\delta]_m, \psi_{\delta,-}^m) = - \int_{I_m} (y_\delta - y_\delta^d, \psi_\delta) \, dt.$$

However, $(q_{\delta,-}^{N_T}, p_{\delta,+}^{N_T})$ does not match the right-hand side of (18), so it is set to zero, i.e. $p_{\delta,+}^N = 0$. Now, we use $a_h^s(\psi_\delta, p_\delta) = a_h^a(p_\delta, \psi_\delta)$. Thus, we arrive at (16). We note that *OD* and *DO* approaches lead to the same optimality conditions, which can be observed by differentiating the discrete Lagrangian with respect to u_δ. Therefore, both approaches commute.

5 Error Estimates

In this section, firstly, we give the norms used in the analysis and mention some estimates in the literature. Secondly, the discrete characteristic function which enables us to provide error estimates at arbitrary time points is explained. Then, we prove some useful lemmas and state the main estimate of this study.

We introduce the L^2 inner product on the inflow or outflow boundaries as follows

$$(w, v)_{\Gamma-} = \int_{\Gamma_-} |\boldsymbol{\beta} \cdot n| wv \, ds$$

with analogous definition of $(\cdot, \cdot)_{\Gamma+}$ and associated norms $\| \cdot \|_{\Gamma-}$ and $\| \cdot \|_{\Gamma+}$.

The broken Sobolev space is defined as

$$H^k(\Omega, \mathscr{T}_h) = \left\{ v \, : \, v \,|_K \in H^k(K) \quad \forall K \in \mathscr{T}_h \right\},$$

with the semi-norm defined by

$$|v|_{H^k(\Omega,\mathscr{T}_h)} = \left(\sum_{K\in\mathscr{T}_h} |v|^2_{H^k(K)} \right)^{1/2}, \quad v \in H^k(\Omega,\mathscr{T}_h).$$

The Bochner space of functions whose kth time derivative is bounded almost everywhere on $(0,T)$ with values in X is denoted by $W^{k,\infty}(0,T;X)$. We use the dG energy norm in [33, Sect. 4]

$$|||v|||^2_{DG} = |v|^2_{H^1(\Omega,\mathscr{T}_h)} + J_\sigma(v,v). \tag{19}$$

We give the multiplicative trace inequality for all $K \in \mathscr{T}_h$, for all $v \in H^1(K)$ as follows:

$$\|v\|^2_{L^2(\partial K)} \leq C_M \left(\|v\|_{L^2(K)} |v|_{H^1(K)} + h_K^{-1} \|v\|^2_{L^2(K)} \right), \tag{20}$$

where C_M is a positive constant independent of v, h and K. We refer the reader to the study [12, Lemma 3.1] for the proof.

In addition, the generalization of Poincaré inequality to the broken Sobolev space $H^1(\Omega,\mathscr{T}_h)$ is given as [26, Sect. 3.1.4]

$$\|v\|^2_{L^2(\Omega)} \leq C_S \left(|v|^2_{H^1(\Omega,\mathscr{T}_h)} + \sum_{E\in\mathscr{E}_h} \frac{1}{h_E} \| [\![y]\!] \|^2_{L^2(E)} \right). \tag{21}$$

We proceed with the standard estimates derived for finite element methods [9]. Consider the L^2-projection $\Pi_h : L^2(\Omega) \to V_{h,p}$ so that

$$\|\Pi_h v - v\|_{L^2(K)} \leq C_\Pi h^{p+1} |v|_{H^{p+1}(K)}, \quad |\Pi_h v - v|_{H^1(K)} \leq C_\Pi h^p |v|_{H^{p+1}(K)}, \tag{22}$$

for all $v \in H^{p+1}(K)$, $K \in \mathscr{T}_h$ where C_Π is a positive constant and independent of v and h. In addition, as suggested in [33, Sect. 4], using the study [13], the following estimate holds for all $v \in H^{p+1}(\Omega,\mathscr{T}_h)$

$$|||\Pi_h v - v|||_{DG} \leq (2C_M + 1) C_\Pi h^p |v|_{H^{p+1}(\Omega,\mathscr{T}_h)}, \tag{23}$$

where C_M and C_Π are positive constants from (20) and (22), respectively. In the following we introduce the parabolic projection for $m = 0, \ldots, N_T$ and mention the properties given in [33]. Suppose that $X \subset L^2(\Omega)$ is a Hilbert space. Let us denote the space of polynomial functions depending on time as follows:

$$P^\alpha(I_m, X) = \left\{ v \in L^2(0,T;L^2(\Omega)) \, : \, v = \sum_{s=0}^{\alpha} t^s \phi_{s,m}, t \in I_m, \phi_{s,m} \in X \right\}.$$

A space-time projection π of $y \in C(0,T;H^1(\Omega))$ into $V_{h,p}^{k,q}$ is employed for the convergence estimates. Time projection P of $y \in C(0,T;H^1(\Omega))$ is defined as

$$
\begin{aligned}
&Py \in \left\{ v \in L^2(Q_T) \ : \ v|_{I_m} \in P^q(I_m, L^2(\Omega)) \right\}, \\
&\int_{I_m} (Py - y, t^j v) \, dt = 0, \quad \forall v \in L^2(\Omega), j = 0, \dots, q-1, \\
&(Py)_-^m = y(t^m).
\end{aligned}
$$

In addition, for $m = 0, \dots, N_T$, with $y \in C(0,T;H^1(\Omega))$, $\pi y \in V_{h,p}^{k,q}$ is defined as

$$
\begin{aligned}
&\pi y = \Pi_h(Py) \Longleftrightarrow ((\pi y)(t), v) = ((Py)(t), v), \quad \forall v \in V_{h,p}, \forall t \in I_m, \\
&\int_{I_m} (\pi y - y, v) \, dt = \int_{I_m} ((Py, v) - (y, v)) \, dt = 0, \quad \forall v \in V_{h,p}^{k,q-1}, \qquad (24) \\
&((\pi y)_-^m - y(t^m), v) = (((Py)_-^m, v) - (y(t^m), v)) = 0, \quad \forall v \in V_{h,p}.
\end{aligned}
$$

We note that the definition of the projection π is likewise in the study [28].

We give some estimates from [33, Lemmas 4.3, 4.5], which we need in the proofs.

Lemma 1 *Suppose that* $y \in W^{q+1,\infty}(I_m, H^1(\Omega))$ *such that* $y = 0$ *on* $\partial\Omega$. *Then,*

$$
\begin{aligned}
\|y(t) - Py(t)\| &\le C_P k_m^{q+1} |y|_{W^{q+1,\infty}(I_m, L^2(\Omega))} \quad \forall t \in I_m, \\
|y(t) - Py(t)|_{H^1(\Omega)} &\le C_P k_m^{q+1} |y|_{W^{q+1,\infty}(I_m, H^1(\Omega))} \quad \forall t \in I_m, \qquad (25) \\
|||y(t) - Py(t)|||_{DG} &\le C_P k_m^{q+1} |y|_{W^{q+1,\infty}(I_m, H^1(\Omega))} \quad \forall t \in I_m.
\end{aligned}
$$

Lemma 2 *Suppose that* $y \in W^{q+1,\infty}(I_m, H^1(\Omega)) \cap L^\infty(I_m, H^{p+1}(\Omega))$ *such that* $y = 0$ *on* $\partial\Omega$. *Then,*

$$
\begin{aligned}
\|y(t) - \pi y(t)\| &\le C_\pi (h^{p+1} + k_m^{q+1}) \|y\|_R \quad \forall t \in I_m, \\
|||y(t) - \pi y(t)|||_{DG} &\le C_\pi (h^p + k_m^{q+1}) \|y\|_R \quad \forall t \in I_m, \qquad (26)
\end{aligned}
$$

where $\|y\|_R = \max(|y|_{W^{q+1,\infty}(I_m, H^1(\Omega))}, |y|_{L^\infty(I_m, H^{p+1}(\Omega))})$ *and* C_π *is a positive constant independent of* h, k_m, m *and* y.

Lemma 3 *There exists a positive constant* C_A *which is independent of* h, v_h, w_h, ϵ *such that*

$$
\begin{aligned}
&a^d(y(t) - \Pi_h y(t), v_h) \le C_A \epsilon h^p \|y(t)\|_{H^{p+1}(\Omega)} |||v_h|||_{DG} \\
&\qquad a.e. \ t \in (0,T), y \in L^2(0,T; H^{p+1}(\Omega)), v_h \in V_{h,p}. \qquad (27)
\end{aligned}
$$

Proof The proof in [11, Lemma 3.8] is adopted to the bilinear form (7) using the estimate (23). □

Remark 1 A similar estimate for the bilinear form arising from the nonsymmetric interior penalty Galerkin method can be found in [33, Lemma 4.2].

Lemma 4 *The bilinear form $a^d(\cdot,\cdot)$ satisfies the coercivity inequality*

$$a^d(v_h, v_h) \geq \frac{\epsilon}{2}|||v_h|||_{DG}^2, \quad \forall v_h \in V_{h,p}. \tag{28}$$

Proof The proof in [11, Corollary 3.10] is adopted to the bilinear form (7) using the norm (19). □

5.1 Discrete Characteristic Function

We use the discrete characteristic function in order to provide error estimates at arbitrary time points as suggested in [8]. We can work on $[0,k)$ instead of I_m, since the construction of the discrete characteristic function is invariant under translation. We consider polynomials $s \in \mathscr{P}_q(0,k)$ and the discrete approximation of $\chi_{[0,t)}s$ of s which is a polynomial

$$\tilde{s} \in \left\{\tilde{s} \in \mathscr{P}_q(0,k) \ : \ \tilde{s}(0) = s(0)\right\} \text{ such that } \int_0^k \tilde{s}z = \int_0^t sz, \quad \forall z \in \mathscr{P}_{q-1}(0,k).$$

This definition can be extended from $\mathscr{P}_q(0,k)$ to $V_{h,p}^{k,q}$. The discrete approximation of $\chi_{[0,t)}v$ for $v \in V_{h,p}^{k,q}$ is written as $\tilde{v} = \sum_{i=0}^q \tilde{s}_i(t)v_i$. On account of these inequalities, the following estimate is given in [33]

$$\int_{I_m} |||\tilde{w}|||_{DG}^2 \, dt \leq C_D \int_{I_m} |||w|||_{DG}^2 \, dt, \quad C_D = C_D(q). \tag{29}$$

We mention that a suitable discrete approximation $\chi_{(t,t^n]}v_h$ must be constructed for the adjoint problem, as it is noted in the proof of [7, Theorem 3.8]. The discrete approximation of $\chi_{(t,t^{N_T}]}s$ is a polynomial

$$\tilde{s} \in \{\tilde{s} \in \mathscr{P}_q(t^{N_T-1}, t^{N_T}) : \tilde{s}(t^{N_T}) = s(t^{N_T})\} \text{ such that } \int_{t^{N_T-1}}^{t^{N_T}} \tilde{s}z = \int_t^{t^{N_T}} sz,$$

$\forall z \in \mathscr{P}_{q-1}(t^{N_T-1}, t^{N_T})$. This definition can be extended from $\mathscr{P}_q(t^{N_T-1}, t^{N_T})$ to $V_{h,p}^{k,q}$ and the estimates above can be modified for the adjoint [7, Theorem 3.8].

5.2 A Priori Error Estimates

We proceed with the derivation of the convergence estimates for the optimality system and its space-time dG approximation. We define the auxiliary state and adjoint equation which are needed for a priori error analysis

$$
\begin{aligned}
\sum_{m=1}^{N_T} \int_{I_m} (\partial_t y_\delta^u, v_\delta)\, dt + \int_0^T a_h^s(y_\delta^u, v_\delta)\, dt + \sum_{m=1}^{N_T} ([y_\delta^u]_{m-1}, v_{\delta,+}^{m-1}) \\
= \int_0^T (f_\delta + u, v_\delta)\, dt, \\
y_{\delta,-}^{u,0} = (y_0)_\delta, \\
\sum_{m=1}^{N_T} \int_{I_m} (-\partial_t p_\delta^u, \psi_\delta)\, dt + \int_0^T a_h^a(p_\delta^u, \psi_\delta)\, dt - \sum_{m=1}^{N_T} ([p_\delta^u]_m, \psi_{\delta,-}^m) \\
= -\int_0^T (y_\delta^u - y_\delta^d, \psi_\delta)\, dt, \\
p_{\delta,+}^{u,N} = 0.
\end{aligned}
\tag{30}
$$

Following [15], we assume that the reaction term satisfies $|r| \leq C_r$ a.e. in Ω; the velocity field is bounded by a constant C_β a.e. in Ω.

We prove some useful lemmas before stating the main theorem of this study.

Lemma 5 *Let (y_δ, p_δ) and (y_δ^u, p_δ^u) be the solutions of (16) and (30), respectively. Then, there exists a constant C independent of h and k such that*

$$
\sup_{t\in I_n} \|y_\delta^u(t) - y_\delta(t)\| + \sup_{t\in I_n} \|p_\delta^u(t) - p_\delta(t)\| \leq C \int_0^{t_n} \|u - u_\delta\|\, dt. \tag{31}
$$

Proof Firstly, we study the fully discrete state equation on each subinterval I_m. We subtract (16) from (30) to obtain

$$
\int_{I_m} (\partial_t \theta, v_\delta)\, dt + ([\theta]_{m-1}, v_{\delta,+}^{m-1}) + \int_{I_m} a_h^s(\theta, v_\delta)\, dt = \int_{I_m} (u - u_\delta, v_\delta)\, dt, \tag{32}
$$

where $\theta = y_\delta^u - y_\delta$. We substitute $v_\delta = 2\theta$ in (32). Then,

$$
\int_{I_m} 2(\partial_t \theta, \theta)\, dt + 2([\theta]_{m-1}, \theta_+^{m-1}) = \|\theta_-^m\|^2 - \|\theta_-^{m-1}\|^2 + \|[\theta]_{m-1}\|^2, \tag{33}
$$

is achieved. For the right-hand side, we employ Cauchy-Schwarz, Young inequalities, Poincaré inequality (21) and the definition of dG norm (19). For the left-hand side, we use (28) for diffusion term and follow the technique in (see [15, Theorem 5.1]) for convection and reaction terms. Then, we derive the following estimate in the middle of (34)

$$
\begin{aligned}
&\|\theta_-^m\|^2 - \|\theta_-^{m-1}\|^2 + \frac{\epsilon}{2}\int_{I_m} |||\theta|||_{DG}^2 \, dt + 2C_0 \int_{I_m} \|\theta\|^2 \, dt \\
&\quad + \frac{\epsilon}{2}\int_{I_m} \left(\sum_{K\in\mathcal{T}_h} \left(\|\theta\|^2_{\partial K^- \cap \Gamma^-} + \| [\![\theta]\!] \|^2_{\partial K^- \setminus \Gamma^-} + \|\theta\|^2_{\partial K^+ \cap \Gamma^+} \right) \right) dt \\
&\leq \|\theta_-^m\|^2 - \|\theta_-^{m-1}\|^2 + \frac{\epsilon}{2}\int_{I_m} |||\theta|||_{DG}^2 \, dt + 2C_0 \int_{I_m} \|\theta\|^2 \, dt \\
&\quad + \int_{I_m} \left(\sum_{K\in\mathcal{T}_h} \left(\|\theta\|^2_{\partial K^- \cap \Gamma^-} + \| [\![\theta]\!] \|^2_{\partial K^- \setminus \Gamma^-} + \|\theta\|^2_{\partial K^+ \cap \Gamma^+} \right) \right) dt \\
&\quad \leq C \int_{I_m} \|u - u_\delta\|^2 \, dt.
\end{aligned} \tag{34}
$$

We note that the lower bound on the left-hand side of (34) has been added after deriving the estimate in the middle for the clearance of the proof and will be used later. Now, we proceed by substituting $v_\delta = 2\tilde{\theta}$ into (32). We employ the discrete characteristic function as in the proof of [33, Theorem 5.2] to obtain an estimate at arbitrary points and use the properties given there. With $z = \arg\sup_{\bar{I}_m} \|\theta(t)\|$, the discrete characteristic function defined in Sect. 5.1 leads to

$$
\int_{I_m} (\partial_t \theta, \tilde{\theta}) \, dt = \int_{t_{m-1}}^{z} (\partial_t \theta, \theta) \, dt, \quad \tilde{\theta}_+^{m-1} = \theta_+^{m-1}, \quad [\tilde{\theta}]_{m-1} = [\theta]_{m-1}, \tag{35}
$$

$$
\int_{I_m} 2(\partial_t \theta, \tilde{\theta}) \, dt + 2([\theta]_{m-1}, \tilde{\theta}_+^{m-1}) = \|\theta(z)\|^2 - \|\theta_-^{m-1}\|^2 + \|[\theta]_{m-1}\|^2.. \tag{36}
$$

We use (35) and (36) and the inequality $\|\theta_-^{m-1}\| \leq \sup_{t\in I_{m-1}} \|\theta(t)\|$ to bound the terms arising in the time derivative. We proceed by moving $2\int_{I_m} a_h(\theta, \tilde{\theta}) \, dt$ to the right-hand side. We employ (27) for the diffusion term, the proof of [15, Theorem 5.1] for the convection term. The reaction term and the control on the right-hand side is bounded by using Cauchy-Schwarz and Young inequalities (21) and (19) such that $\| \cdot \|^2 \leq C||| \cdot |||_{DG}^2$ is satisfied for a positive constant C. We eliminate the term $|||\tilde{\theta}|||_{DG}^2$ on the right-hand side by using (29). Then, we obtain the following inequality

$$\begin{aligned}
&\sup_{t\in I_m}\|\theta(t)\|^2-\sup_{t\in I_{m-1}}\|\theta(t)\|^2\\
&\le C_b\int_{I_m}|||\theta|||_{DG}^2\,dt+\int_{I_m}\sum_{K\in\mathcal{T}_h}\left(\|\theta\|^2_{\partial K^+\cap\Gamma^+}+\|\,[\![\theta]\!]\,\|^2_{\partial K^-\setminus\Gamma^-}\right)\,dt\\
&\quad+C\int_{I_m}\|u-u_\delta\|^2\,dt\\
&\le C_b'\int_{I_m}\left(|||\theta|||_{DG}^2+\sum_{K\in\mathcal{T}_h}\left(\|\theta\|^2_{\partial K^+\cap\Gamma^+}+\|\,[\![\theta]\!]\,\|^2_{\partial K^-\setminus\Gamma^-}\right)\right)\,dt\\
&\quad+C\int_{I_m}\|u-u_\delta\|^2\,dt,
\end{aligned}\tag{37}$$

where $C_b=C(1+C_D)(\epsilon C_A+C_S(C_r+C_\beta))$, $C_b'=\max\{1,C_b\}$. In order to eliminate the terms θ on the right-hand side of (37), we use (34) multiplying it by $C_b''=\frac{2}{\epsilon}C_b'$. By adding these inequalities and denoting $\Theta_m=\sup_{t\in I_m}\|\theta(t)\|^2+C_b''\|\theta_-^m\|^2$, we arrive at

$$\Theta_m-\Theta_{m-1}\le C(1+C_b'')\int_{I_m}\|u-u_\delta\|^2\,dt.\tag{38}$$

We sum (38) over $m=1,\ldots,n\le N_T$ and use $\theta=0$ at $t=0$ to derive the estimate

$$\sup_{t\in I_n}\|\theta(t)\|^2=\sup_{t\in I_n}\|y_\delta^u(t)-y_\delta(t)\|^2\le C\int_0^{t_n}\|u-u_\delta\|^2\,dt.\tag{39}$$

Secondly, we proceed with the adjoint equation subtracting (16) from (31) and using $\zeta=p_\delta^u-p_\delta$. A discrete approximation to $\chi_{(t,t_m]}v_h$ specified for the adjoint problem must be used, as we discussed in Sect. 5.1. Then, this leads to

$$\int_{I_m}2(-\partial_t\zeta,\tilde{\zeta})\,dt-2([\zeta]_m,\tilde{\zeta}_-^m)=\|\zeta(z)\|^2-\|\zeta^m\|^2+\|[\zeta]_m\|^2,\tag{40}$$

where $z=\arg\sup_{\bar{I}_m}\|\zeta(t)\|$. In addition, the inequalities $\|\zeta^m\|^2\le\sup_{I_{N_T-m+2}}\|\zeta(t)\|^2$ and $\|\zeta(z)\|^2=\sup_{I_{N_T-m+1}}\|\zeta(t)\|^2$ are needed. Then, we follow the same idea used to derive (39) to reach the inequality

$$\sup_{t\in I_{N_T-m+1}}\|\zeta(t)\|^2-\sup_{t\in I_{N_T-m+2}}\|\zeta(t)\|^2\le Ck_m\int_{t\in I_m}\|u-u_\delta\|^2\,dt.\tag{41}$$

We sum (41) over $m = N_T, \dots, n \geq 1$ and use $\zeta = 0$ at $t = t_{N_T}$. The final result (31) follows from standard algebra, (39) and (41). □

We proceed with the estimate between the exact and the approximate control.

Lemma 6 *Let (y, p, u) and $(y_\delta, p_\delta, u_\delta)$ be the solutions of (2) and (16), respectively. Then, we have*

$$\|u - u_\delta\|_{L^2(0,T;L^2(\Omega))} \leq \frac{1}{\alpha}\|p - p_\delta^u\|_{L^2(0,T;L^2(\Omega))}. \tag{42}$$

Proof We apply the technique used for the steady-state optimal control problem in [21, Sect. 4.2]. We start using the continuous and the fully-discrete optimality conditions (3)–(17) to obtain the following equation

$$\begin{aligned}\alpha\|u - u_\delta\|^2_{L^2(0,T;L^2(\Omega))} &= \alpha\int_0^T (u - u_\delta, u - u_\delta)\, dt \\ = \int_0^T (\alpha u - p, u - u_\delta)\, dt &- \int_0^T (\alpha u_\delta - p_\delta, u - u_\delta)\, dt + \int_0^T (p - p_\delta, u - u_\delta)\, dt \\ = \int_0^T (p - p_\delta^u, u - u_\delta)\, dt &+ \int_0^T (p_\delta^u - p_\delta, u - u_\delta)\, dt = J_1 + J_2. \end{aligned} \tag{43}$$

We use Cauchy-Schwarz and Young inequalities to show that

$$0 \leq J_1 \leq \frac{1}{2\alpha}\|p - p_\delta^u\|^2_{L^2(0,T;L^2(\Omega))} + \frac{\alpha}{2}\|u - u_\delta\|^2_{L^2(0,T;L^2(\Omega))}. \tag{44}$$

We proceed with J_2 and usc the auxiliary state equation (30) to obtain

$$\begin{aligned} J_2 &= \int_0^T (p_\delta^u - p_\delta, u - u_\delta)\, dt \\ &= \sum_{m=1}^{N_T} \int_{I_m} (\partial_t(y_\delta^u - y_\delta), p_\delta^u - p_\delta)\, dt + \int_0^T a_h^s(y_\delta^u - y_\delta, p_\delta^u - p_\delta)\, dt \\ &+ \sum_{m=1}^{N} ([y_\delta^u - y_\delta]_{m-1}, (p_\delta^u - p_\delta)_+^{m-1}). \end{aligned}$$

We proceed applying integration by parts in time and use the auxiliary adjoint equation (30) to arrive at

$$
\begin{aligned}
J_2 = & -\sum_{m=1}^{N_T} \int_{I_m} \big(p_\delta^u - p_\delta, \partial_t(y_\delta^u - y_\delta)\big)\, dt + \sum_{m=1}^{N} \big(y_\delta^u - y_\delta, p_\delta^u - p_\delta\big)\big|_{t_{m-1}}^{t_m} \\
& + \int_0^T a_h^s(y_\delta^u - y_\delta, p_\delta^u - p_\delta)\, dt + \sum_{m=1}^{N} \big([y_\delta^u - y_\delta]_{m-1}, (p_\delta^u - p_\delta)_+^{m-1}\big) \\
= & -\sum_{m=1}^{N_T} \int_{I_m} \big(p_\delta^u - p_\delta, \partial_t(y_\delta^u - y_\delta)\big)\, dt + \int_0^T a_h^s(y_\delta^u - y_\delta, p_\delta^u - p_\delta)\, dt \\
& - \sum_{m=1}^{N} \big((y_\delta^u - y_\delta)_-^m, [p_\delta^u - p_\delta]_m\big) \\
= & -\int_0^T \big(y_\delta^u - y_\delta, y_\delta^u - y_\delta\big)\, dt \le 0. \qquad (45)
\end{aligned}
$$

Then, using (43)–(45), we derive the final result (42). □

Lemma 7 *Let (y, p) and (y_δ^u, p_δ^u) be the solutions of (2) and (30), respectively. Assume that $y, p \in W^{q+1,\infty}(0, T; H^1(\Omega)) \cap L^\infty(0, T; H^{p+1}(\Omega))$. Then, there exists a constant C independent of h and k such that*

$$
\sup_{t\in I_n} \|y - y_\delta^u\| + \sup_{t\in I_n} \|p - p_\delta^u\| \le \mathcal{O}(h^p + k^{q+1}). \qquad (46)
$$

Proof Firstly, we integrate (2) over I_m and subtract the result from (30) in order to obtain the following equation

$$
\begin{aligned}
& \int_{I_m} (\partial_t \xi, v_\delta)\, dt + ([\xi]_{m-1}, v_{\delta,+}^{m-1}) + \int_{I_m} a_h^s(\xi, v_\delta)\, dt \\
& = -\left(\int_{I_m} (\partial_t \eta, v_\delta)\, dt + ([\eta]_{m-1}, v_{\delta,+}^{m-1}) \right) - \int_{I_m} a_h(\eta, v_\delta)\, dt, \qquad (47)
\end{aligned}
$$

where $y - y_\delta^u = (y - \pi y) + (\pi y - y_\delta^u) = \eta + \xi$.

Since we use the same mesh on each time interval, (24) leads to the following identity.

$$
\int_{I_m} (\partial_t \eta, v_\delta)\, dt + ([\eta]_{m-1}, v_{\delta,+}^{m-1}) = 0, \quad \forall v_\delta \in V_h^{k,q}. \qquad (48)
$$

We proceed as in the proof of Lemma 5 and the proof of [15, Theorem 5.1] by inserting the estimate (26) to obtain

$$\begin{aligned}
&\int_{I_m} (\partial_t \xi, v_\delta)\, dt + ([\xi]_{m-1}, v_{\delta,+}^{m-1}) + \int_{I_m} a_h^s(\xi, v_\delta)\, dt \\
&\leq \frac{\epsilon}{4} \int_{I_m} |||v_\delta|||_{DG}^2\, dt + \frac{C_0}{2} \int_{I_m} \|v_\delta\|^2\, dt \\
&+ \frac{1}{2} \int_{I_m} \sum_{K \in \mathcal{T}_h} \left(\|v_\delta\|_{\partial K^+ \cap \Gamma+}^2 + \|\, [\![v_\delta]\!]\, \|_{\partial K^- \setminus \Gamma-}^2 \right) dt \\
&+ k_m C_A C_\pi (h^{2p} + k^{2q+2}) |y|_R^2 + k_m 2 C_\beta C_\pi C_M (h^{2p+1} + k^{2q+2}) |y|_R^2 \\
&+ k_m C_\pi \frac{C_\beta C_r}{C_0} (h^{2p+2} + k^{2q+2}) |y|_R^2,
\end{aligned} \tag{49}$$

where $|y|_R = \max(|y|_{W^{q+1,\infty}(I_m;H^1(\Omega))}, |y|_{L^\infty(I_m;H^{p+1}(\Omega))})$.

Firstly, we shall substitute $v_\delta = 2\xi$ into (49) to obtain

$$\begin{aligned}
&\|\xi_-^m\|^2 - \|\xi_-^{m-1}\|^2 + \frac{\epsilon}{2} \int_{I_m} |||\xi|||_{DG}^2\, dt + C_0 \int_{I_m} \|\xi\|^2\, dt \\
&+ \int_{I_m} \sum_{K \in \mathcal{T}_h} \left(\|\xi\|_{\partial K^- \cap \Gamma-}^2 + \frac{1}{2} \|\, [\![\xi]\!]\, \|_{\partial K^- \setminus \Gamma-}^2 + \frac{1}{2} \|\xi\|_{\partial K^+ \cap \Gamma+}^2 \right) dt \\
&\leq k_m C_b (h^{2p} + h^{2p+1} + h^{2p+2} + k^{2q+2}) |y|_R^2,
\end{aligned} \tag{50}$$

where $C_b = \max\{C_A C_\pi, 2 C_\beta C_\pi C_M, C_\pi \frac{C_\beta C_r}{C_0}\}$.

Secondly, we substitute $v_\delta = 2\tilde{\xi}$ into (49) to obtain

$$\begin{aligned}
&\sup_{t \in I_m} \|\xi(t)\|^2 - \sup_{t \in I_{m-1}} \|\xi(t)\|^2 \\
&\leq C_b' \int_{I_m} |||\xi|||_{DG}^2\, dt + \int_{I_m} \sum_{K \in \mathcal{T}_h} \left(\|\, [\![\xi]\!]\, \|_{\partial K^- \setminus \Gamma-}^2 + \|\xi\|_{\partial K^+ \cap \Gamma+}^2 \right) dt \\
&+ k_m C_b (h^{2p} + h^{2p+1} + h^{2p+2} + k^{2q+2}) |y|_R^2 \\
&\leq C_b'' \int_{I_m} \left(|||\xi|||_{DG}^2 + \sum_{K \in \mathcal{T}_h} \left(\|\, [\![\xi]\!]\, \|_{\partial K^- \setminus \Gamma-}^2 + \|\xi\|_{\partial K^+ \cap \Gamma+}^2 \right) \right) dt \\
&+ k_m C_b (h^{2p} + h^{2p+1} + h^{2p+2} + k^{2q+2}) |y|_R^2,
\end{aligned} \tag{51}$$

where $C_b' = C(1 + C_D)(\epsilon C_A + C_S(C_\beta + C_r))$, $C_b'' = \max\{1, C_b'\}$. Now, we proceed as in the proof of Lemma 5. We multiply (50) by $C_b''' = \frac{2}{\epsilon} C_b''$ in order to eliminate the terms ξ on the right-hand side of (51). Then, we add it to (51) and denote $\Theta_m =$

$\sup_{t\in I_m} \|\xi(t)\|^2 + C_b''' \|\xi_-^m\|^2$ in order to obtain

$$\Theta_m - \Theta_{m-1} \leq k_m 2C_b'''(h^{2p} + h^{2p+1} + h^{2p+2} + k^{2q+2})|y|_R^2. \tag{52}$$

We sum (52) over $m = 1, \ldots, n \leq N_T$ to obtain

$$\sup_{t\in I_n} \|\xi(t)\|^2 \leq \mathcal{O}(h^{2p} + k^{2q+2}). \tag{53}$$

Thirdly, we integrate (2) over I_m and subtract it from (31) and denote $p - p_\delta^u = (p - \pi p) + (\pi p - p_\delta^u) = \varphi + \mu$. Then, we use the idea in the proof of (53) in order to derive

$$\sup_{t\in I_{N-m+1}} \|\mu(t)\|^2 - \sup_{t\in I_{N-m+2}} \|\mu(t)\|^2 \leq Ck_m \sup_{t\in I_m} \|\xi(t)\|^2 \, dt + \mathcal{O}(h^{2p} + k^{2q+2}), \tag{54}$$

for $C > 0$. The resulting inequality is summed over $m = N_T, \ldots, n \geq 1$. Then, it is combined with (53) to derive the final result (46). □

Remark 1 For guaranteeing the assumptions on the exact solution, it is necessary to require a higher regularity of the data of the problem.

We state the main estimate of this study by combining Lemmas 5, 6, and 7.

Theorem 1 *Suppose that* (y, p, u) *and* $(y_\delta, p_\delta, u_\delta)$ *are the solutions of (2) and (16), respectively. We assume that all conditions of Lemmas 5, 6 and 7 are satisfied. Then, there exists a constant C independent of h and k such that*

$$\|y - y_\delta\|_{L^\infty(0,T;L^2(\Omega))} + \|p - p_\delta\|_{L^\infty(0,T;L^2(\Omega))} + \|u - u_\delta\|_{L^2(0,T;L^2(\Omega))} \leq C\left(h^p + k^{q+1}\right). \tag{55}$$

In Theorem 1, the error in the state and control is measured with respect to the norm $L^\infty(0, T; L^2(\Omega))$ and $L^2(0, T; L^2(\Omega))$, respectively. The same norms are used, for example, in the study of Fu [16], too. The former norm is due to the discrete characteristic function which is used to provide error estimates at arbitrary time points. The latter norm arises from the optimality condition which is shown in Lemma 6. On the other hand, we observe that Theorem 1 is optimal in time, suboptimal in space in the $L^\infty(0, T; L^2(\Omega))$ norm for the state and $L^2(0, T; L^2(\Omega))$ for the control, i.e. $\mathcal{O}(h^p, k^{q+1})$, using p-degree spatial, q-degree temporal polynomial approximation. However, for example, optimal spatial convergence rate for SIPG discretization combined with backward Euler is achieved using an elliptic projection in [1]. The first reason behind the order reduction in this study is the estimate (26) for the space-time projection which is employed to bound the continuity estimate of the bilinear form in Lemma 3. The convection term also has an influence on the spatial order reduction since we follow the proof of [15, Theorem 5.1]. After eliminating the effect of the space-time projection in the bilinear form of the diffusion term, this suboptimal estimate can be improved as in [1].

6 Numerical Results

In this section, we present some numerical results. We measure the error in the state and the control in terms of $L^\infty(0,1;L^2(\Omega))$ and $L^2(0,1;L^2(\Omega))$ norm, respectively. We have used discontinuous piecewise linear polynomials in space. In all numerical examples, we have taken $h = k$.

We note that, in the case of dG(0) method, the approximating polynomials are piecewise constant in time and the resulting scheme is a version of the backward Euler method with a modified right-hand side [31, Chap. 12]:

$$(M + kA^s)y_{h,m} = My_{h,m-1} + \frac{k}{2}(f_{h,m} + f_{h,m-1}) + \frac{k}{2}M(u_{h,m} + u_{h,m-1}),$$

$$(M + kA^a)p_{h,m-1} = Mp_{h,m} - \frac{k}{2}M(y_{h,m} + y_{h,m-1}) + \frac{k}{2}(y^d_{h,m} + y^d_{h,m-1}).$$

For dG(1) method, we use piecewise linear polynomials in time. The resulting linear system for the state on each time interval is given as follows [31, Chap. 12]:

$$\begin{pmatrix} M + kA^s & M + \frac{k}{2}A^s \\ \frac{k}{2}A^s & \frac{1}{2}M + \frac{k}{3}A^s \end{pmatrix} \begin{pmatrix} Y_0 \\ Y_1 \end{pmatrix} = \begin{pmatrix} My_{h,m-1} + \frac{k}{2}(f_{h,m} + f_{h,m-1}) + \frac{k}{2}M(u_{h,m} + u_{h,m-1}) \\ \frac{k}{2}(f_{h,m} + Mu_{h,m}) \end{pmatrix}, \tag{56}$$

where $A^s$and M are the stiffness and the mass matrices of the state equation, respectively. We derive the solution at the time step t_m as $y_{h,m} = Y_0 + Y_1$. For the adjoint equation, we have the following linear system:

$$\begin{pmatrix} M + kA^a & M + \frac{k}{2}A^a \\ \frac{k}{2}A^a & \frac{1}{2}M + \frac{k}{3}A^a \end{pmatrix} \begin{pmatrix} P_1 \\ P_0 \end{pmatrix} = \begin{pmatrix} Mp_{h,m} - \frac{k}{2}M(y_{h,m} + y_{h,m-1}) + \frac{k}{2}(y^d_{h,m} + y^d_{h,m-1}) \\ -\frac{k}{2}(My_{h,m-1} - y^d_{h,m-1}) \end{pmatrix}, \tag{57}$$

where A^a is the stiffness matrix for the adjoint equation. We obtain the adjoint at the time step t_{m-1} as $p_{h,m-1} = P_0 + P_1$.

The main drawback of dG time discretization is the solution of large coupled linear systems in block form. Because we are using constant time steps, the coupled matrices on the right-hand side of (56) and (57) have to be decomposed (LU block factorization) at the beginning of the integration. Then, the state and the adjoint equations are solved at each time step by forward elimination and back substitution using the block factorized matrices.

Example 1 The first example is a convection dominated OCP with smooth solutions. It is converted to an unconstrained optimal control problem [17, Ex. 1] by

Table 1 Example 1 by dG(0) and backward Euler(in parenthesis) method

k	$\|y - y_\delta\|$	Rate	$\|u - u_\delta\|$	Rate
$\frac{1}{5}$	5.91e−2(4.88e−2)	–(–)	3.94e−2(2.41e−2)	–(–)
$\frac{1}{10}$	2.42e−2(1.40e−2)	1.29(1.80)	2.14e−2(8.81e−3)	0.88(1.45)
$\frac{1}{20}$	1.17e−2(4.01e−3)	1.04(1.80)	1.12e−2(3.68e−3)	0.92(1.26)
$\frac{1}{40}$	6.01e−3(1.60e−3)	0.96(1.33)	5.81e−3(1.74e−3)	0.95(1.08)
$\frac{1}{80}$	3.07e−3(7.58e−4)	0.97(1.07)	2.95e−3(8.62e−4)	0.98(1.01)

Table 2 Example 1 by Crank-Nicolson method *OD* and *DO* approach(in parenthesis)

k	$\|y - y_\delta\|$	Rate	$\|u - u_\delta\|$	Rate
$\frac{1}{5}$	6.79e−2(6.91e−2)	–(–)	2.26e−2(2.47e−2)	–(–)
$\frac{1}{10}$	1.86e−2(1.89e−2)	1.86(1.87)	6.69e−3(8.12e−3)	1.76(1.61)
$\frac{1}{20}$	4.86e−3(4.89e−3)	1.94(1.95)	1.81e−3(3.30e−3)	1.89(1.30)
$\frac{1}{40}$	1.24e−3(1.24e−3)	1.97(1.98)	4.72e−4(2.19e−3)	1.94(0.59)
$\frac{1}{80}$	3.13e−4(8.27e−4)	1.99(0.59)	1.16e−4(2.04e−3)	2.02(0.10)

adding the reaction term with

$$Q = (0,1] \times \Omega, \ \Omega = (0,1)^2, \ \epsilon = 10^{-5}, \ \beta = (1,0)^T, \ r = 1, \ \alpha = 1.$$

The source function f, the desired state y_d and the initial condition y_0 are computed from (2) using the following exact solutions of the state and the control, respectively,

$$\begin{aligned} y(x_1, x_2, t) &= \exp(-t)\sin(2\pi x_1)\sin(2\pi x_2), \\ u(x_1, x_2, t) &= \exp(-t)(1-t)\sin(2\pi x_1)\sin(2\pi x_2). \end{aligned}$$

In Table 1, errors and converge rates for dG(0) and backward Euler method are shown. We observe that the first order convergence rate is achieved in time, due to the dominance of temporal errors.

In Table 2, errors and converge rates for Crank-Nicolson method obtained by *OD* and *DO* approaches are shown. For Crank-Nicolson method, through *OD* approach, the second order convergence rate is achieved. However, for *DO* approach, discretization of the right-hand side of the adjoint equation (14) by one-step method is reflected to the numerical results and the quadratic order of convergence is not observed.

In Table 3, We present numerical results for dG(1) time discretization. Numerical results indicate a higher order experimental order of convergence, namely $\mathcal{O}(h^2)$, than the one shown in Theorem 1, which is $\mathcal{O}(h)$ with $h = k$. The error in the state is smaller than for Crank-Nicolson method with *OD* approach, while the error in the control is close to one for Crank-Nicolson method with *OD* approach.

Table 3 Example 1 by dG(1)method

k	$\|y - y_\delta\|$	Rate	$\|u - u_\delta\|$	Rate
$\frac{1}{5}$	5.65e−2	–	2.93e−2	–
$\frac{1}{10}$	1.56e−2	1.86	8.50e−3	1.78
$\frac{1}{20}$	4.04e−3	1.95	2.28e−3	1.90
$\frac{1}{40}$	1.03e−3	1.97	5.95e−4	1.93
$\frac{1}{80}$	2.59e−4	1.99	1.46e−4	2.03

Table 4 Example 2 by Crank-Nicolson method *OD* and *DO* approach(in parenthesis)

k	$\|y - y_\delta\|$	Rate	$\|u - u_\delta\|$	Rate
$\frac{1}{5}$	1.90(1.90)	–(–)	2.06e−1(2.03e−1)	–(–)
$\frac{1}{10}$	1.03(1.03)	0.89(0.89)	3.63e−2(3.63e−2)	2.51(2.49)
$\frac{1}{20}$	3.62e−1(3.62e−1)	1.50(1.50)	8.23e−3(8.15e−3)	2.12(2.15)
$\frac{1}{40}$	1.06e−1(1.05e−1)	1.78(1.78)	3.01e−3(6.17e−3)	1.45(0.40)
$\frac{1}{80}$	2.77e−2(2.71e−2)	1.93(1.95)	9.07e−4(4.95e−3)	1.73(0.32)

Example 2 The second example is a convection dominated OCP adapted from [16, Ex. 2] with

$$Q = (0,1] \times \Omega, \ \Omega = (0,1)^2, \ \epsilon = 10^{-5}, \ \beta = (0.5, 0.5)^T, \ r = 3, \ \alpha = 1.$$

The source function f, the desired state y_d and the initial condition y_0 are computed from (2) using the following exact solutions of the control and state, respectively,

$$u(x_1, x_2, t) = \sin(\pi t)\sin(2\pi x_1)\sin(2\pi x_2)\exp\left(\frac{-1 + \cos(t_x)}{\sqrt{\varepsilon}}\right),$$

$$\begin{aligned} y(x_1, x_2, t) &= u\left(\frac{1}{2\sqrt{\varepsilon}}\sin(t_x) + 8\varepsilon\pi^2 + \frac{\sqrt{\varepsilon}}{2}\cos(t_x) - \frac{1}{2}\sin^2(t_x)\right) \\ &\quad - \pi\cos(\pi t)\sin(2\pi x_1)\sin(2\pi x_2)\exp\left(\frac{-1 + \cos(t_x)}{\sqrt{\varepsilon}}\right), \end{aligned}$$

where $t_x = t - 0.5(x_1 + x_2)$. As opposed to the previous example, the exact solution of the PDE constraint depends on the diffusion explicitly and the problem is highly convection dominated. This example cannot be solved properly by using dG(0) and backward Euler method. Therefore, we present numerical results for Crank-Nicolson method in Table 4, where the differences between *OD* and *DO* can be seen clearly. *DO* approach causes order reduction in the control. However, due to the convection dominated nature of the problem, the quadratic convergence rate cannot be achieved with *OD* approach in contrast to *Example 1*. The orders of convergence correspond to those in [5].

Table 5 Example 2 by dG(1) method

k	$\|y - y_\delta\|$	Rate	$\|u - u_\delta\|$	Rate
$\frac{1}{5}$	1.70	–	2.03e−1	–
$\frac{1}{10}$	6.14e−1	1.47	4.35e−2	2.22
$\frac{1}{20}$	1.50e−1	2.03	1.23e−2	1.83
$\frac{1}{40}$	3.40e−2	2.15	2.38e−3	2.36
$\frac{1}{80}$	7.92e−3	2.10	4.77e−4	2.32

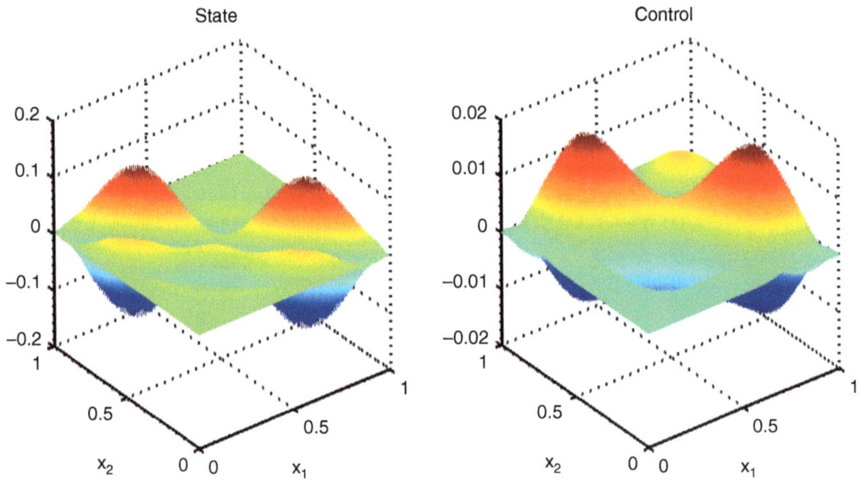

Fig. 1 Example 2: Error at t = 0.5 with Crank-Nicolson(*DO* approach) $h = k = 1/80$

In Table 5, we present numerical results for dG(1) discretization. As opposed to the results in Table 4, the error in the state and the control are smaller than in the case of Crank-Nicolson. Numerical results indicate a better experimental order of convergence, namely $\mathscr{O}(h^2)$, than the theoretical error estimate in Theorem 1. Similar observations are made for nonstationary non-linear diffusion-convection equations for the SIPG spatial discretization in [20]. In Figs. 1 and 2, we present the error between the exact and the approximate solution at $t = 0.5$ obtained using Crank-Nicolson-*DO* approach and dG(1) discretization. These figures also show that dG(1) discretization solves the problem well.

7 Conclusion

For dG time discretization, the numerical results show that linear and quadratic convergence rates are achieved using piecewise discontinuous constant and linear polynomials in time, respectively, and *DO* and *OD* approaches commute. In a future

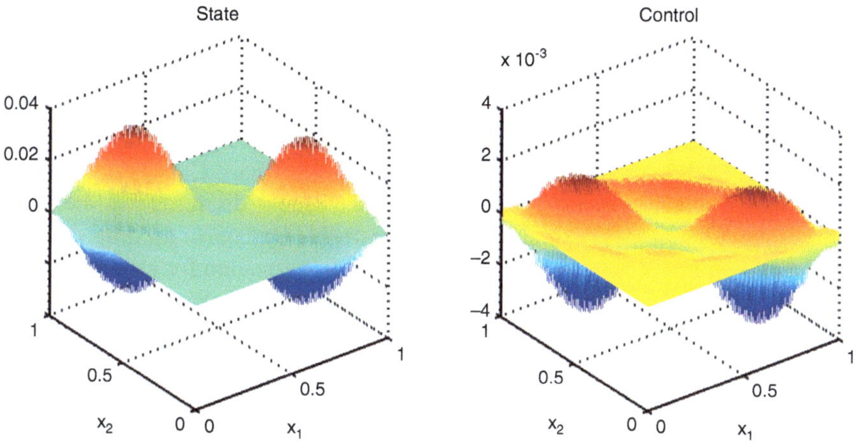

Fig. 2 Example 2: Error at t = 0.5 with dG(1) method $h = k = 1/80$

work, we will study control constrained problem and derive the optimal convergence rates under lower regularity assumptions.

8 Outlook: Efficient Solvers for DG Time Discretization

Discontinuous Galerkin time stepping is used for solving linear and nonlinear OCPs by multiple shooting methods in [6, 18] because of the commutativity property of discretization and optimization. At each subinterval of multiple shooting, a very large system of linear or nonlinear equations has to be solved, which can be handled by iterative methods, such as Krylov subspace method. In the references mentioned above, the first order dG(0) method is used, where for nonlinear problems at each Newton iteration step, a linear system of equations with the same structure of implicit Euler method has to be solved. Higher order dG methods lead to coupled block systems and the number of the unknowns grows linearly with increasing order. Therefore, for OCPs constrained by linear and nonlinear parabolic PDEs in several space dimensions, efficient solution techniques are needed. In the following, we will give an overview of the existing approaches by narrowing our discussion to 2x2 coupled block systems arising from different dG discretizations.

In the last decade, several variational time discretization methods were developed. The test spaces always consist of piecewise discontinuous polynomials. When the solution space consists of continuous piecewise polynomials of degree k and the test functions are piecewise discontinuous polynomials of degree $k-1$, the resulting method is called continuous Galerkin discretization cGP(k). For discontinuous Galerkin dG(k) method, both test and trial spaces are piecewise discontinuous polynomials of degree k. Advantages of variational time discretization are stability, convergence, space-time adaptivity. Both continuous and discontinuous Galerkin methods are A-stable; the discontinuous Galerkin methods are even L-stable (strongly stable). The convergence order of cGP(k) methods is of one order higher than the dG(k) methods. Both of these methods are super-convergent at the nodal points, namely of order $2r + 1$, when the order of the method is r and the solution of the problem is sufficiently regular [31, Chap. 12]. The time-space adaptivity can be easily implemented, because the time is discretized as the space with finite elements. Using a posteriori error estimates, adaptive hp time stepping and dynamic meshes (the use of different spatial discretization for each time step) can directly be incorporated in the discrete formulation [25]. We want to mention that dynamic meshes (meshes changing with time) were used by combining dG(0) time discretization with multiple shooting method for linear and nonlinear OCPs in [18], whereas Carraro et al. [6] use fixed meshes for all discrete time levels.

As we have mentioned, the main disadvantage of variational time discretization is the large system of coupled equations as a result of space-time discretization. To illustrate this, we consider the semilinear parabolic initial value problem

$$\frac{du}{du} = Au + f(u), \quad u(0) = u_0, \tag{58}$$

where A is a linear second order elliptic differential operator and $f(u)$ is locally Lipschitz continuous and monotone.

The 2×2 block system associated to dG discretization of (58) can be written in the following form:

$$\begin{aligned} \alpha_{1,1}MU_n^1 + \alpha_{1,2}MU_n^2 + \Delta t\beta_{1,1}F(U_n^1) + \Delta t\beta_{1,2}F(U_n^2) &= c_1MU_0 + d_1F(U_0), \\ \alpha_{2,1}MU_n^1 + \alpha_{1,2}MU_n^2 + \Delta t\beta_{2,1}F(U_n^1) + \Delta t\beta_{2,2}F(U_n^2) &= c_2MU_0 + d_2F(U_0), \end{aligned} \tag{59}$$

where M is the mass matrix and $F(\cdot)$'s are dGFEM semi-discretized nonlinear terms of the right hand side of (58).

One step of the Newton iteration for solving the coupled system in (59) corresponds to solving the following 2×2 block system:

$$\begin{pmatrix} \Delta t\alpha_{1,1}M + \Delta t\beta_{1,1}\bar{A} & \Delta t\alpha_{1,2}M + \Delta t\beta_{1,2}\bar{A} \\ \Delta t\alpha_{1,2}M + \Delta t\beta_{1,2}\bar{A} & \Delta t\alpha_{2,2}M + \Delta t\beta_{2,2}\bar{A} \end{pmatrix} \begin{pmatrix} W_n^1 \\ W_n^2 \end{pmatrix} = \begin{pmatrix} R_n^1 \\ R_n^2 \end{pmatrix}, \tag{60}$$

where the vectors W_n^i and R_n^i, for $i = 1, 2$, denote the Newton correction and residual for a temporal basis function, respectively [25].

In [35], the linear system of equations associated to dG(k) method, derived from the solution of the linear parabolic equations, are decoupled into complex valued linear systems having the same structure as the implicit Euler discretization. Because the existing finite element codes do not support complex arithmetic, implementation would be difficult and costly. In order to avoid the use of complex arithmetic, Richter et al. [25] developed an inexact Newton method for solving nonlinear parabolic PDEs discretized by dG(k) methods. At each time step, several linear systems of equations are solved with the same structure as for the implicit Euler discretization. Weller and Basting [34] suggest a different solution strategy for linear parabolic PDEs under dG(2) method approximated at Gauss-Radau points. The essential component U_n^2, which is the solution of the problem at the next time step, can be obtained by an inexact factorization of the Schur complement, due to the property $\beta_{1,2} = \beta_{2,1} = 0$ in (59) and (60). Because the Schur complement is of the fourth order, the condition number will be worse than the condition number of the original system. They apply a symmetric preconditioned conjugate gradient method so that a number of linear systems with the same structure arising from implicit Euler discretization must be solved at each step. The nice property of the method is that it can be applied to linear parabolic PDEs with non-self adjoint operators like diffusion-convection-reaction equation, because the Schur complement is symmetric. Efficiency of the solution technique for nonlinear parabolic problems has to be tested. Schieweck [27] introduced a continuous dG method where the solution space consists piecewise continuous polynomials of degree $k \geq 1$ and test space of piecewise discontinuous polynomials of degree $k-1$ approximated at Gauss-Lobatto nodes. They call this technique discontinuous Galerkin-Petrov dGP(k) method. Because the time derivative of the discrete solution is contained in the discrete test space, the method has energy decreasing property so that it can be applied to gradient systems like Allen-Chan and Chan-Hilliard equations. Again, the essential unknown is U_n^2 for dGP(2) method due to $\beta_{11} = 0$ in (59) and (60), and the solution can be determined by fixed point iteration. However, the linear system which must be solved at each time level consists of powers of mass and stiffness matrices, which could be difficult to solve. Instead, a defect correction algorithm was introduced [27], so that at each defect correction step, linear systems like in the implicit Euler discretization have to be solved again.

Acknowledgements The authors thank to Konstantinos Chrysafinos for his explanations regarding error estimates and references. This research was supported by the Middle East Technical University Research Fund Project (BAP-07-05-2012-102).

References

1. Akman, T., Yücel, H., Karasözen, B.: A priori error analysis of the upwind symmetric interior penalty Galerkin (SIPG) method for the optimal control problems governed by unsteady convection diffusion equations. Comput. Optim. Appl. **57**(3), 703–729 (2014)
2. Apel, T., Flaig, T.G.: Crank-Nicolson schemes for optimal control problems with evolution equations. SIAM J. Numer. Anal. **50**(3), 1484–1512 (2012)
3. Ayuso, B., Marini, L.D.: Discontinuous Galerkin methods for advection-diffusion-reaction problems. SIAM J. Numer. Anal. **47**(2), 1391–1420 (2009)
4. Becker, R., Vexler, B.: Optimal control of the convection-diffusion equation using stabilized finite element methods. Numer. Math. **106**(3), 349–367 (2007)
5. Burman, E.: Crank-Nicolson finite element methods using symmetric stabilization with an application to optimal control problems subject to transient advection-diffusion equations. Commun. Math. Sci. **9**(1), 319–329 (2011)

6. Carraro, T., Geiger, M., Rannacher, R.: Indirect multiple shooting for nonlinear parabolic optimal control problems with control constraints. SIAM J. Sci. Comput. **36**(2), A452–A481 (2014)
7. Chrysafinos, K.: Discontinuous Galerkin approximations for distributed optimal control problems constrained by parabolic PDE's. Int. J. Numer. Anal. Model. **4**(3–4), 690–712 (2007)
8. Chrysafinos, K., Walkington, N.J.: Error estimates for the discontinuous Galerkin methods for parabolic equations. SIAM J. Numer. Anal. **44**(1), 349–366 (electronic) (2006)
9. Ciarlet, P.G.: The Finite Element Method for Elliptic Problems. North-Holland, Amsterdam, New York (1978)
10. Collis, S.S., Heinkenschloss, M.: Analysis of the streamline upwind/Petrov Galerkin method applied to the solution of optimal control problems. Tech. Rep. TR02–01, Department of Computational and Applied Mathematics, Rice University, Houston, TX (2002)
11. Dolejší, V., Feistauer, M.: Error estimates of the discontinuous Galerkin method for nonlinear nonstationary convection-diffusion problems. Numer. Funct. Anal. Optim. **26**(3), 349–383 (2005)
12. Dolejší, V., Feistauer, M., Schwab, C.: A finite volume discontinuous Galerkin scheme for nonlinear convection-diffusion problems. Calcolo **39**, 1–40 (2002)
13. Dolejší, V., Feistauer, M., Sobotíková, V.: Analysis of the discontinuous Galerkin method for nonlinear convection-diffusion problems. Comput. Methods Appl. Mech. Eng. **194**(25–26), 2709–2733 (2005)
14. Eriksson, K., Johnson, C., Thomée, V.: Time discretization of parabolic problems by the discontinuous Galerkin method. RAIRO Modél. Math. Anal. Numér. **19**(4), 611–643 (1985)
15. Feistauer, M., Švadlenka, K.: Discontinuous Galerkin method of lines for solving nonstationary singularly perturbed linear problems. J. Numer. Math. **12**(2), 97–117 (2004)
16. Fu, H.: A characteristic finite element method for optimal control problems governed by convection-diffusion equations. J. Comput. Appl. Math. **235**(3), 825–836 (2010)
17. Fu, H., Rui, H.: A priori error estimates for optimal control problems governed by transient advection-diffusion equations. J. Sci. Comput. **38**(3), 290–315 (2009)
18. Hesse, H.K., Kanschat, G.: Mesh adaptive multiple shooting for partial differential equations. I. Linear quadratic optimal control problems. J. Numer. Math. **17**(3), 195–217 (2009)
19. Hinze, M., Yan, N., Zhou, Z.: Variational discretization for optimal control governed by convection dominated diffusion equations. J. Comput. Math. **27**(2–3), 237–253 (2009)
20. Hozman, J., Dolejjí, V.: A priori error estimates for DGFEM applied to nonstationary nonlinear convection-diffusion equation. In: Kreiss, G., et al. (eds.) Numerical Mathematics and Advanced Applications. ENUMATH 2009, pp. 459–467. Springer, Heidelberg (2010)
21. Leykekhman, D.: Investigation of commutative properties of discontinuous Galerkin methods in PDE constrained optimal control problems. J. Sci. Comput. **53**(3), 483–511 (2012)
22. Leykekhman, D., Heinkenschloss, M.: Local error analysis of discontinuous Galerkin methods for advection-dominated elliptic linear-quadratic optimal control problems. SIAM J. Numer. Anal. **50**(4), 2012–2038 (2012)
23. Lions, J.L.: Optimal control of systems governed by partial differential equations. Translated from the French by S. K. Mitter. Die Grundlehren der Mathematischen Wissenschaften, Band 170. Springer, New York (1971)
24. Meidner, D., Vexler, B.: A priori error estimates for space-time finite element discretization of parabolic optimal control problems. I. Problems without control constraints. SIAM J. Control Optim. **47**(3), 1150–1177 (2008)
25. Richter, T., Springer, A., Vexler, B.: Efficient numerical realization of discontinuous Galerkin methods for temporal discretization of parabolic problems. Numer. Math. **124**(1), 151–182 (2013)
26. Rivière, B.: Discontinuous Galerkin Methods for Solving Elliptic and Parabolic Equations: Theory and Implementation. Frontiers in Applied Mathematics, vol. 35. Society for Industrial and Applied Mathematics (SIAM), Philadelphia, PA (2008)
27. Schieweck, F.: *A*-stable discontinuous Galerkin-Petrov time discretization of higher order. J. Numer. Math. **18**(1), 25–57 (2010)

28. Schötzau, D., Schwab, C.: Time discretization of parabolic problems by the *hp*-version of the discontinuous Galerkin finite element method. SIAM J. Numer. Anal. **38**(3), 837–875 (2000)
29. Stoll, M., Wathen, A.: All-at-once solution of time-dependent PDE-constrained optimization problems. Tech. rep., Computational Methods in Systems and Control Theory, Max Planck institude for Dynamics of Complex Technical Systems, Magdeburg (2010). http://www.eprints.maths.ox.ac.uk/1017/1/NA-10-13.pdf
30. Sun, T.: Discontinuous Galerkin finite element method with interior penalties for convection diffusion optimal control problem. Int. J. Numer. Anal. Model. **7**(1), 87–107 (2010)
31. Thomée, V.: Galerkin Finite Element Methods for Parabolic Problems, 2nd edn. Springer Series in Computational Mathematics, vol. 25. Springer, Berlin (2006)
32. Tröltzsch, F.: Optimal Control of Partial Differential Equations: Theory, Methods and Applications. Graduate Studies in Mathematics, vol. 112. American Mathematical Society, Providence, RI (2010). Translated from the 2005 German original by Jürgen Sprekels
33. Vlasák, M., Dolejší, V., Hájek, J.: A priori error estimates of an extrapolated space-time discontinuous Galerkin method for nonlinear convection-diffusion problems. Numer. Methods Partial Differ. Equ. **27**(6), 1456–1482 (2011)
34. Weller, S., Basting, S.: Efficient preconditioning of variational time discretization methods for parabolic partial differential equations. ESIAM Math. Model. Numer. Anal. **49**(2), 331–347 (2015)
35. Werder, T., Gerdes, K., Schötzau, D., Schwab, C.: *hp*-discontinuous Galerkin time stepping for parabolic problems. Comput. Methods Appl. Mech. Eng. **190**(49–50), 6685–6708 (2001)
36. Yücel, H., Heinkenschloss, M., Karasözen, B.: Distributed optimal control of diffusion-convection-reaction equations using discontinuous Galerkin methods. In: Cangiani, A., Davidchack, R.L., Georgoulis, E., Gorban, A.N., Levesley, J., Tretyakov, M.V. (eds.) Numerical Mathematics and Advanced Applications 2011, pp. 389–397. Springer, Berlin/Heidelberg (2013)
37. Yücel, H., Karasözen, B.: Adaptive symmetric interior penalty Galerkin (SIPG) method for optimal control of convection diffusion equations with control constraints. Optimization **63**(1), 145–166 (2014)
38. Zhou, Z., Yan, N.: The local discontinuous Galerkin method for optimal control problem governed by convection diffusion equations. Int. J. Numer. Anal. Model. **7**(4), 681–699 (2010)

Reducing Memory Requirements in Scientific Computing and Optimal Control

Sebastian Götschel, Christoph von Tycowicz, Konrad Polthier, and Martin Weiser

Abstract In high accuracy numerical simulations and optimal control of time-dependent processes, often both many timesteps and fine spatial discretizations are needed. Adjoint gradient computation, or post-processing of simulation results, requires the storage of the solution trajectories over the whole time, if necessary together with the adaptively refined spatial grids. In this paper we discuss various techniques to reduce the memory requirements, focusing first on the storage of the solution data, which are typically double precision floating point values. We highlight advantages and disadvantages of the different approaches. Moreover, we present an algorithm for the efficient storage of adaptively refined, hierarchic grids, and the integration with the compressed storage of solution data.

1 Introduction

The numerical solution and optimal control of time-dependent, nonlinear PDEs often requires fine discretization both of the time interval $[0, T]$ and the—typically three-dimensional—spatial domain Ω to achieve accurate results. For optimization, adjoint gradient computation is often used, see e.g. [29]. There, the solution trajectory over the whole time interval needs to be stored, together with the adaptively refined spatial grids. To be more precise, consider the abstract optimal control problem

$$\min_{y,u} J(y, u) \text{ s.t. } c(y, u) = 0,$$

A pre-print of this article appeared as ZIB-Report 13-64.

S. Götschel (✉) • M. Weiser
Zuse Institute Berlin, Takustr. 7, 14195 Berlin, Germany
e-mail: goetschel@zib.de; weiser@zib.de

C. von Tycowicz • K. Polthier
Freie Universität Berlin, Arnimallee 6, 14195 Berlin, Germany
e-mail: christoph.vontycowicz@fu-berlin.de; konrad.polthier@fu-berlin.de

T. Carraro et al. (eds.), *Multiple Shooting and Time Domain Decomposition Methods*, Contributions in Mathematical and Computational Sciences 9,
DOI 10.1007/978-3-319-23321-5_10

where the relation between the state y and the control u is governed by the equality constraint $c : Y \times U \to Z^\star$, which, for example, may be a parabolic PDE on Hilbert spaces Y, U, Z. Under suitable assumptions, and with $y = y(u)$ the unique solution of the state equation $c(y, u) = 0$, we arrive at the reduced formulation

$$\min_u j(u) := J(y(u), u).$$

The reduced gradient, required for the optimization, is then given by

$$j'(u) = J_u(y, u) - c_u(y, u)^\star \lambda,$$

where the adjoint variable λ fulfills $c_y(y, u)^\star \lambda = J_y(y, u)$. Partial derivatives with respect to the variables y, u are denoted by $J_y, J_u, \ldots$. The adjoint equation is backwards-in-time, and—depending on the objective functional and the state equation—the solution of the forward state equation is needed for the adjoint equation, so the state has to be stored at every timestep.

Of course the storage of simulation results is not only important for optimization, but also for other post-processing algorithms, visualization, archiving of results, and more.

Not only the mere storage size is important, with the ever-growing speed of CPUs, memory access time is more and more becoming a bottleneck for large-scale simulation and optimization. To be able to tackle real-world applications, compression methods are required in order to reduce the amount of data. In this paper, we discuss various techniques to reduce the memory requirements, both in terms of bandwidth and size. An important criterion to judge the quality of compression methods is the *compression factor*, which is defined as the ratio between uncompressed and compressed storage size. Typically—but not in all cases—a reduction of memory size leads also to a similar reduction of the required memory bandwidth. Of course when using lossy compression, where parts of the original information are discarded, the compression factor has to be discussed in relation with the induced error.

This paper is organized as follows. In Sect. 2, we discuss approaches for general floating point compression, before we come to methods specialized for optimal control in Sect. 3. In both sections we mainly focus on the compression of the solution data. In Sect. 4 we describe an algorithm for the efficient storage of the adaptively refined spatial grids, and its integration with the lossy compression approach discussed in Sect. 3.2.

2 General Floating Point Compression

In this section, we discuss approaches for general purpose floating point compression, both lossless and lossy.

2.1 Lossless Methods

For lossless methods, the sole criterion for comparison of different approaches is the compression factor. The comparison depends on the test data sets used, which differ in the literature. Nevertheless, the reported compression ratios are good indicators for the quality and applicability of the algorithms to our problem at hand.

FPC In [7], Burtscher and Ratanaworabhan present the lossless, single-pass, linear-time compression algorithm FPC. It aims at compressing floating-point data with unknown internal structure, with maximizing throughput, i.e. compression speed, as the main objective. Sequences of double-precision floating-point values are processed by predicting a value, determining the prediction error by an XOR operation, and compressing the result.

As predictors, fcm [57] and dfcm [16] are used, such that prediction is essentially a hash-table look-up to determine which value followed the last time a given sequence of values occurred. If the predicted value is close to the true value, the XOR operation produces many leading zeros. The number of leading zeros is encoded in a 3-bit value, which is stored together with a bit specifying the chosen predictor and the remaining non-zero bytes of the prediction error. The reported compression ratios range between 1.02 and up to 15.05 (for one special test data set), the geometric mean compression ratio is 1.2–1.9 depending on the size of the look-up table for the predictors.

fpzip While FPC uses no information about the structure of the data, the algorithm fpzip by Lindstrom and Isenburg, based on [48], traverses the data in some coherent order, and uses the Lorenzo predictor [30] to estimate values based on a subset of the already encoded data. Row-by-row traversal of the data works especially well for data on structured, cartesian grids. The predicted and true value is mapped from floating-point to an integer representation. While fpzip is foremost a lossless compression algorithm, it can be run in a lossy mode. Then, during the mapping stage, the least significant bits are discarded, reducing the precision to 48, 32 or 16 bits/value, without control of the quantization error. The integer residual is stored using range coding [49], a variant of arithmetic coding. Lossless compression ratios of 1.4–2.7 for a double precision test data set are reported in [48], with an average ratio of approximately 1.6.

2.2 Lossy Methods

As expected, lossless methods cannot reduce the amount of data significantly, due to many quasi-random least significant bits. To achieve good compression ratios, we have to resort to lossy compression techniques. Typically, the accuracy is reduced by *quantization* of the true values, or of predicted values, which is essentially rounding. Here, control of the quantization error is of crucial importance.

Comparison criteria for lossy methods are the compression ratio in relation with the induced error. The different test data sets given in the literature, together with the different error norms used to report the quantization errors, makes it difficult to give a quantitative comparison of the algorithms described below.

Adaptive Coarsening/Sub-sampling This method, presented in [59, 72], is closely related to adaptive mesh refinement. Starting from simulation results on some fine, uniform mesh, the mesh is tentatively coarsened. After reconstructing the solution, grid points are removed where the data is represented on the coarser mesh with sufficient accuracy. This procedure is carried out recursively on the new coarser meshes, until no further coarsening is possible without violating the error bound. The result is an adaptive mesh representing the data up to a specified accuracy. As no quantization is used, compression is solely achieved by adaptivity. If the simulations are carried out using standard adaptive mesh refinement during the solution process, data reduction is only possible, if the necessary accuracy for solution and post-processing differ, like for adjoint gradient computation. In [72] the reported compression factors range between 7.44 (3D data) and 25.1 (2D data) for a pointwise relative ℓ^∞ error of 10^{-3}.

Graph Decomposition In a recent work, Iverson, Kamath and Karypis [35] propose a compression algorithm based on the decomposition of the computational grid into so-called ε-bounded sets. The method works on structured and unstructured meshes, which are modeled via a graph. The nodes of the graph are the grid vertices for which values are computed. These vertices are partitioned into non-overlapping sets V_i, such that each set contains only vertices with values differing at most by a specified ε. In each set V_i, the values are replaced by the mean value of the set, such that the point-wise absolute error is bounded by ε. If there is local smoothness in the data, this substitution increases the redundancy of the data, which is afterwards compressed using standard lossless compression methods. For a testset consisting of data on structured and unstructured grids with between 486,015 and 100,663,296 vertices, they report average compression ratios between 20 and 50 for pointwise relative ℓ^∞ errors of orders 10^{-2} to 10^{-3}.

ISABELA Lakshminarasimhan et al. [45, 46] propose a method for "In situ Sort-And-B-spline Error-bounded Lossy Abatement" (ISABELA), specifically designed for spatio-temporal scientific data that is inherently noisy and random-like. In the spatial domain, data is sorted from an irregular signal to a smooth monotonous curve. Then a B-spline is fitted to the sorted data, the difference between data and fitted curve is quantized and stored, together with the information necessary to invert the sorting process. Their experience suggests that the ordering of the sorted data is similar between adjacent timesteps such that delta-encoding can be used to compress the ordering information. The accuracy of the reconstructed data is reported by two quantities, the normalized root mean squared error (NRMSE), and Pearson's correlation coefficient defined by

$$\text{NRMSE} = \frac{(\sum_i (D_i - \tilde{D}_i)^2)^{\frac{1}{2}}}{\max(D) - \min(D)}, \qquad \rho = \frac{\text{cov}(D, \tilde{D})}{\sigma(D)\sigma(\tilde{D})},$$

where D denotes the original data, $\tilde{D}$ the de-compressed data, and σ the standard deviation. In [46] they report compression factors between 3.8 and 5.6 for error bounds $\rho > 0.99$ and NRMSE < 0.01 and five different data sets.

FEMZIP FEMZIP [67, 68] is a commercial tool for the compression of finite element solutions created by certain FE-programs. After a quantization step with prescribed relative or absolute tolerance, a prediction step follows. In space, a hierarchic approximation of the FE functions is performed, using coarsening of the computational grid by algebraic multigrid techniques [68]. In time, prediction based on rigid body movements is used. As a final step, the approximation residual is compressed using arithmetic encoding. Compression factors of up to 13.3 are reported [67], but without quantitative specification of the accuracy.

2.3 Geometry Compression

Compression of geometric data is a vital factor in many computer graphics and visualization applications. Here we will briefly discuss techniques developed for the compression of polygonal surface meshes. For a more detailed overview and comparisons of various schemes we refer to the excellent surveys [2, 54].

A wealth of compression schemes have been developed for single-rate coding (compressing the whole mesh in a region-growing fashion) as well as progressive coding (encoding the model from coarse to fine). For triangle meshes, the most prominent single-rate coders are the Edgebreaker [56] and the method of Touma and Gotsman [69] which both spawned numerous descendants. In particular, the early-split coder of Isenburg and Snoeyink [33] and the optimized Edgebreaker encoding of Szymczak [66] are among the best-performing variants and are able to achieve bit rates well below the Tutte limit [71] of roughly 3.24 bits per vertex. In addition, many triangle mesh compression schemes have been generalized to polygonal meshes, such as Martin Isenburg's method [31] which extends the Touma-Gotsman coder.

Bits rates can be improved even further by accessing already encoded geometry data when encoding connectivity and vice versa, hence exploiting the correlation between connectivity and geometry. Based on this approach, FreeLence [38] is especially performant in the triangular case, while Angle Analyzer [47] is optimized for quadrilateral meshes.

Progressive coders follow a different approach: the coder starts from a coarse mesh and then successively encodes refinement data for finer representations of the model. This approach allows the application of multiresolution analysis to decorrelate high- and low-frequency components of the geometry and/or attribute data such as colors and texture coordinates. Details in the data are thus represented as wavelet-coefficients which typically feature a smaller entropy than the original representation.

Wavelet transforms have been presented for both (unstructured) hierarchical and irregular grids. The latter group employs mesh coarsening methods that progressively remove vertices causing the smallest distortion. Prominent coders in this category are [1, 32, 73]. The best results for geometry compression however have been achieved for hierarchical meshes where efficient wavelet transforms have been derived based on the notion of subdivision. The best known scheme in this group is the progressive geometry compression (PGC) codec by Khodakovsky et al. [42] adapting the established zerotree coding scheme [61] from image compression. Numerous variants have been proposed extending PGC to different types of meshes [41], resolution scalability [3], and efficient embedded quantization [53]. Using context-based entropy coding to account for intraband correlations of the wavelet coefficients, Denis et al. [12] and von Tycowicz et al. [75] are able to further improve the compression performance. In addition, [75] incorporates strategies to encode adaptively refined hierarchies independently of the geometry or attribute data. We utilize these strategies in our coding technique presented in Sect. 4.

In the field of geometry compression the accuracy is typically measured in terms of the *root mean square error* as reported by METRO [9] which is based on a point-to-surface distance and thus neglects tangential errors. For an accuracy of orders 10^{-4} to 10^{-5} w.r.t. the bounding box diameter, FreeLence reports average compression factors of 21 for irregular triangle meshes whereas [75] achieves average factors of 29 and 122 for adaptive and uniform hierarchical grids, respectively.

3 Specialized Methods for Optimization with Differential Equations

In the remainder of this paper, we focus on methods tailored to the needs of optimal control problems governed by time-dependent differential equations.

3.1 Checkpointing

So-called "checkpointing methods" are a tool for the computation of the reduced gradient using the adjoint equation. Instead of keeping track of the whole forward trajectory, only the solution at some intermediate timesteps is stored. During the integration of the adjoint equation, the required states are re-computed starting from the snapshots. In the analysis of checkpointing methods, fixed spatial grids are usually considered, such that each checkpoint has the same size.

3.1.1 Fixed Timesteps

During the forward simulation, the algorithm has to decide when to create a checkpoint. In the simplest setting, the temporal mesh is fixed as well as the spatial grid, and the checkpoint distribution can be computed beforehand ("offline checkpointing"). In the following we denote by c the total number of checkpoints, and by n_t the total number of timesteps of the time discretization.

One obvious strategy would be to place checkpoints uniformly over the time interval, a technique also known as *windowing*, see e.g. [6]. Recursive application of this strategy to each group of timesteps between two checkpoints results in a *multilevel checkpointing* strategy [6, 22]. Neither technique yields optimal distributions, i.e. distributions leading to a minimal amount of re-computations. *Binomial checkpointing* [20, 21] is based on the fact that the maximal number of timesteps $\beta(c, r)$ that can be reversed fulfills

$$\beta(c, r) = \binom{c + r}{c},$$

when c checkpoints and at most r re-computations of each timestep are allowed. Via dynamic programming one can evaluate the minimal extra number of forward steps $t(n_t, c)$ necessary to compute the adjoint using c checkpoints as

$$t(n_t, c) = rn_t - \beta(c + 1, r - 1),$$

where r is the unique integer satisfying $\beta(c, r - 1) < n_t \leq \beta(c + 1, r - 1)$, see e.g. [21, 22]. An implementation called revolve by Griewank and Walther [21] is available.

The achieved compression factor for storage space is given by n_t/c. However, due to multiple read- and write-accesses of checkpoints during the re-computations for the adjoint equation, the reduction in memory bandwidth is significantly smaller. An evaluation of the number of times a snapshot is written or read can be found in [63]. There Stumm and Walther analyze a multistage approach, where some checkpoints are kept in RAM, others written to a hard disk or tape. Evaluating the write counts for instance for $n_t = 1000$ timesteps, and $c = 50$ checkpoints, i.e. compression factor 20, shows that only about 5 % reduction of memory bandwidth is achieved for these parameters [78]. In this example we get $r = 2$, and the computational overhead amounts to $1,948$ additional forward steps.

Here, we assumed that each timestep has the same computational cost; in case of non-uniform timestep cost, optimal checkpoint distributions can be evaluated in

$\mathcal{O}(cn_t^2)$ if the timestep costs are known a priori [76]. Alternatively, heuristics can be used for checkpoint placement [62].

3.1.2 Adaptive Timesteps

If the number of timesteps is not known beforehand, the optimal checkpoint distribution cannot be computed. Thus in practical applications, the user has to resort to "online" placement of checkpoints during the state integration.

An extension of the revolve algorithm, named a-revolve, is proposed in [27, 62], and applied to optimal control of the Navier-Stokes equations. There, a heuristic strategy to overwrite the contents of a previously recorded checkpoint is developed, based on estimates of the computational cost for the current and the updated snapshot distribution. While the resulting scheme is not proven to be optimal, numerical experiments indicate that the generated checkpoint distribution is close to the corresponding offline one.

Other work on online checkpointing was started in [26], with extensions and theoretical foundations in [64]. The approach presented there is proven to be optimal in terms of re-computations for repetition number $r = 2$ and $n_t \leq \beta(c, 2)$. For $r = 3$ and $\beta(c, 2) < n_t \leq \beta(c, 3)$ optimal checkpoint distributions cannot be computed, but for a wide range of timesteps n_t, the resulting algorithm is close to optimal. The method works by continuously overwriting certain previously set checkpoints, until the end of the state integration. For re-computations during the adjoint integration, intermediate snapshots are stored using optimal offline checkpointing.

A different strategy for choosing which checkpoints to replace is devised in [77]. Although their algorithm, called *dynamic checkpointing*, works for an arbitrary number of timesteps n_t, the resulting distribution has just an optimal repetition number r, but is not optimal in terms of the total number of re-computations.

For all three methods the reduction in memory bandwidth is drastically smaller than the reduction in storage space. In fact, due to the frequent overwriting of snapshots, it is questionable if a reduction of bandwidth can be achieved at all.

3.1.3 Discussion

Checkpointing is a compression method, which was originally developed for computation of gradients via the reverse mode of automatic differentiation, where a huge number n_t of arithmetic operations has to be reversed. In that context, two features are particularly important: checkpointing is lossless, and, for a constant number of checkpoints c, the additional computational work, governed by the repetition number r, grows slowly for an increasing number of operations, $r \approx n_t^{1/c}$ [22]. For optimization with time-dependent differential equations as constraints, we are not interested in adjoining every arithmetic operation, but in the solution of the backward-in-time adjoint PDE. This requires access only to the solution of the state equation at every timestep, which is re-computed by solving the state equation with the closest checkpoint as initial data. As the number of timesteps is typically small compared to the number of arithmetic operations in automatic differentiation, the actual additional work—for typical settings two up to four additional solves of the state equation—carries more weight than the limit behavior for increasing n_t. Moreover, in terms of data transferred, only a small reduction of bandwidth can be achieved, in particular with online checkpointing.

When using second order optimization methods, like Newton-CG, the state trajectory is needed in each CG iteration to evaluate Hessian-times-vector products, leading to higher computational work, as checkpoints are overwritten during adjoint integration, and thus their original information is lost for the subsequent CG iterations and has to be re-computed as well.

For non-uniform timestep cost which is not known a-priori, checkpoint distributions have to be chosen heuristically. With adaptive mesh refinement, also the sizes of the snapshots are unknown a priori. For this case, no optimal checkpoint distributions are known, not even heuristics.

3.2 *Lossy Compression*

Checkpointing methods pay for the storage reduction with an increase in runtime, but reconstruct the solution data exactly. However, due to discretization of the state equation by finite elements, quadrature, and iterative solution of the resulting linear equation systems, the solution is inherently inexact. Thus a trade-off between storage demand and accuracy is natural.

3.2.1 Point-Wise Error Bounds

In [78] we propose using the general principle of transform coding for the compression of finite element solution trajectories. It consists of a prediction step, quantization, and entropy coding of the prediction errors, see Fig. 1. To fix the setting we consider spatial discretization by a nested family $\mathcal{T}_0 \subset \cdots \subset \mathcal{T}_l$ of simplicial triangulations, constructed from an initial triangulation $\mathcal{T}_0$ of a polygonal domain $\Omega \subset \mathbb{R}^d$. This grid hierarchy can be created either by uniform or adaptive refinement. The set of vertices on level j is denoted by $\mathcal{N}_j$. The time grid for the time interval $[0, T]$ is given by $0 = t_0 < \cdots < t_f = T$. For brevity, we restrict the attention to piecewise linear finite elements.

Quantization For a given accuracy $\delta > 0$ we define the *quantization* $Q_\delta : \mathbb{R} \to \mathbb{Z}$ as

$$Q_\delta(y) := \left\lfloor \frac{y + \delta}{2\delta} \right\rfloor,$$

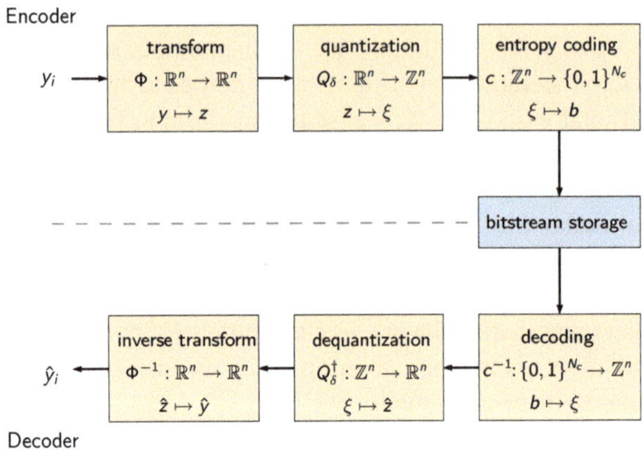

Fig. 1 Principle of transform coding

the *reconstruction* $Q_\delta^\dagger : \mathbb{Z} \to \mathbb{R}$ is then given by $Q_\delta^\dagger(\xi) := 2\delta\xi$. This guarantees the quantization error bound

$$|y - Q_\delta^\dagger(Q_\delta(y))| \le \delta.$$

This implies an ℓ^∞ error bound of δ on the coefficient vector, and hence an L^∞ bound on the FE function.

Prediction in Space Values y_k associated with coarse level vertices are quantized directly to $\xi_k = Q_\delta(y_k)$, yielding a reconstructed value $\hat{y}_k := Q_\delta^\dagger(\xi_k)$. For new vertices $x_k \in \mathcal{N}_j \setminus \mathcal{N}_{j-1}$ on level $j > 0$, we make use of the grid hierarchy and quantize and store only the deviation of y_k from a prediction $P_k(\hat{y}_m : m \in \mathcal{N}_{j-1})$ obtained from reconstructed values $\hat{y}_m$ from lower level vertices. There are several algorithmic choices for the predictor. One possibility is a change of basis from the nodal basis to the hierarchic basis [79]. This is easily implemented, as it is essentially the application of prolongation matrices between grid levels, which is easily accessible in most FE codes.

A priori estimates for the compression factors were derived in [78]. To achieve L^∞-interpolation error accuracy for the reconstructed FE function, 2.9 bits/vertex in 2D and 2.5 bits/vertex in 3D are sufficient. This amounts to compression factors of 22.1 and 25.6, respectively, compared to storing double precision floating point values at 64 bits/vertex.

Prediction in Time Additionally, temporal correlations can be used to further reduce the entropy of the data. If gradient based methods like steepest-descent or quasi-Newton methods are used, the state solution is only accessed backwards in time, and no random access is required. Thus, we can use delta-encoding, and store for a timestep $t_n < t_f$ only the difference

$$\xi_k^\Delta(t_n) = \begin{cases} \xi_k(t_n) - \xi_k(t_{n+1}), & k \in \left(\mathcal{N}_j(t_n)\backslash\mathcal{N}_{j-1}(t_n)\right) \cap \left(\mathcal{N}_j(t_{n+1})\backslash\mathcal{N}_{j-1}(t_n)\right) \\ \xi_k(t_n), & \text{otherwise} \end{cases}.$$

At the final timestep t_f,

$$\xi_k^\Delta(t_f) = \xi_k(t_f) \quad \forall k \in \mathcal{N}_j(t_f)\backslash\mathcal{N}_{j-1}(t_f).$$

Delta-encoding the quantized coefficients avoids error accumulation. Of course higher order prediction can be used instead of the constant predictor described above, at the expense of keeping more timesteps in RAM.

Entropy Coding The quantized and possibly delta-encoded coefficients ξ_k^Δ are written to disk using range coding [49]. They are encoded with variable-size symbols, where fewer bits are assigned to the more frequent coefficients.

More details and numerical examples can be found in [78].

3.2.2 Adaptive Error Control

A crucial algorithmic choice is the quantization tolerance δ. To choose the error bound as large as possible without impeding the convergence of the optimization algorithm, we need to estimate the induced error in the reduced gradient $j'(u) = J_u(y,u) - c_u(y,u)^\star\lambda$. Typically, J_u and c_u are independent of y, such that the error is determined by the error of the adjoint. If additionally c_y does not depend on the state, e.g. for linear equations, the error in the adjoint e_λ satisfies the equation $c_y(y,u)^\star e_\lambda = -J_{yy}(y,u)\varepsilon_y$, where ε_y denotes the error in the reconstructed state solution. For nonlinear equations, the error additionally depends on $c_{yy}(y,u)\varepsilon_y$ and the solution of the adjoint equation with inexact data. Computationally available estimates of the gradient error can be used to determine the quantization tolerance according to the progress of the optimization procedure. A detailed discussion can be found in [17].

For second order methods, errors in the reduced Hessian have to be considered as well. A derivation of error estimates and the influence on a Newton-CG method can be found in [19] specialized to the application in optimal control of cardiac defibrillation, and more general in [17].

3.2.3 H^{-1} Error Bounds

While bounding the pointwise ℓ^∞ error in the coefficients of the reconstructed FE solution trajectory is easily implemented and yields good compression factors, considering other error measures is reasonable in the optimal control setting. In particular, the reconstructed solution enters into the right-hand side of the adjoint equation. Due to smoothing properties of parabolic equations, a quantization error with high spatial frequency is preferable, such that the H^{-1}-norm is more appropriate.

Controlling the H^{-1} error can be achieved by using a wavelet transform to represent the finite element solution $y(x, t_n)$ at some fixed timestep t_n as

$$y(x,t_n) = \sum_{k\in\mathcal{N}_0} y_{0,k}\phi_{0,k}(x) + \sum_{j=0}^{l-1} \sum_{m\in\mathcal{N}_{j+1}\setminus\mathcal{N}_j} z_{j,m}\psi_{j,m}(x).$$

For this we assume again a level partitioning of the grid vertices $\mathcal{N} = \mathcal{N}_0 \cup \cdots \cup \mathcal{N}_l$, and denote by the subscript j, k values belonging to vertex k on grid level j. We use the abbreviations $n(j,k) = \{m \in \mathcal{N}_{j+1} \setminus \mathcal{N}_j \mid m \text{ is a child of } k\}$ and $N(j,m) = \{k \in \mathcal{N}_j \mid k \text{ is a parent of } m\}$. Here, a vertex $m \in \mathcal{N}_{j+1}$ is a child of $k_1 \in \mathcal{N}_j$, if m was created by splitting an edge $[k_1, k_2]$. The *scaling functions* $\phi_{j,k}$ satisfy the refinement relation

$$\phi_{j,k} = \phi_{j+1,k} + \sum_{m\in n(j,k)} \frac{1}{2}\phi_{j+1,m},$$

the *wavelets* are of the form

$$\psi_{j,m} = \phi_{j+1,m} - \sum_{k\in N(j,m)} s_{j,k,m}\phi_{j,k}.$$

The lifting coefficients $s_{j,k,m}$ are determined to impose vanishing moments on the wavelets, see e.g. [8, 65]. In particular, one vanishing moment is easily obtained on unstructured grids if the mass matrix is available.

The modified coarse grid values $y_{0,k}$ and wavelet coefficients $z_{j,m}$ are computed using the fast wavelet transform with lifting [58, 65], for grid levels $l-1,\ldots,0$:

$$z_{j,m} = y_{j+1,m} - \frac{1}{2}\sum_{k\in N(j,m)} y_{j,k} \qquad \forall m \in \mathcal{N}_{j+1}\setminus\mathcal{N}_j$$

$$y_{j,k} = y_{j+1,k} + \sum_{m\in n(j,k)} s_{j,k,m} z_{j,m} \qquad \forall k \in \mathcal{N}_j.$$

Norm equivalences, e.g. [11], suggest that the error bound $\|y - \hat{y}\|_{H^{-1}} < \varepsilon$ holds, if the wavelet coefficients $z_{j,k}$ are quantized using a level-dependent tolerance $\delta_j \sim 2^{j(d/2+1)}\varepsilon$.

To compare a first, simple implementation of this approach with the hierarchical basis (HB) prediction of Sect. 3.2.1, we use the three functions

$$f_1(x) = \sin(12(x_0 - 0.5)(x_1 - 0.5)), \quad x \in [0,1]^2$$

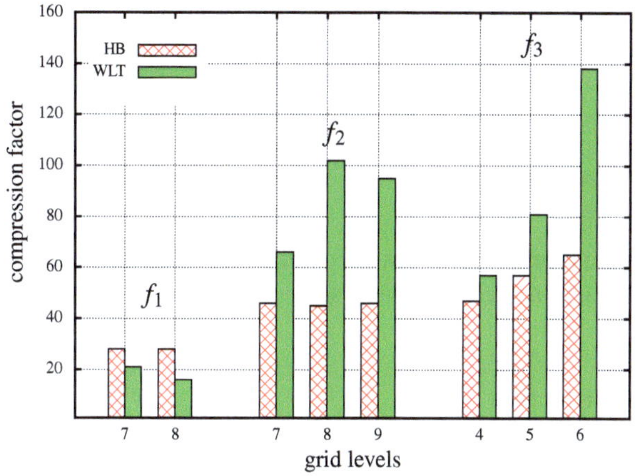

Fig. 2 Compression factors for hierarchical basis prediction/point-wise error bounds (HB) and wavelet-based compression (WLT), for three test functions and different mesh sizes

$$f_2(x) = \sin(50(x_0 - 0.5)(x_1 - 0.5)), \quad x \in [0, 1]^2$$

$$f_3(x) = \|x\|^2 + \sin(12(x_0 - 0.5)(x_1 - 0.5)), \quad x \in [0, 1]^3.$$

For the HB approach, quantization tolerances were chosen to achieve L^∞-interpolation error accuracy. For the wavelet approach the tolerance was set to achieve the same H^{-1} error as the corresponding HB result. The resulting compression factors shown in Fig. 2 indicate that on average a wavelet approach might indeed give better compression factors when H^{-1} reconstruction errors are used..

3.2.4 Discussion

The lossy compression technique sketched in this section offers significant reduction of storage space as well as memory bandwidth, as only the compressed data is transferred to storage media. The computational cost of the basic method is negligible: Quantization, delta-encoding in time, and entropy coding consist only of cheap elementary arithmetic operations; in space, the prediction step amounts to the computation of products between prolongation matrices and FE coefficient vectors. Prolongation matrices are often available from multigrid preconditioners, or can otherwise be computed inexpensively on the fly.

As a downside, information is discarded in the quantization step, and the FE solution cannot be reconstructed exactly. If used in optimal control of differential equations, adaptive control of the quantization error ensures that the inexactness has no influence on the convergence of the optimization. For other post-processing tasks, like data analysis or visualization, the error norms and tolerances can be chosen according to the particular needs of the application, offering a flexible way to balance data size and accuracy.

3.3 Other Techniques

In this section we briefly discuss two techniques for the solution of optimal control problems, with memory reduction as a side effect.

3.3.1 Model Reduction

Model reduction techniques focus mainly on the reduction of computational complexity via approximation of large-scale problems by smaller ones. First developed for handling parameter-dependent differential equations, in the last years this algorithm class has been applied to optimal control and inverse problems as well. One popular method for the construction of reduced models is proper orthogonal decomposition (POD). There, a basis is computed from the solution of the state equation at a number of given timesteps. For many problems, only a few basis vectors are necessary to get sufficiently accurate approximations. A detailed analysis of POD methods for parabolic PDEs can be found in [43], see e.g. [28] for the use of POD in optimal control. In terms of memory requirements, only the snapshots of the solution of the large-scale problem need to be stored.

Due to the reduced-order model, only sub-optimal controls can be computed. To judge the quality of the approximate solution, a-posteriori error estimators were developed. In [70], such an estimator is derived for the linear-quadratic case, and extended to semilinear equations in [40]. For the evaluation of the error estimate, state and adjoint solutions of the full problem are needed, posing the same requirements for storage space as the original large-scale problem. A different technique is suggested in [37]: they use the full model to compute the gradient and only use reduced models to find a suitable steplength for the control update. Again, no reduction in memory size is achieved. While both methods reduce memory bandwidth, a combination with lossy trajectory compression for evaluation of error estimators or gradient computation appears attractive.

3.3.2 Multiple Shooting

Multiple shooting is a well established method for the solution of ODE boundary value problems. The time interval $[0, T]$ is decomposed in a number of sub-intervals, with auxiliary variables for the interfaces ensuring continuity of the solution. The resulting cyclic, nonlinear system of equations is typically solved using Newton's method. Details and a short overview of the history of shooting methods can be found in [13], for instance. In the last years, this principle was applied to solve optimal control problems governed by time-dependent partial differential equations, e.g. [10, 23–25]. The decomposition of the time domain leads to optimization problems on the sub-intervals, where locally state and adjoint are implicit functions of the control and the auxiliary variables [10]. Sequential solution of the local problems leads to a storage reduction, as only the trajectory on the respective sub-interval is needed. The coupling of the sub-problems via the auxiliary variables ("matching conditions") avoids the disadvantage of moving horizon techniques, where only sub-optimal controls can be computed (see e.g. [34]). Combination with adaptive mesh refinement is discussed e.g. in [24, 25], where a dual weighted residual (DWR) method [5] is used for error estimation.

Although the resulting algorithms are easily parallelizable due to the splitting in local sub-problems, significant storage reduction is only achieved in sequential computations, or if the number of sub-intervals is considerably larger than the number of CPUs. Each CPU then has to provide storage only for the currently processed local problem, plus additional storage for the auxiliary variables. Again, a combination with lossy trajectory storage is an attractive possibility.

4 Compression of Hierarchical, Unstructured Grids

For problems with spatially localized solution features, it is beneficial to use adaptively refined spatial meshes to reduce the computational cost and memory demand of simulation and optimization. As a downside, in the context of trajectory storage discussed here, this incurs the need to store the mesh together with the trajectory data. For applying our lossy compression approach, we even need the complete hierarchy, not just the leaf level. In this section we discuss an efficient algorithm for mesh storage [39, 75], as well as the integration of this method with the lossy compression approach described above.

4.1 Connectivity Compression

Numerous strategies for the adaptive refinement of grids have been presented in the literature. Exploiting the particular structure inherent to a given strategy is paramount in the construction of an efficient compression scheme. Here we present a method that is tailored to hierarchies based on split operations for which the resulting grid is independent of the order in which the operations are applied. In particular, we confine our attention to the well-known and established *red–green refinement* [4] on two-dimensional grids. However, the ideas presented here can also be adapted to refinement schemes for three-dimensional grids and/or other types of elements. For example, [75] additionally provides results for quadrilateral hierarchies.

Typically, the root grid is described by a small, carefully laid out mesh that can be compressed well using single-rate coders. Explicitly, our implementation uses FreeLence [38] to losslessly encode the triangular base mesh. Starting from the root grid, it is sufficient to encode which elements (including those on finer levels) have been refined to reconstruct the adaptive hierarchy. Thus, the hierarchy can be represented as a forest where each node corresponds to an element in the grid and the parent-child relation reflects which triangles resulted through refinement of a particular coarse one.

We convert this representation into a bit stream by traversing the forest breadth-first and writing `true` only if the node has children, i.e., was refined. If a geometric criterion is used to resolve non-conforming situations between elements of differing refinement grade, we can uniquely determine the connectivity of the grid from the root grid together with the bit stream. However, if the conformization is determined exclusively by local indexing, additional symbols have to be coded whenever there is freedom of choice, e.g. a coarse triangle with two refined neighbors (see Fig. 3 middle). We entropy code these symbols, but found that they where virtually incompressible without knowing the exact implementation of the grid manager.

Before compressing the bit stream we remove nodes whose state can be implicitly reconstructed. In particular, no symbols are written in the following cases:

1-Regularity In balanced grids, neighboring triangles must not differ in more than one level of refinement to ensure a certain level of shape regularity. Thus, elements adjacent to coarse green triangles cannot be refined and can therefore be culled from the bit stream.

Stream Truncation Due to breadth-first traversal, nodes at the finest level are visited last. The corresponding `false` symbols can be left from the stream since they cause no further refinements. In fact, we discard all trailing `false` entries.

Uniform Refinement We store a separate byte that encodes the degree of uniform refinement, allowing the coder to skip all nodes on coarser levels.

Overall, for the test set of adaptive hierarchies used in [75], the above steps allow to nearly halve the number of bits in the binary representation, without even looking at the characteristics of the particular input grid. However, grids do show certain characteristics in practice and we use *context groups* as a simple measure to account for the conditional entropy (see [60]) of the bit stream. Just like two adjacent pixels in a digital photograph are likely to be similar, the refinement grades in hierarchical meshes typically tend to be locally similar. We call two nodes within one level of the hierarchy adjacent, if their corresponding triangles share an edge. This notion of locality allows us to define the context of a node based on the refinement status of its

Fig. 3 *Red-green* conformizations of non-conforming configurations due to adaptive refinement

neighbors. Naturally, we may only assume knowledge of neighbors whose status is also available during decoding. Thus, we define the context of a node by the number of refined, not refined, and unknown neighbors. The latter category is made up of nodes whose status cannot be implied and have not been traversed so far. We write (x, y, z) to denote the context with x refined, y not refined, and z unknown adjacent triangles. The symbols of different context groups are kept in separate arrays, which are entropy coded independently. With arithmetic coding, each context group will then compress to its conditional entropy in the limit, allowing us to exploit the correlation in the refinement status of adjacent triangles.

However, the mutual information inherent to each context group varies drastically. For example, context (0,0,3) with all neighbors unknown is virtually incompressible as no advantage can be taken of mutual information. The same holds for context groups where the extra information is rather ambiguous, for example (1,1,1), (1,2,0), and (2,1,0). At the contrary, the other context groups perform very well in the experiments. These observations motivate an optimization of the traversal scheme used within each level since the iteration of nodes can be arbitrary as long as encoder and decoder agree on a common one. Instead of iterating each node using a standard breadth-first in-order traversal of the forest, we determine an ordering that attempts to maximize the mutual information. The idea is to prioritize each node by the entropy of its current context. Learning these entropies, however, is expensive in terms of compression performance as well as computational cost. As shown in [75], this approach typically leads to a fixed prioritization of context groups once the learning phase is settled. Therefore, the effects of the learning process of the contexts' entropies can be remedied by using fixed priorities. Explicitly, context groups are assigned higher priorities with decreasing number of unknown neighbors, where ties are resolved by prioritizing contexts with a higher number of known refined neighbors. While the (culled) binary representation of the hierarchy is almost incompressible when entropy coded directly, the proposed context groups together with the improved traversal reduced the code length by more than 50 % for the test data in [75].

Furthermore, the proposed context-based coding can easily be extended to time-varying series. When coding the status of a node in a sequence of frames we can extend the context groups to account for its status in a previous frame. The previous state of a node can either be refined, not refined, or it did not exist, hence we triple each context. If the refinements between frames does not vary much, the contexts corresponding to previously refined nodes will mainly comprise `true` symbols whereas the other contexts will primarily contain `false`. Therefore, grids which equal their preceding frame can be stored at no cost (except for a small overhead due to the increased number of context groups). On the contrary, if there is no correlation between the frames, the compression will be as good as in the static case since the entropy of the individual context groups cannot increase.

4.2 Numerical Example

The lossy compression method discussed in Sect. 3.2 is implemented in the C++ finite element toolbox Kaskade 7 [18]. An implementation of the algorithm for connectivity compression is available in JavaView [36], a toolkit for mathematical geometry processing and visualization. Both packages have been combined using the Java Native Interface to allow a fully adaptive solution of optimal control problems with compression of both meshes and solution data.

As an illustrative example we use an optimal control problem for the monodomain equations describing the electrical activity in the heart (see e.g. [44, 51]) on a simple 2D unit square domain $\Omega = (0,1)^2$. As membrane model, we use the Rogers-McCulloch variant of the Fitzhugh-Nagumo model [55]. The state equations for the *transmembrane voltage* v and the *gating variable* w are given by

$$v_t = \nabla \cdot \sigma \nabla v - gv\big(1 - \frac{v}{v_{th}}\big)\big(1 - \frac{v}{v_p}\big) + \eta_1 v w + \chi_{\Omega_c} u(t)$$

$$w_t = \eta_2\big(\frac{v}{v_p} - \eta_3 w\big),$$

together with homogeneous Neumann boundary conditions, and initial values

$$v(x,0) = \begin{cases} 101.0 & \text{in } \Omega_{\text{exi}} \\ 0 & \text{otherwise} \end{cases}$$

$$w(x,0) = 0 \quad \text{in } \Omega.$$

Here, Ω_{exi} is a circle with radius 0.04 and midpoint $(0.5, 0.5)$. The state variable is $y = (v, w)$; $\sigma : \mathbb{R}^2 \to \mathbb{R}^{2\times 2}$ and $g, \eta_i, v_p, v_{th} \in \mathbb{R}_+$ are given parameters. For details, see e.g. [50]. The control u is spatially constant on the control domain $\Omega_c = [0.37, 0.4] \times [0.45, 0.55] \cup [0.6, 0.63] \times [0.45, 0.55]$. The objective is to dampen out the excitation wave front induced by the initial values,

$$J(y,u) = \frac{1}{2}\, \|v\|^2_{L^2(\Omega_{\text{obs}} \times (0,T))} + \frac{\alpha}{2}\, \|u\|^2_{L^2(0,T)} \to \min,$$

where $\Omega_{\text{obs}} = \Omega \setminus \big([0.35, 0.42] \times [0.43, 0.57] \cup [0.58, 0.65] \times [0.43, 0.57]\big)$, and $\alpha = 3 \times 10^{-6}$. Optimality conditions and more details can be found in [19]. We use adjoint gradient computation and the BFGS-Quasi-Newton method [52] for optimization. Spatial adaptivity is performed individually for state and adjoint using the hierarchical DLY error estimator [15]. For time stepping, a linearly implicit extrapolated Euler method [14] is used, with fixed timestep sizes for ease of implementation.

First, we consider just one iteration, i.e. one state and adjoint solve, on the time interval $[0, 6]$ with timestep size 0.04. In space, we restrict the number of vertices

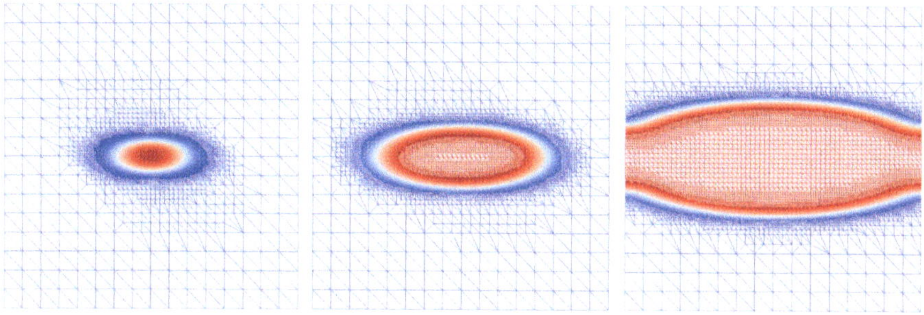

Fig. 4 Uncontrolled solution v at 1, 3 and 6 ms. The adaptively refined meshes have 37,344, 41,729 and 38,346 vertices

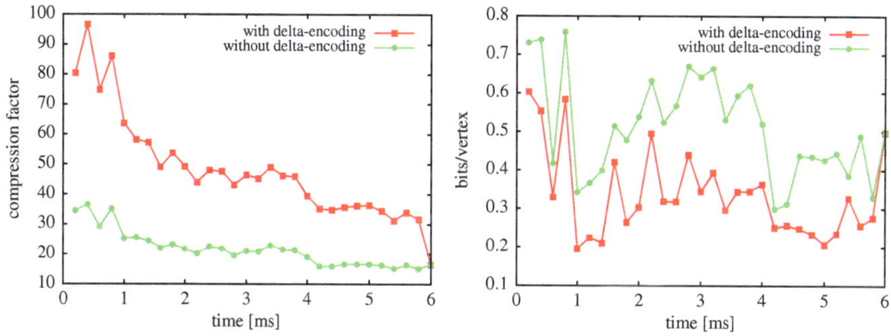

Fig. 5 Compression factor of the state values using lossy compression with $\delta = 10^{-2}$ compared to double precision floating point values at 64 bits/vertex (*left*), bits/vertex for connectivity encoding (*right*), both with and without delta-encoding between timesteps

Table 1 CPU times (in seconds) for one state and adjoint solve, averaged over five test runs

		State values			Grid		
	Solve	Setup	Encode	Decode	Encode	Decode	Transfer
State	3026.2	31.4	11.3	–	90.6	–	–
Adjoint	1473.1	31.4	–	3.7	–	59.3	183.6

Times are measured without delta-encoding of trajectory and mesh

to be less than 60,000. We choose a fixed quantization tolerance $\delta = 10^{-2}$, yielding a relative absolute error bound of 10^{-4} for v. In Fig. 4 we show the v component of the state variable at selected times. Compression factors for the state values, and the number of bits/vertex necessary for connectivity encoding is shown in Fig. 5. Using delta-encoding in time more than doubles the achieved overall compression factor for the state values. The bits/vertex for connectivity encoding are reduced to 66 % of the storage size obtained by compressing each timestep separately. Detailed CPU times are shown in Table 1. *solve* consists of time for assembly, adaptivity, and solution of the linear systems using BiCGStab [74] with an ILU preconditioner. For state value compression, during *setup*, the prolongation matrices required for the spatial prediction are generated, which is more expensive than the actual encoding and decoding. The overall computational overhead for state value compression amounts to 1.4 % in the state equation, and 2.4 % in the adjoint. Encoding and decoding the mesh take up 3 % and 4 % in state and adjoint, respectively. As state and adjoint equations are solved on independently adapted grids, the de-compressed state trajectory has to be interpolated on the adjoint mesh (last column in Table 1) which takes up 12.5 % CPU time of an adjoint solve; this overhead occurs also if the trajectory is stored uncompressed. In our current preliminary implementation, we have to re-create the mesh hierarchy in the Java code for encoding, and in the C++ code after decoding. Additionally, as different data structures are used in the two combined software toolboxes, re-assignment of the degrees of freedom is necessary after the encoding step. This significant overhead is not included in the CPU times reported here, as it can be avoided by improving the implementation. For delta-encoding, at each timestep t_n, additional work is required for checking if a vertex k already existed in the previous timestep t_{n-1}, determining its corresponding quantized residual $\xi_k(t_{n-1})$, and computing the difference $\xi_k(t_{n-1}) - \xi_k(t_n)$. For a suitable implementation, this increases the computation times for encoding and decoding by approximately 50 %.

Second, the complete optimization is performed on the time interval $[0, 4]$, with timestep size 0.04, and a restriction to at most 25,000 vertices in space. Figure 6 shows the progress of the optimization method. For trajectory compression, different fixed quantization tolerances were used. We estimate the spatial discretization error in the reduced gradient by using a solution on a finer mesh as a reference. Clearly, lossy compression has no influence on the optimization progress, up to discretization error accuracy.

Remark Numerical results are given here for one specific example. The proposed lossy compression scheme for the state values was applied to several other optimal control problems, both linear and nonlinear, with comparable results [17, 78]. Good compression factors are achieved even for highly nonlinear dynamics and adaptively

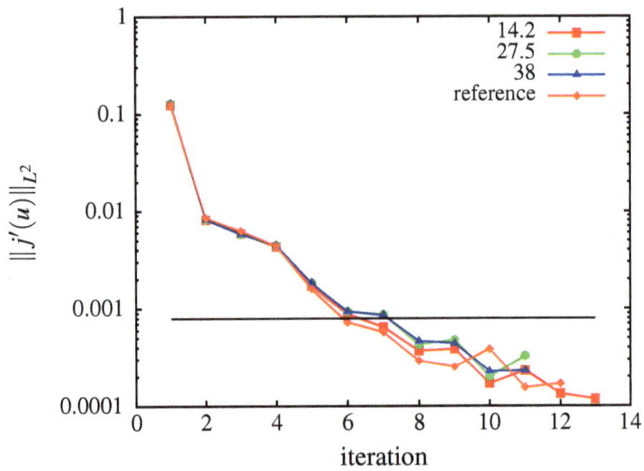

Fig. 6 Optimization progress for different quantization tolerances for the state trajectory. No delta-encoding between timesteps was used

refined grids [19]. Compression of hierarchical meshes yielded excellent results for a variety of examples [75].

5 Conclusion

To reduce the memory requirements of scientific data, essentially two classes of algorithms are available: methods like checkpointing, which reduce storage space at the cost of computation time, and lossy compression techniques, where the trade-off is between memory requirements and accuracy. While general purpose floating point compression methods can be used for many different applications, good compression results can only be achieved with structure-exploiting techniques, like checkpointing, FEMZIP, or our lossy compression approach.

Optimal control problems pose specific requirements for accuracy, which can be satisfied using quantitative error estimates to choose suitable quantization tolerances. The combination of lossy state values compression and compressed storage of adaptively refined meshes yields significant reduction of storage space, at small computational cost. As only the compressed data has to be transferred to and from mass storage, memory bandwidth requirements are reduced by the same factor.

Acknowledgements The authors gratefully acknowledge support by the DFG Research Center MATHEON, project F9.

References

1. Alliez, P., Desbrun, M.: Progressive compression for lossless transmission of triangle meshes. In: Proceedings of 28th Annual Conference on Computer Graphics and Interactive Techniques, pp. 195–202. ACM (2001)
2. Alliez, P., Gotsman, C.: Recent advances in compression of 3d meshes. In: Dodgson, N., Floater, M., Sabin, M. (eds.) Advances in Multiresolution for Geometric Modelling, Mathematics and Visualization, pp. 3–26. Springer, Berlin/Heidelberg (2005). doi:10.1007/3-540-26808-1_1. http://www.dx.doi.org/10.1007/3-540-26808-1_1
3. Avilés, M., Morán, F., García, N.: Progressive lower trees of wavelet coefficients: efficient spatial and SNR scalable coding of 3D models. In: Advances in Mulitmedia Information Processing-PCM 2005 pp. 61–72 (2005)
4. Bank, R.E., Sherman, A.H., Weiser, A.: Some refinement algorithms and data structures for regular local mesh refinement. In: Stepleman, R., et al. (eds.) Scientific Computing. Applications of Mathematics and Computing to the Physical Sciences, vol. I. IMACS/North-Holland, Amsterdam (1983)
5. Becker, R., Rannacher, R.: An optimal control approach to a posteriori error estimation in finite element methods. Acta Numerica **10**(1), 1–102 (2001)
6. Becker, R., Meidner, D., Vexler, B.: Efficient numerical solution of parabolic optimization problems by finite element methods. Optim. Methods Softw. **22**(5), 813–833 (2007). http://www.dx.doi.org/10.1080/10556780701228532

7. Burtscher, M., Ratanaworabhan, P.: FPC: a high-speed compressor for double-precision floating-point data. IEEE Trans. Comput. **58**(1), 18–31 (2009)
8. Castrillón-Candás, J.E., Amaratunga, K.: Spatially adapted multiwavelets and sparse representation of integral equations on general geometries. SIAM J. Sci. Comput. **24**(5), 1530–1566 (2003)
9. Cignoni, P., Rocchini, C., Scopigno, R.: Metro: measuring error on simplified surfaces. Tech. Rep., Paris, France (1996)
10. Comas, A.: Time-domain decomposition preconditioners for the solution of discretized parabolic optimal control problems. Ph.D. thesis, Rice University (2005)
11. Dahmen, W.: Wavelet methods for PDEs – some recent developments. J. Comput. Appl. Math. **128**(1), 133–185 (2001)
12. Denis, L., Satti, S., Munteanu, A., Cornelis, J., Schelkens, P.: Scalable intraband and composite wavelet-based coding of semiregular meshes. IEEE Trans. Multimedia **12**(8), 773–789 (2010)
13. Deuflhard, P., Bornemann, F.: Scientific Computing with Ordinary Differential Equations, vol. 42. Springer, Berlin (2002)
14. Deuflhard, P., Nowak, U.: Extrapolation integrators for quasilinear implicit ODEs. In: Deuflhard, P., Engquist, B. (eds.) Large Scale Scientific Computing. Progress in Scientific Computing, vol. 7, pp. 37–50. Birkhäuser, Basel (1987)
15. Deuflhard, P., Leinen, P., Yserentant, H.: Concepts of an adaptive hierarchical finite element code. IMPACT Comput. Sci. Eng. **1**(1), 3–35 (1989)
16. Goeman, B., Vandierendonck, H., De Bosschere, K.: Differential FCM: increasing value prediction accuracy by improving table usage efficiency. In: The 7th International Symposium on High-Performance Computer Architecture, 2001. HPCA, pp. 207–216. IEEE (2001)
17. Götschel, S., Weiser, M.: Lossy compression for PDE-constrained optimization: Adaptive error control. ZIB Report, pp. 13–27 (2013)
18. Götschel, S., Weiser, M., Schiela, A.: Solving optimal control problems with the Kaskade 7 finite element toolbox. In: Dedner, A., Flemisch, B., Klöfkorn, R. (eds.) Advances in DUNE, pp. 101–112. Springer, New York (2012)
19. Götschel, S., Nagaiah, C., Kunisch, K., Weiser, M.: Lossy compression in optimal control of cardiac defibrillation. J. Sci. Comput. **60**(1), 35–59 (2014)
20. Griewank, A.: Achieving logarithmic growth of temporal and spatial complexity in reverse automatic differentiation. Optim. Methods Softw. **1**(1), 35–54 (1992)
21. Griewank, A., Walther, A.: Algorithm 799: revolve: an implementation of checkpointing for the reverse or adjoint mode of computational differentiation. ACM Trans. Math. Softw. **26**(1), 19–45 (2000)
22. Griewank, A., Walther, A.: Evaluating Derivatives: Principles and Techniques of Algorithmic Differentiation. SIAM, Philadelphia (2008)
23. Heinkenschloss, M.: A time-domain decomposition iterative method for the solution of distributed linear quadratic optimal control problems. J. Comput. Appl. Math. **173**(1), 169–198 (2005)
24. Hesse, H.K.: Multiple shooting and mesh adaptation for PDE constrained optimization problems. Ph.D. thesis, University Heidelberg (2008)
25. Hesse, H.K., Kanschat, G.: Mesh adaptive multiple shooting for partial differential equations. part I: linear quadratic optimal control problems. J. Numer. Math. **17**(3), 195–217 (2009)
26. Heuveline, V., Walther, A.: Online checkpointing for parallel adjoint computation in PDEs: application to goal-oriented adaptivity and flow control. In: Euro-Par 2006 Parallel Processing, pp. 689–699. Springer, Berlin (2006)
27. Hinze, M., Sternberg, J.: A-revolve: an adaptive memory-reduced procedure for calculating adjoints; with an application to computing adjoints of the instationary Navier-Stokes system. Optim. Methods Softw. **20**(6), 645–663 (2005)
28. Hinze, M., Volkwein, S.: Error estimates for abstract linear-quadratic optimal control problems using proper orthogonal decomposition. Comput. Optim. Appl. **39**, 319–345 (2008)
29. Hinze, M., Pinnau, R., Ulbrich, M., Ulbrich, S.: Optimization with PDE Constraints. Springer, Berlin (2009)

30. Ibarria, L., Lindstrom, P., Rossignac, J., Szymczak, A.: Out-of-core compression and decompression of large n-dimensional scalar fields. In: Computer Graphics Forum, vol. 22, pp. 343–348. Wiley Online Library (2003)
31. Isenburg, M.: Compressing polygon mesh connectivity with degree duality prediction. In: Graphics Interface Conference Proceedings, pp. 161–170 (2002)
32. Isenburg, M., Snoeyink, J.: Mesh collapse compression. In: Proceedings of SIBGRAPI'99, pp. 27–28 (1999)
33. Isenburg, M., Snoeyink, J.: Early-split coding of triangle mesh connectivity. In: Graphics Interface Proceedings, pp. 89–97. Canadian Information Processing Society, Toronto, ON (2006)
34. Ito, K., Kunisch, K.: Receding horizon optimal control for infinite dimensional systems. ESAIM Control Optim. Calc. Var. **8**(1), 741–760 (2002)
35. Iverson, J., Kamath, C., Karypis, G.: Fast and effective lossy compression algorithms for scientific datasets. In: Euro-Par 2012 Parallel Processing, pp. 843–856. Springer, Berlin (2012)
36. JavaView Homepage: www.javaview.de (2014)
37. Jörres, C., Vossen, G., Herty, M.: On an inexact gradient method using proper orthogonal decomposition for parabolic optimal control problems. Comput. Optim. Appl. **55**(2), 1–10 (2013)
38. Kälberer, F., Polthier, K., Reitebuch, U., Wardetzky, M.: Freelence - coding with free valences. Comput. Graph. Forum **24**(3), 469–478 (2005)
39. Kälberer, F., Polthier, K., von Tycowicz, C.: Lossless compression of adaptive multiresolution meshes. In: Proceedings of Brazilian Symposium on Computer Graphics and Image Processing (SIBGRAPI), vol. 22 (2009)
40. Kammann, E., Tröltzsch, F., Volkwein, S.: A method of a-posteriori error estimation with application to proper orthogonal decomposition. Tech. Rep. (2011)
41. Khodakovsky, A., Guskov, I.: Compression of normal meshes. In: Geometric Modeling for Scientific Visualization, pp. 189–206. Springer, Berlin (2003)
42. Khodakovsky, A., Schröder, P., Sweldens, W.: Progressive geometry compression. In: SIGGRAPH'00 Proceedings, pp. 271–278 (2000)
43. Kunisch, K., Volkwein, S.: Galerkin proper orthogonal decomposition methods for parabolic problems. Numer. Math. **90**, 117–148 (2001)
44. Kunisch, K., Wagner, M.: Optimal control of the bidomain system (I): The monodomain approximation with the Rogers-McCulloch model. Nonlinear Anal. Real World Appl. **13**(4), 1525–1550 (2012). doi:10.1016/j.nonrwa.2011.11.003. http://www.sciencedirect.com/science/article/pii/S1468121811003099
45. Lakshminarasimhan, S., Shah, N., Ethier, S., Klasky, S., Latham, R., Ross, R., Samatova, N.F.: Compressing the incompressible with ISABELA: in-situ reduction of spatio-temporal data. In: Euro-Par 2011 Parallel Processing, pp. 366–379. Springer, Berlin (2011)
46. Lakshminarasimhan, S., Shah, N., Ethier, S., Ku, S.H., Chang, C.S., Klasky, S., Latham, R., Ross, R., Samatova, N.F.: ISABELA for effective in situ compression of scientific data. Concurr. Comput. Pract. Exper. **25**, 524–540 (2013)
47. Lee, H., Alliez, P., Desbrun, M.: Angle-analyzer: a triangle-quad mesh codec. In: Drettakis, G., Seidel, H.-P. (eds.) Eurographics 2002 Conference Proceedings, pp. 383–392 (2002)
48. Lindstrom, P., Isenburg, M.: Fast and efficient compression of floating-point data. IEEE Trans. Vis. Comput. Graph. **12**(5), 1245–1250 (2006). http://www.dx.doi.org/10.1109/TVCG.2006.143
49. Martin, G.: Range encoding: an algorithm for removing redundancy from a digitised message. In: Presented at Video & Data Recording Conference, Southampton (1979)
50. Nagaiah, C., Kunisch, K., Plank, G.: Numerical solution for optimal control of the reaction-diffusion equations in cardiac electrophysiology. Comput. Optim. Appl. **49**, 149–178 (2011). doi:10.1007/s10589-009-9280-3. http://www.dx.doi.org/10.1007/s10589-009-9280-3
51. Nielsen, B.F., Ruud, T.S., Lines, G.T., Tveito, A.: Optimal monodomain approximations of the bidomain equations. Appl. Math. Comput. **184**(2), 276–290 (2007)
52. Nocedal, J., Wright, S.J.: Numerical Optimization. Springer, New York (2006)

53. Payan, F., Antonini, M.: An efficient bit allocation for compressing normal meshes with an error-driven quantization. Comput. Aided Geom. Des. **22**(5), 466–486 (2005)
54. Peng, J., Kim, C.S., Jay Kuo, C.C.: Technologies for 3d mesh compression: a survey. J. Vis. Commun. Image Represent. **16**(6), 688–733 (2005). doi:10.1016/j.jvcir.2005.03.001. http://www.dx.doi.org/10.1016/j.jvcir.2005.03.001
55. Rogers, J.M., McCulloch, A.D.: A collocation-Galerkin finite element model of cardiac action potential propagation. IEEE Trans. Biomed. Eng. **41**, 743–757 (1994)
56. Rossignac, J.: Edgebreaker: connectivity compression for triangle meshes. IEEE Trans. Vis. Comput. Graph. **5**, 47–61 (1999)
57. Sazeides, Y., Smith, J.E.: The predictability of data values. In: Proceedings of 30th Annual IEEE/ACM International Symposium on Microarchitecture, 1997, pp. 248–258. IEEE (1997)
58. Schröder, P., Sweldens, W.: Spherical wavelets: efficiently representing functions on the sphere. In: SIGGRAPH '95 Proceedings of the 22nd Annual Conference on Computer Graphics and Interactive Techniques, pp. 161–172. ACM (1995)
59. Shafaat, T.M., Baden, S.B.: A method of adaptive coarsening for compressing scientific datasets. In: Kågström, B., Elmroth, E., Dongarra, J., Wasniewski, J. (eds.) Applied Parallel Computing. State of the Art in Scientific Computing. 8th International Workshop, PARA 2006, Umeå, Sweden, 18–21 June 2006, Revised Selected Papers. Lecture Notes in Computer Science, vol. 4699, pp. 774–780. Springer, New York (2007)
60. Shannon, C.E.: A mathematical theory of communication. Bell Syst. Tech. J. **27**, 379–423 (1948)
61. Shapiro, J.M.: Embedded image coding using zerotrees of wavelet coefficients. IEEE Trans. Signal Process. **41**, 3445–3462 (1993)
62. Sternberg, J., Hinze, M.: A memory-reduced implementation of the Newton-CG method in optimal control of nonlinear time-dependent PDEs. Optim. Methods Softw. **25**(4), 553–571 (2010)
63. Stumm, P., Walther, A.: Multi-stage approaches for optimal offline checkpointing. SIAM J. Sci. Comput. **31**(3), 1946–1967 (2009)
64. Stumm, P., Walther, A.: New algorithms for optimal online checkpointing. SIAM J. Sci. Comput. **32**(1), 836–854 (2010)
65. Sweldens, W.: The lifting scheme: a construction of second generation wavelets. SIAM J. Math. Anal. **29**(2), 511–546 (1998)
66. Szymczak, A.: Optimized edgebreaker encoding for large and regular triangle meshes. In: DCC '02 Proceedings, p. 472. IEEE Computer Society, Washington, DC (2002)
67. Teran, R.I., Thole, C.A., Lorentz, R.: New developments in the compression of LS-DYNA simulation results using FEMZIP. In: 6th European LS-DYNA Users' Conference (2007). https://www.dynalook.com/european-conf-2007/new-developments-in-the-compression-of-ls-dyna.pdf
68. Thole, C.A.: Compression of LS-DYNA3D™ simulation results using FEMZIP©. 3. LS-DYNA Anwenderforum (2004)
69. Touma, C., Gotsman, C.: Triangle mesh compression. In: Graphics Interface Conference Proceedings, pp. 26–34 (1998)
70. Tröltzsch, F., Volkwein, S.: POD a-posteriori error estimates for linear-quadratic optimal control problems. Comput. Optim. Appl. **44**, 83–115 (2009)
71. Tutte, W.: A census of planar triangulations. Can. J. Math. **14**, 21–38 (1962)
72. Unat, D., Hromadka, T., Baden, S.: An adaptive sub-sampling method for in-memory compression of scientific data. In: Data Compression Conference, 2009 (DCC '09), pp. 262–271. IEEE (2009)
73. Valette, S., Prost, R.: Wavelet-based progressive compression scheme for triangle meshes: wavemesh. IEEE Trans. Vis. Comput. Graph. **10**(2), 123–129 (2004)
74. van der Vorst, H.A.: Bi-CGSTAB: a fast and smoothly converging variant of bi-cg for the solution of nonsymmetric linear systems. SIAM J. Sci. Stat. Comput. **13**, 631–644 (1994)

75. von Tycowicz, C., Kälberer, F., Polthier, K.: Context-based coding of adaptive multiresolution meshes. Comput. Graph. Forum **30**(8), 2231–2245 (2011). doi:10.1111/j.1467-8659.2011.01972.x. http://www.dx.doi.org/10.1111/j.1467-8659.2011.01972.x
76. Walther, A.: Program reversal schedules for single-and multi-processor machines. Ph.D. thesis, Institute of Scientific Computing, Technical University Dresden, Germany (1999)
77. Wang, Q., Moin, P., Iaccarino, G.: Minimal repetition dynamic checkpointing algorithm for unsteady adjoint calculation. SIAM J. Sci. Comput. **31**(4), 2549–2567 (2009)
78. Weiser, M., Götschel, S.: State trajectory compression for optimal control with parabolic PDEs. SIAM J. Sci. Comput. **34**(1), A161–A184 (2012)
79. Yserentant, H.: On the multi-level splitting of finite element spaces. Numer. Math. **49**(4), 379–412 (1986)

Optimal Control of Heat and Fluid Flow for Efficient Energy Utilization

Yosuke Hasegawa

Abstract Application of the optimal control theory to turbulent flows and associated transport phenomena opens up a unique possibility of seeking an optimal set of control inputs (design parameters) without relying on researchers' subjective insights. As an example, it is shown that heat transfer enhancement and skin friction drag reduction is simultaneously achieved in wall turbulence, where it has been considered to be difficult to achieve such dissimilar heat transfer enhancement due to the strong similarity in the governing equations of heat and fluid flow. The control input is assumed to be zero-net-mass-flux wall blowing/suction and its spatio-temporal distribution is optimized so as to minimize a prescribed cost functional defined within a finite time horizon. Surprisingly, the resultant control input exhibits a streamwise traveling wave-like property. Although increase in the time horizon significantly enhances the resultant control performance, time horizons employed in previous studies are commonly limited due to the strong nonlinearity of turbulent flows. Applying a multiple shooting method would be promising to further increase the time horizon, and thereby improve the resultant control performance.

1 Background

1.1 New Horizon for Optimizing Thermo-Fluids Systems

Towards achieving the future sustainable society, prediction and control of interfacial phenomena play curtail roles. For example, the turbulent momentum transfer at solid-fluid interfaces governs the energy losses in high-speed transport applications, such as aircrafts, marine vessels, trains, automobiles, pipelines, ventilation systems, to name a few. Enhancing heat and mass transfer across solid-fluid or fluid-fluid interfaces is essential for improving energy efficiencies in air conditioning systems, heat recovery systems, chemical reactors, and so forth. Optimal design

Y. Hasegawa (✉)
Institute of Industrial Science, The University of Tokyo, 4-6-1 Komaba, Meguro-ku, Tokyo 153-8505, Japan
e-mail: ysk@iis.u-tokyo.ac.jp

T. Carraro et al. (eds.), *Multiple Shooting and Time Domain Decomposition Methods*, Contributions in Mathematical and Computational Sciences 9,
DOI 10.1007/978-3-319-23321-5_11

of three-dimensional complex porous structure is particularly important to promote electrochemical reactions in electrodes of solid-oxide fuel cells and lithium ion batteries. Due to the multi-scale and highly non-linear nature of fluid flow and associated transport phenomena, however, optimal design of these energy devices are not trivial. In the present paper, we focus on control of turbulent transport phenomena as a typical example of non-linear problems.

Conventionally, academic researchers have been extracting essential elements from practical problems, and trying to understand the underlying physics. It has been believed that such fundamental knowledge will eventually be useful for designing innovative thermo-fluids systems. Since the first direct numerical simulation (DNS) of wall turbulence by Kim et al. [12], with the aid of rapid development of computational resources, numerical simulation has grown as a powerful tool alternative to existing experimental techniques in deepening our understanding and modeling of complex turbulent transport phenomena. Indeed, the range of application of numerical simulation has been significantly expanded, and this enables to obtain much more detailed flow statistics which cannot be measured experimentally. Despite these progresses, optimization of thermo-fluids systems remain a difficult task. There exists no established approach for predicting how a finite change in a certain design parameter influences resultant drag, heat/mass transfer or chemical reactions in thermo-fluid systems, due to their highly non-linear and multi-scale nature. Optimal control theory opens up a new horizon for seeking an optimal set of design parameters based on the governing equations of underlying physics.

1.2 Overview of Turbulence Control Research for Skin Friction Drag Reduction

During the past several decades, a huge amount of effort of the turbulence research community has been devoted to advance our understanding of turbulent dynamics both experimentally and numerically. Based on this knowledge, various types of control strategies have been proposed. Although flow control has a wide range of applications, such as modifying momentum/heat/mass transfer, suppressing noise, enhancing lift and so forth, we hereafter focus on skin friction drag reduction, which is one of the most active topics in the flow control community.

Existing control schemes are roughly classified into two categories, i.e., active and passive controls. Passive control typified by a riblet surface is advantageous in the sense that it does not require additional energy consumption for flow control. However, their control performance is generally smaller than that of active controls. In addition, they are effective under limited flow conditions close to a design point.

In contrast, active control is generally more flexible and effective, although additional energy consumption for driving actuators is required. Active control is further classified into predetermined and feedback controls. In the former, a control input with spatio-temporal coherence is specified *a priori* and it is applied without sensing the instantaneous flow field. Starting from spanwise wall oscillation [9], various types of predetermined control schemes have been developed. They are, for example, spanwise traveling wave of body force[5], streamwise traveling wave of wall blowing/suction [14] and deformation [8], standing and traveling waves of spanwise wall forcing [15, 16]. Although significant drag reduction rate is achieved in the predetermined controls, they commonly suffer from relative large energy consumption for applying control. Finding a control input leading to a larger drag reduction rate with smaller control energy input is a major challenge in the predetermined control.

The feedback control generally offers better control performance with small power consumption than the predetermined control. However, it requires a complex sensor-actuator system, possibly fabricated by Micro Electro Mechanical System (MEMS) [10] in order to detect an instantaneous flow state, of which signals are used to trigger actuators. One of the most widely-accepted feed-back control strategy is the so-called opposition control proposed by Choi et al. [4]. In this strategy, wall blowing/suction is applied in order to oppose the wall-normal velocity fluctuation at a certain distance away from the wall. By optimizing the sensing location, they demonstrated 15–20 % drag reduction in DNS of a low Reynolds number turbulent flow. In this study, the control input is determined based on the sensing information inside the fluid domain. In real systems, however, the available information is considered to be limited to wall quantities. Accordingly, Lee et al. [13] developed a control algorithm using wall information based on the suboptimal control theory. In the suboptimal control theory, the control input is optimized so as to minimized a prescribed cost functional in the next computational time step. Their algorithm achieves 12 % drag reduction by using the spanwise wall shear stress or wall pressure. These quantities, however, are in most cases difficult to measure using small sensors distributed on the wall [10]. Hence, Fukagata and Kasagi [6] redefined the cost functional based on the near-wall Reynolds shear stress, and achieve drag reduction by using streamwise wall shear stress, which is the easiest quantity to measure with a relatively small error. In the above studies, the control inputs are optimized by taking into account only short-term dynamics in the suboptimal control framework. The "real" optimal control with a finite, but non-vanishing time horizon was first conducted by Bewley et al. [3], where more than 60 % drag reduction is obtained, and an initial turbulent flow is eventually relaminarized due to the applied control. The significant enhancement of control performance from suboptimal to optimal controls implies the importance of taking into account the future dynamics in determining a control input.

1.3 Dissimilar Control of Momentum and Heat Transfer: Less Friction and More Heat Transfer

In many practical problems, one often encounters a significant challenge to not only minimizing drag, but also enhance heat and mass transfer. Indeed, the analysis [2] based on the second law of thermodynamics shows that one of ultimate goals in controlling heat and fluid flow is to achieve an infinitely large heat transfer rate with minimum drag. However, such dissimilar heat transfer enhancement should be a difficult task due to strong similarity between the governing equations of fluid flow and heat in most of shear flows. Namely, on the one hand, turbulence has to be suppressed to reduce drag, but at the same time, mixing has to be promoted in order to enhance heat/mass transfer.

Recently, Kasagi et al. [11] revisited the governing equations and boundary conditions of heat and fluid flow in order to clarify possible scenarios for dissimilar heat transfer control. Among these scenarios, a control strategy based on the fundamental difference between the divergence-free velocity vector and the conservative scalar is considered to be most promising. Based on this idea, Hasegawa and Kasagi [7] first demonstrated dissimilar heat transfer enhancement in a fully developed turbulent channel flow by applying the suboptimal control theory. More recently, Yamamoto et al. [17] applied the optimal control theory to the same problem and higher control performance was obtained. Specifically, they first achieved simultaneous drag reduction and heat transfer enhancement. In the following, we summarize the recent advancement on dissimilar heat transfer enhancement control in wall turbulence.

2 Numerical Configurations

2.1 Numerical Schemes and Conditions

We consider a fully developed turbulent flow between two parallel plates as shown in Fig. 1. The streamwise, wall-normal and spanwise directions are denoted by x_1, x_2 and x_3, whereas the corresponding velocity components are u_1, u_2 and u_3, respectively. The origin of x_2 is located at the center of the channel so that the locations of the two walls are $x_2 = 1$ and -1, respectively. The total volume of the computational domain is V_Ω, whereas the domain boundary is expressed by Γ, the subscript of which represents the normal direction. In all cases, the horizontal channel dimensions are set to $2.5\pi\delta$ and $\pi\delta$ in x_1 and x_3 directions, respectively. These are sufficiently large to reproduce the reliable turbulent statistics in a low Reynolds number flow considered here.

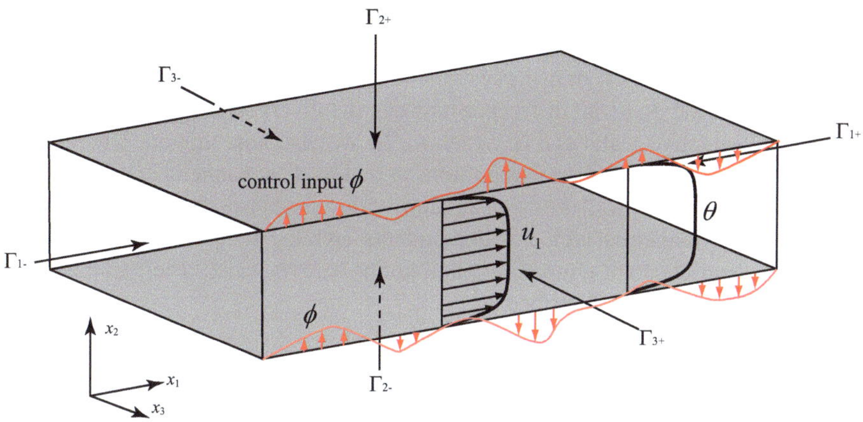

Fig. 1 Computational domain and coordinate system

The governing equations of incompressible fluid flow are given by the following Navier-Stokes and continuity equations:

$$\frac{\partial u_i}{\partial t} + u_j \frac{\partial u_i}{\partial x_j} = -\frac{\partial p}{\partial x_i} + \frac{1}{Re} \frac{\partial^2 u_i}{\partial x_j \partial x_j}, \tag{1}$$

$$\frac{\partial u_i}{\partial x_i} = 0. \tag{2}$$

Here, all variables are normalized by the bulk mean velocity U_b defined later in Eq. (7) and the channel half depth δ, so that the dimensionless channel height is two (see, Fig. 1), whereas p is the static pressure and t is time. The Reynolds number is defined as $Re = U_b \delta / \nu$, where ν is the kinematic viscosity of fluid.

The temperature is treated as a passive scalar, so that any buoyancy effects do not arise. Consequently, the transport equation of heat is given by

$$\frac{\partial \theta}{\partial t} + u_j \frac{\partial \theta}{\partial x_j} = Q + \frac{1}{PrRe} \frac{\partial^2 \theta}{\partial x_j \partial x_j}. \tag{3}$$

Here, the temperature is also non-dimensionalized by the temperature difference between the bulk fluid and the wall, $\Theta_b - \Theta_w$. The Prandtl number is the ratio of ν and the thermal diffusivity α, i.e., $Pr = \nu/\alpha$, whilst the heat source term Q is generally a function of time and space.

In the present study, the heat source is assumed to be time-independent and spatially uniform throughout the computational domain, and identical to the mean pressure gradient:

$$Q = -\overline{\frac{\partial p}{\partial x_1}}, \tag{4}$$

where the over-bar represents averaging in the homogeneous directions, i.e., x_1 and x_3, and also time t. In addition, Pr is also set to be unity. This particular condition is chosen, since it makes the transport equations and boundary conditions for u_1 and θ similar, so that the essential difference between the divergence-free velocity vector and the conservative scalar can be analyzed[7]. We also note that the present ideal condition is close to the thermal conditions in real heat exchangers[17].

We consider local wall blowing/suction with zero-net-mass-flux as a control input. For the tangential velocity components and the temperature, we impose the no-slip and constant-temperature conditions at two walls. The resultant wall boundary conditions are described as

$$u_i\Big|_{\Gamma_{2\pm}} = \phi n_i, \tag{5}$$

$$\theta\Big|_{\Gamma_{2\pm}} = 0. \tag{6}$$

Here, the control input, i.e., the wall-normal velocity component imposed at the wall, is denoted by ϕ, the sign of which is defined to be positive when the applied control input is directed to the outer normal vector n_i at the boundaries of the fluid domain. In the horizontal directions x_1 and x_3, we apply periodic boundary conditions.

The governing equations (1)–(3) for the velocity and thermal fields are solved by DNS with a second-order finite volume method. More detailed description of the numerical scheme can be found in Yamamoto et al. [17]. All calculations are conducted under a constant bulk mean velocity and the Reynolds number is mostly set to be $\mathrm{Re} = 2293$. Due to the similarity in the mathematical form between physical and adjoint problems (see, Eqs. (16) and (23)), the essentially same numerical method is used for solving the adjoint velocity and thermal fields introduced later.

2.2 Control Performance Indices

Following Hasegawa and Kasagi [7], the bulk velocity U_b and the bulk temperature Θ_b are respectively defined as the following cross-sectional average of flow rate and temperature:

$$U_b = \frac{1}{V_\Omega}\int_\Omega u_1 dV, \tag{7}$$

$$\Theta_b = \frac{1}{V_\Omega}\int_\Omega \theta dV. \tag{8}$$

As the indices of heat transfer and pressure loss, the following Stanton number St and the friction coefficient C_f are defined:

$$St = \frac{q_w}{U_b(\Theta_b - \Theta_w)}, \tag{9}$$

$$C_f = \frac{\tau_w}{\frac{1}{2}{U_b}^2}, \tag{10}$$

where

$$q_w = -\frac{1}{PrRe}\frac{\partial\overline{\theta}}{\partial y}n_2\bigg|_{\Gamma_2}, \tag{11}$$

$$\tau_w = -\frac{1}{Re}\frac{\partial\overline{u_1}}{\partial y}n_2\bigg|_{\Gamma_2} \tag{12}$$

are the dimensionless wall heat flux and skin friction, respectively.

If the profiles of the averaged streamwise velocity and temperature are similar, $2St$ is exactly equal to C_f at $Pr = 1$. Therefore, we define an analogy factor as

$$A = \frac{2\text{St}}{C_f}. \tag{13}$$

Physically, A represents heat transfer per unit pumping power. The main objective in dissimilar control is to increase A from unity by manipulating turbulence.

2.3 *Optimization Procedure*

In applying optimal control theory to flow problems, a cost functional is first defined, and then a control input is iteratively updated so as to minimize the cost function within a prescribed time horizon. The correction of a control input in each iteration is obtained by solving the adjoint velocity and thermal fields backward in time. Ideally, the time horizon should be long enough to cover the whole life-time of turbulence dynamics, but it is not computationally trackable. Therefore, it is common to choose an intermediate finite time horizon T as shown in Fig. 2. Once a control input converges, the time horizon is advanced by T_a, and then a new optimization procedure in the next time horizon starts. In the following, we define the cost functional, derive the adjoint equations, and describe the optimization procedures without getting into the mathematical details, which can be found in other literatures [1, 3]. In the present study, the time horizon is set to be $T/(\delta/U_b) = 10$, which is almost identical to the maximal value used in the drag reduction control by Bewley et al. [3], while $T_a = T/10$ is employed.

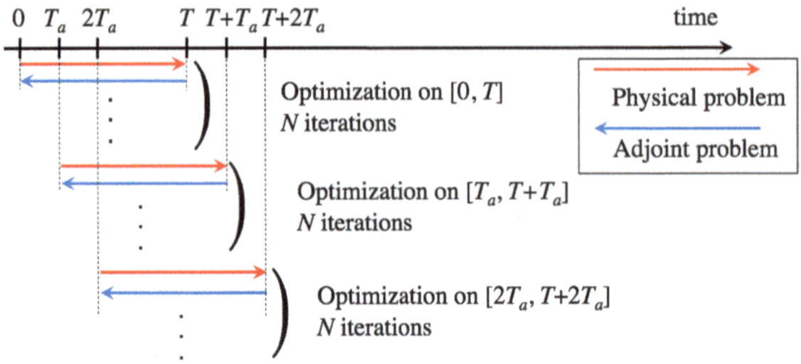

Fig. 2 Schematic of optimization procedure. In each time horizon, the evolution of velocity and thermal fields are solved under a preliminary control input as shown by the *red arrows*, which is followed by adjoint computation depicted by the *blue arrows*

2.3.1 Defining the Cost Functional

We define a cost functional as follows:

$$J = \kappa \int_0^T \int_{\Gamma_{2\pm}} \frac{1}{2}\phi^2 dSdt - A$$

$$\approx \kappa \int_0^T \int_{\Gamma_{2\pm}} \frac{1}{2}\phi^2 dSdt - \frac{\displaystyle\int_0^T \int_{\Gamma_{2\pm}} -\frac{1}{PrRe}\frac{\partial\theta}{\partial n} dSdt}{\displaystyle\int_0^T \int_{\Gamma_{2\pm}} -\frac{1}{Re}\frac{\partial u}{\partial n} dSdt}, \tag{14}$$

where $t = 0$ corresponds to the beginning of the time horizon. The first term represents the cost of control, while the second term is exactly a quantity we attempt to enhance, i.e., the analogy factor. Hence, under this cost functional, the control input is optimized so as to maximize A with the least intensity of wall blowing/suction. Ideally, A has to be determined by the ratio of $2St$ and C_f integrated over a sufficiently long period. Since the optimal control theory takes into account only flow dynamics within a finite time horizon, however, A is approximated by the integrals within the time horizon as shown in the second line of Eq. (14). The weight coefficient κ corresponds to the relative cost of the control input. In the present study, κ is specified so that the intensity of the control input ϕ is 5 % of the bulk mean velocity.

2.3.2 Optimal Control Theory

For ease of notation, the flow state $\boldsymbol{\psi}$, the flow perturbation state $\boldsymbol{\psi}'$ and the adjoint state $\boldsymbol{\psi}^*$ are expressed as the following vector forms:

$$\boldsymbol{\psi} = \begin{pmatrix} p \\ u_i \\ \theta \end{pmatrix}, \boldsymbol{\psi}' = \begin{pmatrix} p' \\ u_i' \\ \theta' \end{pmatrix}, \boldsymbol{\psi}^* = \begin{pmatrix} p^* \\ u_i{}^* \\ \theta^* \end{pmatrix}. \tag{15}$$

The governing equations (1)–(3) for the velocity and thermal fields can be written in a functional form as

$$N(\boldsymbol{\psi}) = \begin{pmatrix} \dfrac{\partial u_i}{\partial x_i} \\ \dfrac{\partial u_i}{\partial t} + \dfrac{\partial u_j u_i}{\partial x_j} - \dfrac{1}{\mathrm{Re}} \dfrac{\partial^2 u_i}{\partial x_j{}^2} + \dfrac{\partial p}{\partial x_i} \\ \dfrac{\partial \theta}{\partial t} + \dfrac{\partial u_j \theta}{\partial x_j} - \dfrac{1}{\mathrm{RePr}} \dfrac{\partial^2 \theta}{\partial x_j^2} - Q \end{pmatrix} = \mathbf{0}. \tag{16}$$

Then, we consider the perturbation field $\boldsymbol{\psi}'$ of velocity and thermal fields induced by a small change of a control input ϕ. Following Bewley et al. [3], the perturbation is defined by the Frechét differential of the original flow state $\boldsymbol{\psi}$ as

$$\boldsymbol{\psi}' \triangleq \lim_{\epsilon \to 0} \frac{\boldsymbol{\psi}(\phi + \phi'\epsilon) - \boldsymbol{\psi}(\phi)}{\epsilon}, \tag{17}$$

where ϵ is an infinitesimal constant.

Since both the original and perturbed flow states satisfy Eq. (16), the following linear equations for $\boldsymbol{\psi}'$ is obtained:

$$N'(\boldsymbol{\psi}') = \begin{pmatrix} \dfrac{\partial u_i'}{\partial x_i} \\ \dfrac{\partial u_i'}{\partial t} + \dfrac{\partial}{\partial x_j}\left(u_j u_i' + u_j' u_i\right) - \dfrac{1}{\mathrm{Re}} \dfrac{\partial^2 u_i'}{\partial x_j{}^2} + \dfrac{\partial p'}{\partial x_i} \\ \dfrac{\partial \theta'}{\partial t} + \dfrac{\partial}{\partial x_j}\left(u_j \theta' + u_j' \theta\right) - \dfrac{1}{\mathrm{RePr}} \dfrac{\partial^2 \theta'}{\partial x_j^2} \end{pmatrix} = \mathbf{0}, \tag{18}$$

where the wall boundary conditions are given by

$$u_i' = -\phi' n_i, \ \theta' = 0 \quad \text{on } \Gamma_{\pm 2}, \tag{19}$$

$$u_i' = \mathbf{0}, \ \theta' = 0 \quad \text{at } t = 0. \tag{20}$$

In Eq. (18), the products between perturbations are all neglected since the perturbation is assumed to be sufficiently small. Although Eqs. (18)–(20) indicate the linear relationship between ϕ' and $\boldsymbol{\psi}'$, it is not straightforward to derive the explicit relationship between ϕ' and the resultant change of the cost functional J'. In order to overcome this difficulty, the adjoint velocity and thermal fields are introduced.

The flow optimization can generally be viewed as a minimization problem of a cost functional J under the constraints on the flow states, i.e., the governing equations and the boundary conditions of flow and thermal fields. This is equivalent to minimizing the following Hamiltonian H:

$$H = J - \langle N(\boldsymbol{\psi}), \boldsymbol{\psi}^* \rangle \,, \tag{21}$$

where the adjoint state $\boldsymbol{\psi}^*$ corresponds to the Lagrangian multiplier.

The Frechét differential of Eq. (21) leads to

$$\begin{aligned} \frac{\mathscr{D}H}{\mathscr{D}\phi}\phi' &= J' - \left\langle N'(\boldsymbol{\psi}'), \boldsymbol{\psi}^* \right\rangle \\ &= J' - \left\langle \boldsymbol{\psi}', N^*(\boldsymbol{\psi}^*) \right\rangle - b, \end{aligned} \tag{22}$$

where N^* is the adjoint operator of N'. We impose the following relationship for the adjoint field:

$$N^*(\boldsymbol{\psi}^*) = \begin{pmatrix} -\dfrac{\partial u_i^*}{\partial x_i} \\ -\dfrac{\partial u_i^*}{\partial t} - u_j \left(\dfrac{\partial u_i^*}{\partial x_j} + \dfrac{\partial u_j^*}{\partial x_i} \right) - \dfrac{1}{\mathrm{Re}} \dfrac{\partial^2 u_i^*}{\partial x_j{}^2} - \dfrac{\partial p^*}{\partial x_i} - \theta \dfrac{\partial \theta^*}{\partial x_i} \\ -\dfrac{\partial \theta^*}{\partial t} - u_j \dfrac{\partial \theta^*}{\partial x_j} - \dfrac{1}{\mathrm{RePr}} \dfrac{\partial^2 \theta^*}{\partial x_j^2} \end{pmatrix} = \mathbf{0}, \tag{23}$$

so that the second term on the right-hand-side of Eq. (22) vanishes. The first term on the right-hand-side of Eq. (22) is the Frechét differential of the cost functional, and can be written as

$$\begin{aligned} J' &= \kappa \int_0^T \int_{\Gamma_{2\pm}} \phi\phi' dSdt \\ &+ \frac{A}{TS_{\Gamma_{2\pm}} \tau_w} \int_0^T \int_{\Gamma_{2\pm}} -\frac{\partial u'}{\partial n} dSdt - \frac{A}{TS_{\Gamma_{2\pm}} q_w} \int_0^T \int_{\Gamma_{2\pm}} -\frac{\partial \theta'}{\partial n} dSdt, \end{aligned} \tag{24}$$

where $S_{\Gamma 2\pm}$ represents the boundary area of $\Gamma_{2\pm}$. The third term on the right-hand-side of Eq. (22) is called a boundary term, since it includes only boundary integrals as shown below:

$$
\begin{aligned}
b = & \int_{\Omega} (u'_j u^*_j + \theta'\theta^*)\Big|_{t^+=0}^{t^+=T} dV \\
& + \int_0^T \int_{\Gamma_{2\pm}} n_j \Big[p^* u'_j + u^*_j p' + u^*_i (u_j u'_i + u_i u'_j) - \frac{1}{\mathrm{Re}} (u^*_i \frac{\partial u'_i}{\partial x_j} - u'_i \frac{\partial u^*_i}{\partial x_j}) \\
& + (u'_j \theta + u_j \theta')\theta^* - \frac{1}{\mathrm{RePr}} (\theta^* \frac{\partial \theta'}{\partial x_j} - \theta' \frac{\partial \theta^*}{\partial x_j}) \Big] dSdt. \qquad (25)
\end{aligned}
$$

The terminal and boundary conditions for the adjoint state are given by

$$\boldsymbol{\psi}^*\Big|_{t=T} = \mathbf{0} \qquad (26)$$

$$u^*_i\Big|_{\Gamma_{2\pm}} = A \frac{\mathrm{Re}}{T S_{\Gamma_{2\pm}} \tau_w} \delta_{1i} \qquad (27)$$

$$\theta^*\Big|_{\Gamma_{2\pm}} = -A \frac{\mathrm{RePr}}{T S_{\Gamma_{2\pm}} q_w}, \qquad (28)$$

so that the integrand of Eq. (22) is eventually factorized by ϕ' as follows

$$
\begin{aligned}
\frac{\mathscr{D}H}{\mathscr{D}\phi}\phi' &= J' - \langle \boldsymbol{\psi}', N^*(\boldsymbol{\psi}^*) \rangle - b \\
&= \kappa \int_0^T \int_{\Gamma_{2\pm}} \phi \phi' dSdt - \frac{A}{\tau_w} \int_0^T \int_{\Gamma_{2\pm}} \frac{\partial u'}{\partial n} dSdt + \frac{A}{q_w} \int_0^T \int_{\Gamma_{2\pm}} \frac{\partial \theta'}{\partial n} dSdt \\
&\quad - \int_0^T \int_{\Gamma_{2\pm}} \Big[p^* \phi' - \frac{A}{\tau_w} \frac{\partial u'_i}{\partial n} + \frac{A}{q_w} \frac{\partial \theta'}{\partial n} \Big] dSdt \\
&= \int_0^T \int_{\Gamma_{2\pm}} (\kappa \phi - p^*) \phi' dSdt. \qquad (29)
\end{aligned}
$$

The final form of Eq. (29) guarantees that correcting the control input by $\phi' = -(\kappa\phi - p^*)$ decreases H. Therefore, after solving the adjoint field, the control input is updated as follows:

$$\phi^{n+1} = \phi^n - \beta(\kappa\phi^n - p^*), \qquad (30)$$

where the superscript represents the number of iteration, while β is a relaxation coefficient. In the present study, β is determined so that both $|\phi^{n+1} - \phi^n| < 3.0 \times 10^{-3}$. This increases β as the control input converges. Hence, $\beta < 8$ is also imposed throughout the optimization procedure.

2.3.3 Suboptimal Control Theory

One of major obstacles in applying the optimal control theory to fluid flow and associated transport phenomena is to solve physical and adjoint problems iteratively within a time horizon as shown in Fig. 2. In addition, solving the adjoint equations requires complete information of the physical field during the time horizon (see, Eq. (23)), so that large memory capacity is needed. In order to mitigate the computational load, the suboptimal control theory was developed. In the suboptimal control, a control input minimizing a cost functional within a vanishingly small time horizon is considered. Neglecting the response of the non-linear terms appearing in Eqs. (1)–(3) to an infinitesimal change of control input, the short-term response of the velocity and thermal fields to a control input can be obtained analytically by taking into account linear processes only.

The suboptimal control input for dissimilar heat transfer enhancement was derived in Hasegawa and Kasagi [7]. The resultant control inputs at bottom and top walls are given by

$$\hat{\phi}\Big|_{\Gamma_{2-}} = \gamma \left[\int_{-1}^{1} y \left\{ \frac{\sinh\{k(y-1)\}}{\sinh(2k)} \hat{u}_1(y) - \frac{ik_1 \cosh\{k(y-1)\}}{k \cdot \sinh(2k)} \hat{u}_2(y) \right\} dy \right.$$
$$\left. - \int_{-1}^{1} y \left\{ \frac{\sinh\{k(y-1)\}}{\sinh(2k)} \hat{\theta}(y) \right\} dy \right], \tag{31}$$

$$\hat{\phi}\Big|_{\Gamma_{2+}} = \gamma \left[\int_{-1}^{1} y \left\{ -\frac{\sinh\{k(y+1)\}}{\sinh(2k)} \hat{u}_1(y) + \frac{ik_1 \cosh\{k(y+1)\}}{k \cdot \sinh(2k)} \hat{u}_2(y) \right\} dy \right.$$
$$\left. + \int_{-1}^{1} y \left\{ \frac{\sinh\{k(y+1)\}}{\sinh(2k)} \hat{\theta}(y) \right\} dy \right]. \tag{32}$$

Here, $\hat{\cdot}$ represents Fourier coefficient for a particular combination of the streamwise and spanwise wave numbers, i.e., k_x and k_z, and $k = \sqrt{k_x^2 + k_z^2}$. It should be emphasized that the control inputs (31), (32) are expressed with the information of the physical field at the same instant. Therefore, the iterative computation of the adjoint field is not required in the suboptimal control.

The proportional constant γ is determined so that the intensity of ϕ is equal to 5 % of the bulk mean velocity as is the case for the optimal control introduced in the previous subsection.

3 Results

In this section, the control performances achieved by the optimal and suboptimal control theories are compared. Note that the control performance is generally

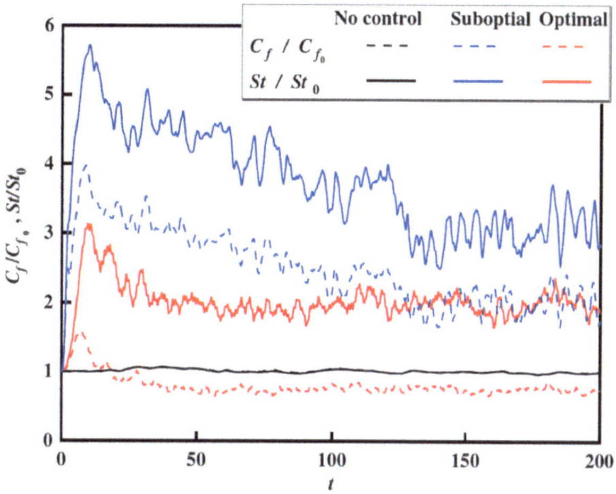

Fig. 3 Time traces of C_f and St achieved in the optimal and suboptimal controls normalized by the values of the uncontrolled flow

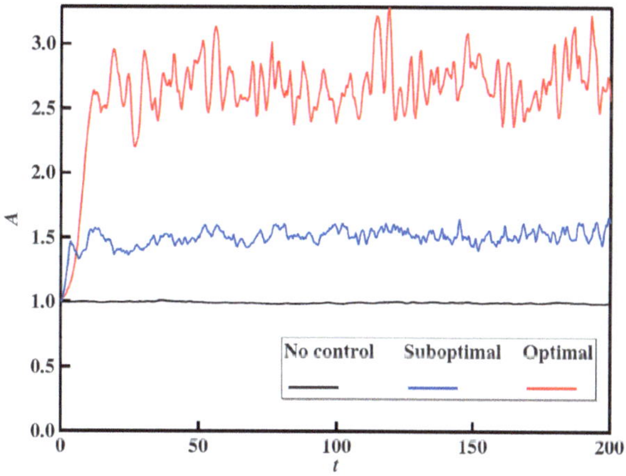

Fig. 4 Time traces of analogy factor achieved in the optimal and suboptimal controls

enhanced with increasing the intensity of the control input. Accordingly, the intensity of the control input is fixed to 5 % of the bulk mean velocity in both controls for fair comparison.

In Fig. 3, the time traces of C_f and St obtained in the suboptimal and optimal controls normalized by the values in the uncontrolled flow are shown. In the case of the suboptimal control, C_f and St are both increased due to the control. However, St is enhanced more than C_f. Specifically, St is increased three times from the uncontrolled value, whereas C_f remains only doubled. In the case of optimal control, more significant control performance can be confirmed. Namely, St is doubled, whilst C_f is decreased by 30 % from the uncontrolled value. This is a surprising result, since 30 % drag reduction rate is larger than that obtained in the opposition control [4]. In addition, the opposition control causes drag reduction only, but does not enhance heat transfer. The time trace of the analogy factor A is shown in Fig. 4.

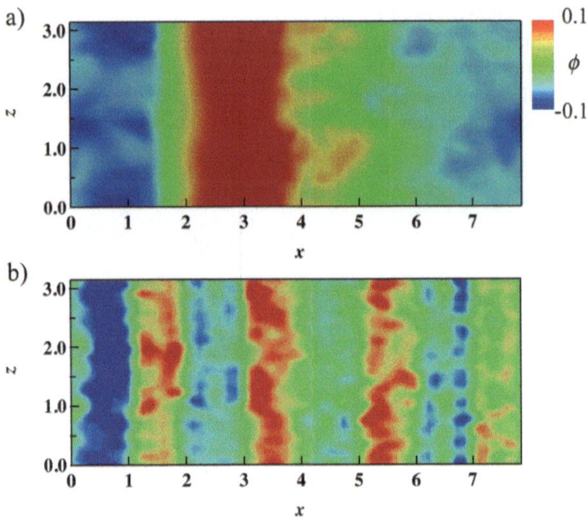

Fig. 5 Instantaneous snapshot of the control input obtained in (**a**) suboptimal and (**b**) optimal control theories. The flow direction is from left to right

It is found that $A \approx 1.5$ is achieved in the suboptimal control, whereas A reaches as high as 2.7 in the optimal control.

The top views of the instantaneous control inputs at the bottom wall obtained in the suboptimal and optimal control theories are shown in Fig. 5. The red and blue colors correspond to regions of wall blowing and suction, respectively. Interestingly, both the control inputs are characterized by wavy distributions in the streamwise direction. In addition, visualization of time traces of these waves (not shown here) reveals that they travel downstream at a constant phase speed, which is around 20–30 % of the bulk mean velocity [7, 17]. These results indicate that the streamwise traveling wave of wall blowing/suction is promising for enhancing heat transfer with minimum pressure penalty.

4 Discussions: Possible Application of Multiple Shooting Method

The advantage of applying the optimal control theory to flow problems is that a control input is optimized based on the governing equations of heat and fluid flow. As shown in Fig. 5, the present control inputs obtained by the optimal and suboptimal control theories commonly exhibit a streamwise traveling-wave like property. Despite their simplicity, it is quite difficult to derive such a control strategy only from researchers' physical insight.

It is also interesting to compare the control performances obtained in the suboptimal and optimal control theories. Obviously, the control performance achieved in the optimal control theory with a finite time horizon T is better than that achieved in the suboptimal control theory, where a vanishingly small time horizon is assumed. Similar trend is also observed in the drag reduction control by Bewley et al. [3].

Although there is tendency that a larger T results in better control performance up to $T \approx 10$, it is found that further increase of the time horizon makes the adjoint computation diverge. This could be attributed to the fact that the mathematical derivation of the optimal control input is based on the linearized perturbation equation (18), where the perturbation of a flow state induced by a small change of a control input is assumed to remain sufficiently small during the time horizon, so that all non-linear terms can be neglected. However, it is well-known that a small disturbance grows very rapidly in turbulent flow due to its nonlinear nature. This implies that the perturbation equation (18) is invalid for a large time horizon. Obviously, a different approach is necessary to extend the time horizon further. In this respect, a multiple shooting technique would be an interesting option. Its application to flow problems remains the future work.

5 Summary

Although significant progresses have been made in understanding and modeling turbulent flows in the last few decades, control of turbulence and associated transport phenomena remains a challenging task due to their highly nonlinear and multi-scale nature. The optimal control theory provides a unique opportunity to optimize a control input without relying on researchers' subjective insights. In the present article, we apply two different approaches, i.e., the optimal and suboptimal control theories, to wall turbulence with heat transfer. In the former, the control input is determined so as to minimize a prescribed cost functional defined within a finite time horizon, whereas in the latter, the time horizon is assumed to be infinitesimal, so that the response of the non-linear terms are all neglected. Although the suboptimal control theory has advantage that it does not require iterative computations of the physical and adjoint equations, there exists a general trend that the control performance is enhanced with increasing the time horizon. This implies significance of taking into consideration the future dynamics in determining the control input. However, the time horizons employed in previous studies are commonly limited due to strong non-linearity of turbulent flows. Applying a multiple shooting method would be one promising option for further increasing the time horizon, and thereby achieving higher control performances.

References

1. Abergel, F., Temam, R.: On some control problems in fluid mechanics. Theor. Comput. Fluid Dyn. **1**, 303–325 (1990)
2. Bejan, A.E.: General criterion for rating heat-exchanger performance. Int. J. Heat Mass Transfer **21**, 655–658 (1978)

3. Bewley, T., Moin, P., Temam, R.: DNS-based predictive control of turbulence: an optimal benchmark for feedback algorithms. J. Fluid Mech. **447**, 179–225 (2001)
4. Choi, H., Moin, P., Kim, J.: Active turbulence control for drag reductio in wall-bounded flows. J. Fluid Mech. **262**, 75–110 (1994)
5. Du, Y., Karniadakis, G.: Suppressing wall turbulence by means of a transverse traveling wave. Science **288**, 1230 (2000)
6. Fukagata, K., Kasagi, N.: Suboptimal control for drag reduction via suppression of near-wall Reynolds shear stress. Int. J. Heat Fluid Flow **25**, 341—350 (2004)
7. Hasegawa, Y., Kasagi, N.: Dissimilar control of momentum and heat transfer in a fully developed turbulent channel flow. J. Fluid Mech. **683**, 57–93 (2011)
8. Hoepffner, J., Fukagata, K.: Pumping or drag reduction? J. Fluid Mech. **635**, 171–187 (2009)
9. Jung, W., Mangiavacchi, N., Akhavan, R.: Suppression of turbulence in wall-bounded flows by high-frequency spanwise oscillation. Phys. Fluids A **4**, 1605–1607 (1992)
10. Kasagi, N., Suzuki, Y., Fukagata, K.: Microelectromechanical systems-based feedback control of turbulence for skin friction reduction. Annu. Rev. Fluid Mech. **41**, 231–251 (2009)
11. Kasagi, N., Hasegawa, Y., Fukagata, K., Iwamoto, K.: Control of turbulent transport: less friction and more heat transfer. Trans. ASME J. Heat Transf. **134**, 031009 (2012)
12. Kim, J., Moin, P., Moser, R.: Turbulent statistics in fully developed channel flow at a low Reynolds number. J. Fluid Mech. **177**, 133–166 (1987)
13. Lee, C., Kim, J., Choi, H.: Suboptimal control of turbulent channel flow for drag reduction. J. Fluid Mech. **358**, 245–258 (1998)
14. Min, T., Kang, J., Kim, J.: Sustained sub-laminar drag in a fully developed channel flow. J. Fluid Mech. **558**, 309–318 (2006)
15. Quadrio, M., Ricco, P., Viotti, C.: Streamwise-travelling waves of spanwise wall velocity for turbulent drag reduction. J. Fluid Mech. **627**, 161–178 (2009)
16. Viotti, C., Quadrio, M., Luchini, P.: Streamwise oscillation of spanwise velocity at the wall of a channel for turbulent drag reduction. Phys. Fluids **21**, 115109 (2009)
17. Yamamoto, A., Hasegawa, Y., Kasagi, N.: Optimal control of dissimilar heat and momentum transfer in a fully developed turbulent channel flow. J. Fluid Mech. **733**, 189–220 (2013)

A Unified Approach to Integration and Optimization of Parametric Ordinary Differential Equations

Daniel Kaschek and Jens Timmer

Abstract Parameter estimation in ordinary differential equations, although applied and refined in various fields of the quantitative sciences, is still confronted with a variety of difficulties. One major challenge is finding the global optimum of a log-likelihood function that has several local optima, e.g. in oscillatory systems. In this publication, we introduce a formulation based on continuation of the log-likelihood function that allows to restate the parameter estimation problem as a boundary value problem. By construction, the ordinary differential equations are solved and the parameters are estimated both in one step. The formulation as a boundary value problem enables an optimal transfer of information given by the measurement time courses to the solution of the estimation problem, thus favoring convergence to the global optimum. This is demonstrated explicitly for the fully as well as the partially observed Lotka-Volterra system.

1 Introduction

Ordinary differential equation (ODE) models play a key role for understanding and predicting the behavior of dynamic systems originating from various disciplines like physics, chemistry or the life sciences. In many cases, these dynamic models depend on parameters that are not known beforehand but need to be determined from measurement data by means of statistical methods. Inference of parameters of dynamic systems from measurement data is commonly realized by optimization of the likelihood function. Optimization is a broad field and many different algorithms have come up over the last decades [1, 7, 11, 13, 14], each of them with problem specific advantages and disadvantages. One characteristic distinction between optimizers is whether they include stochasticity or not. Stochastic optimizers, e.g. evolutionary algorithms [6], particle swarms [9, 12] or simulated annealing [18] are especially valuable for discontinuous likelihood functions where gradient

D. Kaschek (✉) • J. Timmer
Institute of Physics, Freiburg University, Freiburg, Germany
e-mail: daniel.kaschek@physik.uni-freiburg.de; jeti@fdm.uni-freiburg.de

T. Carraro et al. (eds.), *Multiple Shooting and Time Domain Decomposition Methods*, Contributions in Mathematical and Computational Sciences 9,
DOI 10.1007/978-3-319-23321-5_12

information is not available or not defined. On the other hand, many deterministic optimizers employ information about the differentiable structure of the likelihood, i.e. gradient and Hessian information. For differentiable likelihood functions this has the advantage that convergence is achieved much faster. However, this approach has to struggle with other difficulties. If the likelihood function has local optima, the outcome of the optimization procedure depends on the starting point. Once the optimizer is approaching a local optimum, the algorithm will not leave this optimum disregarding the existence of better optima.

The problem of local optima has been addressed by several approaches. It has been shown that a combination of deterministic and stochastic optimization can help escaping local optima and finding the global optimum [15]. Other approaches modify the dynamic system by homotopy transformations [16] introducing a factor λ that allows for a continuous transition between the modified, convex problem and the original problem. Hence another approach is the multiple-shooting method [2]. Most optimizers follow a single-shooting approach, i.e. model trajectories are computed based on given initial values and the outcome is compared to the data. In contrast, the multiple-shooting approach introduces a grid of time-points and initial condition parameters. The optimizer is initialized with discontinuous trajectories and constraints are defined guaranteeing that all trajectories become continuous in the course of optimization.

In our work, we present a reformulation of the optimization problem as a boundary value problem (BVP). The motivation for this approach is twofold. The first argument follows from the history of gradient-based single-shooting optimization for parameter estimation in ordinary differential equations. The performance and accuracy of this method has been enormously increased by solving the ODE together with its' derivatives with respect to the parameters, i.e. the sensitivity equations, in one integration run. This augmentation step allows a fast and accurate computation of the gradient but still evaluation and optimization of the objective function are separate steps. Our aim is to take the next logical step and incorporate even optimization into the ODE integration. The second argument takes up the multiple-shooting idea: the possibility to initialize BVP solvers with prior knowledge like approximate trajectories from measurement data. If the optimization problem is equivalently expressed as a boundary value problem then a good initialization should increase the solver's ability to find the correct solution.

In the following, we show how both objectives can be matched. Our augmentation of the ODE is based on continuation of the log-likelihood function to a differentiable function of time. The resulting system constitutes a BVP. By construction, the solution of this BVP is optimal with respect to the log-likelihood function and it can be obtained by standard numerical BVP solvers. The initialization of the BVP solver allows for an efficient transfer of information provided by the observation data.

2 Methods

We consider a dynamic system defined by ordinary differential equations (ODE),

$$\frac{\mathrm{d}}{\mathrm{d}t}x = f(x, p), \quad x(0) = x_0, \tag{1}$$

with time t, states $x \in \mathbb{R}^n$ and parameters $p \in \mathbb{R}^r$. The extension of the system by $\frac{\mathrm{d}}{\mathrm{d}t}p = 0$ transforms the parameters into usual state variables. For the augmented states $\xi = (x, p)$, parameter estimation becomes an estimation of initial conditions. Furthermore, let $x_{\mathrm{obs}} = (x_1, \ldots, x_m)$, with $m \leq n$, be the observed states and let $\{x_1^D(t_j), \ldots, x_m^D(t_j)\}_j$ denote the time-discrete observation data. The observation data can be approximated by a continuous data function $x_{\mathrm{obs}}^D(t) = (x_1^D(t), \ldots, x_m^D(t))$, e.g. by linear interpolation or spline interpolation. On the other hand, we assume that measurement events for different time points are statistically independent, consequently, the likelihood function

$$L(\xi_0 | \{x_{\mathrm{obs}}^D(t_j)\}_j) = \prod_j L_j(\xi_0 | x_{\mathrm{obs}}^D(t_j)) \tag{2}$$

factorizes and the negative log-likelihood

$$\ell(\xi_0 | \{x_{\mathrm{obs}}^D(t_j)\}_j) = \sum_j -\log L_j(\xi_0, x_{\mathrm{obs}}^D(t_j)) \tag{3}$$

$$\approx \frac{1}{T}\int_0^T r(\xi_0, t)\mathrm{d}t \tag{4}$$

can be approximated by the integral. Here, $r(\xi_0, t)$ denotes the continuous approximation of $-\log L_j(\xi_0, x_{\mathrm{obs}}^D(t_j))$. For standard normally distributed noise, $r(\xi_0, t)$ becomes $\left(x_{\mathrm{obs}}(t) - x_{\mathrm{obs}}^D(t)\right)^2$ which will be used in the following. The argumentation also holds for other noise distributions.

An initial condition vector $\hat{\xi}_0 = (\hat{x}_0, \hat{p}_0) \in \mathbb{R}^{n+r}$ is a local optimum if $\nabla\ell(\hat{\xi}_0, t = T) = 0$ vanishes at the latest observed time point T. Since $\ell(\xi_0, t = 0) = 0$ for all values of ξ_0 at initial time, the gradient $\nabla\ell(\hat{\xi}_0, t = 0) = 0$ vanishes, too. This observation constitutes the boundary condition that needs to be matched for a local optimum, i.e.

$$\nabla\ell(\hat{\xi}_0, 0) = \nabla\ell(\hat{\xi}_0, T) = 0. \tag{5}$$

Each line of Eq. (5) has the potential to determine one parameter value. In order to include this condition into the dynamic system (1), $\nabla\ell$ is derived with respect to time. At this point it is crucial having approximated the negative log-likelihood by an integral expression:

$$\frac{\mathrm{d}}{\mathrm{d}t}\nabla\ell(\xi_0|x^D_{\mathrm{obs}}(t)) = \frac{1}{T}\frac{\mathrm{d}}{\mathrm{d}t}\int_0^t \nabla r(\xi_0,\tau)\mathrm{d}\tau \tag{6}$$

$$= \frac{1}{T}\nabla r(\xi_0, t) \tag{7}$$

$$= \frac{2}{T}\big(x_{\mathrm{obs}}(t) - x^D_{\mathrm{obs}}(t)\big)^* \mathrm{D}_{\xi_0} x_{\mathrm{obs}}(t). \tag{8}$$

Here, * indicates the transpose and $\mathrm{D}_{\xi_0} x_{\mathrm{obs}}(t)$ denotes the Jacobian of $x_{\mathrm{obs}}(t)$ with respect to the initial conditions ξ_0, also known as the sensitivities of the solution trajectory $x_{\mathrm{obs}}(t)$. The sensitivities are determined by an ODE, too, hence the complete systems reads

$$\frac{\mathrm{d}}{\mathrm{d}t}\xi = f(\xi) \tag{9}$$

$$\frac{\mathrm{d}}{\mathrm{d}t}\mathrm{D}_{\xi_0}\xi = \mathrm{D}_\xi f\,\mathrm{D}_{\xi_0}\xi \tag{10}$$

$$\frac{\mathrm{d}}{\mathrm{d}t}\nabla\ell = \frac{2}{T}\big(x_{\mathrm{obs}} - x^D_{\mathrm{obs}}\big)^* \mathrm{D}_{\xi_0} x_{\mathrm{obs}}. \tag{11}$$

The sensitivity equation (10) have fixed initial conditions, $\mathrm{diag}(\mathbb{1}_{n+r})$, with the identity matrix $\mathbb{1}_{n+r} \in \mathbb{R}^{(n+r)\times(n+r)}$. The gradient equations (11) have both zero initial *and* final condition, see Eq. (5), a boundary constraint that fully determines the initial values of the augmented states in Eq. (9). On the other hand, these are the parameters and initial conditions we seek to estimate. Hence, the desired values $\hat{\xi}_0$ optimizing the negative log-likelihood are part of the solution of the two-point boundary value problem, Eqs. (9)–(11). Compared to gradient based single-shooting methods, Eq. (11) represents the pivotal difference. It translates optimization into the ambit of integration. This is the principle behind our optimization approach.

The solution of the two-point boundary value problem is obtained by the Fortran 77 code TWPBVP [4, 5], available from the Netlib repository. The method used in TWPBVP is a deferred correction method based on mono-implicit Runge-Kutta formulas and adaptive mesh refinement. The deferred correction algorithm uses the trapezoidal rule to obtain a first approximation to the required solution. Finite difference approximations to the local truncation error are then added onto this low order solution, increasing the accuracy of the solution repeatedly [3].

3 Example

In the following, we examine the Lotka-Volterra equations [8, 10, 17]. They give a basic description of the predator and prey population dynamics. The system is defined by two differential equations

$$\frac{d}{dt}A = A(\alpha - \beta B), \tag{12}$$

$$\frac{d}{dt}B = -B(\gamma - \delta A), \tag{13}$$

where A and B correspond to prey and predator respectively. The parameters α and β describe prey reproduction and reduction, γ and δ describe predator extinction and reproduction. For non-zero initial condition, the solution of Eqs. (12)–(13) is a sustained oscillation.

In the first part of the study, it is assumed that both populations are observed. Observation data is simulated by numerically integrating the ODE system with $A_0 = 1$, $B_0 = 1.2$, $\alpha = 0.45$, $\beta = 0.5$, $\gamma = 0.3$, $\delta = 0.1$ and adding Gaussian noise with $\sigma = 0.15$ to the solution. Subsequently, the parameter values are sought to be recovered from randomly chosen initial parameter guesses. The initialization of the BVP solver, i.e. the grid of initially assumed values for each of the variables in Eqs. (9)–(11), is obtained by integrating the sensitivity equation (10) with the initial parameter guess and the data interpolations as input trajectories. The gradient is assumed to vanish over the entire range.

Independently of the initial parameter values, the BVP solver converges to the same solution. The result for one representative data set is shown in Fig. 1. The *States* panel shows the solutions of the state variables A and B together with the data points and error bars. As expected, the predator and prey trajectories hit about 67 % of the error bars. In the *Parameters* panel, the solutions of the state variables α, β, γ and δ are shown on a logarithmic scale, i.e. the dynamic parameters. The dots indicate the values that have been used for simulation. The *Sensitivities* panel shows the sensitivity trajectories which are typical for oscillating systems, i.e. oscillations with increasing amplitude. Finally, in the *Negative log-likelihood gradient* panel, the gradient solution is plotted. The time scale of gradient changes is determined by the sampling density of the simulated time course. It hits the ground line in the end point as desired, guaranteeing an optimum.

The BVP method has been tested systematically against a single-shooting approach based on the Levenberg-Marquardt algorithm, implemented in the MINPACK Fortran 77 package. For different simulated data sets, both, BVP method and single-shooting method have been applied to the same random set of initial parameter guesses. In order to avoid that, by chance, parameter vectors are too similar, we employed Latin hypercube sampling with a hypercube covering 4 orders of magnitude around the true parameter values. Both optimization approaches failed convergence a number of times in which case 10^6 was assigned as value to the negative log-likelihood. Figure 2 shows the first 200 sorted negative log-likelihood values of the total 300 initial guesses for both approaches. The single-shooting approach gets stuck in different local optima and finds the global optimum only in 4 % of the cases. In contrast, the BVP method proves to be robust against different initial guesses for the parameter values. The solution converges either to the global

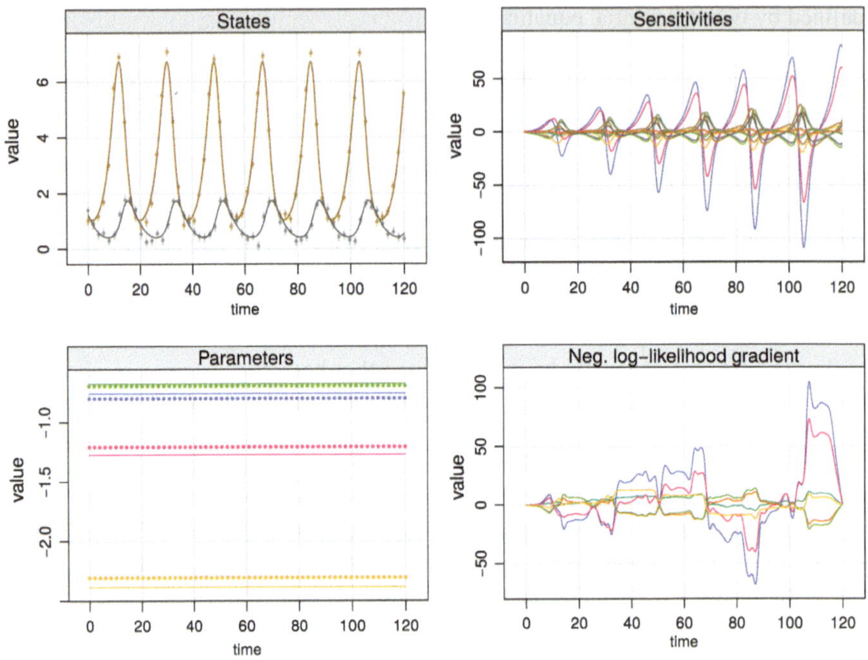

Fig. 1 Solution generated by the BVP solver. The four panels show state solutions with simulated data points, parameter solutions on a logarithmic scale with true values as *dots*, sensitivity solutions and the gradient of the negative log-likelihood

minimum or it fails convergence. It is 6 times more efficient in finding the best optimum. Figure 2 also gives some indication about parameter convergence regions. For the BVP method, the set of initial parameters that finally converged to the best parameter value covers almost the total range. However, fewer initial guesses with α and δ larger than 1 lead to a successful reconstruction of the BVP solution. The broad plateau of local optima for the single-shooting method is reflected in a clear shift of final parameter values and a certain number of randomly distributed final parameters.

In a second step, the observation of B is omitted and β is fixed to 1 in order to keep the system identifiable. Analogously to the fully observed system, data sets have been simulated and Gaussian noise has been added. The comparison between the single-shooting method and the BVP method is shown in Fig. 3a. The plots indicate that the situation becomes more intricate if only one state is observed. The convergence rate drops below 2 % for both approaches and the BVP method reconstructs a variety of local optima, each of them with an almost identical negative log-likelihood value. From the scatter plots in Fig. 3, two conclusions can be drawn for the BVP method: First, whenever we found the global minimum, the parameters were initially situated in the negative orthant, and second, the final parameters

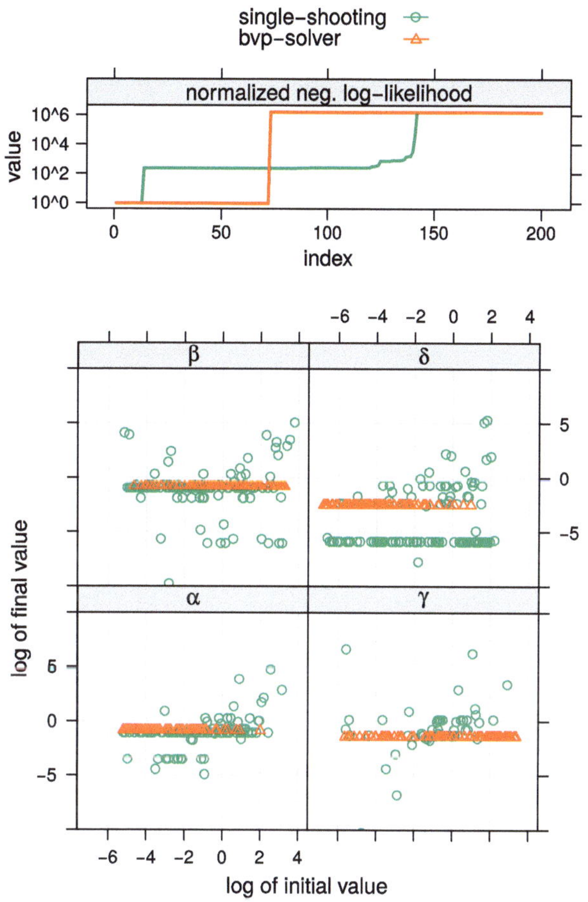

Fig. 2 Comparison of single-shooting and BVP method tested on the fully observed Lotka-Volterra system. Each method has been applied to simulated data sets and for each data set, initial parameter vectors have been generated by Latin hypercube sampling covering a range of 4 orders of magnitude around the true parameter values. The resulting negative log-likelihood values were sorted, normalized by the smallest value and plotted on a logarithmic scale. In the scatter plots, initial parameter vectors are plotted against final parameter values for each optimization that resulted in a negative log-likelihood value smaller than 10^3

corresponding to the local optima form a submanifold with boundary in parameter space.

Figure 3b shows the same picture for initial guesses starting from the negative orthant only. In agreement with the expectation, the number of BVP solutions corresponding to the global optimum increases considerably and exceeds the success rate of the single-shooting method by a factor of 5.

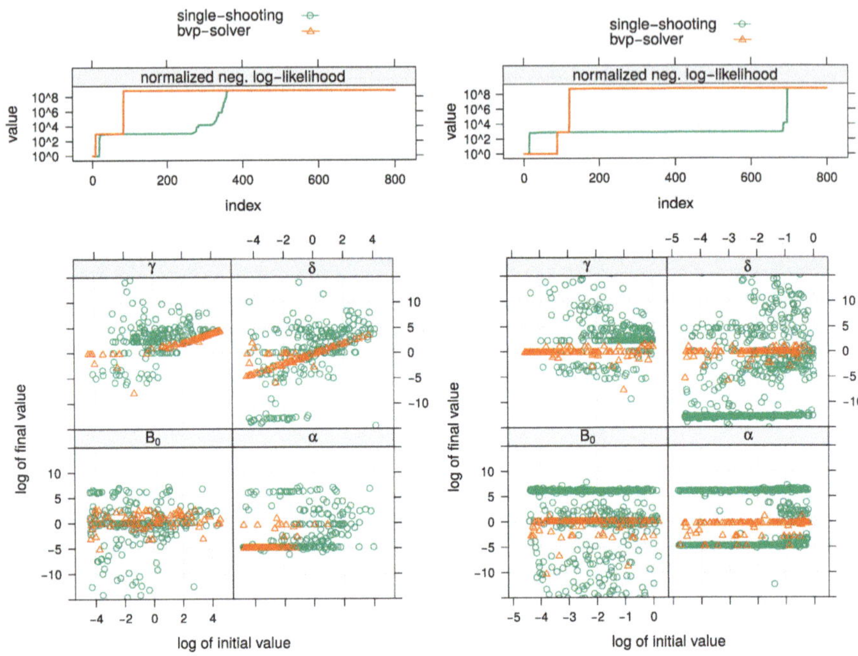

Fig. 3 Comparison of single-shooting and BVP method tested on the partially observed Lotka-Volterra system. Each method has been applied to simulated data sets and for each data set, initial parameter vectors have been generated by Latin hypercube sampling. Column (**a**) shows the results for initial parameter vectors covering a range of 4 orders of magnitude around the true parameter values. For column (**b**), initial parameters were restricted to values smaller than 10^0. In both cases, the resulting negative log-likelihood values were sorted, normalized by the smallest value and plotted on a logarithmic scale. In the scatter plots, initial parameter vectors are plotted against final parameter values for each optimization that resulted in a negative log-likelihood value smaller than 10^3

4 Conclusion

In case of a partially observed system, the success rate of the BVP method depends on favorable initial conditions. Whereas the single-shooting algorithm performed equally badly over the entire parameter space, for the boundary value approach, it was possible to identify an attractive basin resulting in a considerably increased convergence rate.

From the fully observed Lotka-Volterra system, we conclude that the boundary value approach is excellently suited as optimization approach if numerous observables are available. In this case, it clearly outperforms the single-shooting Levenberg-Marquardt algorithm in terms of convergence to the global optimum. The strength of the presented optimization approach is its ability to exploit the measured time courses in a natural way. This favors convergence to the global optimum. Unlike single-shooting approaches, convergence to local optima or false convergence claims are efficiently reduced. On the other hand, the deferred correction algorithm seemed to be very sensitive to the grid initialization by initial parameter and state guesses, manifesting in a large number of non-convergent attempts. This problem increased with the size of the time domain. At this point, a multiple shooting algorithm, being based on time-domain decomposition, is expected to be more stable and to provide a higher convergence rate.

In summary, we presented a reformulation of the estimation problem as a boundary value problem which, in turn, is enabled by continuation of the negative log-likelihood function to a time-differentiable function. This restatement elegantly incorporates optimization and ODE solution in one task. By nature of the boundary value problem, an initial guess for all state variables needs to be presented to the numerical solver. The initialization by measured time-courses carries exactly the information that is necessary to make the algorithm converge to the best optimum.

Acknowledgements This work was supported by the MIP-DILI project, Innovative Medicines Initiative Joint Undertaking under grant agreement No. 115336. We thank our colleague Marcus Rosenblatt for supporting the computational implementation.

References

1. Amritkar, R.E.: Estimating parameters of a nonlinear dynamical system. Phys. Rev. E **80**(4), 047,202 (2009). doi:10.1103/PhysRevE.80.047202
2. Bock, H.G., Kostina, E., Schlöder, J.P.: Numerical methods for parameter estimation in nonlinear differential algebraic equations. GAMM-Mitt. **30**(2), 376–408 (2007)
3. Cash, J.: A variable order deferred correction algorithm for the numerical solution of nonlinear two point boundary value problems. Comput. Math. Appl. **9**(2), 257–265 (1983)
4. Cash, J., Mazzia, F.: A new mesh selection algorithm, based on conditioning, for two-point boundary value codes. J. Comput. Appl. Math. **184**(2), 362–381 (2005)
5. Cash, J., Wright, M.H.: A deferred correction method for nonlinear two-point boundary value problems: implementation and numerical evaluation. SIAM J. Sci. Stat. Comput. **12**(4), 971–989 (1991)
6. Goswami, G., Liu, J.S.: On learning strategies for evolutionary monte carlo. Stat. Comput. **17**(1), 23–38 (2007). doi:10.1007/s11222-006-9002-y
7. Horbelt, W., Timmer, J., J. Bünner, M., Meucci, R., Ciofini, M.: Identifying physical properties of a CO_2 laser by dynamical modeling of measured time series. Phys. Rev. E **64**(1), 016,222 (2001). doi:10.1103/PhysRevE.64.016222
8. Lotka, A.J.: Contribution to the theory of periodic reactions. J. Phys. Chem. **14**(3), 271–274 (1909). doi:10.1021/j150111a004
9. Mendes, R., Kennedy, J., Neves, J.: The fully informed particle swarm: simpler, maybe better. IEEE Trans. Evol. Comput. **8**(3), 204–210 (2004). doi:10.1109/TEVC.2004.826074
10. Parker, M., Kamenev, A.: Extinction in the lotka-volterra model. Phys. Rev. E **80**(2), 021,129 (2009). doi:10.1103/PhysRevE.80.021129
11. Parlitz, U.: Estimating model parameters from time series by autosynchronization. Phys. Rev. Lett. **76**(8), 1232–1235 (1996). doi:10.1103/PhysRevLett.76.1232
12. Peng, H., Li, L., Yang, Y., Liu, F.: Parameter estimation of dynamical systems via a chaotic ant swarm. Phys. Rev. E **81**(1), 016,207 (2010). doi:10.1103/PhysRevE.81.016207
13. Sitz, A., Schwarz, U., Kurths, J., Voss, H.U.: Estimation of parameters and unobserved components for nonlinear systems from noisy time series. Phys. Rev. E **66**(1), 016,210 (2002). doi:10.1103/PhysRevE.66.016210

14. Sohl-Dickstein, J., Battaglino, P.B., DeWeese, M.R.: New method for parameter estimation in probabilistic models: Minimum probability flow. Phys. Rev. Lett. **107**(22), 220,601 (2011). doi:10.1103/PhysRevLett.107.220601
15. Villaverde, A.F., Egea, J.A., Banga, J.R.: A cooperative strategy for parameter estimation in large scale systems biology models. BMC Syst. Biol. **6**(1), 75 (2012). doi:10.1186/1752-0509-6-75
16. Vyasarayani, C., Uchida, T., McPhee, J.: Single-shooting homotopy method for parameter identification in dynamical systems. Phys. Rev. E **85**(3) (2012). doi:10.1103/PhysRevE.85.036201
17. Wang, M.X., Lai, P.Y.: Population dynamics and wave propagation in a lotka-volterra system with spatial diffusion. Phys. Rev. E **86**(5), 051,908 (2012). doi:10.1103/PhysRevE.86.051908
18. Xiang, Y., Gong, X.G.: Efficiency of generalized simulated annealing. Phys. Rev. E **62**(3), 4473–4476 (2000). doi:10.1103/PhysRevE.62.4473

A Variational Approach for Physically Based Image Interpolation Across Boundaries

Matthias Klinger

Abstract In this contribution we present an optimal control approach for physics-based optical flow estimation and image interpolation. The aim of the developed process is to identify appropriate boundary data of an underlying physical model describing the transport field, which reason the movement of an initial brightness distribution. Thereby, the flow field as solution of the time-dependent non-linear Navier-Stokes equations is coupled to a transport dominant convection-diffusion equation describing the brightness intensity. Thus, we have to deal with a weakly coupled PDE system as state equation of a PDE constrained optimisation problem. The data is given in form of consecutive images, with a sparse temporal resolution, representing the brightness distribution at different time points. We will present the mathematical theory of the resulting optimisation problem, which is based on a Robin-type boundary control. We describe the numerical solution process and present by means of synthetical test cases the functionality of the method. Finally we discuss the application of multiple shooting techniques for the considered problem, since we observed that the employed Newton-type method is very sensitive with respect to the chosen initial value.

1 Introduction

In many fields of research scientists are interested in fluid motion, e.g. weather forecast, circulation around obstacles, microfluidic flows. However, to evaluate accurate flow fields is often a hard task, regardless if we use measurement techniques to document an observed flow or numerical modelling to simulate a similar flow situation. A common methodology to document a fluid motion is to observe the movement of a passive tracer in a given fluid flow by consecutive images. These images represent a spatial and temporal discretisation of the evolution of a brightness distribution given by a camera apparatus, which detects light signals transmitting through the transparent fluid and the light absorbing passive tracer.

M. Klinger (✉)
Institut für Angewandte Mathematik, Im Neuenheimer Feld 294, 69120 Heidelberg, Germany
e-mail: matthias.klinger@iwr.uni-heidelberg.de

T. Carraro et al. (eds.), *Multiple Shooting and Time Domain Decomposition Methods*, Contributions in Mathematical and Computational Sciences 9,
DOI 10.1007/978-3-319-23321-5_13

The transport of a brightness value in the image domain is caused by a velocity field, which is called the optical flow. In some situations the optical flow is closely connected to the underlying physical flow. Heitz et al. [20] for example mention that there exists a straightforward connection between the optical flow and the fluid flow of a laser sheet visualisation of a two dimensional incompressible flow, when the laser sheet is perfectly aligned with the flow. Thus the optical flow $\mathbf{w}$ is proportional to the velocity $\mathbf{u}$ and satisfies a convection-diffusion equation. This connection between optical flow and fluid flow can also be assumed for the observation of three dimensional flows by two dimensional images, as long as the flow in z direction is negligible. However, the optical flow field $\mathbf{w}$ serves then only as an approximation of the planar flow field.

Hence, it is a good opportunity to work with optical flow estimation techniques to obtain approximations of fluid flow fields. Approaches toward this direction are already presented in the literature, as for example in the mentioned article of Heitz et al. [20]. Another approach basing on the reformulation of the optical flow functional by means of the underlying flow model was considered by Nakajima et al. [29]. Another promising class of approaches was presented by Ruhnau and coworkers [35–37], who regularised the classical Horn and Schunck cost functional (cf. Horn et al. [22]) by applying physical models as PDE side condition in the optimisation framework.

Especially, for environmental sciences this technique is attractive for the investigation of local wind systems in areas where a dense grid of measurement stations is unavailable, but a tracer is transported in the atmosphere, which can be observed by satellite remote sensing. For example Héas et al. [19] and Papadakis et al. [32] considered the estimation of wind field information from satellite image sequences observing cloud formations by image processing approaches. Another example for environmental fluid flows is the movement of dust aerosols in northern Africa in the Sahara desert. The measurements of the aerosol density in a certain area on the earths surface at different time points, obtained by an instrument installed on a geosynchronous satellite lead to a sequence of brightness distributions of a passive tracer, the aerosols, which is transported by an optical flow field, which we assume to be a approximation of the planar flow field in the ground-based atmospheric layers. First attempts to use this image sequences to obtain these optical flow fields were presented in the work of Bachl et al. [1, 2].

However, we aim to introduce a novel approach for physical-based fluid flow estimation from observations of a passive tracer by combining the single features of the above mentioned techniques in one approach, which place the emphasis on the coupling of a high fidelity physical flow model, namely the incompressible, non-linear and time-dependent Navier-Stokes equations, to the optical flow equation by applying so called boundary controls. Then our focus is on the theoretical justification and the numerical realisation of the method. Furthermore, we are interested in the quality of the reconstructed underlying transport field by our approach. Unfortunately, the mentioned example of a real world application has too many complications (e.g. measurement errors, model uncertainties, occlusions, varying illumination in the images) to tackle them all at the same time. Furthermore, appropriate reference data for a qualitative comparison of our results are not available. Hence, we will consider a prototypical example with synthetic images and some simplifications to present our approach, which employs beside the image information also informations about the physical flow model and allows also movement across the image domain boundaries, since images often represent only an aperture of the real scenery.

Therefore, we assume that we observe a plane motion described by the time-dependent (Navier-) Stokes equations, which transport a brightness distribution $I(\mathbf{x}, t)$ due to the following system of equations

$$\begin{aligned} \partial_t I - \varepsilon \Delta I + \mathbf{u} \cdot \nabla I &= 0, \\ \partial_t u - \nu \Delta \mathbf{u} + \mathbf{u} \cdot \nabla \mathbf{u} + \nabla p &= \mathbf{f}, \qquad \text{in } \Omega \times (0, T], \\ \nabla \cdot \mathbf{u} &= 0, \end{aligned} \tag{1}$$

with appropriate initial and boundary data. The image sequences are then obtained by setting $\mathscr{I}_k = I(\mathbf{x}, t_k)$ at discrete times t_k.

Figures 1, 2 and 3 show three examples of such artificial image sequences. The first two sequences are obtained with very simple flow fields but with flow across the boundaries. In the third test case we observe a flow field which exhibits the time-dependent character of the system (1). Our aim is then to identify the underlying flow fields and the movement of the (bulb) signal. For this purpose we present an optimisation problem with system (1) as PDE constraint. Thus, our methodology can be interpreted as an optimal control problem with parabolic PDE constraints. Such parabolic optimal control problems and their numerical treatment are widely

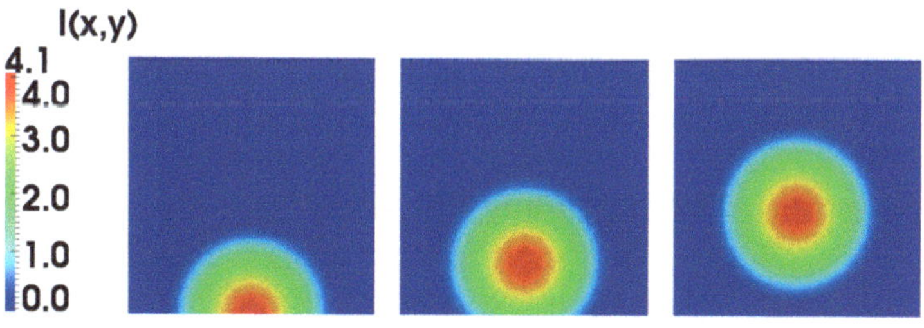

Fig. 1 *Left*: $\mathscr{I}_1$ at $t = 0$. *Middle*: $\mathscr{I}_2$ at $t = 0.1$. *Right*: $\mathscr{I}_3$ at $t = 0.2$. The transport field is given by $\mathbf{u} = (2, 0)^T$

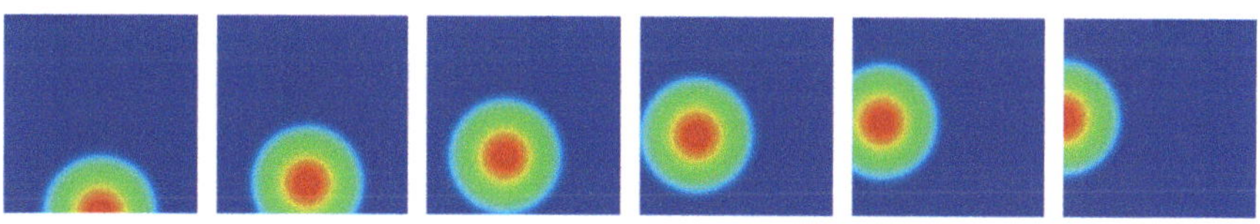

Fig. 2 $(\mathscr{I}_k)_{k=1}^6$ at the time points $t_k = 0.04(k-1)$. The transport field is $\mathbf{u} = \kappa(-y, x)^T$ with $\kappa = \frac{5}{2}\pi$

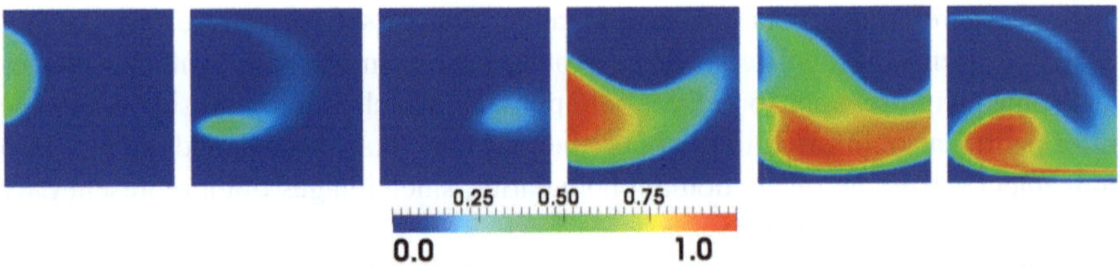

Fig. 3 $(\mathscr{I}_k)_{k=1}^6$ at six different time points transported by a time-dependent solution of the Navier-Stokes equations $\mathbf{u}(\mathbf{x}, t)$

discussed in the literature. We will mention a few example which are related to the presented topics in this article. For example Kolmbauer et al. [26] present an approach for time-periodic eddy current optimal control problems. Pearson et al. [33] consider the numerical treatment of optimal control problems constrained by convection-reaction problems while the article of Stoll et al. [38] considers the time-dependent Stokes equation as side condition. Finally Gunzburger et al. [18] discuss the use of space-time adaptive methods for optimal control problems with parabolic evolution equations as side condition.

The article is organised in the following way. At first we describe variational optical flow estimation techniques for the above mentioned synthetic image sequences. Furthermore we describe their enhancement to optimal control problems. Then we enhance the techniques to formulations which can even deal with flows across the boundaries. Afterwards we discuss the mathematical theory for the presented approach before we talk about the numerical techniques we need for a solution of the presented approach. The sixth section is devoted to the presentation of some numerical results for the mentioned artificial sequences. In the final section we discuss the reformulation of our abstract problem as temporal boundary value problem and the possible advantages of applying a multiple shooting method.

2 Physics-Based Optical Flow Equation

In the following $\Omega \subset \mathbb{R}^2$ denotes the image domain, which is in general a rectangle $\Omega = (0, a) \times (0, b)$. The brightness function is given by

$$I : \Omega \times [0, T] \to \mathbb{R}^+, \qquad \{\mathbf{x}, t\} \mapsto I(\mathbf{x}, t).$$

Our observations, the images, are spatial and temporal discretisations of $I(\mathbf{x}, t)$:

$$\mathscr{I}(i, j, k) = I(\mathbf{x}_{ij}, t_k)$$

on a certain space-time grid. However we assume throughout the article that the spatial discretisation is fine, while the temporal discretisation is coarse, so that the data is given by a temporally discrete sequence

$$(\mathscr{I}_k)_{k=1}^N = (I(\mathbf{x}, t_k))_{k=1}^N.$$

We consider throughout the article a two-dimensional incompressible flow with the field $\mathbf{u} = (u, v)^T$. This flow field satisfies the two-dimensional non-stationary Navier-Stokes equations

$$\begin{aligned} \partial_t \mathbf{u} - \nu \Delta \mathbf{u} + \mathbf{u} \cdot \nabla \mathbf{u} + \nabla p = \mathbf{f} \qquad & \text{in } \Omega \times (0, T] \\ \nabla \cdot \mathbf{u} = 0 \qquad & \text{in } \Omega \times (0, T] \end{aligned}$$

for appropriate initial values and boundary data. Since we do not consider a real world application we conjecture $\mathbf{w} = \mathbf{u}$.

Remark 1 (Relation Between Optical Flow and Fluid Flow) As we already mentioned in the introduction their is a close relationship between the optical flow in an image sequence and the fluid flow, which is documented by these images observing the movement of a passive tracer. For a detailed description of this relationship we refer the interested reader to Sect. 2.1 of Heitz et al. [20]. Especially, for laser sheet visualisation of two-dimensional incompressible flows the connection between optical and fluid flow is straightforward, as long as the laser sheet is aligned with the flow field.

The brightness intensity function $I(\mathbf{x}, t)$ fulfills then the physics-based optical flow equation

$$\partial_t I + \mathbf{u} \cdot \nabla I = \varepsilon \Delta I \qquad \text{in } \Omega \times (0, T]$$

with certain boundary conditions and an appropriate initial condition.

The coupled system of Eq. (1) describes then the evolution of an initial brightness distribution $\mathscr{I}^0(\mathbf{x})$. For further considerations we assume that no domain forces cause the flow in the image domain, which means $\mathbf{f} = 0$. Thus the boundary conditions describe the flow scenario completely.

The aim of our work is to describe a method which recovers appropriate boundary conditions for the flow field and the connected brightness function only by the following available information:

(a) A temporal sparse image sequence: $(\mathscr{I}_k)_{k=1}^N$,
(b) Model parameters: ε and ν,
(c) An estimate for the initial flow field: $\mathbf{u}^0$.

3 Optimisation for Physics-Based Optical Flow

The recovering of appropriate boundary conditions is realised by a PDE constrained optimisation problem. First ideas towards this direction, especially in the image processing framework, are given in the work of Borzi et al. [8], Chen et al. [13] and Klinger [25]. We start with the definition of the optimisation problem.

Definition 1 Find $q_I \in Q_I$, $\mathbf{q_u} \in Q_\mathbf{u}$ and $(\mathbf{u}, p, I) \in V_\mathbf{u} \times V_p \times V_I$ so that the functional

$$J(\{q_I, \mathbf{q_u}\}, I) = \frac{1}{2}\sum_{k=1}^{N} \|I(t_k) - \mathscr{I}_k\|_2^2 + \frac{\alpha_1}{2}\int_0^T \|q_I(t)\|_{Q_I}^2 dt + \frac{\alpha_2}{2}\int_0^T \|q_\mathbf{u}(t)\|_{Q_\mathbf{u}}^2 dt$$

is minimised subject to an appropriate weak formulation of system (1), with $\mathbf{f} = 0$, $I(0) = \mathscr{I}_1$ and $\mathbf{u}(0) = \mathbf{u}^0$ and so called control functions $\mathbf{q_u}$ and q_I for I and $\mathbf{u}$ on the boundary.

The choices of $Q_\mathbf{u}$, Q_I and the type of boundary conditions are crucial for the well-posedness of this optimisation problem as well as for the computational realisation of the problem. We discuss a promising compromise in the next section.

3.1 Treatment of the Boundary Control Formulation

The first idea is to use Dirichlet controls

$$\begin{aligned} I &= q_I && \text{on } \partial\Omega \times (0, T] \\ \mathbf{u} &= \mathbf{q_u} && \text{on } \partial\Omega \times (0, T], \end{aligned} \tag{2}$$

where we have to choose appropriate control spaces Q_I and $Q_\mathbf{u}$. For the existence theory of solutions of the presented optimisation problem a main ingredient is the existence theory of the state equation. In our case for the coupled system (1). This system consists of a convection-diffusion equation, which is weakly coupled to the incompressible Navier-Stokes equations. Both are parabolic PDEs. The natural choice of functions to prescribe Dirichlet boundaries for parabolic PDEs is the space $H^{\frac{1}{2}}(\partial\Omega)$. This space is defined in the following way

$$H^{\frac{1}{2}}(\partial\Omega) = \{\varphi \in L^2(\partial\Omega) : \exists \omega \in H^1(\Omega),\ \varphi = \omega|_{\partial\Omega}\},$$

tributing to the fact, that L^2-functions on the boundary exist which have no H^1-extensions to the interior of the domain.

However, the implementation of an appropriate norm for this space in the cost functional is complicated. One way of doing this was suggested by Of et al. [31], by defining the semi-norm:

$$|q|^2_{H^{\frac{1}{2}}(\partial\Omega)} = \langle \mathcal{S} q, q \rangle$$

with the Steklov-Poincare operator

$$\mathcal{S} : H^{\frac{1}{2}}(\partial\Omega) \to H^{-\frac{1}{2}}(\partial\Omega), \quad \text{with} \quad q \mapsto \partial_{\mathbf{n}}\omega,$$

where ω is the solution of the elliptic PDE

$$-\Delta\omega = 0, \quad \text{in } \Omega, \quad \omega = q, \quad \text{on } \partial\Omega.$$

Hence for each component controlled via a Dirichlet boundary condition we have to solve an additional PDE problem, which increases the computational costs drastically for fine spatial and temporal grids.

Other possibilities of implementing an $H^{\frac{1}{2}}$-norm or an $H^{\frac{1}{2}}$-semi-norm are based on the calculation of complicated boundary integrals, which can hardly be treated in the context of optimisation problems.

From the numerical point of view the most attractive choice of the control space is the $L^2(\partial\Omega)$. As mentioned before the theoretical justification for Dirichlet controls is now cumbersome. However, there is a conceptual access to the problem, which we want to briefly describe by means of the very simple PDE-constrained optimisation problem

$$\min_{u \in V, q \in \mathcal{Q}} J(u,q) = \frac{1}{2}\|u - \hat{u}\|_2^2 + \frac{\alpha}{2}\|q\|_2^2$$

subject to the time-independent Poisson problem

$$-\Delta u = f \quad \text{in } \Omega, \quad u = q \quad \text{on } \partial\Omega.$$

The idea is now to work with the very weak formulation of the state equation as constraint (see May et al. [27])

$$-(u, \Delta\varphi) + \langle q, \partial_n\varphi \rangle = (f, \varphi) \quad \forall \varphi \in H^2(\Omega) \cap H_0^1(\Omega),$$

and the spaces $V = L^2(\Omega)$ and $\mathcal{Q} = L^2(\partial\Omega)$. The so formulated problem admits a unique solution pair (u, q). Belgacem et al. [7] showed that μ-dependent solutions (u_μ, q_μ) of the optimisation problem

$$\min_{\substack{u_\mu \in H^1(\Omega), \\ q_\mu \in L^2(\partial\Omega)}} J(u_\mu, q_\mu)$$

subject to

$$\left(\nabla u_\mu, \nabla \varphi\right) + \frac{1}{\mu}\left\langle u_\mu - q_\mu, \varphi\right\rangle = (f, \varphi)\,, \quad \forall \varphi \in H^1(\Omega)$$

with $\mu > 0$, converge to a solution (u, q) of the L^2-Dirichlet control problem with the very weak formulation of the Poisson problem as side condition as μ tends to zero.

Remark 2 It was shown by Hou et al. [23] that this concept of approximating Dirichlet controls by using the penalised Neumann conditions for small choices of μ works also for optimisation problems with the time-independent Navier-Stokes equations as PDE side condition and a certain choice of the cost functional.

However, for system (1) the presented theoretical background cannot easily be carried over due to the fact that a very weak solution of the time-dependent Navier-Stokes equations is only L^4-regular in space (see Farwig et al. [16]), which is neither sufficient for the presented cost functional nor for the existence theory of the convection-diffusion equation describing the evolution of the brightness function.

We are not limited to use Dirichlet controls. For us the choice of the boundary conditions is only a tool for the estimation of a reliable flow field which transports the brightness distribution in an appropriate manner. Thus, we can work with the above described Robin-type boundary conditions anyway. We state the optimisation problem after introducing appropriate vector spaces.

Definition 2 (Solenoidal Vector Spaces) We define the following vector spaces

$$\begin{aligned} H^1_{\text{div}}(\Omega)^2 &:= \{\boldsymbol{\varphi} \in H^1(\Omega)^2 : \nabla \cdot \boldsymbol{\varphi} = 0 \text{ in a weak sense}\}, \\ L^2_{\text{div}}(\Omega)^2 &= \overline{H^1_{\text{div}}(\Omega)^2}^{\|\cdot\|_2}. \end{aligned}$$

Definition 3 (Robin-Type Control for Image Interpolation) Find

$$\{\mathbf{u}, I\} \in L^2\left(0, T; H^1_{\text{div}}(\Omega)^2\right) \times L^2\left(0, T; H^1(\Omega)\right)$$

and

$$\{\mathbf{q_u}, q_I\} \in L^2\left(0, T; L^2(\partial\Omega)^2\right) \times L^2\left(0, T; L^2(\partial\Omega)\right)$$

so that the functional

$$J(\{q_I, \mathbf{q_u}\}, I)$$

in Definition 1 is minimised, subject to the following weak formulation of the above mentioned coupled system of equations.

Definition 4 (Weak Formulation of the State Equation) For initial values $\mathbf{u}^0 \in L^2_{\text{div}}(\Omega)^2$ and $\mathscr{I}_1 \in L^2(\Omega)$ find a pair

$$\{\mathbf{u}, I\} \in L^2\left(0, T; H^1_{\text{div}}(\Omega)^2\right) \times L^2\left(0, T; H^1(\Omega)\right)$$

so that

$$\begin{aligned} \int_0^T \big(-(I, \partial_t \psi) + a_I(\mathbf{u}; I, \psi) + b_I(\mathbf{u}; q_I; I, \psi)\big) dt &= (\mathscr{I}_1, \psi(0)) \\ \int_0^T \big(-(\mathbf{u}, \partial_t \boldsymbol{\varphi}) + a_{\mathbf{u}}(\mathbf{u})(\boldsymbol{\varphi}) + b_{\mathbf{u}}(\mathbf{q}_{\mathbf{u}}; \mathbf{u})(\boldsymbol{\varphi})\big) dt &= \left(\mathbf{u}^0, \boldsymbol{\varphi}(0)\right) \end{aligned} \tag{3}$$

is fulfilled for all test functions

$$\begin{aligned} \psi &\in \left\{\psi \in L^2\left(0, T; H^1(\Omega)\right) \text{ and } \partial_t \psi \in L^2\left(0, T; \left(H^1(\Omega)\right)'\right)\right\}, \\ \boldsymbol{\varphi} &\in \left\{\boldsymbol{\varphi} \in L^2\left(0, T; H^1_{\text{div}}(\Omega)^2\right) \text{ and } \partial_t \boldsymbol{\varphi} \in L^2\left(0, T; \left(H^1_{\text{div}}(\Omega)^2\right)'\right)\right\}. \end{aligned} \tag{4}$$

The bi- and semi-linear forms are defined as follows

$$\begin{aligned} a_I(\mathbf{u}; I, \psi) &:= \varepsilon (\nabla I, \nabla \psi) + (\mathbf{u} \cdot \nabla I, \psi), \\ a_{\mathbf{u}}(\mathbf{u})(\boldsymbol{\varphi}) &:= \nu (\nabla \mathbf{u}, \nabla \boldsymbol{\varphi}) + (\mathbf{u} \cdot \nabla \mathbf{u}, \boldsymbol{\varphi}), \\ b_I(\mathbf{u}; q_I; I, \psi) &:= \frac{1}{\mu_1} \langle I - q_I, \psi \rangle_{\partial\Omega} - \frac{1}{2} \langle (\mathbf{u} \cdot \mathbf{n}) I, \psi \rangle_{\partial\Omega}, \\ b_{\mathbf{u}}(\mathbf{q}_{\mathbf{u}}; \mathbf{u})(\boldsymbol{\varphi}) &:= \frac{1}{\mu_2} \langle \mathbf{u} - \mathbf{q}_{\mathbf{u}}, \boldsymbol{\varphi} \rangle_{\partial\Omega} - \frac{1}{2} \langle (\mathbf{u} \cdot \mathbf{n}) \mathbf{u}, \boldsymbol{\varphi} \rangle_{\partial\Omega} \end{aligned} \tag{5}$$

Remark 3 (Temporal Regularity) The above formulation has on first glance not enough regularity for a well-defined cost functional and meaningful initial conditions in I and $\mathbf{u}$. However, assume for a moment enough regularity to achieve the equivalent weak formulation

$$(\partial_t I, \tilde{\psi}) + a_I(\mathbf{u}; I, \tilde{\psi}) + b_I(\mathbf{u}; q_I; , \tilde{\psi}) = 0 \qquad \forall \tilde{\psi} \in H^1(\Omega)$$

after partial integration and using a test function $\psi(\mathbf{x}, t) = \bar{\psi}(t)\tilde{\boldsymbol{\psi}}(\mathbf{x})$. It is straightforward to show $\partial_t I \in L^2(0, T; H^1(\Omega)')$ for the assumed regularity of $I, \mathbf{u}$ and q_I in Definition 3 outgoing from the last equation. Thus, due to the Gelfand tripel

$$H^1(\Omega) \subset L^2(\Omega) \subset H^1(\Omega)'$$

we obtain $I \in C\left([0,T];L^2(\Omega)\right)$ by a standard result (cf. Evans [15]). The same thing can also be proven for the solenoidal spaces in the Navier-Stokes case (cf. Temam [39]).

Remark 4 (Strong Formulation of the Boundary Conditions) Under the assumption of sufficient regularity of the functions we can extract from the above stated weak formulations the following corresponding Robin boundary conditions:

$$\begin{aligned} \varepsilon \partial_n I &= \frac{1}{\mu_1}(q_I - I) + \frac{1}{2}(\mathbf{u}\cdot\mathbf{n})\, I && \text{on } \partial\Omega \times (0,T], \\ \nu \partial_{\mathbf{n}} \mathbf{u} - p\mathbf{n} &= \frac{1}{\mu_2}(\mathbf{q_u} - \mathbf{u}) + \frac{1}{2}(\mathbf{u}\cdot\mathbf{n})\,\mathbf{u} && \text{on } \partial\Omega \times (0,T]. \end{aligned}$$

In the next section we will prove the existence of minimisers of the optimisation problem in Definition 3.

4 Mathematical Theory of the Optimisation Problem

A first step towards a proof of the existence of minimisers of the optimisation problem in Definition 3 is to prove unique solvability of the weak formulation in Definition 4. Therefore, we consider at first the following result:

Theorem 1 *For $\mu_2 \in (0,1]$, $u^0 \in L^2(\Omega)^2$ and a fixed boundary function*

$$\mathbf{q_u} \in L^2\left(0,T;L^2(\partial\Omega)^2\right)$$

the Navier-Stokes system in the second equation of (3) has a unique solution

$$\mathbf{u} \in L^\infty\left(0,T;L^2_{div}(\Omega)^2\right) \cap L^2\left(0,T;H^1_{div}(\Omega)^2\right).$$

Proof We can obtain the result by a few simple modifications of the standard Galerkin technique as presented in Temam [39]. At first we obtain the a-priori bound

$$\int_0^T \left(\frac{d}{dt}\|\mathbf{u}(t)\|_2^2 + \nu\|\nabla\mathbf{u}(t)\|_2^2 + \frac{1}{\mu_2}\|\mathbf{u}(t)\|^2_{L^2(\partial\Omega)^2} \right) dt \le \frac{c}{\mu_2}\int_0^T \|\mathbf{q_u}(t)\|^2_{L^2(\Omega)^2}\, dt, \tag{6}$$

since

$$(\mathbf{u}\cdot\nabla\mathbf{u},\mathbf{u}) = \frac{1}{2}\int_{\partial\Omega}(\mathbf{u}\cdot\mathbf{n})\mathbf{u}^2 ds, \tag{7}$$

in contrast to the usual proof. With inequality (6) we obtain the usual (weak and strong) convergence properties of a certain subsequence.

The properties are then used to prove the convergence of the sequence of approximative solutions $\mathbf{u}^m$ of the Navier-Stokes system. The only difference to the standard proof is the convergence of the semi-linear boundary form

$$\int_0^T \int_{\partial\Omega} ((\mathbf{u}^m \cdot \mathbf{n})\mathbf{u}^m - (\mathbf{u} \cdot \mathbf{n})\mathbf{u})\, \varphi ds dt \to 0.$$

Therefore, we set $\mathbf{w}^m = \mathbf{u}^m - \mathbf{u}$ and consider

$$\begin{aligned} \left| \int_0^T \int_{\partial\Omega} ((\mathbf{w}^m \cdot \mathbf{n})\, \mathbf{u}^m + (\mathbf{u} \cdot \mathbf{n})\, \mathbf{w}^m)\, \varphi ds dt \right| \leq & \left| \int_0^T \langle (\mathbf{w}^m \cdot \mathbf{n})\, \mathbf{u}^m, \varphi \rangle_{\partial\Omega}\, dt \right| \\ & + \left| \int_0^T \int_{\partial\Omega} (\mathbf{u} \cdot \mathbf{n})\, \mathbf{w}^m \varphi ds dt \right|. \end{aligned} \tag{8}$$

As test functions φ we choose functions from a subset, which is sufficiently smooth on the whole boundary and the time interval. The first term of the right hand side in (8) is transformed in domain integrals by

$$\int_0^T \langle (\mathbf{w}^m \cdot \mathbf{n})\, \mathbf{u}^m, \varphi \rangle_{\partial\Omega}\ dt = \int_0^T (\mathbf{w}^m \cdot \nabla \mathbf{u}^m, \varphi)\ dt + \int_0^T (\mathbf{w}^m \cdot \nabla \varphi, \mathbf{u}^m)\ dt,$$

since $\mathbf{u}^m$ and $\mathbf{w}^m$ are solenoidal. Due to the smoothness of the test function it is easy to obtain that the terms on the right hand side vanish in the limit, since $\mathbf{u}^m$ converges strongly in $L^2(0, T; L^2(\Omega)^2)$. The second term of the right hand side in (8) can be estimated

$$\left| \int_0^T \int_{\partial\Omega} (\mathbf{u} \cdot \mathbf{n})\, \mathbf{w}^m \varphi\ ds dt \right| \leq \sup_{(\mathbf{x},t)\in\partial\Omega\times[0,T]} |\varphi(\mathbf{x}, t)| \int_0^T |\, \langle \mathbf{u}, \mathbf{w}^m \rangle_{\partial\Omega} \,|\ dt$$

Weak convergence in $L^2(0, T; L^2(\partial\Omega)^2)$ yields that this term is also vanishing. Thus, the whole convergence of the Galerkin approximations in the weak formulation can be obtained by standard continuity arguments.

For the uniqueness we assume the existence of two different solutions for the same initial and boundary data. The difference is given by $\mathbf{w}^m = \mathbf{u} - \mathbf{v}$ and we find the identity

$$\frac{1}{2}\frac{d}{dt}\|\mathbf{w}^m\|_2^2 + \nu\|\nabla \mathbf{w}^m\|_2^2 + \frac{1}{\mu_2}\|\mathbf{w}^m\|^2_{L^2(\partial\Omega)^2} = -\frac{1}{2}\left((\mathbf{w}^m \cdot \nabla \mathbf{u}, \mathbf{w}^m) - (\mathbf{w}^m \cdot \nabla \mathbf{w}^m, \mathbf{u})\right),$$

after standard manipulations and by using Eq. (7). The Ladyzhenskaya inequality in two space dimensions

$$\|\mathbf{w}^m\|_{L^4(\Omega)^2} \leq c\|\mathbf{w}^m\|_2^{\frac{1}{2}}\, \|\nabla \mathbf{w}^m\|_2^{\frac{1}{2}}$$

is used in combination with Young's inequality to find

$$\frac{d}{dt}\|\mathbf{w}^m(t)\|_2^2 \leq \beta(t)\| \|\mathbf{w}^m(t)\|_2^2.$$

with

$$\beta(t) := c(\nu, \Omega) \left(\|\mathbf{u}(t)\|_{H^1(\Omega)^2}^2 + \|\mathbf{u}(t)\|_2^2 \|\mathbf{u}(t)\|_{H^1(\Omega)^2}^2 \right)$$

Gronwall's inequality and $\mathbf{w}^m(0, \mathbf{x}) = 0$ yields the uniqueness, since

$$\int_0^t \beta(s)\, ds \leq c(\nu, \Omega) \left(\int_0^t \|\mathbf{u}(s)\|_{H^1(\Omega)^2}^2\, ds + \operatorname*{ess\,sup}_{s\in[0,t]} \|\mathbf{u}(s)\|_2^2 \int_0^t \|\mathbf{u}(s)\|_{H^1(\Omega)^2}^2\, ds \right)$$

stays bounded due to $\mathbf{u} \in L^\infty\left(0, T; L^2_{\text{div}}(\Omega)^2\right) \cap L^2\left(0, T; H^1_{\text{div}}(\Omega)^2\right)$, which was obtained by the a-priori bound in formula (6). □

Theorem 2 *For fixed parameters μ_1 and μ_2 in $(0, 1]$ and boundary functions $q_I \in L^2\left(0, T; L^2(\partial\Omega)\right)$ and $\mathbf{q_u} \in L^2\left(0, T; L^2(\partial\Omega)^2\right)$ there exists a unique solution pair*

$$\{I_\mu, \mathbf{u}_\mu\} \in L^2(0, T; H^1(\Omega)) \times L^2(0, T; H^1_{div}(\Omega)^2).$$

Proof We consider the two aspects existence and uniqueness. We skip the index μ_i with $i = 1, 2$ for abbreviation:

(Existence)

We use the standard Galerkin technique. Since the convection-diffusion equation for the brightness I is not coupling back to the Navier-Stokes system we can argue in the following way. By Theorem 1 we have the existence of a unique transport field

$$\mathbf{u} \in L^\infty\left(0, T; L^2_{\text{div}}(\Omega)^2\right) \cap L^2\left(0, T; H^1_{\text{div}}(\Omega)^2\right).$$

Hence, we find in the standard way the a-priori bound

$$\int_0^T \left(\frac{d}{dt}\|I(t)\|_2^2 + \varepsilon\|\nabla I(t)\|_2^2 + \frac{1}{\mu_1}\|I(t)\|_{L^2(\partial\Omega)}^2 \right) dt \leq \frac{c}{\mu_1} \int_0^T \|q_I(t)\|_{L^2(\partial\Omega)}^2\, dt, \tag{9}$$

since

$$(\mathbf{u} \cdot \nabla I, I) - \frac{1}{2}\langle(\mathbf{u} \cdot \mathbf{n})I, I\rangle_{\partial\Omega} = 0, \tag{10}$$

for the transport field $\mathbf{u}$. By the standard Galerkin technique and modifications mentioned in the proof of Theorem 1 we find easily the existence of the solution $I \in L^2\left(0,T;H^1(\Omega)\right)$.

(Uniqueness)

The uniqueness of a solution pair $\{I,\mathbf{u}\}$ can be achieved by assuming as usual the existence of two solution pairs $\{I_1,\mathbf{u}_1\}$ and $\{I_2,\mathbf{u}_2\}$ for the same data and building the difference for the systems. We use the notation $K = I_1 - I_2$ and $\mathbf{w}^m = \mathbf{u}_1 - \mathbf{u}_2$. Due to the independence of the Navier-Stokes part of the system from K and the uniqueness of the Navier Stokes solution we have $\mathbf{u}_1 = \mathbf{u}_2$. Thus, the convection-diffusion part is a linear equation, since we can use Eq. (10) for $I = K$. The rest of the argumentation is standard and yields $K = 0$ by Gronwall's lemma. □

Now we are able to prove the existence of a minimiser of the optimisation problem in Definition 3.

Theorem 3 (Solution of the Optimisation Problem) *For $\mu := \mu_1 = \mu_2 \in (0,1]$ fixed we have the existence of at least one minimiser*

$$\begin{aligned}
I_\mu &\in L^2\left(0,T;H^1(\Omega)\right)\\
\mathbf{u}_\mu &\subset L^2\left(0,T;H^1_{div}(\Omega)^2\right)\\
q_{I,\mu} &\in L^2\left(0,T;L^2(\partial\Omega)\right)\\
\mathbf{q}_{\mathbf{u},\mu} &\in L^2\left(0,T;L^2(\partial\Omega)^2\right)
\end{aligned}$$

of the optimisation problem in Definition 3.

Proof Thanks to the previous theorem we have the existence of solutions of the state equation and therefore the admissible set is not empty.

We skip the index μ for abbreviation, collect the controls in the overall vector $\mathbf{q} := (q_I, \mathbf{q}_\mathbf{u}) \in L^2(\partial\Omega)^3$ and choose then a minimising sequence $\{I^{(k)}, \mathbf{u}^{(k)}, \mathbf{q}^{(k)}\}$ in this set with the property

$$\lim_{k\to\infty} J(\mathbf{q}^{(k)}, I^{(k)}) = \inf_{\{I,\mathbf{q}\}} J(\mathbf{q}, I) =: \theta.$$

By using Young's inequality we obtain a uniform bound for $\mathbf{q}$:

$$\|\mathbf{q}^{(k)}\|_{L^2(0,T;L^2(\partial\Omega)^3)} \le \frac{1}{\alpha} J(\mathbf{q}^{(k)}, I^{(k)}) + \frac{1}{2} \le B.$$

Thus, the controls q_I and $\mathbf{q}_\mathbf{u}$ are bounded in $L^2\left(0,T;L^2(\partial\Omega)\right)$ and $L^2\left(0,T;L^2(\partial\Omega)^2\right)$. Hence, by the energy estimates (6) and (9) we receive all necessary uniform bounds for $I^{(k)}$ and $\mathbf{u}^{(k)}$. Finally we find

$$I^{(k)} \in L^\infty(0,T;L^2(\Omega)) \cap L^2(0,T;H^1(\Omega)) \cap L^2(0,T;L^2(\partial\Omega)) \tag{11}$$

$$\mathbf{u}^{(k)} \in L^\infty(0,T;L^2(\Omega)^2) \cap L^2(0,T;H^1_{\mathrm{div}}(\Omega)^2) \cap L^2(0,T;L^2(\partial\Omega)^2) \tag{12}$$

We can then extract the subsequences

$$I^{(k')} \rightharpoonup I \quad \begin{array}{ll} \text{weakly} & \text{in } L^2(0,T;H^1(\Omega)), \\ \text{weakly-}\star & \text{in } L^\infty(0,T;L^2(\Omega)), \\ \text{weakly} & \text{in } L^2(0,T;L^2(\partial\Omega)), \end{array} \quad \text{as } k' \to \infty,$$

and

$$\mathbf{u}^{(k')} \rightharpoonup \mathbf{u} \quad \begin{array}{ll} \text{weakly} & \text{in } L^2(0,T;H^1_{\text{div}}(\Omega)^2), \\ \text{weakly-}\star & \text{in } L^\infty(0,T;L^2(\Omega)^2), \\ \text{weakly} & \text{in } L^2(0,T;L^2(\partial\Omega)^2), \end{array} \quad \text{as } k' \to \infty,$$

By compactness results we obtain also the strong convergence properties

$$I^{(k')} \to I \quad \text{in } L^2(0,T;L^2(\Omega)), \quad \mathbf{u}^{(k')} \to \mathbf{u} \quad \text{in } L^2(0,T;L^2(\Omega)^2),$$

of the subsequences. Thus, passing to the limit in the state equation is a standard task.

It remains to show that the pair $\{I,\mathbf{u},\mathbf{q}\}$ is in fact a minimum of $J(\cdot,\cdot)$. We use the obtained convergence properties to compute

$$\begin{aligned} \theta &= \lim_{k\to\infty} J(I^{(k)},\mathbf{q}^{(k)}) = \lim_{k\to\infty}\left(\frac{1}{2}\sum_{j=1}^{N}\|I^{(k)}(t_j)-\mathscr{I}_j\|_2^2 + \frac{\alpha}{2}\int_0^T \|\mathbf{q}^{(k)}(t)\|^2_{L^2(\partial\Omega)^3}\,dt\right) \\ &= \frac{1}{2}\sum_{j=1}^{N}\|I(t_j)-\mathscr{I}_j\|_2^2 + \liminf_{k\to\infty}\frac{\alpha}{2}\int_0^T \|\mathbf{q}(t)\|^2_{L^2(\partial\Omega)^3}\,dt. \end{aligned}$$

Due to the continuity and convexity of the norm $\|\cdot\|_{L^2(0,T;L^2(\partial\Omega)^3)}$ the norm is also weakly lower semicontinuous. Thus, we find

$$\theta \geq \frac{1}{2}\sum_{j=1}^{N}\|I(t_j)-\mathscr{I}_j\|_2^2 + \frac{\alpha}{2}\int_0^T \|\mathbf{q}(t)\|^2_{L^2(\partial\Omega)^3}\,dt = J(I,\mathbf{q}),$$

which proves the optimality of the pair $\{I,\mathbf{u},\mathbf{q}\}$. □

5 Numerical Solution Process and Further Specifics

5.1 Optimisation Algorithm

We briefly present the optimisation algorithm. All following aspects are very well summarised in the thesis of Meidner [28]. By Theorem 1 we know that the state equation of our PDE constrained optimisation problem is uniquely solvable. Hence we can introduce the solution operator and find $\{\mathbf{u}, I\} = S(\mathbf{q}_\mathbf{u}, q_I)$. By means of this operator we can transform the original problem into an unconstrained problem

$$j(\tilde{\mathbf{q}}) = J(S(\tilde{\mathbf{q}}), \tilde{\mathbf{q}}), \qquad \text{with } \tilde{\mathbf{q}} = \{\mathbf{q}_\mathbf{u}, q_I\}.$$

The first-order necessary condition is then given by

$$j'(\mathbf{q})(\delta\mathbf{q}) = 0 \qquad \forall \delta\mathbf{q} \in \mathscr{Q},$$

where $\mathscr{Q}$ denotes the vector space for the controls.

We use now a Newton-type algorithm to find a solution of the last equation. Therefore we represent the first and second variationals derivative of $j(\cdot)$ by auxiliary variables, which have to be evaluated by solving additional PDE problems. The key for this representation is the identity

$$j(\mathbf{q}) = J(\mathbf{q}, \mathbf{u}) = \mathscr{L}(\mathbf{q}, \mathbf{u}, \mathbf{z}), \tag{13}$$

where $\mathscr{L}(\cdot)$ denotes the Lagrangian.

The first derivative can be expressed by

$$j'(\mathbf{q})(\delta\mathbf{q}) = \mathscr{L}'_\mathbf{q}(\mathbf{q}, \mathbf{u}, \mathbf{z})(\delta\mathbf{q}) \qquad \forall \delta\mathbf{q} \in \mathscr{Q}.$$

Therefore we have to compute u and z by solving the primal and the adjoint equations

$$\mathscr{L}'_\mathbf{z}(\mathbf{q}, \mathbf{u}, \mathbf{z})(\varphi) = 0 \qquad \forall \varphi \in V \qquad \text{(Primal Eq.)},$$
$$\mathscr{L}'_\mathbf{u}(\mathbf{q}, \mathbf{u}, \mathbf{z})(\varphi) = 0 \qquad \forall \varphi \in V \qquad \text{(Adjoint Eq.)}.$$

The second derivative is given by

$$j''(\mathbf{q})(\delta\mathbf{q}, \tau\mathbf{q}) = \mathscr{L}''_{\mathbf{qq}}(\mathbf{q}, \mathbf{u}, \mathbf{z})(\delta\mathbf{q}, \tau\mathbf{q}) + \mathscr{L}''_{\mathbf{uq}}(\mathbf{q}, \mathbf{u}, \mathbf{z})(\delta\mathbf{u}, \tau\mathbf{q}) + \mathscr{L}''_{\mathbf{zq}}(\mathbf{q}, \mathbf{u}, \mathbf{z})(\delta\mathbf{z}, \tau\mathbf{q}).$$

Beside the primal and the adjoint solutions u and z we need now two further variables δu and δz, which can be obtained by solving the following equations

$$\begin{aligned}
\mathscr{L}''_{\mathbf{qz}}(\mathbf{q},\mathbf{u},\mathbf{z})(\delta\mathbf{q},\varphi) + \mathscr{L}''_{\mathbf{uz}}(\mathbf{q},\mathbf{u},\mathbf{z})(\delta\mathbf{u},\varphi) &= 0 \qquad \text{(Tangent Eq.)}\\
\mathscr{L}''_{\mathbf{qu}}(\mathbf{q},\mathbf{u},\mathbf{z})(\delta\mathbf{q},\varphi) + \mathscr{L}''_{\mathbf{uu}}(\mathbf{q},\mathbf{u},\mathbf{z})(\delta\mathbf{u},\varphi) & \\
+ \mathscr{L}''_{\mathbf{zu}}(\mathbf{q},\mathbf{u},\mathbf{z})(\delta\mathbf{z},\varphi) &= 0 \qquad \text{(Additional Adjoint Eq.)}
\end{aligned}$$

Algorithm 1 summarise theses principles.

Algorithm 1 Newton-CG Algorithm

1: Choose an initial $q^0 \in \mathscr{Q}_h$, $\mu_0 \in R \cup \{+\infty\}$ and set $k = 0$.
2: Solve the State Eq. : $\mathscr{L}'_z(q,u,z)(\varphi) = 0 \quad \forall\varphi \in V$
3: Evaluate the cost functional $J(q,u)$
4: Solve the Dual Eq.: $\mathscr{L}'_u(q,u,z)(\varphi) = 0 \quad \forall\varphi \in V$
5: Evaluate the residual : $f := -j'(q^k)(\delta q^k)$
6: If $\|f\| <$ TOL then STOP
7: Solve the system

$$j''(q^{(k)})(\delta q_i^{(k)}, \tau q_j^{(k)}) = -j'(q^{(k)})(\tau q_j^{(k)}) \tag{14}$$

with a CG-method. For each iteration of the CG method we perform

7.1: Solve the Tangent Eq.: $\mathscr{L}''_{qz}(q,u,z)(\delta q,\varphi) + \mathscr{L}''_{uz}(q,u,z)(\delta u,\varphi) = 0$.
7.2: Solve the Additional Adjoint Eq.:

$$\begin{aligned}
\mathscr{L}''_{qu}(q,u,z)(\delta q,\varphi) + \mathscr{L}''_{uu}(q,u,z)(\delta u,\varphi)& \\
+ \mathscr{L}''_{zu}(q,u,z)(\delta z,\varphi) &= 0
\end{aligned}$$

7.3: STOP the CG-method if the residual of the linear system drops below a given tolerance.

8: Update the control

$$q^{(k+1)} = q^{(k)} + \lambda_k q^{(k)},$$

The relaxation parameter λ_k is used to globalise the Newton method.
9: Go back to Step 2.

Remark 5 (Application of the CG Method in the Algorithm) The CG-method allows a matrix-free solution process, which means that we do not have to assemble the Hessian matrix, which saves computational effort. The desired number of iterations to solve the system (14) for a given tolerance depends on the condition of the Hessian matrix. In the case of ill-posed problems the condition number increases, when the regularisation parameter decreases. The convergence behaviour of the CG-method is then very bad. Usually we stop the iteration after a certain amount of steps, unless the threshold for the residuum was reached. Then we work with a so called inexact Newton method.

5.2 PDE Subproblems

The PDE subproblems mentioned in Algorithm 1 are solved by a Rothe method, which means that the problems are first discretised in the time variable and then in the space variable. We will start with the time discretisation.

5.2.1 Time Discretisation

In the used Software library RoDoBo [6] we have the possibility to use either a discontinuous Galerkin $dG(r)$ method or a continuous Galerkin $cG(r)$ method for the time discretisation. The use of Galerkin discretisation was preferred by the authors of the library, since it is necessary to preserve the property that 'discretise-then-optimise' and 'optimise-then-discretise' commute.

The $cG(r)$ method consists of continuous trial functions of degree r and discontinuous test functions of degree $r-1$, while the $dG(r)$ method is based on the use of discontinuous trial and test functions of degree r. Using a $dG(0)$ method, where all occurring integrals are evaluated with the box rule, leads directly to the standard backward Euler scheme. The $cG(1)$ method, where all occurring temporal integrals are approximated with the trapezoidal rule, generates the Crank-Nicolson scheme. For the following calculations we use these two methods only.

For the spatial discretisation of the resulting quasi-stationary equations we use the well-known finite element method.

5.2.2 Spatial Discretisation

To describe the conceptual features of our finite element approach, we consider the following abstract weak formulation for the quasi-stationary equation:

For a given q, find a suitable $u \in V$ such that

$$a(u, \varphi) + b(q; u, \varphi) = l(\varphi) \quad \forall \varphi \in V \tag{15}$$

with the bilinear forms $a(u, \varphi), b(q; u, \varphi)$ and the linear form $l(\varphi)$.

By choosing a conforming ansatz space V_h we derive the following Galerkin equations:

$$a(u_h, \varphi_h) + b(q; u_h, \varphi_h) = l(\varphi_h) \qquad \forall \varphi_h \in V_h. \tag{16}$$

Due to an appropriate choice of the space V_h and the linearity of the forms we end up with the following system of equations

$$A_h x_h = b_h, \qquad \text{with } u_h = \sum_{i=1}^{N} x_{h,i} e^{(i)},$$

which is then solved by a multigrid method.

For all our calculations we used the ansatz space V_h containing continuous functions, which are piecewise bilinear polynomials on a decomposition of the computational domain into regular quadrilaterals:

$$\hat{V}_h := \{u_h : \bar{\Omega} \to \mathbb{R} \mid u_h \in C(\bar{\Omega}),\ u_h|_T \in Q_1\} \tag{17}$$

For a detailed description of this methodology see the monographs of Braess [9] and Brenner et al. [10].

Remark 6 (Nonlinear PDEs) In the case of nonlinear PDEs, e.g. the Navier-Stokes system, we have to linearise the equation. In this case we use also a Newton method and solve then in each step of the Newton method a PDE which fits in the above presented setting.

5.3 Comments on Boundary Conditions

Since we consider boundary identification problems a special focus of this work is on the treatment of boundary conditions. The suggested penalised Neumann approach depends on the choice of the parameter μ. Decreasing the parameter μ the condition of the discrete system is getting worse. A way out of this dilemma was presented by Juntunen et al. [24] for the Poisson problem. This concept can easily be carried over to convection-diffusion equations and the Navier-Stokes system.

5.3.1 Convection-Diffusion Equation

We consider the following convection-diffusion-reaction equation

$$\begin{aligned} -\nu\Delta u + \beta \cdot \nabla u + cu &= f && \text{in } \Omega, \\ \nu \partial_n u &= \frac{1}{\mu}(q_D - u) + \frac{1}{2}(\boldsymbol{\beta} \cdot \mathbf{n})\, u && \text{on } \partial\Omega. \end{aligned} \tag{18}$$

We use the bilinear forms

$$a(u, \varphi) := \nu\,(\nabla u, \nabla\varphi) + (\beta \cdot \nabla u, \varphi) + (cu, \varphi)$$

for the domain and

$$\begin{aligned} b(q_D;u,\varphi) :=& -\frac{\nu\delta}{\mu+\delta}\left(\langle\partial_{\mathbf{n}}u,\varphi\rangle_{\partial\Omega} + \langle u-q_D,\partial_{\mathbf{n}}\varphi\rangle_{\partial\Omega}\right) + \frac{1}{\mu+\delta}\langle u-q_D,\varphi\rangle_{\partial\Omega} \\ & - \frac{\nu\mu\delta}{\mu+\delta}\langle\partial_{\mathbf{n}}u,\partial_{\mathbf{n}}\varphi\rangle_{\partial\Omega} - \frac{\mu}{2(\mu+\delta)}\langle(\boldsymbol{\beta}\cdot\mathbf{n})u,\varphi\rangle_{\partial\Omega} \\ & + \frac{\mu\delta}{2(\mu+\delta)}\langle(\boldsymbol{\beta}\cdot\mathbf{n})u,\partial_{\mathbf{n}}\varphi\rangle_{\partial\Omega}\,, \end{aligned} \tag{19}$$

for the boundary to formulate the problem weakly

$$a(u,\varphi) + b(q_D;u,\varphi) = (f,\varphi) \quad \forall\varphi\in V. \tag{20}$$

Lemma 1 *A solution of problem (18) also satisfies Eq. (20).*

Proof We integrate the first equation in (18) over the domain after multiplying with an arbitrary test function $\varphi \in V$. Integration by parts yields

$$a(u,\varphi) - \nu\,\langle\partial_{\mathbf{n}}u,\varphi\rangle_{\partial\Omega} = (f,\varphi)\,. \tag{21}$$

We multiply now the boundary part of Eq. (18) with the same test function and integrate over the boundary. Then we multiply both sides by $\frac{\mu}{\mu+\delta}$ and obtain

$$\frac{\mu\nu}{\mu+\delta}\langle\partial_{\mathbf{n}}u,\varphi\rangle_{\partial\Omega} = \frac{1}{\mu+\delta}\langle q_D-u,\varphi\rangle_{\partial\Omega} + \frac{\mu}{2(\mu+\delta)}\langle(\boldsymbol{\beta}\cdot\mathbf{n})u,\varphi\rangle_{\partial\Omega} \tag{22}$$

Doing the same again with the test function $\partial_{\mathbf{n}}\varphi$ and the factor $-\frac{\delta\mu}{\mu+\delta}$ we get

$$\begin{aligned} -\frac{\nu\mu\delta}{\mu+\delta}\langle\partial_{\mathbf{n}}u,\partial_{\mathbf{n}}\varphi\rangle_{\partial\Omega} =& -\frac{\delta}{\mu+\delta}\langle q_D-u,\partial_{\mathbf{n}}\varphi\rangle_{\partial\Omega} \\ & -\frac{\mu\delta}{2(\mu+\delta)}\langle(\boldsymbol{\beta}\cdot\mathbf{n})u,\partial_{\mathbf{n}}\varphi\rangle_{\partial\Omega} \end{aligned} \tag{23}$$

Equation (20) is now the sum of Eqs. (21)–(23). □

The bilinear form $b(q_D;u,\varphi)$ can even be evaluated for the case $\mu = 0$. We obtain then a Nitsche type formulation (Nitsche [30]) for the convection-diffusion equation

$$b_{\mathrm{Ni}}(q_D;u,\varphi) = -\nu\left(\langle\partial_{\mathbf{n}}u,\varphi\rangle_{\partial\Omega} + \langle u-q_D,\partial_{\mathbf{n}}\varphi\rangle_{\partial\Omega}\right) + \frac{1}{\delta}\langle u-q_D,\varphi\rangle_{\partial\Omega}\,. \tag{24}$$

Remark 7 (Linear Transport Equation) In the case of pure transport, $\nu = 0$, the boundary form $b(q_D; u, \varphi)$ reads

$$b(q_D; u, \varphi) = \frac{1}{\mu + \delta}\left(\langle u - q_D, \varphi\rangle_{\partial\Omega} - \frac{\mu}{2}\langle(\boldsymbol{\beta}\cdot\mathbf{n})\,u, \varphi\rangle_{\partial\Omega} + \frac{\delta\mu}{2}\langle(\boldsymbol{\beta}\cdot\mathbf{n})\,u, \partial_{\mathbf{n}}\varphi\rangle_{\partial\Omega}\right)$$

For an appropriate choice $\delta \geq \delta_0 > 0$ we obtain by setting $\mu = 0$ a Nitsche-type term for the realisation of Dirichlet boundary data for the pure transport equation

$$b(q_D; u, \varphi) = \frac{1}{\delta}\langle u - q_D, \varphi\rangle_{\partial\Omega} .$$

We are able to set Dirichlet conditions only on the inflow boundary Γ_{In}, that means all $\mathbf{x} \in \partial\Omega$ with $\boldsymbol{\beta}\cdot\mathbf{n} < 0$, in the following way

$$0 < \delta := -\frac{1}{(\boldsymbol{\beta}\cdot\mathbf{n})},$$

we obtain

$$b(q_D; u, \varphi) := -\langle(\boldsymbol{\beta}\cdot\mathbf{n})(u - q_D), \varphi\rangle_{\Gamma_{In}} . \tag{25}$$

This is consistent with a suggestion for Nitsche-type inflow presented in the work of Freund et al. [17].

5.3.2 Navier-Stokes System

For $V = H_0^1(\Omega)^n$ the Navier-Stokes system

$$\begin{aligned} -\nu\Delta\mathbf{u} + \mathbf{u}\cdot\nabla\mathbf{u} + \nabla p &= f && \text{in } \Omega \\ \nabla\cdot\mathbf{u} &= 0 && \text{in } \Omega \end{aligned} \tag{26}$$

can be weakly stated by

$$\begin{aligned} \nu\,(\nabla\mathbf{u}, \nabla\boldsymbol{\varphi}) + (\mathbf{u}\cdot\nabla\mathbf{u}, \boldsymbol{\varphi}) - (p, \nabla\cdot\boldsymbol{\varphi}) &= (\mathbf{f}, \boldsymbol{\varphi}) && \forall\boldsymbol{\varphi} \in V \\ (\nabla\cdot\mathbf{u}, \psi) &= 0 && \forall\psi \in M \end{aligned} \tag{27}$$

with $M = L^2(\Omega)$. In the case $V = H^1(\Omega)^n$ the additional boundary bilinear form

$$b(\{\mathbf{u}, p\}, \boldsymbol{\varphi}) := -\nu\,\langle\partial_{\mathbf{n}}\mathbf{u}, \boldsymbol{\varphi}\rangle_{\partial\Omega} + \langle p\mathbf{n}, \boldsymbol{\varphi}\rangle_{\partial\Omega} , \tag{28}$$

occurs in the above weak formulation. For the following strongly formulated Robin-type boundary conditions

$$\nu \partial_{\mathbf{n}} \mathbf{u} - p\mathbf{n} = \frac{1}{\mu} (\mathbf{q}_D - \mathbf{u}) + \frac{1}{2} (\mathbf{u} \cdot \mathbf{n}) \mathbf{u} \tag{29}$$

we obtain a well posed problem formulation with the boundary semi-linear form

$$b(\mathbf{q}_D; \mathbf{u})(\boldsymbol{\varphi}) := \frac{1}{\mu} \langle (\mathbf{u} - \mathbf{q}_D), \boldsymbol{\varphi} \rangle_{\partial \Omega} - \frac{1}{2} \langle (\mathbf{u} \cdot \mathbf{n}) \mathbf{u}, \boldsymbol{\varphi} \rangle_{\partial \Omega} . \tag{30}$$

Remark 8 (Nonlinear Term and Solvability Theory) Since

$$(\mathbf{u} \cdot \nabla \mathbf{u}, \mathbf{u}) = \frac{1}{2} \langle (\mathbf{u} \cdot \mathbf{n}) \mathbf{u}, \mathbf{u} \rangle_{\partial \Omega}$$

we find similar uniform bounds for the Galerkin technique, which is used in the solvability theory. This justifies the specific form of $b(\mathbf{q}_D; \{\mathbf{u}, p\})(\boldsymbol{\varphi})$.

Analogously to the convection-diffusion-reaction equation we can formulate a stabilised semi-linear form by

$$\begin{aligned} b_\mu^\delta(\mathbf{q}_D; \{\mathbf{u}, p\})(\boldsymbol{\varphi}) := & -\frac{\delta}{\mu + \delta} \left(\langle \nu \partial_{\mathbf{n}} \mathbf{u} - p\mathbf{n}, \boldsymbol{\varphi} \rangle_{\partial \Omega} + \langle \mathbf{u} - \mathbf{q}_D, \nu \partial_{\mathbf{n}} \boldsymbol{\varphi} + \psi \mathbf{n} \rangle_{\partial \Omega} \right) \\ & + \frac{1}{\mu + \delta} \langle \mathbf{u} - \mathbf{q}_D, \boldsymbol{\varphi} \rangle_{\partial \Omega} - \frac{\mu}{2(\mu + \delta)} \langle (\mathbf{u} \cdot \mathbf{n}) \mathbf{u}, \boldsymbol{\varphi} \rangle_{\partial \Omega} \\ & - \frac{\mu \delta}{\delta + \mu} \langle \nu \partial_{\mathbf{n}} \mathbf{u} - p\mathbf{n}, \nu \partial_{\mathbf{n}} \boldsymbol{\varphi} + \psi \mathbf{n} \rangle_{\partial \Omega} \\ & + \frac{\mu \delta}{2(\delta + \mu)} \langle (\mathbf{u} \cdot \mathbf{n}) \mathbf{u}, \nu \boldsymbol{\varphi} + \psi \mathbf{n} \rangle_{\partial \Omega} \end{aligned} \tag{31}$$

While $b_\mu^0(\cdot;\cdot)(\cdot)$ corresponds to the penalty formulation the parameter choice $\mu = 0$ and $\delta = \gamma(h)$ results in a Nitsche-type formulation for the Navier-Stokes system

$$\begin{aligned} b_0^{\gamma(h)}(\mathbf{q}_D; \{\mathbf{u}, p\})(\boldsymbol{\varphi}) := & -\langle \partial_{\mathbf{n}} \mathbf{u} - p\mathbf{n}, \boldsymbol{\varphi} \rangle_{\partial \Omega} - \langle \mathbf{u} - \mathbf{q}_D, \partial_{\mathbf{n}} \boldsymbol{\varphi} + \psi \mathbf{n} \rangle_{\partial \Omega} \\ & + \frac{1}{\gamma(h)} \langle \mathbf{u} - \mathbf{q}_D, \boldsymbol{\varphi} \rangle_{\partial \Omega} . \end{aligned} \tag{32}$$

This form is almost the same as the one proposed in Becker [3].

Lemma 2 *A strong solution pair $\{\mathbf{u}, p\}$ of the Navier Stokes system (26) with the boundary condition (29) satisfies also the equation*

$$\begin{aligned} \nu\,(\nabla\mathbf{u},\nabla\boldsymbol{\varphi}) + (\mathbf{u}\cdot\nabla\mathbf{u},\boldsymbol{\varphi}) + b^{\delta}_{\mu}(\mathbf{q}_D;\{\mathbf{u},p\})(\boldsymbol{\varphi}) - (p,\nabla\cdot\boldsymbol{\varphi}) &= (\mathbf{f},\boldsymbol{\varphi}) \qquad \forall\boldsymbol{\varphi}\in V,\\ (\nabla\cdot\mathbf{u},\psi) &= 0 \qquad \forall\psi\in M. \end{aligned} \tag{33}$$

Proof By multiplying the classical Navier-Stokes equations with arbitrary test functions $\boldsymbol{\varphi} \in V$ and $\psi \in M$, integrating over the domain Ω and partial integration we obtain Eq. (27) with an additional semi-linear form $b(\mathbf{q}_D; \{\mathbf{u}, p\})(\boldsymbol{\varphi})$ like in Eq. (30).

By multiplying the boundary condition (29) with $\boldsymbol{\varphi}$ and integration over the boundary we obtain after multiplying with $\frac{\mu}{\mu+\delta}$

$$\begin{aligned} \frac{\mu}{\mu+\delta}\,\langle \nu\partial_{\mathbf{n}}\mathbf{u} - p\mathbf{n},\boldsymbol{\varphi}\rangle_{\partial\Omega} + \frac{1}{\mu+\delta}\,\langle \mathbf{u}-\mathbf{q}_D,\boldsymbol{\varphi}\rangle_{\partial\Omega}\\ - \frac{\mu}{2(\mu+\delta)}\,\langle (\mathbf{u}\cdot\mathbf{n})\mathbf{u},\boldsymbol{\varphi}\rangle_{\partial\Omega} = 0 \end{aligned} \tag{34}$$

Furthermore we multiply the boundary condition with the test function

$$\nu\partial_{\mathbf{n}}\boldsymbol{\varphi} + \psi\mathbf{n},$$

and integrate again over the boundary. Afterwards we multiply with the factor $-\frac{\mu\delta}{\mu+\delta}$ and obtain

$$\begin{aligned} -\frac{\mu\delta}{\mu+\delta}\,\langle \nu\partial_{\mathbf{n}}\mathbf{u} - p\mathbf{n}, \nu\partial_{\mathbf{n}}\boldsymbol{\varphi}+\psi\mathbf{n}\rangle_{\partial\Omega} - \frac{\delta}{\mu+\delta}\,\langle \mathbf{u}-\mathbf{q}_D, \nu\partial_{\mathbf{n}}\boldsymbol{\varphi}+\psi\mathbf{n}\rangle_{\partial\Omega}\\ + \frac{\mu\delta}{2(\mu+\delta)}\,\langle (\mathbf{u}\cdot\mathbf{n})\mathbf{u}, \nu\partial_{\mathbf{n}}\boldsymbol{\varphi}+\psi\mathbf{n}\rangle_{\partial\Omega} = 0 \end{aligned} \tag{35}$$

Adding the weak formulation and Eqs. (34) and (35) we obtain directly Eq. (33). □

5.4 Comments on Stabilisation Techniques

The presented numerical scheme is unstable with respect to two different sources, namely the convection dominance of the convection-diffusion equation or the Navier-Stokes system for large Reynolds numbers and the lack of inf-sub stability, due to the specific choice of the finite element spaces for the velocity and pressure components in the Navier-Stokes system. We can overcome both issues by using the so called local projection stabilisation. We briefly comment on the technique and cite further literature.

5.4.1 Convection Stabilisation

We use a method for the convection stabilisation, which was introduced by Becker et al. [5]. For the model problem

$$(u_h, \varphi_h) + ((\boldsymbol{\beta} \cdot \nabla)\, u_h, \varphi_h) = (f, \varphi) \qquad \forall \varphi \in V_h,$$

we add hereby the stabilisation term

$$s(u_h, \varphi_h) = s_{\text{LPS}}(u_h, \varphi_h) := \sum_{T \in \mathcal{T}_h} \delta_T \left(\pi_h\left(\boldsymbol{\beta} \cdot \nabla u_h\right), \pi_h\left(\boldsymbol{\beta} \cdot \nabla \varphi\right)\right).$$

The operator $\pi_h = I - \mathcal{P}_{2h}$ consists of the difference of the identity operator and a projection operator

$$\mathcal{P}_{2h} : V_h \to V_{2h}$$

defining a mapping of the current trial function space V_h onto the coarser one V_{2h}. The so defined mapping measures fluctuation of the convection term. The parameter δ_T is chosen as follows

$$\delta_T = \delta_0 \frac{h}{k\theta|\boldsymbol{\beta}|}.$$

This stabilisation scheme admits in principle the same properties in terms of stability and accuracy as the SUPG scheme. An import advantage is that for this scheme again the principles ‘optimise-then-discretise’ and ’discretise-then-optimise’ are interchangeable.

5.4.2 Pressure Stabilisation

For computational simplicity we work with bilinear finite elements for both the pressure and the velocity approximation. That means $V_h = \hat{V}_h^2$ for the discrete velocity space and $M_h = \hat{V}_h$ for the discrete pressure space with $\hat{V}_h$ defined as in Eq. (17). It is well known that the inf-sup condition

$$\min_{\mu_h \in M_h} \left(\max_{\varphi_h \in V_h} \frac{-(\mu_h, \nabla \cdot \boldsymbol{\varphi}_h)}{\|\boldsymbol{\varphi}_h\|_{V_h} \|\mu_h\|_{M_h}} \right) \geq \gamma_h \geq \gamma > 0$$

for this particular choice is violated.

We can also use the mentioned local projection stabilisation technique to stabilise our approach related to inf-sup-instability as it is presented in the work of Becker et al. [4].

We add therefore the bilinear form

$$s_{\mathrm{LPS}}(\{\mathbf{u}_h, p_h\}, \{\boldsymbol{\varphi}_h, \mu_h\}) = \sum_{T \in \mathscr{T}_h} \alpha_T \left(\nabla(\pi_h p_h), \nabla(\pi_h \mu_h)\right)_T$$

to the left hand side of the divergence equation

$$(\mu_h, \nabla \cdot \mathbf{u}_h) = 0 \qquad \forall \mu_h \in M_h,$$

where π_h denotes again the fluctuation operator defined as in the case of the convection stabilisation. The parameter is chosen as suggested in the literature as

$$\alpha_T = \frac{h_T^2}{\nu}.$$

6 Numerical Results

We will use a few test cases to exemplify the functionality of the presented method. Although the theory was already developed for the use of the fully nonlinear Navier-Stokes equations, we avoid at first to work with high Reynolds-numbers and suppress the time-dependent character of the flow equations in the first two test cases. The reason for this is that our main objective is to prove that we can estimate reliable boundary conditions for the flow field and the intensity function by only sparsely given data. Afterwards, we will in a third test case apply the method to identify a fluid flow, which is described by the fully non-linear and time-dependent Navier-Stokes system, which emphasises especially the usage of the high-fidelity physical model in the proposed identification method.

In the first two examples we use in principle the same setting. The computational domain and the time horizon are given by

$$\Omega = (0,1) \times (0,1), \qquad \mathscr{T} = (0, 0.2].$$

Furthermore we set

$$\varepsilon = 10^{-10}, \qquad \text{and } \nu = 1.$$

All numerical calculations are performed on a mesh with 289 spatial nodes and 40 time steps.

6.1 First Test Data Set

The first test data set was already presented in the introduction in Fig. 1. It consists of three intensity functions $\mathscr{I}_1$, $\mathscr{I}_2$ and $\mathscr{I}_3$ representing a brightness distribution in form of a bulb signal. The first function shows the bulb signal at the initial time $t = 0$, the second function shows the signal after a movement with the constant flow field $\beta = (0, 2)^T$ at the intermediate time point $t = 0.1$ and the third function shows the transported signal at the end time point.

We performed then the approach from the Definitions 3 and 4 for the parameter choice

$$\mu_1 = \mu_2 = h$$

with the cell diameter h.

The method is for small values of the regularisation parameter α very unstable unless we have a good initial value for the time dependent boundary functions q_I and $\mathbf{q_u}$. The reason for this is that we solve in general an inverse problem that requires a certain regularisation (see Engl et al. [14]).

Thus we use a homotopy method in α, by solving the optimisation problem for a large α_k and then taking $q_I^{(k)}$ and $\mathbf{q_u}^{(k)}$ as initial values for a further solution of the optimisation problem with a reduced $\alpha_{k+1} = \sigma\alpha_k$ with $\sigma \in (0, 0.5)$. After a few steps of this technique we obtain the reconstruction $I_{k,h}$ presented in Fig. 4. The regularisation parameter was around 10^{-2} and the final Newton residual was approximately 10^{-5}.

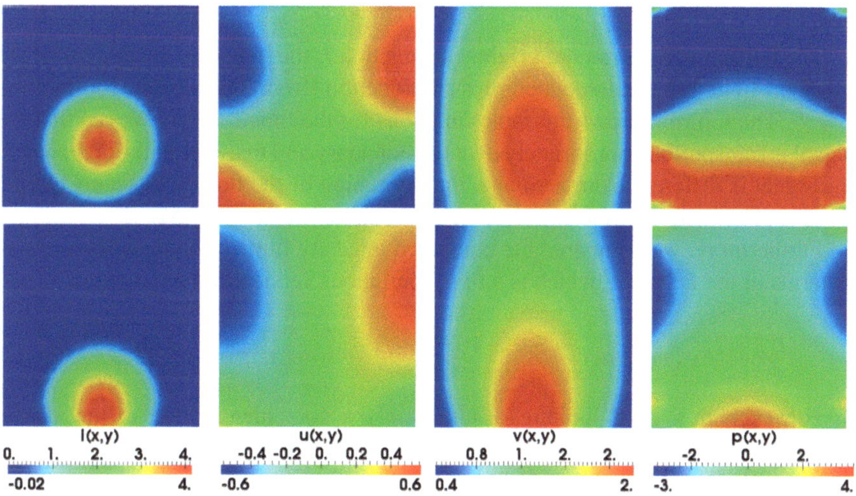

Fig. 4 From *left* to *right*: $I(\mathbf{x})$, $u(\mathbf{x})$, $v(\mathbf{x})$ and $p(\mathbf{x})$. *Top*: $t = \frac{3T}{4}$. *Bottom*: $t = \frac{T}{4}$. Regularisation parameter: $\approx \alpha = 10^{-2}$. $\|I_{k,h} - \hat{I}\|^2_{L^2(0,T;L^2(\Omega))} = 5.01 \cdot 10^{-4}$

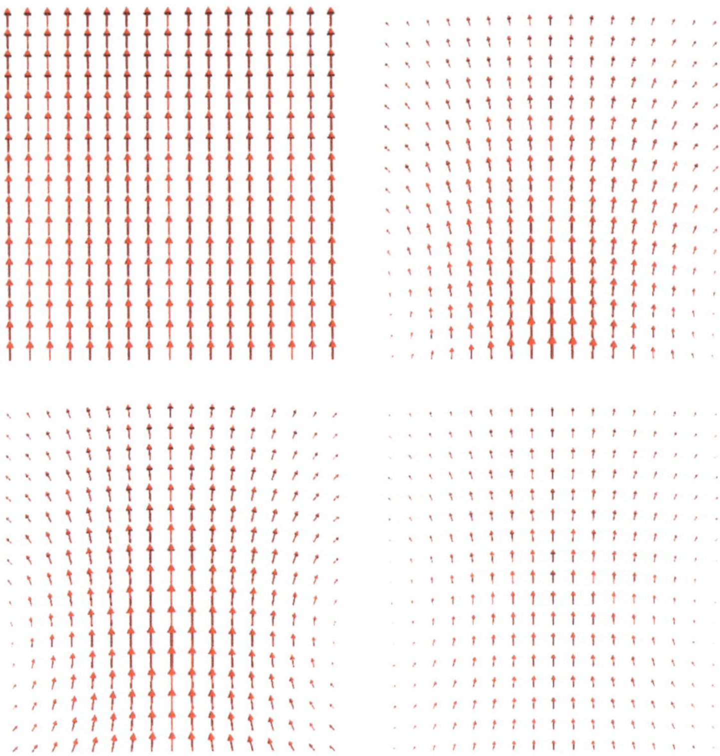

Fig. 5 *Top left*: $\boldsymbol{\beta} = (0, 2)^T$. *Top right*: $\mathbf{u}_h$ at $t = \frac{T}{4}$. *Bottom left*: $\mathbf{u}_h$ at $t = \frac{3T}{4}$. *Bottom right*: $\mathbf{u}_h$ at $t = T$

The bottom row shows the brightness function I, the x- and y-component of the transport field and the associated pressure function (from left to right) for the time $t = \frac{T}{4}$. The top row shows the same functions for the time $t = \frac{3T}{4}$. We see that the signal is transported appropriately over the boundary and that within the area, where the signal is different from zero also the estimation of the flow field seems to be a good approximation of the expected field (see therefore also Fig. 5).

Furthermore, we compared the computed intensity function to the expected one in terms of the L^2-norm, which is a time continuous version of the data term:

$$\|I_{k,h} - \hat{I}\|^2_{L^2(0,T;L^2(\Omega))} = 5.01 \cdot 10^{-4} \qquad (\|\hat{I}\|^2_{L^2(0,T;L^2(\Omega))} = 2.48 \cdot 10^{-1}).$$

However, this good looking result is only a coincidence, due to the simple situation of our test case. Since we assumed the underlying transport field to be constant the special choice of our regularisation term leads to an appropriate recovering of the time-dependent function $I(\mathbf{x}, t)$. In general there are infinitely many solenoidal flow fields that generate the three given intensity functions $\mathscr{I}_1$, $\mathscr{I}_2$ and $\mathscr{I}_3$ (e.g. $\beta_c(\mathbf{x}, t) = \sin^c\left(\frac{t\pi}{T}\right)(0, 2)^T$), which we used as data in the optimisation problem. All of these flow fields lead to completely different intensity functions (e.g. $I_{\beta_c}(\mathbf{x}, t)$). Thus, it is impossible to reconstruct a specific flow field, without introducing further a priori knowledge about the flow field in the optimisation process. This consideration emphasises again the ill-posed character of the problem and shows that we have to be careful in trusting the reconstruction of fluid flow fields, without additional information about this fields.

Nevertheless, we can use the method as an interpolator between discrete intensity functions as the next test case indicates.

6.2 *Second Test Data Set*

The second test case was also mentioned in the introduction and consists of the six brightness functions presented in Fig. 2 at the time points $t = 0.04(k-1)$ with $k = 1, \ldots, 6$. The underlying transport field for the movement is

$$\boldsymbol{\beta}(\mathbf{x}, t) = \frac{5\pi}{2}(-y, x)^T.$$

Obviously the signal is throughout the image sequence transported from the lower boundary to the left boundary. Thus, we initialise our computational method this time by the following assumption of the transport field

$$\mathbf{u}^0(\mathbf{x}) = (-2, 2)^T,$$

since this field describes the principle direction of the flow. Afterwards, we solve as in the first test case the optimisation problem with

$$\mu_1 = \mu_2 = h.$$

The result of this calculation is visualised in Fig. 6. We want to emphasise that we choose time points for the visualisation in which no data is available to make clear that this method interpolates intermediate brightness distributions. Our next aim is to compare different parameter choices for the boundary control formulation. We use therefore four different cases. The first three cases are given as follows

$$\mu_1 = \mu_2 = 1, \qquad \text{(Case 1)}$$
$$\mu_1 = \mu_2 = 0.1, \qquad \text{(Case 2)}$$
$$\mu_1 = \mu_2 = h, \qquad \text{(Case 3)}$$

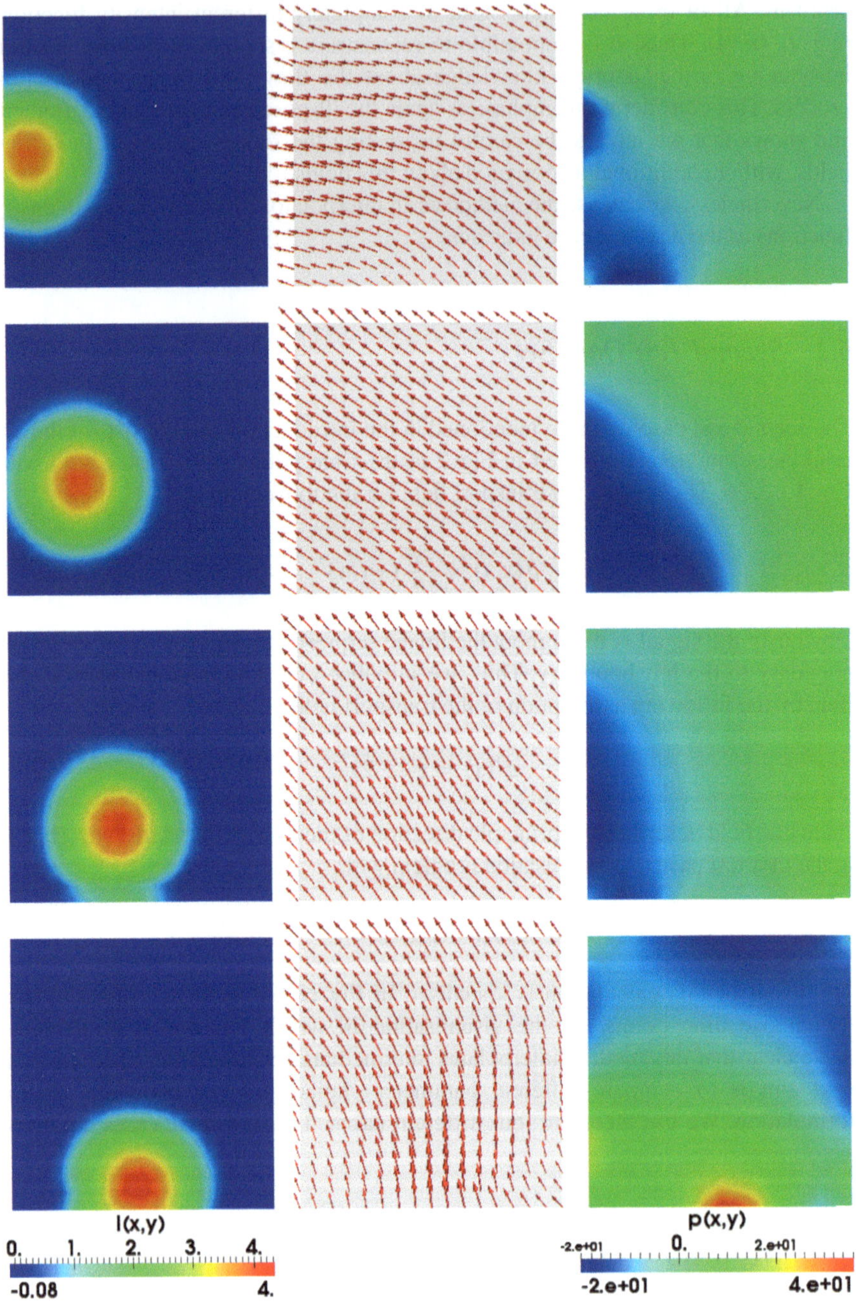

Fig. 6 Calculated solution for $\alpha \approx 10^{-3}$. *Left column*: from *bottom to top*: $I(\mathbf{x}, t_k)$ with $t_k = \frac{kT}{8}$ and $k = 1, 3, 5, 7$. *Middle column*: corresponding transport field. *Right column*: corresponding pressure function

Table 1 We indicate the error between the expected brightness function in time and the calculation by $e_{h,k} := \|I - I_{h,k}\|^2_{L^2(\Omega\times[0,2])}$. The expected brightness function has the following norm $g := \|I\|^2_{L^2(\Omega\times[0,2])} = 2.454 \cdot 10^{-1}$

Case	$e_{h,k}$	Rel. Error ($\frac{e_{h,k}}{g}$) (%)
1	$1.735 \cdot 10^{-3}$	0.71
2	$7.976 \cdot 10^{-4}$	0.33
3	$6.205 \cdot 10^{-4}$	0.25
4	$5.616 \cdot 10^{-4}$	0.23
5	$5.346 \cdot 10^{-4}$	0.22

for the parameters in the Definitions 3 and 4. For the final test case we want to set $\mu = 0$. Thus, we use instead of the usual boundary bilinear forms in the two definitions the stabilised ones in Eq. (32) for the Navier-Stokes part and Eq. (25) for the part of the physic-based optical flow equation

$$\mu_1 = \mu_2 = 0, \qquad \text{Eq. (32) with } \gamma(h) := \frac{h}{5} \text{ and Eq. (25)} \qquad \text{(Case 4)}.$$

We want to emphasise that these two conditions implement Dirichlet boundary controls for which the developed theory could not be applied.

The results of the calculations are shown in Table 1. Here we document the ability of the method to recover the expected movement of the bulb signal during the whole time horizon in terms of the L^2-error. The table shows that for $\mu = 1$ the method is worse than for small choices of μ or even for a implemented weak Dirichlet condition. The reason for this is that artefacts on the inflow boundary occur for the recovery of the signal. Thus, it seems to be a good idea to work with the weak Dirichlet conditions.

Unfortunately, the solution has then another drawback since the transport field is no longer continuous in time as Fig. 7 indicates. The left column represents the transport field in the case of the weak Dirichlet controls. As we can see the field is immediately changing after the time stepping scheme passing a time point, where a brightness function information is available. Hence, the field has some kind of jump with respect to the time variable. We want to emphasise that this is nothing unexpected, since the theory is not working in this case and we cannot expect the same temporal regularity as in the case of Robin controls.

As a result of the discussion we propose a "trade-off" version of the used methods. Here we use the Robin-type control for the Navier-Stokes part with $\mu_1 = h$ and the Dirichlet control in Eq. (25) for the brightness function. This leads to a method which smoothly transports the signal across the boundary by a temporally smooth transport field. We document the error of this combination by "Case 5" in Table 1.

However, the last two examples rely on temporal constant solenoidal fields, which could also be solutions of the steady Stokes equations. Thus, the spent effort for interpolating discrete image intensities with our approach basing on a

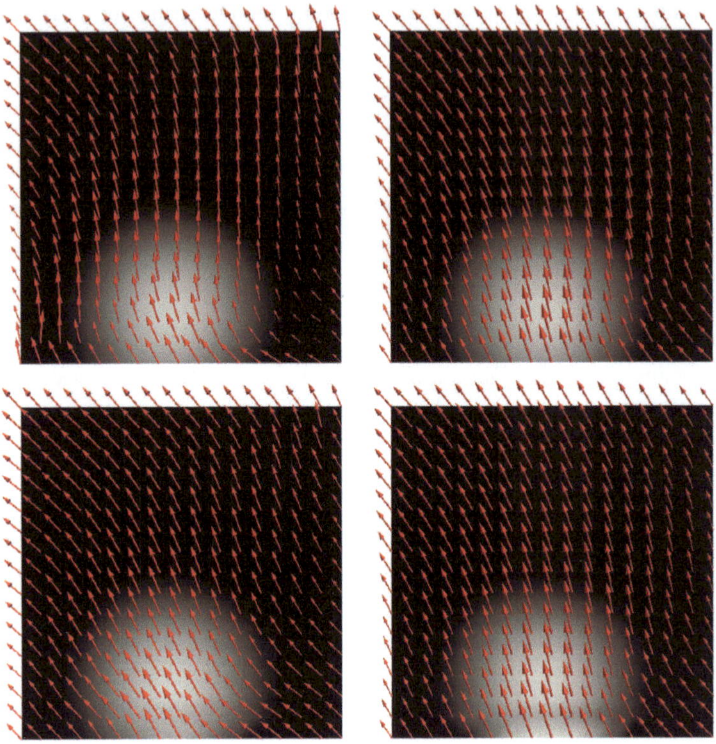

Fig. 7 *Top*: $t = 0.04$. *Bottom*: $t = 0.0425$. *Left column*: transport field for the weak Dirichlet boundary control. *Right column*: transport field for the Robin-type control approach with $\mu_1 = \mu_2 = 0.1$. The left transport field is immediately changing and has therefore a kink in the time variable

regularisation with a fully time-dependent and non-linear physical model for the fluid flow is questionable. To justify the reliability of our consideration we present a final test case, which based on observations of a dynamic flow.

6.3 Third Test Data Set

The following example relies on the sequence $\mathscr{I}_k$ presented in Fig. 3, which was obtained after executing a forward calculation of system (1). We will therefore shortly present the generation of the image sequence. The computational domain is (Fig. 8)

$$\Theta = (-0.01m, 0.19m) \times (0, 0.1m).$$

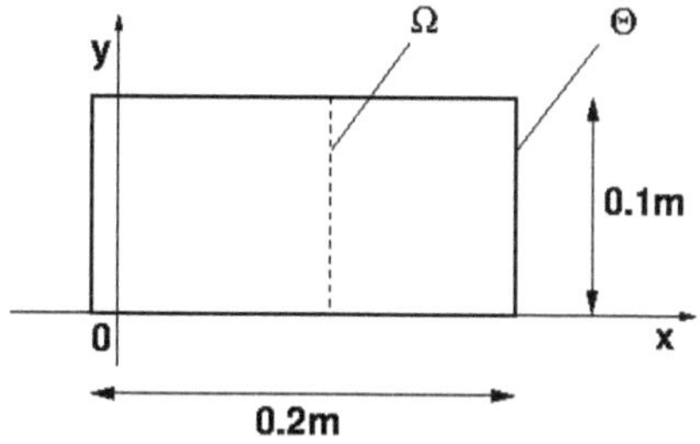

Fig. 8 The image domain $\Omega = (0, 0.1m)^2$ is an aperture of the original computational domain $\Theta = (-0.01m, 0.19m) \times (0, 0.1m)$

The time horizon is $(0, 3s]$. We prescribe no-slip boundary conditions on the upper and lower boundary. The left boundary is an inflow boundary, while on the right boundary the outflow occurs. The inflow is described by the function

$$\mathbf{u}(\mathbf{x}, t) := \left(10 \min(t, 1) \max\left(0, \omega(x, y)\right), 0\right)^T$$

with

$$\omega(x, y) := \frac{1}{4}\left(y - \left(\frac{1}{4} + \frac{1}{5}\sin(\pi t)\right)\right) \cdot \left(\left(\frac{3}{4} + \frac{1}{5}\sin(\pi t)\right) - y\right) \frac{m}{s}.$$

Furthermore, we choose a kinematic viscosity $\nu = 0.01\frac{m^2}{s}$. Finally, the passive tracer is also introduced by a Dirichlet-type boundary condition on the left side of the domain Θ:

$$I(\mathbf{x}, t) := \max\left(\cos(\pi t), 0\right).$$

For the diffusivity parameter in the convection-diffusion equation of the tracer we choose $\varepsilon = 1e - 10$.

Then we perform a calculation with 60 steps of an implicit Euler method in time and with a spatial discretisation into 2145 grid points of a regular mesh. Thus, we obtain the reference solutions $\mathbf{u}_r$ and I_r. By using $I_r(\mathbf{x}, t_k)$ for $\mathbf{x} \in \Omega = (0, 0.1m)^2$ we generate afterwards three different "image sequences" on the aperture domain Ω. Each sequence represents a different sampling rate for the images:

$$\begin{aligned}
\mathscr{I}^{(1)} &:= I_r(\mathbf{x}, t_k), &&\text{with } t_k = 0.05k, && k = 0, \ldots, 60,\\
\mathscr{I}^{(2)} &:= I_r(\mathbf{x}, t_k), &&\text{with } t_k = 0.25k, && k = 0, \ldots, 12,\\
\mathscr{I}^{(3)} &:= I_r(\mathbf{x}, t_k), &&\text{with } t_k = 0.5k, && k = 0, \ldots, 6.
\end{aligned}$$

We have a clear difference between sampling rate and time discretisation in the sequences $\mathscr{I}^{(2)}$ and $\mathscr{I}^{(3)}$, while for $\mathscr{I}^{(1)}$ the sampling rate and the time discretisation coincides.

These sequences are used for the reconstruction of the flow on the aperture domain Ω. Therefore, we use our Robin-boundary control approach with the following settings. From the image sequences we observe a starting phase of the flow and a movement from left to right of the image patterns. Thus, we set

$$\mathbf{g}(\mathbf{x}, t) := (\min(t, 1)u_{\max}y(1-y), 0)^T \, \frac{m}{s}$$

and using instead the modified function

$$\tilde{\mathbf{q}}_{\mathbf{u}}(\mathbf{x}, t) := \mathbf{q}_{\mathbf{u}}(\mathbf{x}, t) + \mathbf{g}(\mathbf{x}, t)$$

in the Robin-boundary formulation [cf. fourth equation in (5)]. The parameter $u_{\max}$ can also be roughly estimated by the given images. We choose $u_{\max} = 2$.

To initialise also the inflow of signals I across the inflow boundary on the left we choose a linear interpolation between the given discrete image signals of the sequences on the boundary. We denote this interpolated function by $\mathfrak{I}(\mathscr{I}^{(j)})(\mathbf{x}, t)$ with $i = 1, 2, 3$ and have thus the modified function

$$\tilde{q}_I := \mathfrak{I}(\mathscr{I}^{(j)})(\mathbf{x}, t) + q_I$$

in the Robin-boundary formulation [cf. third equation in (5)]. For the parameter in the Navier-Stokes part we choose this time $\mu_1 = 1$ and perform a weak Dirichlet control in the fashion of Eq. (25) as in the test case before for the convection-diffusion equation part.

This time we will work with two different values for α_1 and α_2 in the cost functional (see Definition 1). The reason for this is that we want to penalise too big changes of the image function on the boundary since we assume that the linear interpolated boundary conditions of the image sequence data is already a qualitative approximation of the effective boundary conditions. Thus, the α_1-term has more the character of a correction. However, since we have no clue about the boundary conditions for the fluid flow we want to choose α_2 as small as possible. As in the test cases before we apply therefore a homotopy-type method that stops, when α_2 drops below a threshold Tol. Our experiences shows that $\alpha_1 = 100$ and $\mathrm{Tol} = 10^{-3}$ is an appropriate choice, when we additionally weighting the data term in the cost functional by a factor 100.

The results for the identification by this settings are presented in Fig. 9 (Magnitude of the flow field) and Fig. 10 (Intensity function) for the time points $t_1 = 0.75$ (Row A), $t_2 = 1.5$ (Row B), $t_3 = 2.25$ (Row C) and $t_4 = 3$ (Row D). From left to right we visualise in the both figures the expected solution (Column 1), the results for sequence $\mathscr{I}^{(1)}$ (Column 2), $\mathscr{I}^{(2)}$ (Column 3) and $\mathscr{I}^{(3)}$ (Column 4). For a more qualitative comparison we investigate the following quantity of interest

$$\eta(t) := \|\mathbf{u}(t) \cdot \nabla I(t)\|_2^2.$$

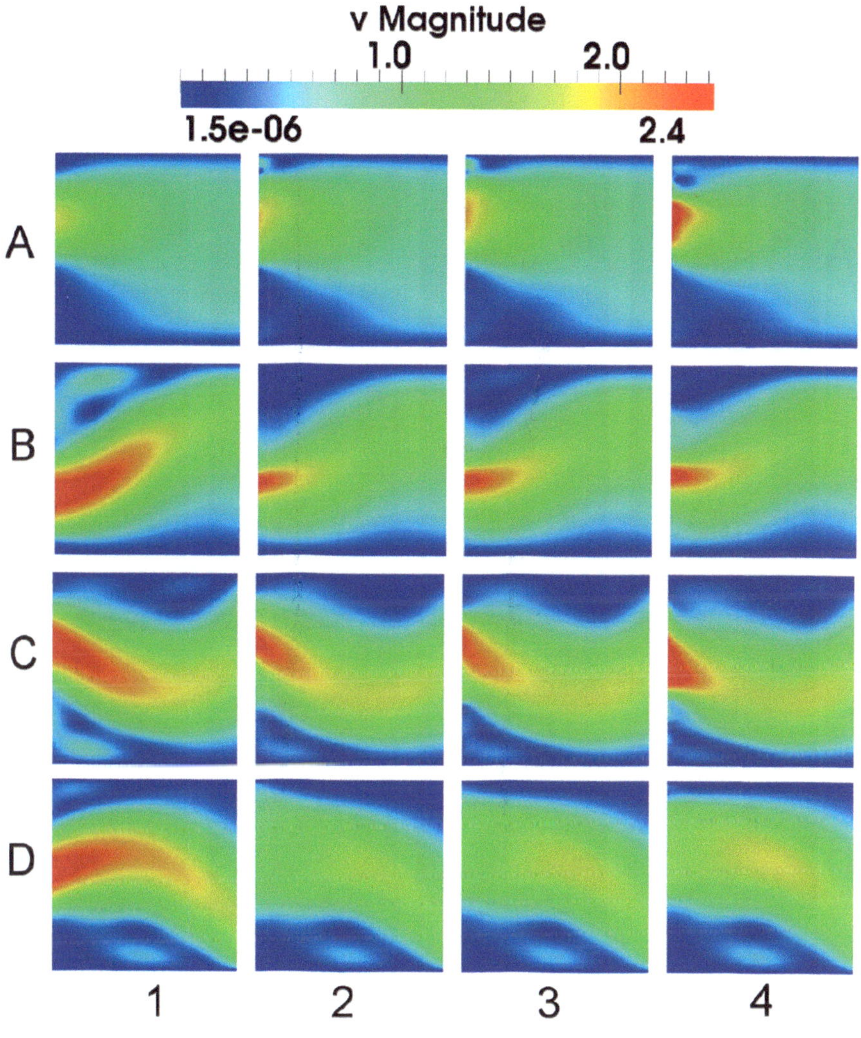

Fig. 9 Comparison of the identified fluid flow in terms of the magnitude of the flow field. Row A to row D: $t_k = 0.75k$ with $k = 1, \ldots, 4$. Columns from *left* to *right*: expected (1), $\mathscr{I}^{(1)}$ (2), $\mathscr{I}^{(2)}$ (3) and $\mathscr{I}^{(3)}$ (4)

This quantity has the feature that it depreciate the recovery in areas where the intensity is nearly constant. The evolution of the $\eta(t)$ across the mentioned time interval is documented in Fig. 11. The left picture shows the expected curve (blue), the estimation with the data from sequence $\mathscr{I}^{(1)}$ (green dashed) and the results of a forward calculation with $\mathbf{g}(\mathbf{x}, t)$ and $\mathfrak{I}(\mathscr{I}^{(j)})$ (red dash-dotted). We see a clear improvement by the usage of our identification process. In the left picture of

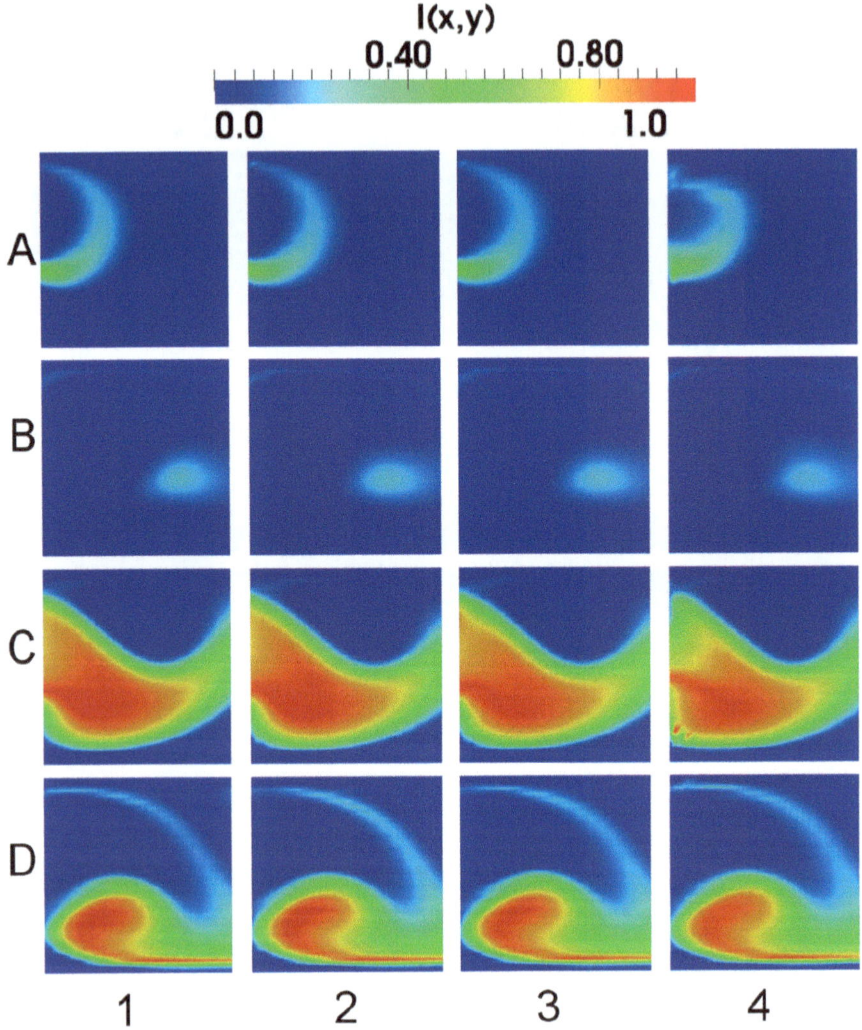

Fig. 10 Comparison of the reconstructed intensity functions. Row A to row D: $t_k = 0.75k$ with $k = 1, \dots, 4$. Columns from *left* to *right*: expected (1), $\mathscr{I}^{(1)}$ (2), $\mathscr{I}^{(2)}$ (3) and $\mathscr{I}^{(3)}$ (4)

Fig. 11 we compared the quality of the reconstruction in terms of the three data sequences $\mathscr{I}^{(1)}$ (green), $\mathscr{I}^{(2)}$ (blue dashed) and $\mathscr{I}^{(3)}$ (red dash-dotted). As expected we see a clear improvement by using as much data as possible. However, the result also indicates that we are not able to fit the curve even if we have image data in every time step of our temporal discretisation. The reason for this is the ill-posed character of the problem.

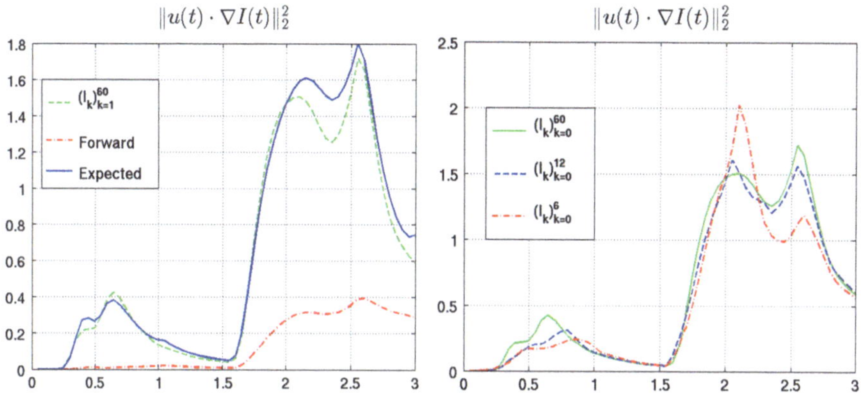

Fig. 11 Recovery of the quantity of interest $\eta(t)$ by our approach. *Left*: comparison of the expected curve (*blue*), identification with intensity informations in each time step (*green*) and a forward calculation (*red*). *Right*: different data sequences: $\mathscr{I}^{(1)}$ (*green*), $\mathscr{I}^{(2)}$ (*blue*) and $\mathscr{I}^{(3)}$ (*red*)

7 Discussion on the Application of Multiple Shooting Methods

The numerical solution of the presented optimisation problem relies on a Newton-type method developed by Becker et al. [6] to find a solution of the optimality system, which can be introduced by means of the Lagrangian [see Eq. (13)].

We observed for all our calculations that the method's performance is very sensitive with respect to the choice of the initial values for the control variables q_I, $\mathbf{q_u}$, the length of the time horizon and the chosen regularisation parameter.

As mentioned in the last section we used a homotopy method in α to be able to solve the problem in a stable way for small values of the regularisation parameter. We want to remark that for each step of the homotopy method a whole optimisation problem must be solved, which increases the computational costs immensely.

However, the unstable behaviour is probably a result of the structure of the PDE-constrained optimisation problem. We remember therefore the cost functional in Definition 1

$$J(q_I, \mathbf{q_u}, I)$$

which is minimised with respect to the weak formulation of the state equation in Definition 4. The latter is now converted into the following abstract form

$$((\partial_t I, \psi)) + \bar{a}(\mathbf{u}; I, \psi) + \bar{b}(q_I, \psi) + (I(0) - \mathscr{I}_1, \psi(0)) = 0 \qquad \forall \psi \in \mathscr{X}_I$$
$$((\partial_t \mathbf{u}, \boldsymbol{\varphi})) + \bar{c}(\mathbf{u})(\boldsymbol{\varphi}) + \bar{d}(\mathbf{q_u}, \boldsymbol{\varphi}) + \big(\mathbf{u}(0) - \mathbf{u}^0, \boldsymbol{\varphi}(0)\big) = 0 \qquad \forall \boldsymbol{\varphi} \in \mathscr{X}_{\mathbf{u}}$$

with the linear forms $\bar{a}(\cdot;\cdot,\cdot)$, $\bar{b}(\cdot,\cdot)$, $\bar{c}(\cdot)(\cdot)$ and $\bar{d}(\cdot)(\cdot)$ containing the terms of the forms in Eq. (5) integrated over the time interval $[0,T]$. Here we separated the control terms from the rest of the equation.

We concretise the Lagrangian

$$\begin{aligned}\mathscr{L}(q_I,\mathbf{q_u},I,\mathbf{u},K,\mathbf{z}) := \; & J(q_I,\mathbf{q_u},I) + ((\partial_t I, K)) + \bar{a}(\mathbf{u};I,K) + \bar{b}(q_I,K) \\ & + (I(0)-\mathscr{I}_1, K(0)) + ((\partial_t \mathbf{u},\mathbf{z})) + \bar{c}(\mathbf{u})(\mathbf{z}) + \bar{d}(\mathbf{q_u},\mathbf{z}) \\ & + \left(\mathbf{u}(0)-\mathbf{u}^0,\mathbf{z}(0)\right).\end{aligned}$$

With the Lagrangian we can obtain the mentioned optimality system, which consists of the state equations

$$\begin{aligned}\mathscr{L}_K(q_I,\mathbf{q_u},I,\mathbf{u},K,\mathbf{z})(\delta K) &= 0, \qquad \forall \delta K \in \mathscr{X}_I,\\ \mathscr{L}_{\mathbf{z}}(q_I,\mathbf{q_u},I,\mathbf{u},K,\mathbf{z})(\delta \mathbf{z}) &= 0, \qquad \forall \delta \mathbf{z} \in \mathscr{X}_{\mathbf{u}},\end{aligned}$$

adjoint equations

$$\begin{aligned}\mathscr{L}_I(q_I,\mathbf{q_u},I,\mathbf{u},K,\mathbf{z})(\delta I) &= 0, \qquad \forall \delta I \in \mathscr{X}_I,\\ \mathscr{L}_{\mathbf{u}}(q_I,\mathbf{q_u},I,\mathbf{u},K,\mathbf{z})(\delta \mathbf{u}) &= 0, \qquad \forall \delta \mathbf{u} \in \mathscr{X}_{\mathbf{u}},\end{aligned}$$

and control equations

$$\begin{aligned}\mathscr{L}_{q_I}(q_I,\mathbf{q_u},I,\mathbf{u},K,\mathbf{z})(\delta q_I) &= 0, \qquad \forall \delta q_I \in \mathscr{Q}_I,\\ \mathscr{L}_{\mathbf{q_u}}(q_I,\mathbf{q_u},I,\mathbf{u},K,\mathbf{z})(\delta \mathbf{q_u}) &= 0, \qquad \forall \delta \mathbf{q_u} \in \mathscr{Q}_{\mathbf{u}}.\end{aligned}$$

The last two conditions can be used to substitute the control variables by the adjoint variables in the state equation. Collecting afterwards the state and the adjoint variables by the overall vector $\mathbf{x} := \{I,\mathbf{u},K,\mathbf{z}\}$ we can summarise the optimality system by the abstract system of equations

$$\begin{aligned}\partial_t \mathbf{x}(t) &= \mathscr{A}(\mathbf{x}(t)), \qquad \forall t \in [0,T],\\ \hat{\mathbf{g}}(\mathbf{x}(0),\mathbf{x}(T)) :&= B_0\mathbf{x}(0) + B_T\mathbf{x}(T) - (\mathscr{I}_1,\mathbf{u}^0, I(T)-\mathscr{I}_N, 0)^T = 0\end{aligned} \tag{36}$$

with

$$B_0 = \begin{pmatrix}\mathbf{I} & \mathbf{0}\\ \mathbf{0} & \mathbf{0}\end{pmatrix}, \qquad B_T = \begin{pmatrix}\mathbf{0} & \mathbf{0}\\ \mathbf{0} & \mathbf{I}\end{pmatrix}, \qquad \text{and } \mathbf{I} = \begin{pmatrix}1 & 0\\ 0 & 1\end{pmatrix},$$

and an abstract differential operator $\mathscr{A}$. It is known that the above abstract problem represents a boundary value problem (BVP) for the state and adjoint variables, with respect to given values for the state variables I and $\mathbf{u}$ at the initial time and for the adjoint variables K and $\mathbf{z}$ at the end time point (cf. Carraro et al. [12] for optimal control problems with parabolic PDE constraints).

This is a possible explanation for the instability of our sophisticated PDE-constrained optimisation problem, especially if we consider long time horizons, as it was figured out in the article of Weiss [40] for a much simpler configuration. To follow this argumentation we discretise system (36) on a fixed spatial grid by the method of lines:

$$\mathbf{x}_h'(t) = \hat{\mathbf{f}}_h\left(t, \mathbf{x}_h(t)\right), \qquad \forall t \in [0, T],$$
$$\hat{\mathbf{g}}_h\left(\mathbf{x}_h(0), \mathbf{x}_h(T)\right) = 0.$$

The resulting system is a classical ODE-based BVP. Weiss showed that for such problems multiple shooting increases the size of the domain within which we may choose the initial value for a successful performance of Newton's method. However, whether working with multiple shooting is reliable requires a further detailed mathematical and numerical investigation. First steps towards the application of multiple shooting methods to parabolic optimal control problems were presented by Carraro et al. [11, 12], Hesse et al. [21] and Potschka [34].

Since our PDE-constrained optimisation problem could also be interpreted as a parabolic boundary control problem, it would be an interesting topic for further research to apply the rather technical multiple shooting method to our highly complex problems. We conjecture that it is possible, due to the stabilising effect of the multiple shooting method, to reduce the effort of the homotopy method in α.

Another big advantage of multiple shooting methods is that they are well suited for parallelisation. This could also clearly accelerate the solution process in terms of needed computational time.

8 Conclusion

In this contribution we suggested a PDE-constrained optimisation problem based on the estimation of boundary functions, which can be used in a certain physical setting to evaluate a flow field transporting a passive tracer across the computational domain boundaries. The only available data is a temporally sparse sequence of measurements of the intensity function of a passive tracer. The motivation for the investigation is the increasing use of image processing, especially optical flow estimation, techniques in fluid flow evaluation.

However our main objective was to formulate a mathematically well-posed problem, which is also easy to handle from the computational point of view. We suggested therefore a Robin-type control formulation, since in simple linear settings a close connection to Dirichlet control problems was observed. Although our Robin-type approach is well-posed and has a straightforward computational treatment, we have not enhanced the theory for approximating Dirichlet controls by Robin-type controls to our system. On the one hand, it is not necessary to work with Dirichlet controls for our identification purposes. On the other hand we have no appropriate very weak formulation of the problem. Here we see a possible direction for further theoretical research to connect Robin-type and Dirichlet controls also for such sophisticated coupled systems of equations.

We proved the functionality of the method based on synthetic test cases. We were able to interpolate a sparsely given image sequence, without the knowledge of the underlying flow field, even when the information on the intensity function moved across the boundary. However the examples were too simple to prove the necessity of using the fully nonlinear Navier-Stokes equations. Applying the method to more realistic prototypical examples or real world problems would be a further step for our work.

For all our numerical calculations we enhanced a generalisation for weakly imposed boundary conditions suggested by Juntunen et al. [24] to the class of PDEs we have to deal with and used them for our numerical calculations.

Acknowledgements This article is based on collaboration with Prof. Dr. Dr. h.c. R. Rannacher (IAM, Heidelberg University) and Priv.-Doz. Dr. C.S. Garbe (IPM, Heidelberg University). Furthermore M. Geiger and Dr. T. Carraro (both IAM, Heidelberg University) supported the author to connect the content to the time decomposition topic in the conclusions. The author gratefully acknowledges all mentioned persons for many fruitful discussions and for their support.

References

1. Bachl, F., Garbe, C., Fieguth, P.: A Bayesian approach to apaceborn hyperspectral optical flow estimation on dust aerosols. In: Geoscience and Remote Sensing Symposium (IGARSS), pp. 256–259 (2012)
2. Bachl, F., Garbe, C., Fieguth, P.: Baysian inference on integrated continuity fluid flows and their application to dust aerosols. In: Geoscience and Remote Sensing Symposium (IGARSS), pp. 2246–2249 (2013)
3. Becker, R.: Mesh adaptation for Dirichlet flow control via Nitsche's method. Commun. Numer. Meth. Eng. **18**(9), 669–680 (2002)
4. Becker, R., Braack, M.: A finite element pressure gradient stabilization for the Stokes equations based on local projections. Calcolo **38**(4), 173–199 (2001)
5. Becker, R., Braack, M.: A two-level stabilization scheme for the Navier-Stokes equations. In: Proceedings of ENUMATH (2003)
6. Becker, R., Meidner, M., Vexler, B.: RoDoBo: A C++ Library for Optimization with Stationary and Nonstationary PDEs. Institute of Applied Mathematics, Heidelberg University, Heidelberg. http://www.rodobo.org (2005)
7. Belgacem, F., Fekih, H., Metoui, H.: Singular pertubation for the Dirichlet boundary control of elliptic problems. ESAIM **37**(5), 833–850 (2003)
8. Borzi, A., Ito, K., Kunisch, K.: Optimal control formulation for determining optical flow. SIAM J. Sci. Comput. **24(3)**, 818–847 (2002)
9. Braess, D.: Finite Elemente, vol. 4. Springer, Berlin (2007)
10. Brenner, S., Scott, L.: The Mathematical Theory of the Finite Element Method, vol. 3. Springer, Berlin (2007)

11. Carraro, T., Geiger, M.: Direct and indirect multiple shooting for parabolic optimal control problems. In: Thomas, C., Michael, G., Stefan, K., Rolf, R. (eds.) Multiple Shooting and Time Domain Decomposition Methods. Springer, Heidelberg (2015, in this issue)
12. Carraro, T., Geiger, M., Rannacher, R.: Indirect multiple shooting for nonlinear parabolic optimal control problems with control constraints. SIAM J. Sci. Comput. **36**(2), 452–481 (2014)
13. Chen, K., Lorenz, D.: Image sequence interpolation based on optical flow, image segmentation and optimal control. IEEE Trans. Image Process. **21**(3), 1020–1030 (2012)
14. Engl, H., Hanke, M., Neubauer, A.: Regularization of Inverse Problems. Kluwer Academic, Dordrecht (2000)
15. Evans, L: Partial Differential Equations, 2nd edn. American Mathematical Society, Providence, RI (2010)
16. Farwig, R., Kozono, H., Sohr, H.: Very weak, weak and strong solutions to the instationary Navier-Stokes system. In: Kaplicky, P., Necasova, S. (eds.) Topics on Partial Differential Equations, vol. 2, pp. 1–54. Publishing House of the Faculty of Mathematics and Physics Charles University, Prague (2007)
17. Freund, J., Stenberg, R.: On weakly imposed boundary conditions for second order problems. In: Proceedings of the International Conference on Finite Elements in Fluids, pp. 327–336 (1995)
18. Gunzburger, M., Kunoth, A.: Space-time adaptive wavelet methods for optimal control problems constrained by parabolic evolution equation. SIAM J. Control Optim. **49**(3), 1150–1170 (2011)
19. Héas, P., Mémin, E., Papadakis, N., Szantai, A.: Layered estimation of atmospheric mesoscale dynamics from satellite imagery. IEEE Trans. Geosci. Remote Sens. **45**(12), 4087–4104 (2007)
20. Heitz, D., Memin, E., Schnörr, C.: Variational fluid flow measurements from image sequences: synopsis and perspectives. Exp. Fluids. Published Online (2009)
21. Hesse, H.K., Kanschat, G.: Mesh adaptive multiple shooting for partial differential equations. Part I: linear quadratic optimal control problems. J. Numer. Math. **17**(3), 195–217 (2009)
22. Horn, B.K.P, Schunck, B.G.: Determining optical flow. Artif. Intell. **14**, 185–203 (1981)
23. Hou, L., Ravindran, S.: A penalized Neumann control approach for solving an optimal Dirichlet control problem for the Navier-Stokes equations. SIAM J. Control Optim. **36**(5), 1795–1814 (1998)
24. Juntunen, M., Stenberg, R.: Nitsche's method for general boundary conditions. Math. Comput. **78**(5), 1353–1373 (2009)
25. Klinger, M.: Parameter estimation problems in physically based image processing. In: Proceedings of ENUMATH 2011, pp. 191–199 (2013)
26. Kolmbauer, M., Langer, U.: Efficient solvers for some classes of time-periodic eddy current optimal control problems. In: Numerical Solution of Partial Differential Equations: Theory, Algorithms, and Their Applications, Springer, New York, pp. 203–216 (2013)
27. May, S., Rannacher, R., Vexler, B.: A priori error analysis for the finite element approximation of elliptic Dirichlet boundary control problems. In: Proceedings of ENUMATH 2007, pp. 637–644 (2007)
28. Meidner, D.: Adaptive space-time finite element methods for optimization problems governed by nonlinear parabolic systems. Ph.D. thesis, Heidelberg University (2008)
29. Nakajima, Y., Inomata, H., Nogawa, H., Sato, Y., Tamura, S., Okazaki, K., Torii, S.: Physics-based flow estimation of fluids. Pattern Recogn. **36**, 1203–1212 (2003)
30. Nitsche, J.: Über ein Variationsprinzip zur Lösung von Dirichlet-problemen bei Verwendung von Teilräumen, die keinen Randbedingungen unterworfen sind. Abhandlungen aus dem Mathematischen Seminar der Universität Hamburg **36**, 9–15 (1971)
31. Of, G., Phan, T., Steinbach, O.: An energy space finite elliptic Dirichlet boundary control problem. SFB-Report No. 2009-001, Uni Graz (2012)
32. Papadakis, N., Mémin, E.: Variational assimilation of fluid motion from image sequences. SIAM J. Image Sci. **1**(4), 343–363 (2008)

33. Pearson, J., Stoll, M.: Fast iterative solution of reaction-diffusion control problems arising from chemical processes. SIAM **35**(5), 987–1009 (2013)
34. Potschka, A.: A Direct Method for the Numerical Solution of Optimization Problems with Time-Periodic PDE Constraints. Springer, Berlin (2014)
35. Ruhnau, P.: Variational fluid motion estimation with physical priors. Ph.D. thesis, Mannheim University (2006)
36. Ruhnau, P., Schnörr, C.: Optical Stokes flow estimation: an imaging-based control approach. Exp. Fluids **42**, 61–78 (2007)
37. Ruhnau, P., Stahl, A., Schnörr, C.: Variational estimation of experimental fluid flows with physics-based spatio-temporal regularization. Meas. Sci. Technol. **18**(3), 755–763 (2007)
38. Stoll, M., Wathen, A.: All-at-once solution of time-dependent Stokes control. J. Comput. Phys. **232**, 498–515 (2013)
39. Temam, R.: Navier-Stokes Equations. AMS Chelsea Edition, North-Holland (1977)
40. Weiss, R.: The convergence of shooting methods. BIT **13**, 470–475 (1973)

Statistics for Model Calibration

Clemens Kreutz, Andreas Raue, and Jens Timmer

Abstract Mathematical models of dynamic processes contain parameters which have to be estimated based on time-resolved experimental data. This task is often approached by optimization of a suitably chosen objective function. Maximization of the likelihood, i.e. *maximum likelihood estimation*, has several beneficial theoretical properties ensuring efficient and accurate statistical analyses and is therefore often performed for identification of model parameters.

For nonlinear models, optimization is challenging and advanced numerical techniques have been established to approach this issue. However, the statistical methodology typically applied to interpret the optimization outcomes often still rely on linear approximations of the likelihood.

In this review, we summarize the maximum likelihood methodology and focus on nonlinear models like ordinary differential equations. The profile likelihood methodology is utilized to derive confidence intervals and for performing identifiability and observability analyses.

1 Introduction

Establishing a mathematical model of a process of interest typically comprise the identification of an appropriate model structure as well as calibration of the model's parameters. For phenomenological models, the structure of the model is typically chosen as simple as possible. Often, the models are linear with respect to the parameters, like in the linear regression case, and in this case there is a well-established methodology for estimating the parameters as well as for statistical interpretations.

C. Kreutz (✉) • A. Raue
Institute of Physics, University of Freiburg, Hermann-Herder Str. 3, 79104 Freiburg, Germany
e-mail: ckreutz@fdm.uni-freiburg.de; andreas.raue@fdm.uni-freiburg.de

J. Timmer
BIOSS Centre for Biological Signalling Studies and Institute of Physics, University of Freiburg, Hermann-Herder Str. 3, 79104 Freiburg, Germany
e-mail: jeti@fdm.uni-freiburg.de

T. Carraro et al. (eds.), *Multiple Shooting and Time Domain Decomposition Methods*, Contributions in Mathematical and Computational Sciences 9,
DOI 10.1007/978-3-319-23321-5_14

In some applications, however, the models are mechanistic, i.e. the variables and parameters of the model have their counterparts in the process of interest. In chemical engineering or in systems biology, for example, ordinary differential equation models are applied to describe chemical or biochemical reactions. Here, the dynamics is given by chemical rate laws. The dynamic variables typically represent concentrations of compounds and parameters are used for unknown rate constants and initial concentrations.

Estimating parameters for such nonlinear models, requires advanced technique for optimization techniques like trust-region approaches [8, 32] or multiple shooting [5, 6]. In addition, the methodology required for reliable statistical interpretation of the optimization result has to be adapted. Confidence intervals, as an example, are, in general, not appropriately represented by the covariance or Fisher information matrix. In addition, commonly performed tasks like model predictions or identifiability analyses of parameters are getting more complex to be done appropriately from the statistical point of view.

2 Measurement Errors

The central limit theorem of statistics [4, 12] states that the distribution of a sum

$$\varepsilon = \sum_i^m \rho_i \tag{1}$$

of independent, identically distributed sources of noise ρ_i converges to a normal distribution $\varepsilon \sim N(\mu, \sigma^2)$ for $m \to \infty$. This theorem holds independent of the specific form of the distribution ρ_m. It is only required that the distributions of $\rho_1, \ldots, \rho_m$ have finite expectations and variances. Due to the central limit theorem, the observational noise ε is very commonly assumed as normally distributed or briefly as *Gaussian*. Additive noise yields measurements

$$y_i = f_i + \varepsilon_i \ , \ \varepsilon_i \sim N(0, \sigma^2) \tag{2}$$

where f_i denotes the model response, i.e. represents the true values of an underlying process. The different experimental conditions are labeled by $i = 1, \ldots, n$. Each experimental condition i is characterized by the values for all predictor variables in the model, e.g. time, position, experimental treatment, or the sample entity.

Experimental biomedical research often boils down to quantification of a compound of interest. Typically, an experimental protocol comprises a sequence of steps, each introducing a measurement error with magnitude relative to the quantity of interest. In analogy to the central limit theorem, one can show that a sequence

$$\eta_1 \propto f \rho_1 \tag{3}$$

$$\eta_m \propto \eta_{m-1} \rho_m \tag{4}$$

of relative errors $\eta_m - \eta_{m-1}$ rapidly converges to a lognormal distribution.

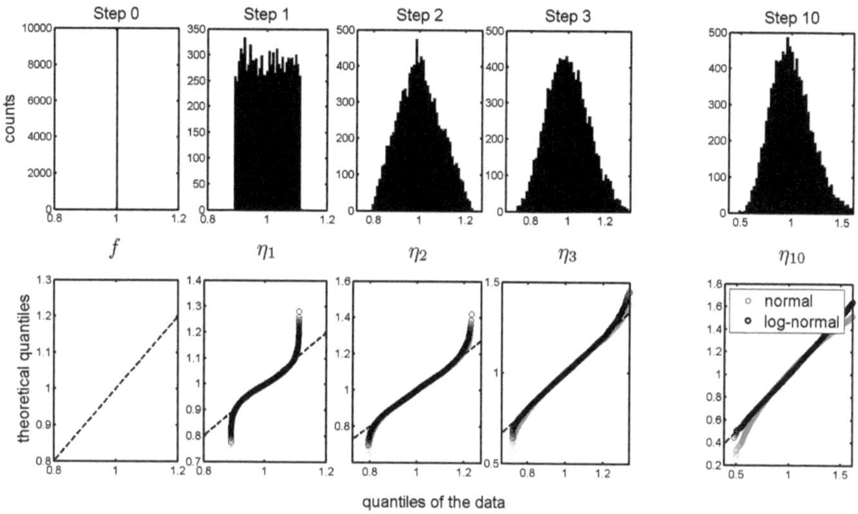

Fig. 1 Illustration showing that a sequence of relative errors yield lognormally distributed errors. The *upper row* shows histograms, the *lower row* depicts quantile-quantile plots of the observations vs. the lognormal distribution (*black*) and vs. the normal distribution (*gray*). Step 0 shows the underlying truth to be measured which has been chosen as 1 in this example. Step 3 already yields a distribution which is close to a lognormal distribution. For increasing number of relative errors, the observed distribution converges to a lognormal distribution

In Fig. 1, this is illustrated for uniformly distribution of ρ_i. After around three steps, the observation η_3 has almost lognormal distribution. For η_{10} there is hardly a difference to the lognormal distribution. In general, the rate of convergence of the central limit theorems depends on the shape of ρ. However, lognormally distributions are very often observed in practice since (1) often holds for the individual relative errors. Another hint for lognormally distributed errors is the fact that raw measurements, i.e. before a background is subtracted, are often strictly positive. In such cases, normally distributed errors are obtained after a log-transformation of the data and the transformed data can be analyzed by the classical statistical methodology established for Gaussian errors.

3 Likelihood

The *likelihood* is defined as the probability $\rho(y|\theta)$ of the data y for a given model with parameters θ [2]. For independent noise realizations, the probability is a product

$$\rho(y|\theta) = \prod_{i}^{N} \rho(y_i|\theta) \tag{5}$$

of terms $\rho(y_i|\theta)$ for the individual data points y_i. The *maximum likelihood estimator*

$$\hat{\theta} = \arg\max_{\theta} L(y|\theta) \tag{6}$$

is given by the parameters maximizing the likelihood. Maximum likelihood estimation is asymptotically unbiased, i.e. for a sufficiently large number N of data points $y_1, \ldots, y_N$ the expectation

$$\lim_{N\to\infty} E(\hat{\theta}) = \theta_{\text{true}} \tag{7}$$

coincides with the true parameters θ_{true}. Moreover, the maximum likelihood estimator has the minimal variance within all unbiased estimators [9].

For Gaussian errors, the likelihood is given by

$$L(y|\theta) = \prod_i^N \frac{1}{\sqrt{2\pi}\sigma} \exp\left(-\frac{(y_i - f_i)^2}{2\sigma^2}\right) \tag{8}$$

and minus two times the log-likelihood is

$$-2\text{LL}(y|\theta) = \sum_i^N \frac{(y_i - f_i)^2}{\sigma^2} + 2N\log(\sigma) + N\log(2\pi) \ . \tag{9}$$

Since only the first term depends on the parameters and minimization of (9) is equivalent to maximization of (8) *least-squares estimation*

$$\hat{\theta} = \arg\min_{\theta} -2\text{LL}(y|\theta) \tag{10}$$

is equivalent to maximum likelihood estimation for Gaussian measurement errors [27].

4 Estimation for Linear Models

The maximum likelihood methodology is very general, i.e. every distributional assumption for the measurement uncertainty can be considered. In this section, the discussion will be focused on least-squares estimation, i.e. on maximum likelihood for the special case of Gaussian measurement errors since due to the central limit theorems, measurement errors are often Gaussian at the measurement or the logarithmic scale as argued in Sect. 2.

If the model f is linear with respect to its parameters, it can be written as

$$f = X\theta \tag{11}$$

Example1: The design matrix for a model $f(x) = \theta_0 + \theta_1 x + \theta_2 \exp(-x)$ for measurements at $x = \{0, 1, 2, 3\}$ is given by

$$X := \begin{pmatrix} 1 & 0 & 1 \\ 1 & 1 & \exp(1) \\ 1 & 2 & \exp(2) \\ 1 & 3 & \exp(3) \end{pmatrix} \tag{12}$$

since then the matrix product $X\theta$ coincides with $f(x)$. Note that X contains the experimental design [17], i.e. the choice of x determines X.

Example2: For a *one-way analysis of variance (ANOVA)*, i.e. for testing whether there is a significant variation between n_g groups of data points with $N_1, \ldots, N_{n_g}$ replicates in each group, the design matrix is given by

$$X := \underbrace{\begin{pmatrix} 1 & 0 & \ldots & 0 \\ \vdots & \vdots & & \vdots \\ 1 & 0 & \ldots & 0 \\ 1 & 1 & \ldots & 0 \\ \vdots & \vdots & & \vdots \\ 1 & 1 & \ldots & 0 \\ \vdots & \vdots & & \vdots \\ 1 & 0 & \ldots & 1 \\ \vdots & \vdots & & \vdots \\ 1 & 0 & \ldots & 1 \end{pmatrix}}_{n_g \text{ columns}} . \tag{13}$$

Here, there is a parameter for the overall average represented by the first column of X and $n_g - 1$ parameters for each group means. These parameters have a corresponding column in X with ones for data points belonging to the respective group and zeros otherwise.

Fig. 2 Two examples for design matrices

in matrix notation where the N data points $y \in N \times 1$ are given as a vector [29]. The rows of the so-called *design matrix* $X \in N \times n_\theta$ indicate the impact of the parameter on each single data point. In the Fig. 2, two basic examples are presented for illustrating the design matrix.

Least-squares estimates for a linear model are obtained via multiplication

$$\widehat{\theta} = X^\dagger y \tag{14}$$

of the data y with the *pseudo-* or *generalized inverse*

$$X^\dagger := \left(X^\top X\right)^{-1} X^\top \; \in M(n_\theta \times N) \tag{15}$$

of the design matrix X [12, 28]. For the linear case, the parameter estimates are normally distributed

$$\widehat{\theta} \sim N(\theta, \Pi) \tag{16}$$

with covariance matrix

$$\Pi = \sigma^2 (X^\dagger X^{\dagger\top})^\top = \sigma^2 (X^\top X)^{-1} \ . \tag{17}$$

The covariance matrix of the parameter estimates $\hat{\theta}_k$ yields point-wise *confidence intervals* or *standard errors*

$$\mathrm{SE}(\hat{\theta}_k) = \sqrt{\Pi_{ii}} \ . \tag{18}$$

The standard error

$$\mathrm{SE}(\hat{\theta}_k) \propto \sigma \tag{19}$$

is proportional to the standard deviation of the measurement error for linear models. If a set of N measurements is repeated n times, the standard error decreases

$$\mathrm{SE}(\hat{\theta}_k) \propto \frac{1}{\sqrt{n}} \tag{20}$$

proportional to one over the square root of the number n of repetitions.

4.1 Confidence Intervals

A confidence interval CI represents the uncertainty of an estimate [11]. The *confidence level* α controls the probability that the true parameter is inside the respective confidence intervals. If, as an example, $\alpha = 0.95$ is chosen, then the true parameters are expected to be within the confidence interval with a probability of 95 %.

For known σ^2, the *standard score* or *z-score*

$$z = \frac{\hat{\theta}_k}{\mathrm{SE}(\hat{\theta}_k)} \sim N(\theta_{k,\mathrm{true}}, 1) \tag{21}$$

is standard-normally distributed around the true parameter $\theta_{k,\mathrm{true}}$ which can utilized to define a more general confidence interval

$$\mathrm{CI}_{\hat{\theta}_k}(\alpha) = \left\{ \theta_k \mid \alpha/2 \leq \mathrm{cdf}_{N(0,1)} \left(\frac{\hat{\theta}_k}{\mathrm{SE}(\hat{\theta}_k)} - \theta_k \right) \leq 1 - \alpha/2 \right\} \tag{22}$$

for the parameter $\hat{\theta}_k$.

Equation (22) provides confidence intervals for a single parameter. For several parameters, the multivariate normal distribution MVN can be utilized, i.e. a *joint* confidence interval for several parameter components θ_k, $k \in K$, $|K| > 1$ is given by [39]

$$\mathrm{CI}_{\hat{\theta}_{k\in K}}(\alpha) = \left\{\theta_{k\in K} \mid \alpha/2 \leq \mathrm{cdf}_{\mathrm{MVN}(\hat{\theta}_k, \Pi(K,K))}(\theta_{k\in K}) \leq 1-\alpha/2\right\} \ . \tag{23}$$

In the multivariate case, the cumulative density function $\mathrm{cdf}(\theta) \equiv \mathrm{cdf}_\rho(\theta)$ of the probability density ρ at a point θ is defined as the integral

$$\mathrm{cdf}(\theta) = \int_{\{x|\rho(x)<\rho(\theta)\}} \rho(x')\ \mathrm{d}x' \tag{24}$$

over the region where the density $\rho(x)$ is smaller than $\rho(\theta)$.

Equation (22) is based on knowledge about the distribution of the estimated parameters. This concept holds true for linear models. For nonlinear models, it only holds in an asymptotic setting, i.e. for a large number of data points. A more powerful concept which is also applicable in nonlinear settings is based on the distribution of the log-likelihood or of differences of the log-likelihood. Figure 3 illustrates this approach. If N data points are simulated with given parameters, the log-likelihood follows a chi-squared distribution

$$-2\,\mathrm{LL}(y|\theta) \sim \chi^2_N \tag{25}$$

with N degrees of freedom.

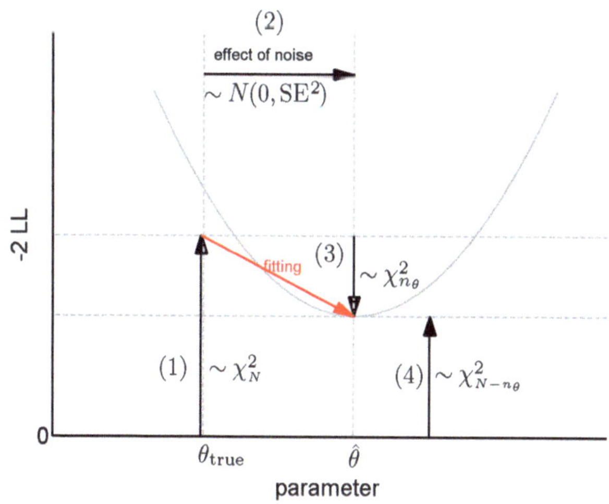

Fig. 3 -2 LL in the simulation setting is χ^2_N distributed (1). For linear models, the log-likelihood is quadratic. Because of measurement errors, the minimum is shifted from the true parameters θ to $\hat{\theta}$ according to a normal distribution (2). Fitting of each model parameter decreases the log-likelihood by a single degree of freedom, i.e. for n_θ parameters -2 LL decreases according to the $\chi^2_{n_\theta}$ distribution (3). Then minimum is then $\chi^2_{N-n_\theta}$-distributed (4)

If n_θ parameters are fitted to such data, the decrease log-likelihood, i.e. the log-likelihood ratio

$$-2\,\mathrm{LR}(\hat{\theta},\theta) := -2\left(\mathrm{LL}(y|\theta) - \mathrm{LL}(y|\hat{\theta})\right) \tag{26}$$

is again given by the chi-squared distribution

$$-2\,\mathrm{LR}(\hat{\theta},\theta) \sim \chi^2_{n\theta} \tag{27}$$

with n_θ degrees of freedom. This decrease can be reversed, i.e. the range in the parameter space where -2LR does not exceed the α-quantile of the $\chi^2_{n\theta}$-distribution can be evaluated for defining confidence intervals. The true parameters are expected to be inside a confidence interval given by

$$\mathrm{CI}_{\hat{\theta}}(\alpha) = \left\{\theta \mid -2\,\mathrm{LR}(\hat{\theta},\theta)) \leq \mathrm{icdf}_{\chi^2_{n\theta}}(\alpha)\right\} \ . \tag{28}$$

with probability α. This equation means, that in -2 LR-"units", $\mathrm{icdf}_{\chi^2_{n\theta}}(\alpha)$ provides a threshold for confidence intervals of n_θ parameters. Equation (25) only holds for Gaussian errors. In contrast, (27) which is the basis for the so-called *likelihood ratio test*, holds under much weaker assumptions as discussed in Sect. 5.

The symbol n_θ for the number of fitted parameters refers to the effective number of parameters in terms of degrees of freedom which are specified by the fitting process. This number can deviate from the dimension of the optimization problem. For multiple-shooting optimization methods, as an example, the time domain is decomposed into several intervals and the initial values within each interval are given by new auxiliary parameters which increase the dimension of the optimization problem. However, since for each auxiliary parameter a single continuity constraints is introduced in parallel, this technical reformulation of the optimization problem has no impact on the estimated parameters and therefore neither contribute to n_θ nor enter the statistical interpretations. A second example where n_θ differ from the dimension of the optimization problem are redundant parameterizations of models as discussed in Sect. 4.3. In such circumstances, n_θ refers to the effective number of fitted parameters, i.e. redundant or non-identifiable parameters are not counted.

If a confidence intervals for a subset $k \in K$ of parameters is of interest, then the likelihood ratios are calculated by optimizing all other parameters. For a single parameter component, this approach yields the so-called *profile likelihood* [34]

$$\mathrm{PL}_k(p) = \min_{\theta \in \{\theta|\theta_k = p\}} -2\,\mathrm{LR}(\theta,\hat{\theta}) \ , \tag{29}$$

which is the log-likelihood ratio (26) as a continuous function of a parameter component $\theta_k = p$. As depicted in Fig. 4, the confidence interval for a single

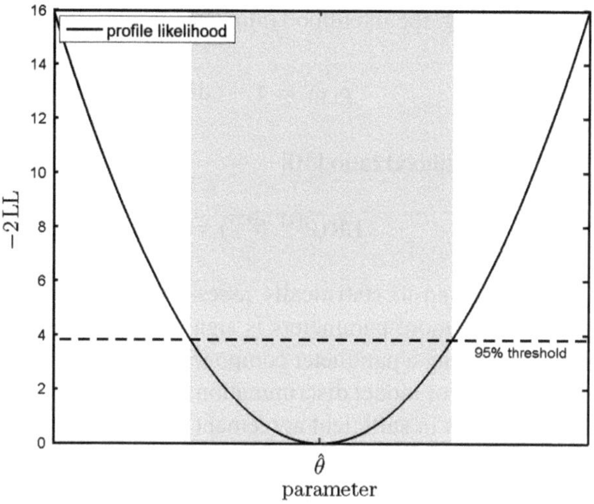

Fig. 4 Applying a threshold given by the χ_1^2-distribution to the profile likelihood yields confidence intervals for single parameter components indicated by the *gray* background color. The minimum $\min_\theta -2\text{LL}$ has been subtracted for illustration purpose. Since the profile likelihood is invariant under nonlinear transformations of the parameters, it is also applicable for nonlinear models

parameter is given by the region [22]

$$\text{CI}_k(\alpha) = \left\{ p \mid -2\,\text{PL}_k(p) \leq \text{icdf}_{\chi_1^2}(\alpha) \right\} \tag{30}$$

where the increase of the profile likelihood is smaller than the α-quantile of the χ_1^2-distribution.

4.2 Model Discrimination for Known Measurement Errors

Statistical assessment of the agreement of two or several models with experimental data is termed *model discrimination* or *model selection* in literature [1, 7, 37, 38]. We restrict the discussion on the setting, where one model is a special case, i.e. a valid simplification of the other, since in this case reliable statistical tests are available.

Let's assume, there are two linear models

$$f^{(1)} = X^{(1)}\theta^{(1)} \tag{31}$$

$$f^{(2)} = X^{(2)}\theta^{(2)} \tag{32}$$

with $n_\theta^{(1)}$ and $n_\theta^{(2)}$ parameters.

If the two models are *nested*, i.e. $f^{(1)}$ is a special case of $f^{(2)}$, the null model $f^{(1)}$ is obtained by setting some parameters $k \in K$ of the alternative model $f^{(2)}$ to specific values

$$\theta_k^{(2)} = \Theta_k \quad \text{for } k \in K\ . \tag{33}$$

In this setting, the likelihood ratio test

$$p_{\mathrm{LRT}} = 1 - \mathrm{cdf}_{\chi^2_{\Delta n}}\left(-2\,\mathrm{LR}(\hat{\theta}^{(1)}, \hat{\theta}^{(2)})\right) \tag{34}$$

with the likelihood ratio [10]

$$\mathrm{LR}(\hat{\theta}^{(1)}, \hat{\theta}^{(2)}) = \mathrm{LL}(y|\hat{\theta}^{(1)}) - \mathrm{LL}(y|\hat{\theta}^{(2)}) \tag{35}$$

can be applied to statistically assess, whether the larger model $f^{(2)}$ with $\Delta n = n_{\theta^{(2)}} - n_{\theta^{(1)}}$ more parameters is significantly better than the simplified model $f^{(1)}$ [9, 23]. Testing a parameter component for being agreement with a specific value is an example for model discrimination, i.e. statistically testing whether a special case of a model is in sufficient agreement with available data. If $p < 1 - \alpha$, the smaller model would be significantly rejected by the test according to a significance level $1 - \alpha$.

The likelihood ratio test (34) is equivalent to profile likelihood based confidence intervals (30). In analogy, also Eq. (22) can be utilized to calculate p-values for statistically testing a specific null hypothesis $\theta_k \equiv \theta_0$. However, since (22) only holds for linear models, the p-value

$$p = 2 \cdot \min\left[\mathrm{cdf}_{N(0,1)}\left(\frac{\hat{\theta}_k}{\mathrm{SE}(\hat{\theta}_k)} - \theta_0\right),\ 1 - \mathrm{cdf}_{N(0,1)}\left(\frac{\hat{\theta}_k}{\mathrm{SE}(\hat{\theta}_k)} - \theta_0\right)\right] \tag{36}$$

based on the cumulative density function of $\mathrm{cdf}_{N(0,1)}$ of the standard normal distribution are also restricted to the linear setting.

4.3 Identifiability

The rank of $X^\top X$ determines the number of identifiable parameters, i.e. if $X^\top X$ has full rank, it is invertible and there is a unique estimate for all parameters. Then, standard errors given by (17) and (18) have finite size. Since standard errors for linear models only depend on the design X, identifiability is independent on the true underlying parameters θ.

For non-identifiable parameters, the profile likelihood, Eq. (29), is entirely flat and there is no unique estimate for the parameters [17, 30]. The existence of such non-identifiable parameters is also called *confounding* for linear models in the statistical literature. Decreasing the measurement error, e.g. by experimental repetitions of the same conditions, does not affect the rank of $X^\top X$ and therefore does not qualitatively change structural identifiability properties.

4.4 Heteroscedastic Data

Up to now, a single, known measurement error σ was assumed for all data points. If the data y_i have different, known measurement errors, i.e. $\sigma \rightarrow \sigma_i$, it is convenient to account for the magnitude of observational noise by rescaling the measurements

$$y_i' := \frac{y_i}{\sigma_i} \tag{37}$$

as well as the model predictions

$$K\theta := \frac{X\theta}{\sigma} \tag{38}$$

i.e. the matrix elements of K are given by

$$(K)_{ik} = \frac{(X)_{ik}}{\sigma_i} \ . \tag{39}$$

Then, equivalently to Eq. (14), least squares parameter estimates are obtained by multiplication of the rescaled data

$$\widehat{\theta} = K^{\dagger} y' \tag{40}$$

with the pseudo inverse of K and the covariance matrix (17) is given by

$$\Sigma = \left(K^{\dagger} K^{\dagger\top}\right)^{\top} = \left(K^{\top} K\right)^{-1} \ . \tag{41}$$

In this so-called *heteroscedastic* case, only the rescaling Eqs. (37) and (39) changes, although the structure of Eqs. (40) and (41) remain unchanged compared to the homoscedastic situation.

4.5 Model Discrimination for Unknown Measurement Error

If there is a single, but unknown variance of the measurement error, σ^2 has been estimated from the same data. Then, the ratio

$$\frac{\hat{\theta}_k}{\mathrm{SE}(\hat{\theta}_k)} \sim t_{\mathrm{df}} \tag{42}$$

is t-distributed with $df = N - n_\theta$ degrees of freedom. In the case of confounding, the number of parameters n_θ is replaced by $rank(X^\top X)$ for the calculation of the degrees of freedom. Note that the maximum likelihood estimate of the measurement error variance

$$\hat{\sigma}^2 = \frac{1}{N}\sum_{i=1}^{N}\left(y_i - f_i(\hat{\theta})\right)^2 \tag{43}$$

is biased in the finite sample case and is replaced by

$$\hat{\sigma}^2 = \frac{1}{df}\sum_{i=1}^{N}\left(y_i - f_i(\hat{\theta})\right)^2 \tag{44}$$

to obtain an unbiased estimate. The estimates $\hat{\theta}$ are not affected by the fact that the sampling variance is unknown and using (44), standard errors are again given by (18).

The cumulative density function $\mathrm{cdf}_{t_{\mathrm{df}}}(\hat{\theta}_k/\mathrm{SE}(\hat{\theta}_k) - \theta_{k,\mathrm{true}})$ of the t-distribution can be evaluated to test whether a parameter θ_k is significantly different from a specific value $\theta_{k,\mathrm{true}}$. p-value for the *two-sided test* of the null hypothesis that $\theta_k = \Theta_k$ are given by the extremes of the lower and upper quantiles

$$p = 2 \cdot \min\left[\mathrm{cdf}_{t_{\mathrm{df}}}\left(\frac{\hat{\theta}_k - \Theta_k}{\mathrm{SE}(\hat{\theta}_k)}\right), 1 - \mathrm{cdf}_{t_{\mathrm{df}}}\left(\frac{\hat{\theta}_k - \Theta_k}{\mathrm{SE}(\hat{\theta}_k)}\right)\right] . \tag{45}$$

More general model discrimination hypotheses can be tested based on the ratio

$$F = \frac{\left(\frac{\left(y - X^{(1)}\hat{\theta}^{(1)}\right)^2}{\sigma^2} - \frac{\left(y - X^{(2)}\hat{\theta}^{(2)}\right)^2}{\sigma^2}\right) / \left(n_\theta^{(2)} - n_\theta^{(1)}\right)}{\left(\frac{\left(y - X^{(2)}\hat{\theta}^{(2)}\right)^2}{\sigma^2}\right) / \left(N - n_\theta^{(2)}\right)} \tag{46}$$

which is known to be F-distributed

$$F \sim f_{n_\theta^{(2)} - n_\theta^{(1)}, n_\theta^{(2)}} . \tag{47}$$

Because a possibly unknown sampling variance σ^2 cancels out, this result is very useful for an unknown measurement error. F-tests are commonly applied in the setting of *analysis of variances (ANOVA)* to test specific null hypothesis. Note, that the distributional statements in Eqs. (42) and (47) are exact, i.e. they hold for any number of repetitions and does not require the *asymptotic* assumption like the likelihood ratio test. Asymptotically, i.e. for large number of data, both Eqs. (42) and (47) become equivalent to the χ^2 and both tests become equivalent to the *likelihood ratio test*.

If the data is heteroscedastic with several unknown measurement errors, so-called mixed effects models have to be applied for the analysis of the data [25]. This topic, however, is beyond the scope of this article.

5 Nonlinear Models

For nonlinear models, the log-likelihood can have arbitrarily complex shape [35]. This alters the following aspects which are discussed in the following sections.

1. Least-squares estimation is not possible by generalized matrix inversion, i.e. minimizing the least-squares objective function has to be performed numerically. A variety of methods have been developed for performing numerical optimization in such a setting.
2. The likelihood can exhibit local minima.
3. The parameter estimates are not normally distributed and confidence intervals are typically not symmetric.
4. A unique maximum likelihood estimates does not guarantee for identifiability since the confidence intervals can have infinite size. This aspect is called *practical non-identifiability* in the literature [30, 31].

5.1 Numerical Optimization

Numerical optimizations techniques are applied if minimization or maximization is not feasible analytically. For nonlinear systems, the computational efficiency and the reliability of numerical convergence is problem-dependent, e.g. dependent on the model and on parameter transformations, and there are no algorithms working in all circumstances. For statistical analysis, the choice of the numerical optimization method, e.g. for solving (10) or (29) only has an impact on the computational efficiency and on robustness of convergence but has no impact on the results because competing numerical algorithms converge to the same parameters.

There is a huge set of numerical approaches for the solution of nonlinear optimization problems [27]. For oscillatory models, *multiple-shooting* approaches [5, 6] have beneficial performance since the problem of local minima is diminished by decomposition of the time domain. In addition, multiple-shooting allows the incorporation of prior knowledge of the dynamics by a reasonable initialization of *multiple-shooting intervals* [3]. This leads to a good initial guess as starting point for optimization which improves the performance of the method. The convergence of multiple shooting is discussed in more detail in [14, 20, 24].

In contrast, if most dynamic variables are unobserved and there is no reliable guess for the dynamics of these variables, initialization of the multiple shooting intervals becomes difficult. This situation occurs, as an example, in systems biology applications since many cellular compounds are often not experimentally accessible. In such applications, the standard *single-shooting* approaches like the *trust-region* [8] or *Levenberg-Marquardt* approaches [21, 27] can be applied, especially if the dynamics is expected to be transient.

If optimization is performed based on ordinary differential equation models, the following aspects should be considered

- Most ordinary differential equations (ODE) can only be solved numerically with a certain accuracy which is controlled to some extent by the tolerances of numerical ODE solvers. Calculating derivatives based on *difference quotients*, is often not feasible since the integration step control depends on the parameters and numerical integration errors are inflated by dividing through a small parameter step size [32]. Then, the ODE system can be augmented by the *sensitivity equations* to calculate derivatives with respect to the parameters [19].
- A strategy has to be chosen to deal with potential local minima as discussed in the following Sect. 5.2.
- Often, optimization problems are numerically more efficiently solved, if the exponents of parameters are fitted, i.e. if optimization is performed at a logarithmic scale. This especially holds if parameters are known to be positive like rate constants or initial concentrations.

5.2 Local Minima

Nonlinearity of a model can yield to local minima of the objective function. For such cases, *stochastic optimization approaches* like *simulated-annealing* [13], *particle-swarm* [26], or *evolutionary algorithms* [18] have been proposed. To mathematically guarantee convergence, these methods combine exploration of the parameter space by random sampling with exploitation of the local geometry of the objective function by approximations of the derivatives of the objective function with respect to the parameters.

In practice, however, stochastic optimization methods often show slow convergence, especially for larger problems with more than 10^1–10^2 parameters [32]. In addition, stochastic optimization methods have hyper-parameters, e.g. to heuristically combine the stochastic sampling strategy with the local descent method. An efficient choice of these hyper-parameters is usually difficult since insights about the optimization problem are required which are typically not available *a priori*.

For systems biology applications, the dynamics is typically smooth at the time scale of the measurements. Then, also the objective function is smooth and deterministic optimization approaches can be efficiently applied to find local minima. Since in most dynamic variables are typically not observed, multiple shooting is difficult to be applied.

A combination with a stochastic initialization approach like *latin-hypercube sampling* works efficiently in a systems biology setting and has been awarded twice as best performing in systems biology benchmark challenges for parameter estimation and network reconstruction [36]. In this challenge, the so-called *Data2Dynamics* framework [33] was utilized which provides an efficient implementation for fitting based on the Matlab's nonlinear least-squares estimation routine "lsqnonlin". This routine, in turn, implements a deterministic trust-region-reflective algorithm of Gauss-Newton optimization.

5.3 Asymmetry of Confidence Intervals

For linear models with Gaussian measurement errors or for nonlinear models in an asymptotic setting, the parameter estimates are normally distributed and the log-likelihood has a quadratic shape as shown in Fig. 5, panel (a). Then, a quadratic approximation of the likelihood at the maximum likelihood estimates $\hat{\theta}$ as provided by the *observed Fisher-Information*

$$F_{ij} = \frac{\partial^2}{\partial\theta_i \partial\theta_j} \left(-2\,\mathrm{LL}(y|\theta)|_{\hat{\theta}}\right) \tag{48}$$

and the related covariance matrix

$$\Pi = F^{-1} \tag{49}$$

of the estimated parameters provides reasonable confidence intervals (see panel (c)).

The effect of a nonlinearity is easily demonstrated by a reparametrization

$$\theta_{\mathrm{nonlinear}} = \log\left(\theta_{\mathrm{linear}}\right) \tag{50}$$

of a linear parameter θ_{linear}. The reparameterization yields an asymmetric distribution of the parameter estimates (see panel (b)) as well as a distortion of the likelihood shape as depicted in panel (d).

Profile likelihood based confidence intervals are independent on monotone transformations of the parameters since the region below a threshold is not affected by distortions in horizontal direction. Therefore, the profile likelihood based confidence intervals indicated in panels (c) and (d) are equivalent. In addition, existence of a monotone transformation yielding a quadratic shape of the profile likelihood is sufficient to guarantee that the threshold can be chosen from the chi-squares distribution according to Eq. (30). The transformation is only required locally, i.e. only for the region in the parameter space which is in statistical agreement with the data. If there are several minima which are in sufficient agreement with the data, or if the confidence interval cover range a boundary of the parameter space, such a transformation does not exist and then Eq. (30) only holds asymptotically.

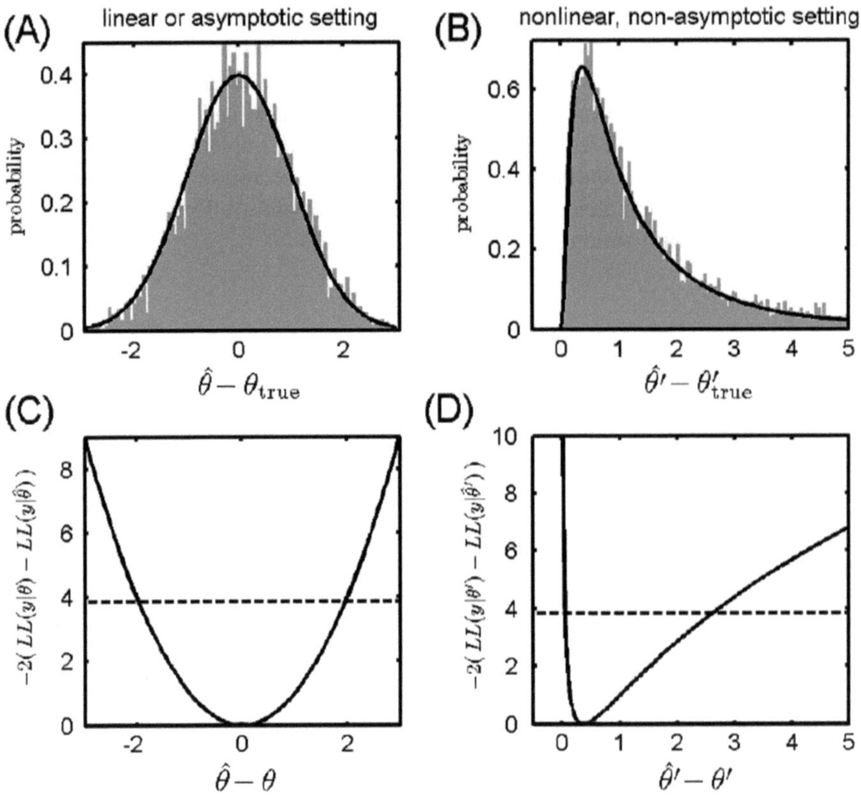

Fig. 5 Illustration of effect of nonlinearity of a model parameter on the distribution of the parameter estimates (*upper row*) and on the shape of the likelihood (*lower row*). Panel (**a**) shows a Gaussian distribution of the maximum likelihood estimates which is obtained for many data realizations in a linear or asymptotic setting. A nonlinear parameter as shown in panel (**b**), in general has an asymmetric distribution in the non-asymptotic case. For this plot, the nonlinear parameter was defined as $\theta' := \exp(\theta)$. In panel (**c**), the likelihood is plotted for the linear parameter. If a model has several parameters, the depicted *curve* corresponds to the profile likelihood. Panel (**d**) shows the profile likelihood for the nonlinear parameter θ'. Confidence intervals are obtained from the curves in (c) and (d) by applying a threshold according to Eq. (30) as indicated by the *dashed lines*

5.4 *Identifiability Analysis*

Non-identifiability refers to situations, where reliable estimation of parameters is not feasible. In the linear setting discussed in Sect. 4.3, the parameters are either identifiable or *structurally non-identifiable*. Structural non-identifiability means that the likelihood does not attain its global optimum at a unique point. In fact, there is a manifold where the likelihood is optimal. If the dimension of this manifold is d, then $n_\theta - d$ parameters are identifiable.

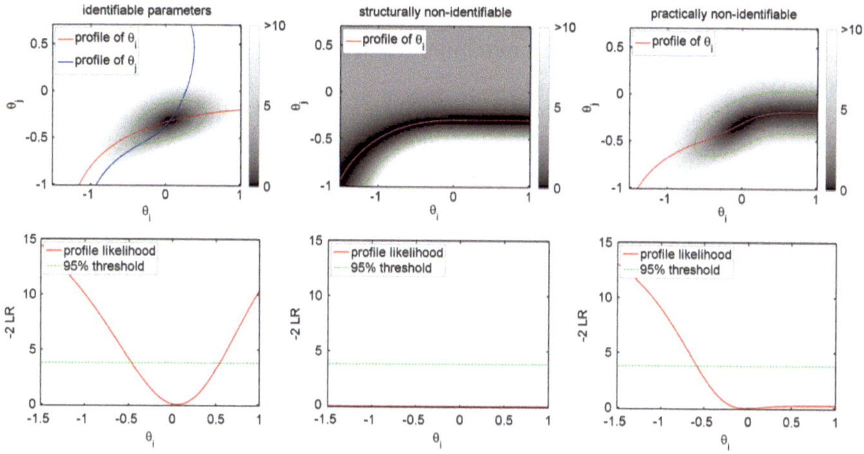

Fig. 6 Illustration of the three qualitative identifiability scenarios. In the *left column*, both parameters θ_i and θ_j are identifiable. Here, the log-likelihood ratios -2 LR has a unique minimum (*upper left panel*) as indicated by the different colors. Then, the profile likelihood (*red line*) shown in the lower left panel exceeds the confidence threshold (*green line*) in upper and lower directions. The *red* and *blue lines* in the upper left panel are the parameters evaluated for calculating the profile likelihood of the two shown parameters θ_i and θ_j. In the *middle*, an example for structural non-identifiability is depicted. Here, there is a flat direction in the likelihood which yields a perfectly flat profile likelihood for parameter θ_i (*lower panel in the middle*). On the *right*, a practical non-identifiability is shown. Although the likelihood has a unique minimum, it is flat for large values of θ_i. Therefore, the confidence interval for θ_i is not restricted in upper direction and has infinite size

For nonlinear models, also several points in the parameter space can have the same optimal likelihood. As a simple example, let's consider a case where only θ_k^2 enters the likelihood, then θ_k and $-\theta_k$ fits the data equally well and if all other parameters are identifiable, there are exactly two point in the parameter space maximizing the likelihood.

As an additional scenario for nonlinear models, there could be a unique maximum, but the confidence interval has infinite size. This happens if the profile likelihood does not exceed the confidence threshold in lower and/or upper direction.

Figure 6 illustrates the identifiable setting (left column) as well as structural (middle) and practical non-identifiability (right column). The colors in the upper panels indicate likelihood ratios and the green lines correspond to point-wise 95 % thresholds.

5.5 *Prediction and Observability Analysis*

A primary purpose of modeling is prediction, i.e. forecasting a system's behavior under specific conditions of interest. A model prediction is any characteristic which is calculated based on the model. Parameter estimation can be seen as a special case of performing a model prediction. Other examples for predictions are extrapolations of the dynamics like predicting a maximum value, a steady state, or the dynamics of unobserved components.

The profile likelihood methodology used to calculated confidence intervals for parameters can be adapted to the more general prediction setting. This can be done by a reparametrization of the model, i.e. by replacing a parameter by the prediction. For ODEs, however, this is usually not feasible since integration is not feasible analytically.

Nevertheless, the likelihood can be maximized in a constrained manner, i.e. a predicted value is fixed and the parameters are optimized under such a constraint. Repeating this procedure for continuous variations of the constraint yields an implicit reparametrization and the prediction profile likelihood [16]. As demonstrated in the next section, the results look similar to the parameter profile likelihood. The major difference is that the profile likelihood curve is calculated for the prediction on the horizontal axis. Confidence intervals for the predictions are calculated by the same thresholds as in the parameter estimation case. The reliability and limitation of the threshold is discussed in [16]. In the prediction setting, a totally flat profile likelihood can be interpreted as structural non-observability. A unique prediction with an infinite size of the confidence interval has been termed as practical non-observability [16].

6 Example

Let's assume for the illustration purpose, a small model consisting of two consecutive reactions

$$A \overset{\theta_1}{\to} B \overset{\theta_2}{\to} C \tag{51}$$

between three compounds A, B, and C. Let's assume further that $C(t)$ is measured at $t = 0, 10, \ldots, 100$. Panel (a) in Fig. 7 shows the dynamics of this system rates $\theta_1 = 0.05, \theta_2 = 0.1$ and initial conditions $A(0) = \theta_3 = 1, B(0) = 0, C(0) = 0.$ as well as a data realization generated with Gaussian noise $\varepsilon \sim N(0, \sigma^2)$ with $\sigma = 0.1$ corresponding to a typical signal-to-noise ratio for applications in cell biology of around 10 %.

Let's assume, the peak location of $B(t)$ is intended to be predicted. Fitting the parameters by the maximum likelihood methodology according to Eq. (10) yields $\hat{\theta}_1 = 0.065, \hat{\theta}_2 = 0.065$, and $\hat{\theta}_3 = 1.03$ and predicts a peak location $t_{\text{peak}} = 15.48$. According to Eqs. (18), (48), and (49), the curvature of the likelihood provides a standard error

$$SE(t_{\text{peak}}) = \sqrt{\left(\frac{\partial^2}{\partial t_{\text{peak}}^2} - 2\,\text{LL}(y|\hat{\theta})\right)^{-1}} = 1.97 \tag{52}$$

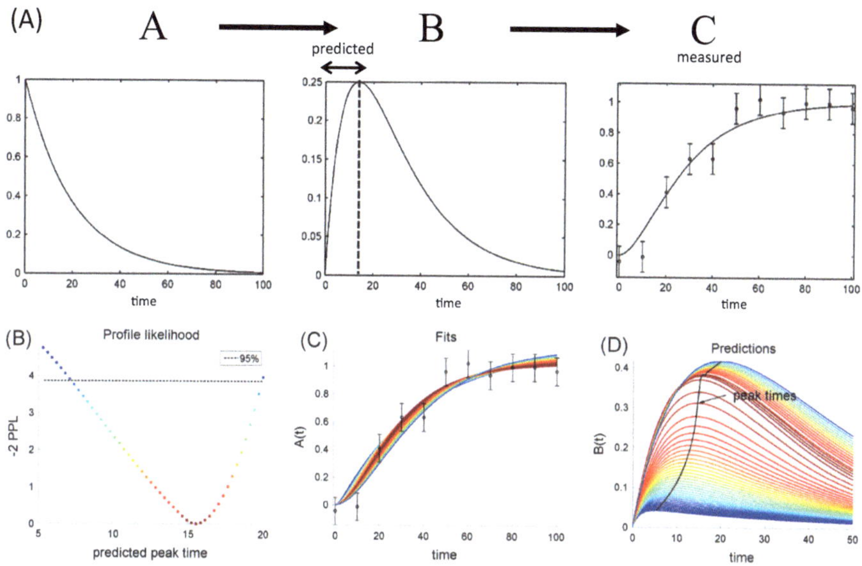

Fig. 7 Panel (**a**) shows the dynamics of the compounds A, B and C as well as the measurements for C. The goal of our illustration analysis is the prediction of the peak time of $B(t)$. The prediction profile likelihood depicted in panel (**b**) for the peak time is obtained by iteratively fitting the data for different constraints for the peak position. Panel (**c**) shows these fits. The *colors* indicate -2 LR. In the last panel (**d**), the predictions for $B(t)$ are plotted indicating that the dynamics of B is only weakly constrained by the data. The peak times are highlighted as a *black line*

with a symmetric confidence interval $\hat{\theta}_{\text{peak}} = 15.46 \pm 1.97$ according to a 68.3 % confidence level and $\hat{\theta}_{\text{peak}} = 15.46 \pm 3.86 = [11.60, 19.32]$ according to a more common confidence level of 95 %.

Panel (b) in Fig. 7 shows the prediction profile likelihood for the peak time as discussed in Sect. 5.5. The colors correspond to the fits shown in panel (c) and the predictions of $B(t)$ depicted in panel (d). The profile likelihood accounts for the nonlinear dependency of the fitted likelihood and the peak time. In this example, the 95 % threshold result in an asymmetric confidence interval [7.22, 19.92].

7 Summary

Establishing a mathematical model comprise model discrimination and model calibration. Both, the selection of an appropriate model structure and the identification of model parameters requires appropriate statistical methodology. Typically, models are calibrated by fitting of experimental data. Then optimization is required to perform maximum likelihood estimation which is the most commonly applied fitting technique. In this review, basic approaches for parameter estimation, confidence interval calculation as well as statistical tests based on maximum likelihood have been presented.

For the maximum likelihood methodology, reliable numerical approaches are required to maximize the likelihood. The increasing complexity of models applied in current research challenges the methodology for optimization as well as for statistical analyses. Although many techniques have been established for optimization as well as for statistical interpretations and their performance have been demonstrated, further improvement of the algorithms in terms of efficiency and robustness especially for large and nonlinear systems is still a topic of current research. Improvements of both, approaches for optimization and for statistical interpretations, are highly relevant for applications in current research.

Acknowledgements The authors acknowledge financial support provided by the BMBF-grants 0315766-VirtualLiver, 0316042GLungSysII.

References

1. Atkinson, A.: Likelihood ratios, posterior odds and information criteria. J. Econ. **16**, 15–20 (1981)
2. Azzalini, A.: Statistical Inference: Based on the Likelihood. Chapman and Hall, London (1996)
3. Baake, E., Baake, M., Bock, H., Briggs, K.M.: Fitting ordinary differential equations to chaotic data. Phys. Rev. A **45**(8), 5524–5529 (1992)
4. Billingsley, P.: Probability and Measure, 3rd edn. Wiley, New York (1995)
5. Bock, H.: Numerical treatment of inverse problems in chemical reaction kinetics. In: Ebert, K., Deuflhard, P., Jäger, W. (eds.) Modeling of Chemical Reaction Systems, vol. 18, pp. 102–125. Springer, New York (1981)
6. Bock, H.: Recent advances in parameter identification for ordinary differential equations. In: Deuflhard, P., Hairer, E. (eds.) Progress in Scientific Computing, vol. 2, pp. 95–121. Birkhäuser, Boston (1983)
7. Box, G.E.P., Hill, W.J.: Discrimination among mechanistic models. Technometrics **9**, 57–71 (1967)
8. Coleman, T., Li, Y.: An interior, trust region approach for nonlinear minimization subject to bounds. SIAM J. Optim. **6**, 418–445 (1996)
9. Cox, D., Hinkley, D.: Theoretical Statistics. Chapman & Hall, London (1994)
10. Feder, P.I.: On the distribution of the log likelihood ratio test statistic when the true parameter is "near" the boundaries of the hypothesis regions. Ann. Math. Stat. **39**(6), 2044–2055 (1968)
11. Hand, D.J.: Understanding the new statistics: effect sizes, confidence intervals, and meta-analysis by geoff cumming. Int. Stat. Rev. **80**(2), 344–345 (2012)
12. Honerkamp, J.: Statistical Physics: An Advanced Approach with Applications. Springer, Heidelberg (2002)
13. Kirkpatrick, S., Gelatt, C., Jr., Vecchi, M.P.: Optimization by simulated annealing. Science **220**(4598), 671–680 (1983)
14. Koch, O., Weinmüller, E.B.: The convergende of shooting methods for singular bondary value problems. Math. Comput. **72**(241), 289–305 (2001)
15. Kreutz, C., Timmer, J.: Systems biology: experimental design. FEBS J. **276**(4), 923–942 (2009)

16. Kreutz, C., Raue, A., Timmer, J.: Likelihood based observability analysis and confidence intervals for predictions of dynamic models. BMC Syst. Biol. **6**, 120 (2012)
17. Kreutz, C., Raue, A., Kaschek, D., Timmer, J.: Profile likelihood in systems biology. FEBS J. **280**(11), 2564–2571 (2013)
18. Kronfeld, H.P.M., Zell, A.: The EvA2 optimization framework. Learn. Intell. Optim. **6073**, 247–250 (2010)
19. Leis, J., Kramer, M.: The simultaneous solution and sensitivity analysis of systems described by ordinary differential equations. ACM Trans. Math. Softw. **14**(1), 45–60 (1988)
20. Lory, P.: Enlarging the domain of convergence for mutiple shooting by homotopy method. Numer. Math. **35**, 231–240 (1980)
21. Marquardt, D.: An algorithm for least-squares estimation of nonlinear parameters. SIAM J. Appl. Math. **11**(2), 431–441 (1963)
22. Meeker, W., Escobar, L.: Teaching about approximate confidence regions based on maximum likelihood estimation. Am. Stat. **49**(1), 48–53 (1995)
23. Neyman, L., Pearson, E.: On the problem of the most efficient tests of statistical hypotheses. Phil. Trans. Roy. Soc. A **231**, 289–337 (1933)
24. Peifer, M., Timmer, J.: Parameter estimation in ordinary differential equations for biochemical processes using the method of multiple shooting. IET Syst. Biol. **1**, 78–88 (2007)
25. Pinheiro, J.C., Bates, D.M.: Mixed-effects models in S and S-plus. In: Statistics and Computing. Springer, New York (2000)
26. Poli, R., Kennedy, J., Blackwell, T.: Particle swarm optimization: an overview. Swarm Intell. J. **1**(1), 33–57 (2007)
27. Press, W., Flannery, B., Saul, S., Vetterling, W.: Numerical Recipes. Cambridge University Press, Cambridge (1992)
28. Puntanen, S.: Projection matrices, generalized inverse matrices, and singular value decomposition by haruo yanai, kei takcuchi, yoshio takane. Int. Stat. Rev. **79**(3), 503–504 (2011)
29. Quinn, G.P., Keough, M.J.: Experimental Design and Data Analysis for Biologists. Cambridge University Press, Cambridge (2002)
30. Raue, A., Kreutz, C., Maiwald, T., Bachmann, J., Schilling, M., Klingmüller, U., Timmer, J.: Structural and practical identifiability analysis of partially observed dynamical models by exploiting the profile likelihood. Bioinformatics **25**, 1923–1929 (2009)
31. Raue, A., Kreutz, C., Maiwald, T., Klingmüller, U., Timmer, J.: Addressing parameter identifiability by model-based experimentation. IET Syst. Biol. **5**(2), 120 130 (2011)
32. Raue, A., Schilling, M., Bachmann, J., Matteson, A., Schelker, M., et al.: Lessons learned from quantitative dynamical modeling in systems biology. PLoS One **8**(9), e74335 (2013). doi:10.1371/journal.pone.0074335
33. Raue, A., Steiert, B., Schelker, M., Kreutz, C., Maiwald, T., Hass, H., Vanlier, J., Tönsing, C., Adlung, L., Engesser, R., Mader, W., Heinemann, T., Hasenauer, J., Schilling, M., Höfer, T., Klipp, E., Theis, F., Klingmüller, U., Schöberl, B., Timmer, J.: Data2Dynamics: a modeling environment tailored to parameter estimation in dynamical systems. Bioinformatics (2015). doi:10.1093/bioinformatics/btv405. First published online 3 July 2015
34. Reid, N., Fraser, D.: Likelihood inference in the presence of nuisance parameters. arXiv:physics/0312079 (2003)
35. Seber, G., Wild, C.: Nonlinear Regression. Wiley, New York (1989)
36. Steiert, B., Raue, A., Timmer, J., Kreutz, C.: Experimental design for parameter estimation of gene regulatory networks. PLoS One **7**(7), e40052 (2012)
37. Steward, W.E., Henson, T.L., Box, G.E.P.: Model discrimination and criticism with single-response data. AIChE J. **42**, 3055–3062 (1996)
38. Steward, W.E., Shon, Y., Box, G.E.P.: Discrimination and goodness of fit of multiresponse mechanistic models. AIChE J. **66**, 1404–1412 (1998)
39. Wald, A.: Tests of statistical hypotheses concerning several parameters when the number of observations is large. Trans. Am. Math. Soc. **54**(3), 426–482 (1943)

On Time Discretizations of Fluid-Structure Interactions

Thomas Richter and Thomas Wick

Abstract In this contribution, time discretizations of fluid-structure interactions are considered. We explore two specific complexities: first, the stiffness of the coupled system including different scales of the Navier-Stokes equations of parabolic type and the structure equation of hyperbolic type and second, the problem of moving domains that is inherent to fluid-structure interactions.

Typical moving mesh approaches, such as the arbitrary Lagrangian-Eulerian framework, give rise to nonlinearities and time-derivatives with respect to the mesh-deformation. We derive different time-stepping techniques of Crank-Nicolson type and analyse their stability and approximation properties. Further, we closely look at the dominant time-scales that must be resolved to capture the global dynamics. Moreover, our discussion is supplemented with an analysis of the temporal discretization of Eulerian fixed-mesh approaches for fluid-structure interactions, where the interface between fluid and solid will change from time-step to time-step. Finally, a formulation of parallel multiple shooting methods for fluid-structure interaction is presented.

1 Introduction

Fluid-structure interactions (fsi) are part of many application problems and appear in mechanical engineering, aeroelasticity, hemodynamics, or as pore-scale modeling in porous media flow. Concretely in this work, we consider the interaction of a laminar, incompressible fluid with an elastic solid governed by the Saint Venant Kirchhoff material law. While the simulation of both single sub-processes is already a difficult task, covering their interaction is further complicated by the two-way

T. Richter (✉)
Institute for Applied Mathematics, INF 294, 69120 Heidelberg, Germany
e-mail: thomas.richter@iwr.uni-heidelberg.de

T. Wick
The Institute for Computational Engineering and Sciences, The University of Texas at Austin, Austin, TX 78712, USA
e-mail: twick@ices.utexas.edu

T. Carraro et al. (eds.), *Multiple Shooting and Time Domain Decomposition Methods*, Contributions in Mathematical and Computational Sciences 9,
DOI 10.1007/978-3-319-23321-5_15

coupling between both systems. In this contribution, we will focus on issues concerning the time-discretization of fully-coupled fluid-structure interactions by means of time-stepping schemes. In every time-step, an approximation to the fully coupled system at a new time-step is to be found. The fully coupled approach is called *monolithic* compared to *partitioned* approaches, where the fluid- and solid-problem are solved separately and coupled via outer iterations. Monolithic schemes always belong to strongly coupled methods whereas partitioned schemes can be either strongly or loosely coupled depending on the number of inner iterations and the desired accuracy of force balance at the interface. For instance, strongly coupled schemes are necessary when the added mass effect takes place as in applications in hemodynamics. On the other hand, partitioned solution schemes are very efficient for problems with a less stiff coupling, as they are common in aeroelasticity.

Time discretization of fluid-structure interactions is mainly governed by two specific complexities. First, the overall stiffness of the coupled problem is by far greater than that of the sub-problems. This is mainly due to the coupling of parabolic-type fluid equations with hyperbolic-type solid equations. Second, using a (most common) moving-mesh approach, time derivatives do not appear separated from spatial differential operators, but they depend nonlinearly on other solution variables and their spatial derivatives, giving rise to terms like

$$\det(I + \nabla \mathbf{u})[I + \nabla \mathbf{u}]^{-1}\nabla \mathbf{v}\partial_t \mathbf{u},$$

where $\mathbf{v}$ is the unknown velocity and $\mathbf{u}$ is the unknown deformation determining the domain motion. Detailed analysis for fluid flows on moving domains has been performed by Formaggia and Nobile [15, 16]. These studies already tackle several important aspects such as stability, order of convergence and the geometric conservation law. In fluid-structure interaction, the fluid-domain movement is caused by the solid deformation. Hence, the analysis of fully coupled fluid-structure interaction is similar but must also include detailed consideration of the solid discretization. This analysis is not present in literature.

In the following section, we will shortly introduce the governing equations for the fully coupled fluid-structure interaction problem in Arbitrary Lagrangian Eulerian (ALE) coordinates. Then, in the central third section, we analyze time-dependent dynamics of a typical fluid-structure interaction benchmark. In Sect. 4 we present and discuss different time-stepping technique and closely analyze their accuracy and stability properties. In Sect. 5, we recapitulate the current state-of-the-art of an alternative (to the ALE approach) monolithic formulation, the fully Eulerian framework and discuss special aspects with respect to its time-discretization. The prospective of applying parallel multiple shooting methods to enhance stability and efficiency is discussed in Sect. 6. Finally, we conclude in Sect. 7.

2 Fluid-Structure Interactions

By $\Omega \subset \mathbb{R}^d$ we denote a two or three-dimensional domain that is split into a fluid-part $\mathscr{F}$ and a solid-part $\mathscr{S}$. The splitting is such, that $\mathscr{F}$ and $\mathscr{S}$ are d-dimensional domains with a common interface $\mathscr{I} = \partial\mathscr{F} \cap \partial\mathscr{S}$. Figure 1 shows a possible configuration of fluid-structure interactions, where the fluid encloses an obstacle with elastic beam. By the arising dynamics, the beam will deform and hence, solid and fluid domain will change such that at time $t > 0$ it holds $\Omega(t) = \mathscr{F}(t) \cup \mathscr{S}(t)$ with interface $\mathscr{I}(t) = \partial\mathscr{F}(t) \cap \partial\mathscr{S}(t)$. In $\mathscr{F}(t)$, fluid's velocity $\mathbf{v}_f$ and pressure p_f are governed by the incompressible Navier-Stokes equations, where in $\mathscr{S}(t)$ the elastic beam's deformation $\mathbf{u}_s$ and velocity $\mathbf{v}_s$ is controlled by a St. Venant Kirchhoff material, see [28].

The coupling between these two sub-problems is controlled by the *kinematic condition* that calls for continuity of the velocities on the common interface $\mathscr{I}(t)$ and the *dynamic condition* that asks for continuity of normal stresses on $\mathscr{I}(t)$.

To cope with the moving fluid-domain, we introduce the *Arbitrary Lagrangian Eulerian* (ALE) formulation that maps the domains $\mathscr{F}(t)$ back to the fixed reference domain $\mathscr{F}$ at time $t = 0$. We introduce by $\mathbf{u}_f$ the deformation field of the fluid-domain, that maps every reference point $\hat{x} \in \mathscr{F}$ to a point in the current configuration $x \in \mathscr{F}(t)$ via

$$x \in \mathscr{F}: \quad T(x,t) = x + \mathbf{u}_f(x,t) \in \mathscr{F}(t).$$

This mapping is not the physically motivated mapping between Lagrangian and Eulerian coordinates that follows the path of a particle, but a mapping between a completely arbitrary reference domain and the current configuration, see [8, 27, 32]. In both subdomains $\mathscr{F}$ and $\mathscr{S}$, we denote by $\mathbf{F} := I + \nabla\mathbf{u}$ the deformation gradient and by $J := \det(\mathbf{F})$ its determinant.

On Ω, $\mathscr{F}$ and $\mathscr{S}$ we introduce the function spaces

$$\mathscr{V} := H_0^1(\Omega;\Gamma^D)^d, \quad \mathscr{L}_f := L^2(\mathscr{F}), \quad \mathscr{V}_f := H_0^1(\mathscr{F};\partial\mathscr{F})^d, \quad \mathscr{V}_s := L^2(\mathscr{S})^d,$$

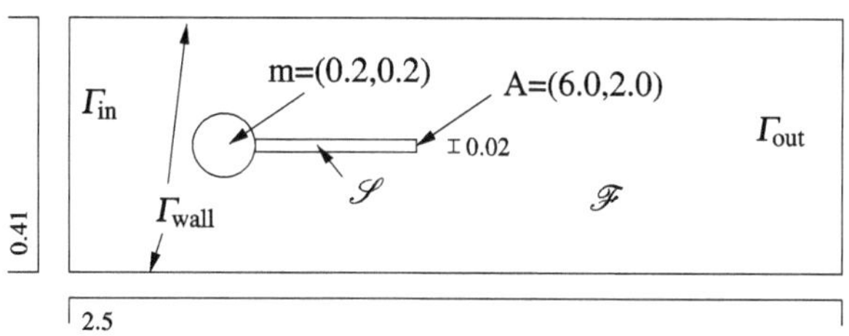

Fig. 1 Configuration of the fsi3-benchmark problem. Flow around obstacle with elastic beam. Inflow boundary Γ_{in}, outflow boundary Γ_{out} and fluid-solid interface $\mathscr{I}$. The cfd-benchmark is recovered by omitting the beam

where by $H^1(\Omega;\Gamma)$ we denote the Sobolev space of Lebesgue integrable functions with the first weak derivative in L^2 and trace zero on the boundary Γ. The full system of coupled fluid-structure interactions in variational formulation is to find

$$v = (\mathbf{v}_f, \mathbf{v}_s) \in \mathscr{V}, \quad \mathbf{u} = (\mathbf{u}_f, \mathbf{u}_s) \in \mathscr{V}, \quad p_f \in \mathscr{L}_f,$$

such that

$$\begin{aligned}
&\Big(\rho_f J(\partial_t \mathbf{v} + \nabla \mathbf{v} \mathbf{F}^{-1}(v - \partial_t \mathbf{u})), \phi\Big)_{\mathscr{F}} + \Big(J\hat{\boldsymbol{\sigma}}_f \mathbf{F}^{-T}, \nabla\phi\Big)_{\mathscr{F}} \\
&\qquad + \Big(\rho_s d_t \mathbf{v}, \phi\Big)_{\mathscr{S}} + \Big(\mathbf{F}\boldsymbol{\Sigma}_s, \nabla\phi\Big)_{\mathscr{S}} \\
&+ \Big(\det(J\mathbf{F}^{-1}\mathbf{v}), \xi_f\Big)_{\mathscr{F}} + \Big(\nabla\boldsymbol{\sigma}_{\text{mesh}}, \nabla\psi_f\Big)_{\mathscr{F}} + \Big(d_t\mathbf{u} - v, \psi_s\Big)_{\mathscr{S}} = (f,\phi)_\Omega,
\end{aligned} \tag{1}$$

for all

$$\phi \in \mathscr{V}, \quad \xi_f \in \mathscr{L}_f, \quad \psi_f \in \mathscr{V}_f, \quad \psi_s \in \mathscr{V}_s.$$

Here, by f we denote a given volume force, by $\boldsymbol{\Sigma}_s$ the 2nd Piola Kirchhoff stress tensor, which for the St. Venant Kirchhoff material takes the form

$$\boldsymbol{\Sigma}_s = 2\mu \mathbf{E}_s + \lambda_s \operatorname{trace}(\mathbf{E}_s) I, \quad \mathbf{E}_s := \frac{1}{2}(\mathbf{F}^T\mathbf{F} - I),$$

where by μ_s and λ_s we indicate the two material parameters. By $\boldsymbol{\sigma}_f$ we denote the fluid's Cauchy stresses expressed in the reference coordinate system

$$\boldsymbol{\sigma}_f = \rho_f \nu_f (\mathbf{F}^{-1}\nabla\mathbf{v}_f + \nabla\mathbf{v}_f^T \mathbf{F}^{-T}) - p_f I,$$

and finally, by $\boldsymbol{\sigma}_{\text{mesh}}$ we denote the mesh-moving operator of harmonic, linear-elastic, or biharmonic type [23, 49, 57].

Kinematic

$$\mathbf{v}_f = \mathbf{v}_s \quad \text{on } \mathscr{I}(t),$$

and dynamic coupling conditions

$$J\boldsymbol{\sigma}_f \mathbf{F}^{-T} n_f = \mathbf{F}\boldsymbol{\Sigma}_s n_s \quad \text{on } \mathscr{I}(t),$$

are embedded in this formulation, as the velocities $v \in \mathscr{V}$ and test-functions $\phi \in \mathscr{V}$ have a well-defined continuous trace on the interface $\mathscr{I}$.

Spatial discretization of this coupled system will be accomplished with conforming finite elements. For details, we refer to [12, 41–43, 45, 57].

3 Analysis of Two Benchmark Problems: cfd and fsi

We start the discussion on time-discretizations of fluid-structure interaction with a literature survey on published results for two benchmark problems in fluid-dynamics: first, the cfd-benchmark *Laminar Flow Around a Cylinder* as published in 1995 by Schäfer and Turek [46] and called cfd-benchmark in the following. Second an extension of this benchmark to fluid-structure interactions, the fsi3-benchmark problem, published in 2006 by Hron and Turek [30] and called fsi3 in the following. Both problems feature the flow around a circular obstacle. In the fsi3 configuration, an elastic beam is attached to the rear of the circular obstacle. See Fig. 1 for a sketch of the configuration. The cfd benchmark is obtained by completely omitting the beam. Both problems are driven by a time-dependent inflow data $v = v^D$ on Γ_{in}. The full set of parameters for both problems is given by

$$\rho_f^{\text{cfd}} = 1, \quad \rho_f^{\text{fsi}} = 10^3, \quad \nu_f^{\text{cfd/fsi}} = 10^{-3}, \quad \mathbf{v}^D(0,y) = 1.5\omega(t)\frac{y(H-y)}{(H/2)^2}\bar{\mathbf{v}},$$

where $\omega(t) = (1 - \cos(\pi t/2))/2$ for $t < 2$ and $\omega(t) = 1$ for $t \geq 2$ is used for regularizing the initial data. As average velocity, $\bar{\mathbf{v}} = 2$ for the fsi3-benchmark and $\bar{\mathbf{v}} = 1$ for the cfd benchmark was considered. With the radius of the circular obstacle $D = 0.1$, the Reynolds number is given by

$$Re_{\text{cfd}} = \frac{\bar{\mathbf{v}}D}{\nu} = 100, \quad Re_{\text{fsi}} = \frac{\bar{\mathbf{v}}D}{\nu} = 200.$$

The description of the problem is closed by providing the material parameters of the elastic solid

$$\rho_s^{\text{fsi}} = 10^3, \quad \mu_s = 2 \cdot 10^6, \quad \lambda_s = 8 \cdot 10^6.$$

As quantity of interest, we consider principal boundary stresses in x- and y-direction on the obstacle with boundary Γ_{obs}:

$$J_{\text{drag}}(v,p) = \frac{2}{\bar{\mathbf{v}}^2 \rho_f L}\int_{\Gamma_{\text{obs}}} \boldsymbol{\sigma}_f \mathbf{n} \mathbf{e}_x \, do, \quad J_{\text{lift}}(v,p) = \frac{2}{\bar{\mathbf{v}}^2 \rho_f L}\int_{\Gamma_{\text{obs}}} \boldsymbol{\sigma}_f \mathbf{n} \mathbf{e}_y \, do.$$

By Γ_{obs} we denote the boundary of the circle with diameter in the case of the cfd-benchmark and the circle with attached beam in the case of the fsi-benchmark problem. Efficient ways for evaluating these functionals are shown in ([5]) and ([43]).

Figure 2 shows the drag-coefficient as function over time $I = [0,5]$ for the two benchmark problems. To avoid confusion, we note, that Hron and Turek [30] also published a new fluid dynamics benchmark, the cfd3-benchmark problem, where the beam was considered as rigid part of the obstacle and the flow was driven with Reynolds number $Re = 200$. Here however, we compare the fsi3-benchmark problem with the published results of the original cfd-benchmark taken from [46]. Both configurations show a similar behavior with a transient initial phase leading to a periodic oscillation with dominant frequencies $f_{\rm cfd} = 13$ Hz for the cfd-benchmark and $f_{\rm fsi} \approx 11$ Hz for the fsi problem. The first obvious difference is the longer transient phase for the fsi-benchmark problem. An insight look into the subinterval $I' = [2.5, 3]$ reveals high frequent oscillations $f_{\rm high} \approx 100$ Hz in the drag-coefficient with a small amplitude $a \approx 10^{-4}$ that is not visible on the large scale. These high frequent oscillations are no numerical artefacts but remain stable under temporal and spatial mesh refinement.

Reviewing the results published by many research groups in the two surveys on the cfd benchmark problem [46] and the fsi3-benchmark [30, 31] a first surprising observation is the choice of discretization parameters that have been necessary to obtain approximations with appropriate accuracy: even though more than a decade lies between both benchmark problems, the dimension of the spatial discretization is very similar. In both cases, different research groups had to use 20.000 to 200.000 spatial degrees of freedom to generate output values with at least 1 % accuracy. The large margin stems from different discretization schemes (lowest order finite element or finite volume schemes, higher order methods) but also from different triangulations of the geometry. The increased difficulty of the fsi3-benchmark problem has been accounted for by a general use of higher order finite elements.

However, observing the temporal discretization, it is found, that the fsi benchmark asks for significantly finer resolution in time. While less than 10 time-steps per period of the oscillation were sufficient in the cfd case, accurate results to the fsi benchmark problem required up to 100 time-steps per period of oscillation resulting in time-steps as small as 10^{-3}. One explanation for this difference in temporal discretization can be found in the high frequent oscillations that are present with small amplitude, see Fig. 2.

Further insight is given by a discrete Fourier analysis of the output functional $J_{\rm drag}(t)$ as function over time. At very fine temporal resolution (down to $k = 10^{-5}$), some complete periods of the fully developed oscillation are analyzed in detail. Figure 3 reveals several dominant frequencies, at about 100 Hz (see also Fig. 2, 500

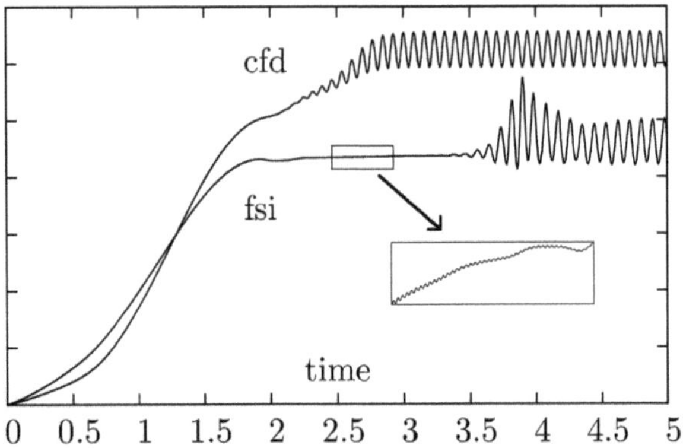

Fig. 2 Comparison of the two benchmark problems, original cfd [46] and fsi3 [30]. We plot the drag coefficient as function over time. For the fsi-problem we show a detailed view of the transient oscillations revealing high frequent modes

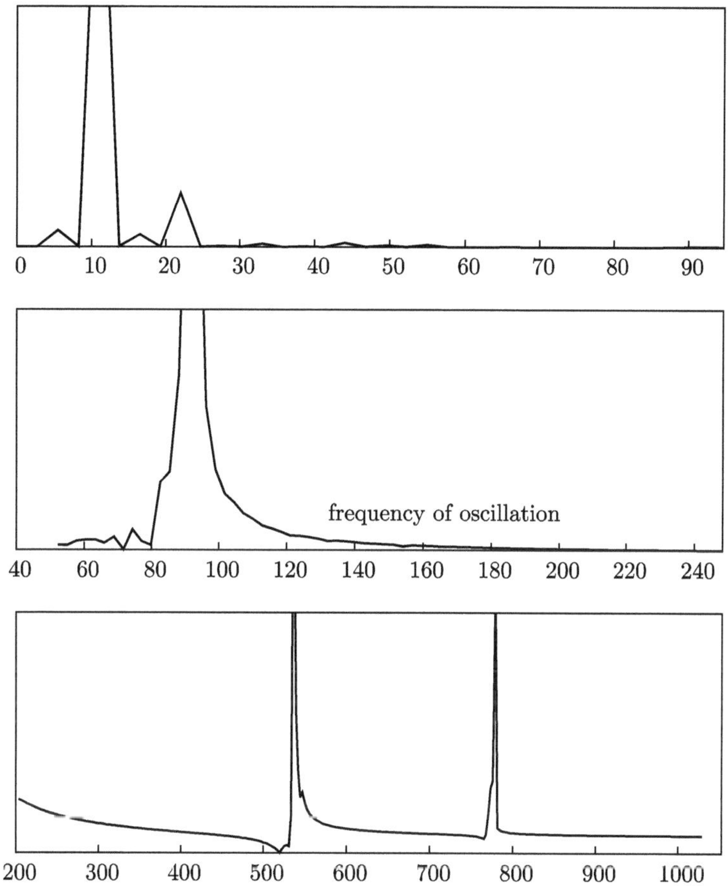

Fig. 3 Discrete Fourier analysis of the output functional (drag) shows the dominant frequency $f_{\text{fsi}} \approx 11$ Hz and further important sub-frequencies at about $f \approx 100$ and 500 Hz as well as 800 Hz. These modes are stable under temporal and spatial mesh refinement

and 800 Hz. These modes are stable under mesh refinement and further decrease of the time step. The results in Fig. 3 are scaled. The modes belonging to higher frequencies carry less energy. But even though the high frequent contributions take place on a much smaller scale as the dominant oscillation $f_{\text{fsi}} \approx 11$ Hz, they must be carefully resolved to capture the overall dynamics of the coupled benchmark problem. The key question in this respect is the origin of these micro-oscillations. They are not present in pure fluid-dynamical simulations. Further, they are no numerical artifact, but stable under discretization of both spatial and temporal discretization. Instead they stem from the coupling to the hyperbolic structure equations.

4 Time-Stepping Schemes for Fluid-Structure Interactions

There is little theoretical background to monolithic time-discretizations of fluid-structure interactions. The main difficulty stems from the motion of the subdomains, that must either be modeled explicitly in partitioned approaches or that must be taken care of by implicit transformations of either the fluid-domain or the solid-domain. Concentrating first on pure fluid problems on moving domains, some crucial aspects with respect to stability and order of convergence are already identified [15, 16]. The equations presented therein can be directly employed in an implementation. In addition, [14] provided stability analysis of fluid-structure interaction problems. Several studies with qualitative comparisons of different time-stepping schemes and their long-time behavior has been reported in [56, 62]. In the primer study and additionally [57, 61], we provide many details for the practical realization and implementation of time-stepping schemes for ALE fluid-structure interaction.

In the following, we focus the attention to the well-established ALE-approach that results from transformation of the moving fluid-domain to a fixed reference domain. The domain motion is hidden in the ALE-map $T_f(x,t)$ and calls for the discretization of non-standard space-time coupled terms like [see (1)]

$$(J(\mathbf{u})\nabla\mathbf{v}\mathbf{F}^{-1}(\mathbf{u})\partial_t\mathbf{u},\phi)_{\mathscr{F}}. \tag{2}$$

Most approaches for the temporal discretization of this term are ad hoc and based on the experience with other types of equations as Navier-Stokes of multiphase fluids, see [29].

Remark 1 An alternative approach to the monolithic formulation of fluid-structure interactions is given by an implicit transformation of the solid-domain to Eulerian coordinates resulting in the Fully Eulerian approach [11, 45]. This method of interface-capturing type must deal with subdomains that move freely through a fixed background mesh from time-step to time-step. We come back to this procedure in Sect. 5.

4.1 Derivation of Second Order Time-Stepping Schemes

The derivation of a second order stable time-stepping scheme is not obvious. Specifically, regarding (2), two immediate reasonable choices for are given by the secant version

$$\left(\left[\frac{J(\mathbf{u}^{m-1})\nabla\mathbf{v}^{m-1}\mathbf{F}^{-1}(\mathbf{u}^{m-1})}{2}+\frac{J(\mathbf{u}^{m})\nabla\mathbf{v}^{m}\mathbf{F}^{-1}(\mathbf{u}^{m})}{2}\right]\frac{\mathbf{u}^{m}-\mathbf{u}^{m-1}}{k_m},\phi\right),$$

and the midpoint-tangent version

$$\left(\left[J(\bar{\mathbf{u}})\nabla\bar{\mathbf{v}}\mathbf{F}^{-1}(\bar{\mathbf{u}})\right]\frac{\mathbf{u}^m-\mathbf{u}^{m-1}}{k_m},\phi\right),\quad \bar{\mathbf{u}}:=\frac{\mathbf{u}^{m-1}+\mathbf{u}^m}{2},\quad \bar{\mathbf{v}}:=\frac{\mathbf{v}^{m-1}+\mathbf{v}^m}{2},$$

of the trapezoidal rule. This idea is explored in [56, 62]. A third version of a time-stepping scheme can be derived by using a temporal $cG(1)/dG(0)$-Galerkin approach on (2) (see [44]):

$$\left(\left[\frac{1}{6}J(\mathbf{u}^{m-1})\nabla\mathbf{v}^{m-1}\mathbf{F}^{-1}(\mathbf{u}^{m-1})+\frac{2}{3}J(\bar{\mathbf{u}})\nabla\bar{\mathbf{v}}\mathbf{F}^{-1}(\bar{\mathbf{u}})\right.\right.$$
$$\left.\left.+\frac{1}{6}J(\mathbf{u}^m)\nabla\mathbf{v}^m\mathbf{F}^{-1}(\mathbf{u}^m)\right]\frac{\mathbf{u}^m-\mathbf{u}^{m-1}}{k_m},\phi\right),$$

where again by $\bar{\mathbf{u}}$ and $\bar{\mathbf{v}}$ we denote the average of old and new approximation. Such a Galerkin-derivation is also possible for more advanced time-stepping schemes like the fractional step theta method, see [37, 38].

Simple truncation error analysis shows second order convergence for $k \to 0$ in all three cases. The leading error constants slightly differ:

$$C_1 \approx \frac{11}{8},\quad C_2 \approx \frac{3}{8},\quad C_3 \approx \frac{3}{4}.$$

In numerical experiments, it is found, that all these variants show a very similar performance. Significant differences in temporal accuracy could not be found.

Finally, we point out, that the Crank-Nicolson scheme applied to the elastic structure equation in mixed formulation is closely related to the Newmark scheme [3], which is one of the most prominent time-discretization techniques in solid mechanics.

4.2 Temporal Stability

Issues of numerical stability are of utter importance for fluid-structure interaction problems, as they consist of the coupled consideration of two different types of equations: the incompressible Navier-Stokes equations which is of parabolic type and that comes with smoothing properties and the hyperelastic solid equation of hyperbolic type, that calls for good conservation properties with very little numerical dissipation. By these considerations, the Crank-Nicolson scheme and its variants like shifted versions [34, 40] or the fractional step theta scheme [7, 54], appear to be ideal candidates that further show second order accuracy.

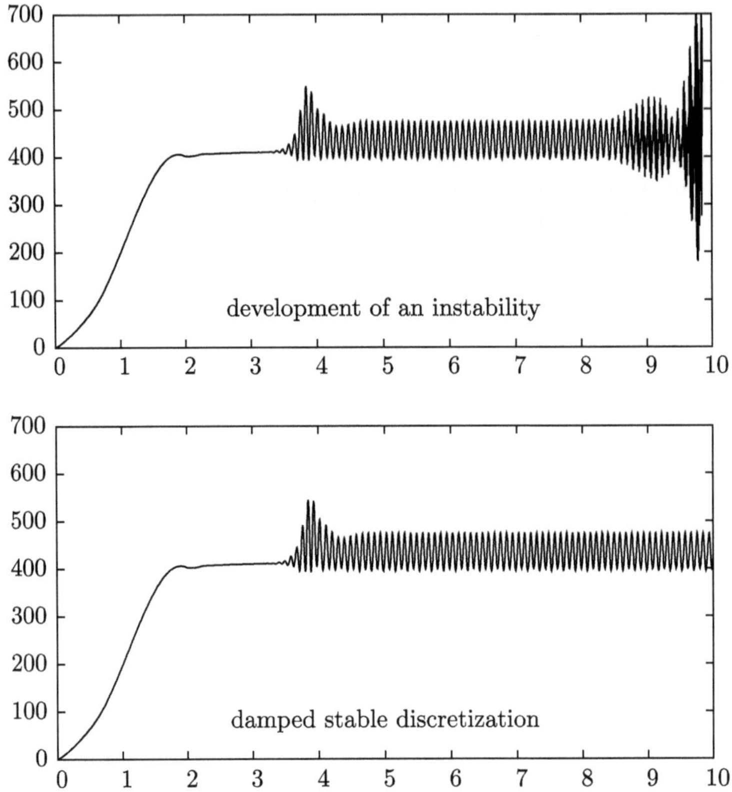

Fig. 4 Simulation for $k = 0.005$. *Top*: undamped Crank-Nicolson scheme develops an instability after $T = 8.5$. *Bottom*: implicitly shifted scheme produces a stable solution on $I = [0, 10]$

Motivated by [16, 26, 34], it is reported in [14, 62], that the discretization of the domain-motion term (2) introduces further stability issues. To investigate this stability problem, we again consult the fsi3-benchmark problem introduced in the previous sections. Figure 4 shows the drag as functional over time for an unstable pair of spatial and temporal discretization parameters. Further, we also show the stable simulation using a damped version of the time-stepping scheme, see Sect. 4.3.

In a first test, we aim at obtaining a stable solution up to $T = 10$. On a sequence of uniform meshes, we identify the largest timestep k that is suited to generate a stable solution. The left part of Table 1 shows the results. Here, we see, that on the coarsest mesh, the large step size $k = 0.02$ is sufficient, while on finer meshes $k < 0.004$ is required. We however cannot identify a further relationship between mesh size and time step if we go to an even finer spatial mesh resolution.

Table 1 Long-term stability of the Crank-Nicolson scheme

Mesh-level	Time step size 0.025	0.02	0.004	$0.00\bar{3}$	Mesh-level	$k = 0.005$	$0.00\bar{3}$
1	×	√	√	√	1	$\gg 10$	$\gg 10$
2	×	×	×	√	2	8.48	10.82
3	×	×	×	√	3	6.04	12.54
					4	3.84	3.84

Left: combination of time-step k and mesh size h, such that the solution is stable in the interval $I = [0, 10]$. We cannot find a strict time-step relation $k \sim h^\alpha$. Right: maximum interval $I = [0, T_{\max}]$, where a solution could be found for $k = 0.005$ and $k = 0.00\bar{3}$, depending on the mesh-size. Here, we also cannot identify an obvious relationship

In a second test-case, we consider the (relatively large) step size $k = 0.005$ and $k = 0.00\bar{3}$ and determine the point in time $T_{\max}$, where the solution gets unstable. Again, we carry out this test-case on different meshes. At first glance, the results in the right part of Table 1 for $k = 0.005$ suggest a stability relationship between time step and mesh size. The results concerning the second configuration with $k = 0.00\bar{3}$ however does not confirm this conjecture. Here, we can even reach a larger final point in time $T_{\max}$ on finer meshes. Further, the simulations on the finest mesh do not cease due to stability problems but due to early failure of the Newton scheme. Altogether, it is not possible to numerically certify a strict time-step restriction. Instead, we find general stability problems for long-term simulation, if we consider the Crank-Nicolson scheme.

4.3 *Stable Time-Discretization and Damping*

By analyzing the fsi3-benchmark problem, it seems, that time-step restrictions due to stability issues are too restrictive and not justified by the needs of approximation accuracy, see Sects. 4.2 and 4.1. It is therefore nearby to search for accurate time-discretization schemes with better stability properties. Different possibilities are either to resort to A stable time-discretization schemes, or to apply modifications to the Crank-Nicolson schemes. Here, two possibilities are often discussed in literature: by slight implicit shifting of the discretization

$$(u^m - u^{m-1}, \phi) + \left(\frac{1}{2} + O(k)\right) a(u^m, \phi) + \left(\frac{1}{2} - O(k)\right) a(u^{m-1}, \phi) = 0,$$

global stability is recovered, see [25, 26, 34]. This is just sufficient for the damping of accumulated errors by truncation, quadrature or inexact solution of the algebraic systems. If the shift depends on the time-step size, the resulting scheme is still second order accurate in time. Similar results are recovered by applying some initial time-steps with the A-stable backward Euler method, see [40]. If these few (usually two are sufficient) backward Euler steps are introduced after every fixed time-interval, e.g. at every $t = j$ for $j = 0, 1, \dots,$ we also recover sufficient stability for long term calculations. This scheme, also referred to as Rannacher time-marching, is second order accurate.

Higher stability, that is also able to cover non-smooth initial data is reached by applying strongly A-stable time-integration techniques. Here, the fractional-step theta method appears to be an optimal choice [7]. This time-stepping scheme consists of three sub-steps, that results in a second order, strongly A-stable scheme that further has very good dissipation properties. It is highly preferable for flow problems [54] and also frequently used in the analysis of fluid-structure interactions problems [29, 55, 56].

An analysis of different damping strategies applied to the fsi-2 benchmark problem (a slightly more difficult test-case) is given in [62], which we briefly summarize in the following: There are only minor differences in the drag evaluation computed with the unstabilized Crank-Nicolson scheme using the different ALE convection term discretizations. Specifically, unstable behavior (blow-up) for computations over long-term intervals is observed. As expected, the shifted Crank-Nicolson scheme and the Fractional-Step-θ scheme do not show any stability problems in long-term computations, even for large time steps $k = 0.01$. This result indicates that the instabilities induced by the ALE convection term have minor consequences, and our observation is in agreement with the statement in [16].

In the following, we compare the three possibilities of a non-damped Crank-Nicolson scheme, with an implicitly shifted version using $\frac{1}{2} + k$ and the Rannacher time-marching algorithm with two steps of the backward Euler method at times $t = 0, t = 1, t = 2$ and so on. In Fig. 5 we compare these three damping strategies. We show the drag-coefficient (see Figs. 2 or 4 for a global view) in the sub-intervals $t \in [3.5, 4.2], t \in [7.95, 8.15]$ and $t \in [9.3, 9.6]$. While all three versions are stable at initial time, Rannacher time-marching develops a first instability after two steps of backward Euler at time $t = 4$, see the left sketch in Fig. 5. This instability will stay during the simulation, but it will not be further developed, as can be seen in the middle and right sketch of the figure. The undamped version of the Crank-Nicolson scheme delivers stable solutions up to a moderate time of about $t = 5$ but develops a strong instability that fill finally lead to a break-down of the scheme, as can be seen in the middle and right sketch. Finally, the implicitly shifted version of the Crank-Nicolson scheme gives stable and good result globally in time.

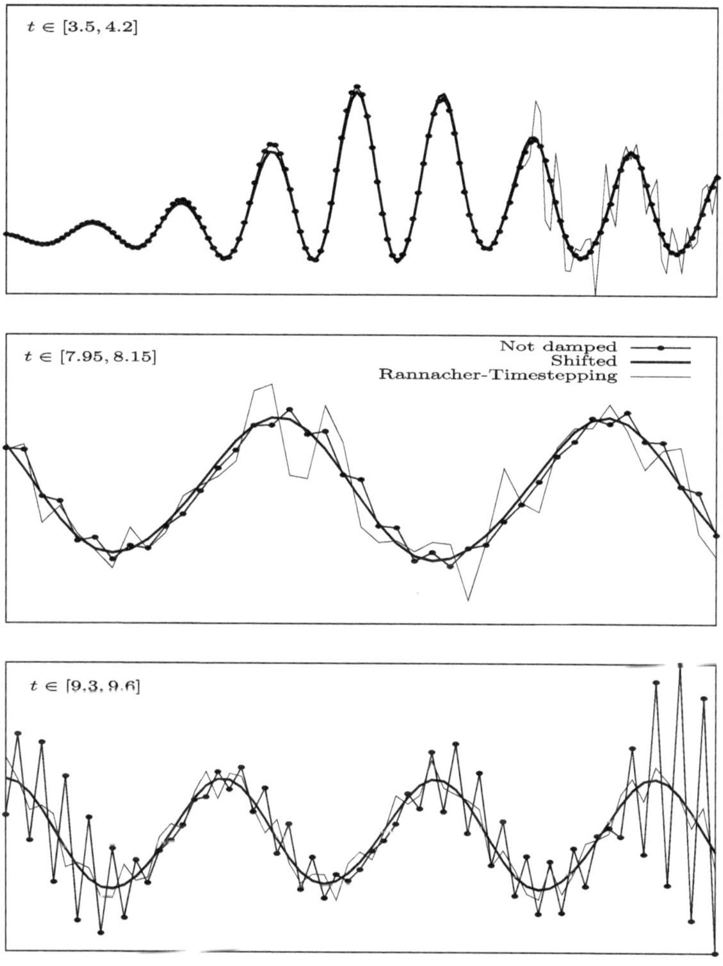

Fig. 5 Comparison of different damping strategies: undamped Crank-Nicolson, shifted version $\frac{1}{2} + k$ and Rannacher time-marching with two backward Euler steps at every time-unit

5 Alternative Formulation in Fully Eulerian Coordinates

As previously mentioned, an interesting alternative to ALE formulations is the fully Eulerian framework [11, 45]. Specific extensions are reported in [43, 58] and combinations with ALE in [59, 60]. Apart from monolithic formulations, other recent studies (but not tested with the fsi-benchmarks) on fully Eulerian formulations are known [9, 22, 50, 51, 64]. As discussed in the previous sections, the ALE formulation of the fluid problems introduced several difficulties for the analysis as well as the implementation. In fact, all kinds of transformation in the fluid equations are avoided in a fully Eulerian description of the equations.

Here, the idea is to have fluid- and solid-problem on the moving domains $\mathscr{F}(t)$ and $\mathscr{S}(t)$. While fluid-system is simply let in Eulerian coordinates, we now need to transform the solid problem to match the Eulerian coordinate system. An illustration is given in Fig. 6.

By the simple observation, that the deformation $\hat{\mathbf{u}}_s(\hat{x},t) = x(\hat{x},t) - \hat{x}$ just maps between Lagrangian points $\hat{x} \in \hat{\mathscr{S}}$ and their Eulerian coordinates $x = x(\hat{x},t) \in \mathscr{S}(t)$, we can defined Eulerian counterparts for deformation $\mathbf{u}_s(x,t) = \hat{\mathbf{u}}_s(\hat{x},t)$ and velocity $\mathbf{v}_s(x,t) = \hat{\mathbf{v}}_s(\hat{x},t)$. Then, following [11, 45], the Eulerian system is similar to a multiphase flow

$$
\begin{aligned}
\rho_f(\partial_t \mathbf{v}_f + \mathbf{v}_f \cdot \nabla \mathbf{v}_f) - \operatorname{div} \boldsymbol{\sigma}_f &= \rho_f f && \text{in} \quad \mathscr{F}(t),\\
\operatorname{div} \mathbf{v}_f &= 0 && \text{in} \quad \mathscr{F}(t),\\
J_s\rho_s(\partial_t \mathbf{v}_s + \mathbf{v}_s \cdot \nabla \mathbf{v}_s) - \operatorname{div} \boldsymbol{\sigma}_s &= J_s\rho_s f && \text{in} \quad \mathscr{S}(t),\\
\partial_t \mathbf{u}_s + \mathbf{v}_s \cdot \nabla \mathbf{u}_s &= \mathbf{v}_s && \text{in} \quad \mathscr{S}(t),\\
\mathbf{v}_f = \mathbf{v}_s, \quad n \cdot \boldsymbol{\sigma}_f &= n \cdot \boldsymbol{\sigma}_s && \text{on} \quad \mathscr{I}(t).
\end{aligned}
\tag{3}
$$

For the exact form of modeling structural stresses $\boldsymbol{\sigma}_s$ in Eulerian coordinates, we refer to the literature [45]. The big advantage of this Eulerian framework is the avoidance of any kind of unphysical (hence arbitrary) mapping of the systems. The transformation between Lagrangian and Eulerian coordinates of the structure system motivated by physical principles and will not be cause for break-down of the scheme, see e.g. [43]. The obvious drawback of an Eulerian model is the front-capturing type of this formulation, where the interface $\mathscr{I}(t)$ will move freely in the domain and through the mesh elements. The ALE technique has a front-tracking character that allows to resolve the interface at all times with a finite element mesh.

To capture the interface, we need to constantly keep record of its location. One classical approach is the Level Set techniques [39, 47], where a scalar function ψ is introduces, that indicates the signed distance to the interface and that is transported with the velocity of the interface. Here, we instead use the Initial Point Set [11, 43, 58], a vector field, that transports the complete reference coordinate system. In the context of fluid-structure interactions, this Initial Point Set is exactly the structure's deformation and its extension to the fluid-domain. Another benefit of the Initial Point Set technique is its ability to depict sharp edges.

As a front-capturing technique, the fully Eulerian formulation is an interface problem, where some mesh elements are cut by the interface and where different equations live on the two sides of the interface. Solutions to such interface problems are usually not regular and standard finite element schemes only give sub-optimal convergence $O(\sqrt{h})$ independent of the approximation degree, see the early works of Babuška [1] or MacKinnon and Carey [35]. Modern techniques to enhance the interface accuracy are to locally fit the mesh in order to recover the optimal approximation order [4, 6, 18, 63] or to enhance the finite element space with special basis functions that can resolve irregularities [2, 20, 21, 36].

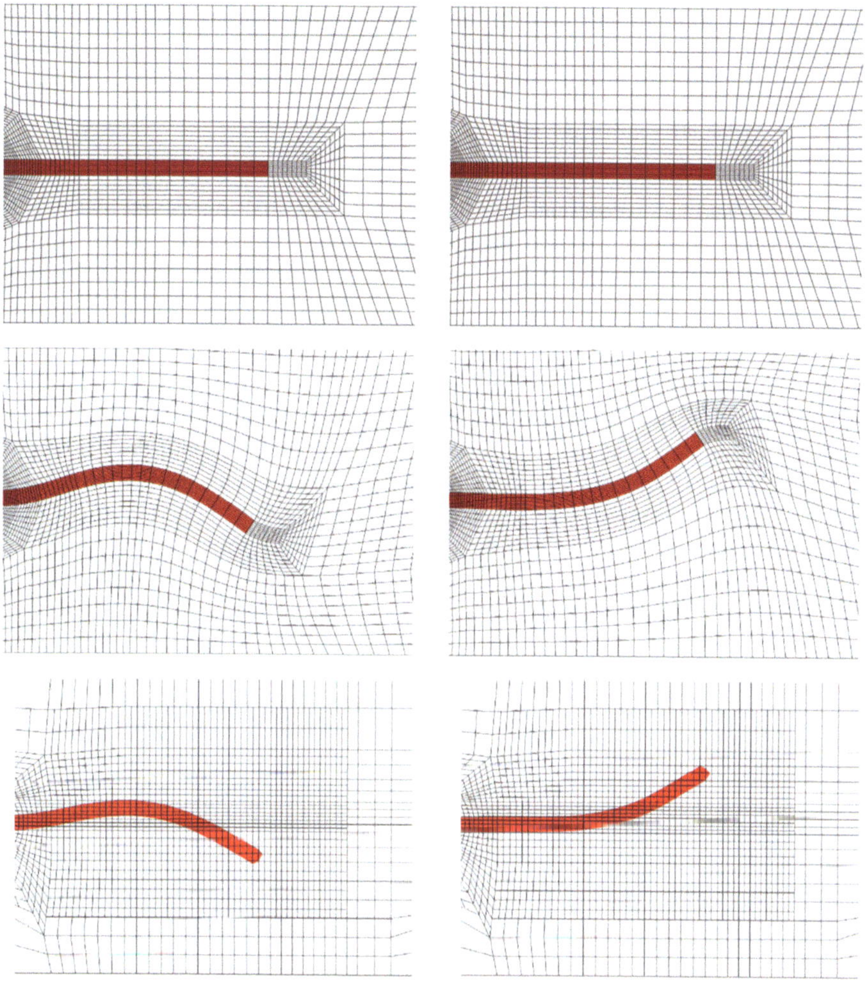

Fig. 6 Comparison of cell occupation and computational domains for two time steps between the ALE (*top* and *middle*) and Eulerian (*bottom*) method. In ALE, all computations are done in the same fixed reference domain Ω (*top*). In particular, a specific cell remains all times the same material (here, an elastic structure in *red*), i.e., Ω_f and Ω_s are time-independent. The mesh movement is hidden in the transformation $\mathbf{F}$ and J. The physical ALE domain $\Omega(t)$ including the mesh movement is displayed in the middle. In contrast, the computation with the Eulerian approach is performed on a fixed (time-independent) mesh Ω_E (*bottom*). However, the two sub-domains for the structure and the fluid $\Omega_{s,E}$ and $\Omega_{f,E}$ change in each time step because the material id of a cell might change since the elastic structure (*red*) moves freely through the mesh. Figures partially taken from [59]

All these techniques are suited to accomplish the problems of limited spatial accuracy. It then remains to derive efficient time-stepping schemes. This problem is still not sufficiently solved. To illustrate this problem, we step back from the coupled fluid-structure interaction problem and instead discuss a simple parabolic equation

$$(\partial_t u, \phi) + \mathscr{A}(u, \phi) = 0, \quad \mathscr{A}(u, \phi) = \begin{cases} \mathscr{A}_1(u, \phi) & \text{in } \mathscr{S}(t) \\ \mathscr{A}_2(u, \phi) & \text{in } \mathscr{F}(t) \end{cases},$$

where $\mathscr{A}_1$ and $\mathscr{A}_2$ represent two different differential operators and where the interface between the two subdomains moves from one time-step to the other. Direct time-stepping approaches result in iterative schemes of the type

$$(u^m - u^{m-1}, \phi) + \theta k \mathscr{A}(u^m, \phi) + (1-\theta) k \mathscr{A}(u^{m-1}, \phi) = 0. \tag{4}$$

It is now possible, that for a given point $x \in \Omega$ it holds $x \in \Omega_1(t_{m-1})$ but $x \in \Omega_2(t_m)$, i.e., that one points belongs to the fluid domain at the old time-step and the solid-domain at the new time. For such a configuration, the expression $u^m(x) - u^{m-1}(x)$ lacks any physical relevance, as there is not immediate relation between the two different phases. This problem also appears in the discretization of multiphase flows, here however it is justifiable to replace the sharp interface by a smoothed one using harmonic averages of the different parameters, see [48, 53]. Smoothing of two entirely different phases like fluid and solid is however no option.

If problem (4) is to be discretized with fitted finite elements, where the mesh locally resolves the interface, or by the extended finite element technique, different time-steps require difference finite element spaces V_h^{m-1} and V_h^m. Then, evaluation of terms like

$$\mathscr{A}(u^{m-1}, \phi^m),$$

with $u^{m-1} \in V_h^{m-1}$ and $\phi^m \in V_h^m$ requires the projection of basis functions from one mesh to the other and numerical quadrature that carefully resolves all possible areas of non-smoothness.

The difficulties of deriving adequate time-discretizations for the Fully Eulerian scheme are even more articulate, if we simply discuss the discretization of the elastic structure equation in Eulerian coordinates. This relates to solving a partial differential equation on a moving domain. For simplicity, we simply consider the parabolic problem:

$$(\partial_t u, \phi)_{\Omega(t)} + (\nabla u, \nabla \phi)_{\Omega(t)} = 0.$$

Here, ad hoc time-discretization with the backward Euler method using changing finite element spaces would results in

$$(u^m - u^{m-1}, \phi^m)_{\Omega(t_m)} + k(\nabla u^m, \nabla \phi^m)_{\Omega(t_m)} = 0.$$

The expression $u^m - u^{m-1}$ is not valid, as the two functions live on different domains. Averaging techniques that may work for multiphase flow problem will not be applicable here.

Efficient and accurate time-discretization schemes for the Eulerian approach will have to consider the motion of the domain. In the following, we will outline two basic ideas. For details, we refer to [17]. The first idea is closely related to the ALE approach flow problems on moving domains and based on old ideas on the method of characteristics [10]. We start by formulating the problem on the time-dependent domain

$$(\partial_t u, \phi)_{\Omega(t)} + (\nabla u, \nabla \phi)_{\Omega(t)} = 0, \quad t \in (t_{m-1}, t_m). \tag{5}$$

Then, by $T_m(t) : \Omega(t_m) \to \Omega(t)$ we denote the transformation between the domains at time t_m and back at time $t \in (t_{m-1}, t_m)$. In the context of fluid-structure interaction or solid problems in Eulerian coordinates, such a mapping is implicitly given by the deformation. By this transformation, we can map Eq. (5) onto the fixed domain $\Omega(t_m)$ similar to the ALE approach:

$$\big(\det(\nabla T_m)(\partial_t u - \partial_t T_m \cdot \nabla u), \phi\big)_{\Omega(t_m)} + \big(\det(\nabla T_m)\nabla T_m^{-1} \nabla u \nabla T_m^{-T}, \nabla\phi\big)_{\Omega(t_m)} = 0. \tag{6}$$

Nonlinearities are introduced, we however shift the motion of the domain into this implicit mapping such that standard time-stepping schemes can be applied.

An alternative approach is given by formulating (5) as a space-time Galerkin approach, based on

$$\int_{t_{m-1}}^{t_m} \Big\{(\partial_t u(t), \phi(t))_{\Omega(t)} + (\nabla u(t), \nabla\phi(t))_{\Omega(t)}\Big\}\, \mathrm{d}t = 0.$$

Then, following the concepts introduced by Eriksson, Estep, Hansbo, and Johnson [13] and Thomée [52], discrete space-time functions are used to approximate this equation. We know for instance, that by combining piece-wise linear (in time) trial functions with piece-wise constant test-functions, we result in a variant of the Crank-Nicolson scheme. To apply this technique to problems with moving interface, we must consider space-time meshes that fit to the interface, see Fig. 7.

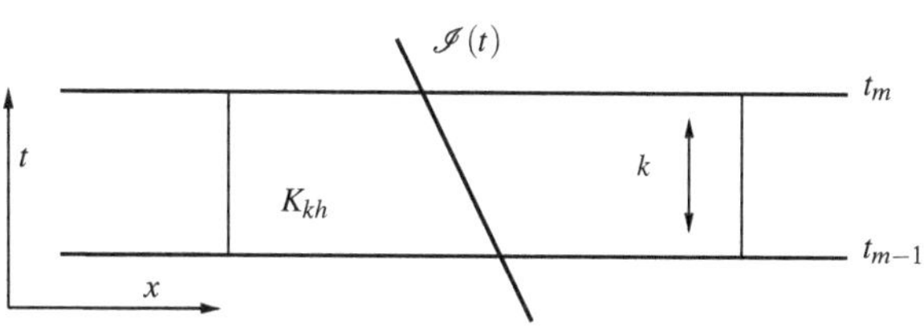

Fig. 7 Triangulation in space and time. The triangulation fits the interface $\mathscr{I}(t)$ in space and time. By K_{kh} we denote one space-time element

A space-time Galerkin approach must now resolve the interface location. To derive a Crank-Nicolson like scheme, the piece-wise linear trial functions must be linear along the interface and not necessarily linear in straight t-direction. This causes an implicit coupling of space- and time-variables. On the element K_{kh} the space-time Galerkin approach is simply the space of bilinear elements in space and time $\text{span}\{1, x, t, xt\}$. This coupling of spacial and temporal variables again leads to an implicit mapping of the space-time slice onto a fixed domain with straight edges which is equivalent to the mapping-approach described in (6). In [17] this technique is analyzed for parabolic problems with moving interfaces. By a proper projection of the old solution to the new finite element space and by applying suitable mapping between the domains at different time-steps, second order schemes in space and time can be derived. First results show however, that the error is not clearly separated into a spatial and a temporal part, but that for elements cut by the interface, simultaneous refinement in space and time is required.

6 Multiple Shooting as Time-Parallel Time-Integration for Fluid-Structure Interaction: Concepts and an Outlook

A further possibility to enhance the stability of time-discretization schemes is to realize them in a time-domain decomposition fashion [19, 33]. Stability issues in the context of fluid-structure interactions mostly appear due to long-time accumulated error contributions. The small time-steps that are necessary to efficiently resolve all these modes are not required for obtaining an adequate accuracy. Here, it might be an option to employ the multiple shooting method for time-discretization. By keeping the sub-intervals small, stability will be under control.

In the following, we briefly explain conceptional issues related to fluid-structure interaction. Once we have set up the basis of a consistent semi-linear form, the solution of the multiple shooting problem follows standard techniques. Specifically, a consistent semi-linear form is provided by a monolithic formulation of the ALE system (1) or the fully Eulerian system (3), respectively.

Find $\mathbf{U} = \{\mathbf{v}_f, \mathbf{v}_s, \mathbf{u}_f, \mathbf{u}_s, p_f\} \in X := \mathcal{V} \times \mathcal{V} \times \mathcal{L}$ such that

$$\int_I A(\mathbf{U})(\Phi) = 0 \quad \forall \Phi \in X$$

with $\Phi = \{\phi, \xi_f, \psi_f, \psi_s\} \in \mathcal{V} \times \mathcal{L} \times \mathcal{V}_f \times \mathcal{V}_s$ and

$$\begin{aligned} A(\mathbf{U})(\Phi) = \Big(\rho_f J(\partial_t \mathbf{v} + \nabla \mathbf{v}\mathbf{F}^{-1}(v - \partial_t \mathbf{u})), \phi\Big)_{\mathcal{F}} + \Big(J\hat{\boldsymbol{\sigma}}_f \mathbf{F}^{-T}, \nabla\phi\Big)_{\mathcal{F}} \\ + \Big(\rho_s d_t \mathbf{v}, \phi\Big)_{\mathcal{S}} + \Big(\mathbf{F}\boldsymbol{\Sigma}_s, \nabla\phi\Big)_{\mathcal{S}} - (f, \phi)_\Omega \qquad (7) \\ + \Big(\det(J\mathbf{F}^{-1}\mathbf{v}), \xi_f\Big)_{\mathcal{F}} + \Big(\nabla\boldsymbol{\sigma}_{\text{mesh}}, \nabla\psi_f\Big)_{\mathcal{F}} + \Big(d_t \mathbf{u} - v, \psi_s\Big)_{\mathcal{S}}. \end{aligned}$$

In fully Eulerian coordinates we have: Find $\mathbf{U} = \{\mathbf{v}_f, \mathbf{v}_s, \mathbf{u}_s, p_f\} \in X := \mathcal{V} \times \mathcal{V}_s \times \mathcal{L}$ such that

$$\int_I A(U)(\Phi) = 0 \quad \forall \Phi \in X$$

with $\Phi = \{\phi, \xi_f, \psi_f, \psi_s\} \in \mathcal{V} \times \mathcal{L} \times \mathcal{V}_f \times \mathcal{V}_s$ and

$$\begin{aligned}
A(\mathbf{U})(\Phi) = (\rho_f(\partial_t \mathbf{v}_f + \mathbf{v}_f \cdot \nabla \mathbf{v}_f), \phi) + (\mathrm{div}\, \boldsymbol{\sigma}_f, \nabla \phi) - (\rho_f f_f, \phi) \\
+(\mathrm{div}\, \mathbf{v}_f, \xi) \\
+(J_s \rho_s(\partial_t \mathbf{v}_s + \mathbf{v}_s \cdot \nabla \mathbf{v}_s), \psi_f) + (\mathrm{div}\, \boldsymbol{\sigma}_s, \nabla \psi_f) - (J_s \rho_s f_s, \psi_f) \\
+(\partial_t \mathbf{u}_s + \mathbf{v}_s \cdot \nabla \mathbf{u}_s - \mathbf{v}_s, \psi_s) \\
= (\rho_f(\partial_t \mathbf{v}_f, \phi) + (J_s \rho_s(\partial_t \mathbf{v}_s, \psi_f) + (\partial_t \mathbf{u}_s, \psi_s) \\
(\rho_f \mathbf{v}_f \cdot \nabla \mathbf{v}_f, \phi) + (\mathrm{div}\, \boldsymbol{\sigma}_f, \nabla \phi) - (\rho_f f_f, \phi) \\
+(\mathrm{div}\, \mathbf{v}_f, \xi) \\
+(J_s \rho_s \mathbf{v}_s \cdot \nabla \mathbf{v}_s, \psi_f) + (\mathrm{div}\, \boldsymbol{\sigma}_s, \nabla \psi_f) - (J_s \rho_s f_s, \psi_f) \\
+(\mathbf{v}_s \cdot \nabla \mathbf{u}_s - \mathbf{v}_s, \psi_s).
\end{aligned} \tag{8}$$

In order to obtain a standard setting for multiple shooting for PDEs [24], we re-arranged and separated the time derivatives from the spatial operators in the previous equation.

In the following, we describe the algorithm for the fully Eulerian case. The formal description (apart from the specific difficulties as described in the previous sections) of the ALE system is analogous. Let $I = (0, T)$ be a decomposition of the time interval into m (not necessarily of the same length) multiple shooting intervals $I_j := (t_j, t_{j+1})$ with

$$0 = t_0 < t_1 < \ldots < t_{m-1} < t_m = T.$$

The multiple shooting formulation asks now for the solution of the matching conditions at the multiple shooting nodes $t_j, j = 0, \ldots, m$ in which we introduce the multiple shooting variables q^j, r^j, s^j (in some Hilbert space) for v_f, v_s, u_s. The variables q^j, r^j, s^j serve as initial value for v_f^j, v_s^j, u_s^j in t_j. Then, the multiple shooting system for the m (separate) interval-wise boundary value problems of fluid-structure interaction read:

$$\int_{I_j} \Big((\rho_f \partial_t \mathbf{v}_f^j, \phi) + (J_s \rho_s \partial_t \mathbf{v}_s^j, \psi_f) + (\partial_t \mathbf{u}_s^j, \psi_s) + A(\mathbf{U}^j)(\Phi) +$$
$$(\mathbf{v}_f^j(t_j) - q^j, \phi(t_j)) + (\mathbf{v}_s^j(t_j) - r^j, \psi_f(t_j)) + (\mathbf{u}_s^j(t_j) - s^j, \psi_s(t_j)) \Big)$$

where the semi-linear form of all 'stationary' terms is given by

$$
\begin{aligned}
A(\mathbf{U}^j)(\Phi) = (\rho_f \mathbf{v}_f \cdot \nabla \mathbf{v}_f, \phi) + (\text{div}\ \boldsymbol{\sigma}_f, \nabla \phi) - (\rho_f f_f, \phi) \\
+(\text{div}\ \mathbf{v}_f, \xi) \\
+(J_s \rho_s \mathbf{v}_s \cdot \nabla \mathbf{v}_s, \psi_f) + (\text{div}\ \boldsymbol{\sigma}_s, \nabla \psi_f) - (J_s \rho_s f_s, \psi_f) \\
+(\mathbf{v}_s \cdot \nabla \mathbf{u}_s - \mathbf{v}_s, \psi_s).
\end{aligned} \tag{9}
$$

Now, the multiple shooting system is solved by finding $q^0, \dots, q^m, r^0, \dots, r^m$, $s^0, \dots, s^m$ such that the matching conditions hold true:

$$
\begin{aligned}
(q^0 - \mathbf{v}_f^0, \psi) &= 0 \quad \forall \text{ admissible } \phi, \\
(s^{j+1} - \mathbf{v}_f^j(t_{j+1}), \psi) &= 0 \quad \forall \text{ admissible } \phi, \quad j = 0, \dots, m-1, \\
(r^0 - \mathbf{v}_s^0, \psi_f) &= 0 \quad \forall \text{ admissible } \psi_f, \\
(r^{j+1} - \mathbf{v}_s^j(t_{j+1}), \psi_f) &= 0 \quad \forall \text{ admissible } \psi_f, \quad j = 0, \dots, m-1, \\
(s^0 - \mathbf{u}_s^0, \psi) &= 0 \quad \forall \text{ admissible } \psi_s, \\
(s^{j+1} - \mathbf{u}_s^j(t_{j+1}), \psi_s) &= 0 \quad \forall \text{ admissible } \psi_s, \quad j = 0, \dots, m-1,
\end{aligned}
$$

These matching conditions form a nonlinear system

$$
F(X) = 0 \quad \text{with } X = (q^0, \dots, q^m, r^0, \dots, r^m, s^0, \dots, s^m),
$$

which can be solved with Newton's method. Details for the general solution algorithm are found in [19]. The implementation and analysis for our fluid-structure interaction systems is planned as next task in our future work. We are specifically interested in the parallel solution capabilities (the link to the parallel algorithm [19, 33]) of this approach because we have to solve for, let us say, ten shooting time intervals for the fsi2-benchmark and $m = 100$ in the case of fsi3-benchmark. This leads to a huge Newton system to solve.

7 Conclusion

We have analyzed implicit time-discretizations of fully coupled monolithic fluid-structure interactions. The difficulties connected to this special application field is two-fold: first, the implicit handling of the mesh motion, captured by the ALE-map, leads to non-standard coupling of temporal and spatial derivatives. For such nonlinear couplings, time discretizations schemes have not been investigated so far. Further, the motion of the mesh, and the coupling of the incompressible Navier-Stokes equations with hyperelastic materials gives rise to stability problems that play an important role for long-time simulations. By analyzing the fsi benchmark problems published by Hron & Turek, these difficulties have been investigated in detail. For running stable long-time simulations, one must either resort to strongly A-stable time-discretization schemes like the fractional step theta method or one must modify standard schemes to improve the stability. A further promising approach to run long-time fluid-structure interaction simulations is the use of the multiple shooting method as a time domain decomposition scheme. Such an approach would allow to use efficient standard schemes like the Crank-Nicolson method by keeping the sub-intervals short.

References

1. Babuška, I.: The finite element method for elliptic equations with discontinuous coefficients. Computing **5**, 207–213 (1970)
2. Babuška, I., Banarjee, U., Osborn, J.E.: Generalized finite element methods: main ideas, results, and perspective. Int. J. Comput. Methods **1**, 67–103 (2004)
3. Bangerth, W., Geiger, M., Rannacher, R.: Adaptive Galerkin finite element methods for the wave equation. Comput. Methods Appl. Math. **10**, 3–48 (2010)
4. Börgers, C.: A triangulation algorithm for fast elliptic solvers based on domain imbedding. SIAM J. Numer. Anal. **27**, 1187–1196 (1990)
5. Braack, M., Richter, T.: Stabilized finite elements for 3 d reactive flows. Int. J. Numer. Math. Fluids **51**, 981–999 (2006)
6. Bramble, J.H., King, J.T.: A finite element method for interface problems in domains with smooth boundaries and interfaces. Adv. Comput. Math. **6**, 109–138 (1996)
7. Bristeau, M.O., Glowinski, R., Periaux, J.: Numerical methods for the Navier-Stokes equations. Comput. Phys. Rep. **6**, 73–187 (1987)
8. Bungartz, H.-J., Schäfer, M. (eds.): Fluid-Structure Interaction II. Modelling, Simulation, Optimisation. Lecture Notes in Computational Science and Engineering. Springer, Berlin (2010)
9. Cottet, G.-H., Maitre, E., Mileent, T.: Eulerian formulation and level set models for incompressible fluid-structure interaction. Math. Model. Numer. Anal. **42**, 471–492 (2008)
10. Douglas, J., Russel, T.F.: Numerical methods for convection dominated diffusion problems based on combining the method of characteristics with finite element or finite difference procedures. SIAM J. Numer. Anal. **19**, 871–885 (1982)
11. Dunne, T.: An Eulerian approach to fluid-structure interaction and goal-oriented mesh refinement. Int. J. Numer. Math. Fluids **51**, 1017–1039 (2006)
12. Dunne, T., Rannacher, R., Richter, T.: Numerical simulation of fluid-structure interaction based on monolithic variational formulations. In: Galdi, G.P., Rannacher, R. (eds.) Comtemporary Challenges in Mathematical Fluid Mechanics. World Scientific, Singapore (2010)
13. Eriksson, K., Estep, D., Hansbo, P., Johnson, C.: Introduction to adaptive methods for differential equations. In: Iserles, A. (ed.) Acta Numerica 1995, pp. 105–158. Cambridge University Press, Cambridge (1995)
14. Fernández, M.A., Gerbeau, J.-F.: Algorithms for fluid-structure interaction problems. In: Formaggia, L., Quarteroni, A., Veneziani, A. (eds.) Cardiovascular Mathematics. MS&A, vol. 1, pp. 307–346. Springer, Milan (2009)
15. Formaggia, L., Nobile, F.: A stability analysis for the arbitrary Lagrangian Eulerian formulation with finite elements. East-West J. Numer. Math. **7**, 105–132 (1999)
16. Formaggia, L., Nobile, F.: Stability analysis of second-order time accurate schemes for ALE-FEM. Comput. Methods Appl. Mech. Eng. **193**(39–41), 4097–4116 (2004)

17. Frei, S., Richter, T.: Time-discretization of parabolic problems with moving interfaces. SIAM J. Numer. Anal. (2015, in preparation)
18. Frei, S., Richter, T.: A locally modified parametric finite element method for interface problems.SIAM J. Numer. Anal. **52**(5), 2315–2334 (2014)
19. Gander, M.J., Vandewalle, S.: Analysis of the parareal time-parallel time-integration method. SIAM J. Sci. Comput. **29**(2), 556–578 (2007)
20. Hansbo, A., Hansbo, P.: An unfitted finite element method, based on Nitsche's method, for elliptic interface problems. Comput. Methods Appl. Mech. Eng. **191**(47–48), 5537–5552 (2002)
21. Hansbo, A., Hansbo, P.: A finite element method for the simulation of strong and weak discontinuities in solid mechanics. Comput. Methods Appl. Mech. Eng. **193**, 3523–3540 (2004)
22. He, P., Qiao, R.: A full-Eulerian solid level set method for simulation of fluid-structure interactions. Microfluid Nanofluid **11**, 557–567 (2011)
23. Helenbrook, B.T.: Mesh deformation using the biharmonic operator. Int. J. Numer. Methods Eng. **56**(7), 1007–1021 (2003)
24. Hesse, H.K., Kanschat, G.: Mesh adaptive multiple shooting for partial differential equations. Part I: linear quadratic optimal control problems. J. Numer. Math. **17**(3), 195–217 (2009)
25. Heywood, J., Rannacher, R.: Finite element approximation of the nonstationary Navier-Stokes problem. iii. smoothing property and higher order error estimates for spatial discretization. SIAM J. Numer. Anal. **25**(3), 489–512 (1988)
26. Heywood, J., Rannacher, R.: Finite element approximation of the nonstationary Navier-Stokes problem. iv. error analysis for second-order time discretization. SIAM J. Numer. Anal. **27**(3), 353–384 (1990)
27. Hirt, C.W., Amsden, A.A., Cook, J.L.: An arbitrary lagrangian-eulerian computing method for all flow speeds. J. Comput. Phys. **14**, 227–469 (1974)
28. Holzapfel, G.A.: Nonlinear Solid Mechanics: A Continuum Approach for Engineering. Wiley-Blackwell, Engelska (2000)
29. Hron, J., Turek, S.: A monolithic FEM/multigrid solver for an ALE formulation of fluid-structure interaction with applications in biomechanics. In: Bungartz, H.-J., Schäfer, M. (eds.) Fluid-Structure Interaction: Modeling, Simulation, Optimization. Lecture Notes in Computational Science and Engineering, pp. 146–170. Springer, Berlin (2006)
30. Hron, J., Turek, S.: Proposal for numerical benchmarking of fluid-structure interaction between an elastic object and laminar incompressible flow. In: Bungartz, H.-J., Schäfer, M. (eds.) Fluid-Structure Interaction: Modeling, Simulation, Optimization. Lecture Notes in Computational Science and Engineering, pp. 371–385. Springer, Berlin (2006)
31. Hron, J., Turek, S., Madlik, M., Razzaq, M., Wobker, H., Acker, J.F.: Numerical simulation and benchmarking of a monolithic multigrid solver for fluid-structure interaction problems with application to hemodynamics. In: Bungartz, H.-J., Schäfer, M. (eds.) Fluid-Structure Interaction II: Modeling, Simulation, Optimization. Lecture Notes in Computational Science and Engineering, pp. 197–220. Springer, Berlin (2010)
32. Huerta, A., Liu, W.K.: Viscous flow with large free-surface motion. Comput. Methods Appl. Mech. Eng. **69**(3), 277–324 (1988)
33. Lions, J.-L., Mayday, Y., Turinici, G.: A parareal in time-discretization of PDEs. C. R. Acad. Sci. Paris Sér. I Math. **332**, 661–668 (2001)
34. Luskin, M., Rannacher, R.: On the smoothing propoerty of the Crank-Nicholson scheme. Appl. Anal. **14**, 117–135 (1982)
35. MacKinnon, R.J., Carey, G.F.: Treatment of material discontinuities in finite element computations. Int. J. Numer. Meth. Eng. **24**, 393–417 (1987)
36. Moës, N., Dolbow, J., Belytschko, T.: A finite element method for crack growth without remeshing. Int. J. Numer. Meth. Eng. **46**, 131–150 (1999)
37. Meidner, D., Richter, T.: A posteriori error estimation for the theta and fractional step theta time-stepping schemes. Comput. Methods Appl. Math. **14**, 203–230 (2014)

38. Meidner, D., Richter, T.:A posteriori error estimation for the fractional step theta discretization of the incompressible Navier-Stokes equations.Comput. Methods Appl. Mech. Eng. **288**, 45–59 (2015)
39. Osher, S., Fedkiw, R.: Level Set Methods and Dynamic Implicit Surfaces. Applied Mathematical Sciences. Springer, Berlin (2003)
40. Rannacher, R.: Finite element solution of diffusion problems with irregular data. Numer. Math. **43**, 309–327 (1984)
41. Richter, T.: Fluid-structure interactions in Fully Eulerian coordinates. In: PAMM, 83st Annual Meeting of the International Association of Applied Mathematics and Mechanics, pp. 827–830 (2012)
42. Richter, T.: Goal oriented error estimation for fluid-structure interaction problems. Comput. Methods Appl. Mech. Eng. **223–224**, 28–42 (2012)
43. Richter, T.: A Fully Eulerian formulation for fluid-structure-interaction problems. J. Comput. Phys. **233**, 227–240 (2013)
44. Richter, T.: Finite Elements for Fluid-Structure Interactions. Lecture Notes in Computational Science and Engineering. Springer, Berlin (2014, submitted)
45. Richter, T., Wick, T.: Finite elements for fluid-structure interaction in ALE and Fully Eulerian coordinates. Comput. Methods Appl. Mech. Eng. (2010). doi:10.1016/j.cma.2010.04.016
46. Schäfer, M., Turek, S.: Benchmark computations of laminar flow around a cylinder. (With support by F. Durst, E. Krause and R. Rannacher). In: Hirschel, E.H. (ed.) Flow Simulation with High-Performance Computers II. DFG priority research program results 1993–1995. Notes Numer. Fluid Mech., vol. 52, pp. 547–566. Vieweg, Wiesbaden (1996)
47. Sethian, J.A.: Level Set Methods and Fast Marching Methods Evolving Interfaces in Computational Geometry. Fluid Mechanics, Computer Vision and Material Science. Cambridge University Press, Cambridge (1999)
48. Shubin, G.R., Bell, J.B.: An analysis of grid orientation effect in numerical simulation of miscible displacement. Comput. Methods Appl. Mech. Eng. **47**, 47–71 (1984)
49. Stein, K., Tezduyar, T., Benney, R.: Mesh moving techniques for fluid-structure interactions with large displacements. J. Appl. Mech. **70**, 58–63 (2003)
50. Sugiyama, K., Li, S., Takeuchi, S., Takagi, S., Matsumato, Y.: A full Eulerian finite difference approach for solving fluid-structure interacion. J. Comput. Phys. **230**, 596–627 (2011)
51. Takagi, S., Sugiyama, K., Matsumato, Y.: A review of full Eulerian mehtods for fluid structure interaction problems. J. Appl. Mech. **79**(1), 010,911 (2012)
52. Thomée, V.: Galerkin Finite Element Methods for Parabolic Problems. Springer Series in Computational Mathematics, vol. 25. Springer, Berlin (1997)
53. Tikhonov, A.N., Samarskii, A.A.: Homogeneous difference schemes. USSR Comput. Math. Math. Phys. **1**, 5–67 (1962)
54. Turek, S., Rivkind, L., Hron, J., Glowinski, R.: Numerical analysis of a new time-stepping theta-scheme for incompressible flow simulations. Ergebnisberichte des Instituts für Angewandte Mathematik, vol. 282. Technical Report, Fakultät für Mathematik, TU Dortmund (2005).
55. Turek, S., Hron, J., Razzaq, M., Wobker, H., Schäfer, M.: Numerical benchmarking of fluid-structure interaction: a comparison of different discretization and solution approaches. In: Bungartz, H.J., Mehl, M., Schäfer, M. (eds.) Fluid Structure Interaction II: Modeling, Simulation and Optimization. Springer, Berlin (2010)
56. Wick, T.: Adaptive finite element simulation of fluid-structure interaction with application to heart-valve dynamics. Ph.D. thesis, Univeristy of Heidelberg (2011). urn:nbn:de:bsz:16-opus-129926
57. Wick, T.: Fluid-structure interactions using different mesh motion techniques. Comput. Struct. **89**(13–14), 1456–1467 (2011)
58. Wick, T.: Coupling of fully Eulerian and arbitrary Lagrangian-Eulerian methods for fluid-structure interaction computations problems. Comput. Mech. **52**(5), 1113–1124 (2013)
59. Wick, T.: Coupling of fully Eulerian and arbitrary Lagrangian-Eulerian methods for fluid-structure interaction computations. Comput. Mech. **52**(5), 1113–1124 (2013)

60. Wick, T.: Flapping and contact FSI computations with the fluid—solid Interface-tracking/Interface-capturing technique and mesh adaptivity. Comput. Mech. **53**(1), 29–43 (2014)
61. Wick, T.: Solving monolithic fluid-structure interaction problems in arbitrary Lagrangian Eulerian coordinates with the deal.ii library. Arch. Numer. Softw. **1**, 1–19 (2013)
62. Wick, T.: Stability estimates and numerical comparison of second order time-stepping schemes for fluid-structure interactions. In: Cangiani, A., Davidchack, R.L., Georgoulis, E., Gorban, A.N., Levesley, J., Tretyakov, M.V. (eds.) Numerical Mathematics and Advanced Applications 2011, Proceedings of ENUMATH 2011, Leicester, September 2011, pp. 625–632 (2013)
63. Xie, H., Ito, K., Li, Z.-L., Toivanen, J.: A finite element method for interface problems with locally modified triangulation. Contemp. Math. **466**, 179–190 (2008)
64. Zhao, H., Freund, J.B., Moser, R.D.: A fixed-mesh method for incompressible flow-structure systems with finite solid deformations. J. Comput. Phys. **227**(6), 3114–1340 (2008)

Parareal Time-Stepping for Limit-Cycle Computation of the Incompressible Navier-Stokes Equations with Uncertain Periodic Dynamics

Michael Schick

Abstract The computation of limit-cycles in time periodic flow problems plays a crucial role in quantifying its dynamical characteristics. Since in many applications model parameters are often subject to uncertainty, the limit-cycle becomes a stochastic quantity itself. In this work we introduce two types of shooting methods based on Polynomial Chaos and the stochastic Galerkin projection. Polynomial Chaos is known to exhibit a convergence breakdown in time, which our proposed algorithms are able to overcome. The first algorithm is a re-interpretation of a Newton-Galerkin method as a single shooting approach. The second one extends this idea using the parareal algorithm on a time domain decomposition. It uses a fine grid propagator, which can be computed in parallel and a coarse grid propagator defined by a low order Polynomial Chaos expansion. We evaluate the convergence properties on a suitable benchmark problem (time periodic vortex shedding) showing promising results in capturing accurately the periodic dynamics of the flow.

1 Introduction

Shooting methods are a popular approach for the computation of stable periodic orbits (limit-cycles), which may arise for example in parabolic type partial differential equations. For example, Duguet et al. [3] introduced a Newton method for determining stable limit-cycles of the unsteady incompressible Navier-Stokes equations, which can be interpreted as a single shooting approach. However, in many fluid flow problems parameters like the kinematic viscosity, boundary and initial conditions as well as forcing terms might be uncertain, for example due to measurement or calibration errors. These uncertainty result in an uncertain

M. Schick (✉)
Heidelberg Institute for Theoretical Studies, Research Group: Data Mining and Uncertainty Quantification, Schloss-Wolfsbrunnenweg 35, 69118 Heidelberg, Germany
e-mail: michael.schick@h-its.org

T. Carraro et al. (eds.), *Multiple Shooting and Time Domain Decomposition Methods*, Contributions in Mathematical and Computational Sciences 9,
DOI 10.1007/978-3-319-23321-5_16

limit-cycle, or equivalently in an almost surely time-periodic flow characterized by a tuple of some non-unique uncertain initial condition and corresponding uncertain period.

In the recent decade, Polynomial Chaos (PC) expansions [6, 23] became increasingly popular to model the dependency of the flow profile on the uncertainty. A finite series of multivariate orthogonal polynomials is used in terms of a vector of independent random variables. It is widely employed if the flow profile is assumed to be square-integrable with respect to the probability space (finite variance) and if no information about the probability distribution of the flow is known a priori. There exist various numerical methods for determining the coefficients in the PC expansion, such as Monte-Carlo or collocation approaches [1, 17]. Alternatively, the stochastic Galerkin projection [6, 14] allows for a best-approximation property in the L^2-sense. However, in time-dependent problems, PC expansions are known to exhibit a convergence breakdown in the presence of oscillatory dynamics [8, 15] making the determination of stochastic limit-cycles a difficult task. The reason for this is that the trajectories corresponding to sample realizations of the uncertainty exhibit a phase-drift. This necessitates a growing polynomial degree in time to maintain accuracy in the approximation, which quickly can become too costly from the numerical point of view. Several methods exist, which allow to extend the point of convergence breakdown in time, see for example [2, 8, 12, 13]. However, non of these address the central difficulty regarding the phase-drift problem.

In this work, we introduce two types of shooting methods for the determination of stochastic limit-cycles of the incompressible Navier-Stokes equations with uncertain parameters. The first approach is a single shooting method based on the developments in our preliminary work [20]. Here, a Newton-Galerkin method was defined in order to compute a tuple of stochastic initial condition and a stochastic period such that each stochastic sample follows a time-periodic trajectory. This was achieved by introduction of an optimality criteria, which stabilizes the phase-drift mentioned earlier. The second approach extends this idea towards multiple shooting/time domain decomposition. It is based on the parareal time integration method [5, 16] adapted to the stochastic computational domain. The parareal algorithm can be obtained by approximating the Jacobian arising in the multiple shooting method by a finite difference scheme on a coarse grid allowing for a parallel computation on fine grid domains. In its classical framework, a coarse grid propagator is defined as an approximation to the fine grid propagator by using either larger time step sizes or lower order time integration methods. In this work, we investigate the use of a coarse propagator based on a low order PC expansion to give a prediction on the limit-cycle, which afterwards is corrected in parallel by a solution obtained by using a higher order PC expansion. Both the single and multiple shooting approach require multiple solves of the (stochastic) Navier-Stokes equations by means of the Finite-Element method. For an overview on existing Finite-Element solvers for the deterministic Navier-Stokes equations see e.g. [4, 7, 22]. For solving and preconditioning the stochastic case using the stochastic Galerkin projection see e.g. [9–11, 18, 19].

This work is structured as follows. Section 2 introduces the unsteady stochastic incompressible Navier-Stokes equations along with the problem definition of finding almost surely time-periodic solutions. Section 3 describes a single shooting approach to compute a tuple of an initial condition and the period. In Sect. 4 we introduce an extension using a parareal time integration method. The convergence properties of the algorithm are evaluated on a benchmark problem in Sect. 5. Finally, Sect. 6 provides a short summary and conclusions.

2 Problem Formulation

2.1 The Incompressible Navier-Stokes Equations with Uncertain Parameters

We consider the unsteady incompressible Navier-Stokes equations in their primitive velocity/pressure (u, p) formulation subject to an uncertain boundary g and initial condition U_0, which we will abbreviate by SNSE:

$$\begin{aligned} \partial_t u(x,t,\xi) + (u(x,t,\xi)\cdot\nabla)u(x,t,\xi)& \\ -\nu(\xi)\Delta u(x,t,\xi) + \nabla p(x,t,\xi) &= 0, && \text{in } \mathscr{D}, && (1)\\ \nabla\cdot u(x,t,\xi) &= 0, && \text{in } \mathscr{D}, && (2)\\ u(x,t,\xi) &= g(x,t,\xi), && \text{on } \Gamma, && (3)\\ u(x,t=0;\xi) &= U_0(x,\xi), && \text{in } \mathscr{D}, && (4)\end{aligned}$$

for $t > 0$ almost surely in Ω. Here, Ω denotes a sample set belonging to some probability space and $\xi : \Omega \to \mathbb{R}^M$ denotes a random vector of dimension M with independent components (parametrization of uncertainty). Note that throughout this work we assume the probability distribution for ξ to be given.

The equations are posed on a computational domain denoted by $\mathscr{D}\subset\mathbb{R}^d$, $d=2,3$ with Dirichlet boundary $\Gamma \subseteq \partial\mathscr{D}$. In general the kinematic viscosity ν is also allowed to be uncertain, i.e., $\nu = \nu(\xi)$.

Note that $u = u(x,t;\xi)$ is a random field due to the explicit dependency on the random vector ξ through the partial differential equations. We employ the Spectral-Stochastic-Finite-Element-Method for discretization of the SNSE (see e.g. [9–11] and references in [14]). To this end, we need to assume that u and p satisfy the following regularity requirements:

$$u \in V \otimes \mathscr{S}, \quad p \in W \otimes \mathscr{S},$$

where for the remainder of this work we define:

$$\mathscr{S} := L^2(\Omega), \quad V := H^1(\mathscr{D}), \quad V_0 := H_0^1(\mathscr{D}), \quad W := L^2(\mathscr{D}).$$

The weak form of the SNSE with respect to the spatial part in $\mathscr{D}$ reads:

$$(\partial_t u, \phi) + ((u \cdot \nabla)u, \phi) + \nu(\nabla u, \nabla \phi) - (p, \nabla \cdot \phi) = 0, \qquad \forall \phi \in V_0, \tag{5}$$

$$(q, \nabla \cdot u) = 0, \qquad \forall q \in W, \tag{6}$$

$$u(x,t,\xi) = g(x,t,\xi), \qquad \text{on } \Gamma, \tag{7}$$

$$u(x,t=0;\xi) = U_0(x,\xi), \qquad \text{in } \mathscr{D}, \tag{8}$$

for $t \geq 0$ almost surely in Ω. Here, $(\cdot,\cdot)$ denotes the inner-product on W. In addition Eqs. (5)–(8) need to be approximated in the stochastic domain Ω. Therefore, we employ a p-th order PC expansion of the velocity and pressure variable:

$$[u(x,t;\xi), p(x,t;\xi)] = \sum_{i=0}^{P} [u_i(x,t), p_i(x,t)]\psi_i(\xi).$$

The truncation parameter P satisfies $(P+1) = (p+M)!/p!M!$, where $p \in \mathbb{N}$ denotes the maximum total polynomial degree of the normalized Chaos Polynomials ψ_i, $i = 0,1,2,\ldots$. Their choice is determined by an orthogonality condition with respect to the probability distribution of ξ, i.e.,

$$\langle \psi_i, \psi_j \rangle := \mathbb{E}(\psi_i \psi_j) = \int_{\mathbb{R}^M} \psi_i(\xi)\psi_j(\xi)w(\xi)\, d\xi = \delta_{ij}.$$

Here, w denotes the probability density function of ξ and δ_{ij} denotes the Kronecker symbol. For example, a Gaussian distribution of ξ corresponds to Hermite polynomials, a Uniform distribution corresponds to Legendre polynomials. We refer the reader to [24] for more details on generalized PC expansions.

We shall denote $\mathscr{S}^P$ the subspace of $\mathscr{S}$ spanned by the PC basis. For deriving the stochastic Galerkin projection of the SNSE, we multiply (5)–(8) by a test function ψ_k and integrate with respect to the probability space:

$$(\partial_t u_k, \phi) + \sum_{i,j=0}^{P} ((u_i \cdot \nabla)u_j, \phi)\langle \psi_i \psi_j, \psi_k \rangle$$

$$+ \langle \nu, \psi_k \rangle (\nabla u_k, \nabla \phi) - (p_k, \nabla \cdot \phi) = 0, \qquad \forall \phi \in V_0, \tag{9}$$

$$(q, \nabla \cdot u_k) = 0, \qquad \forall q \in W, \tag{10}$$

$$u_k = \langle g, \psi_k \rangle, \qquad \text{on } \Gamma, \tag{11}$$

$$u_k = \langle U_0, \psi_k \rangle, \qquad \text{in } \mathscr{D}, \tag{12}$$

for $k = 0,\ldots,P$. In the following, we refer to (9)–(12) as the standard weak formulation of the SNSE.

2.2 Periodic Orbits (Limit-Cycles)

A stable periodic orbit (limit-cycle) can be characterized by a tuple (U_0, T) of a stochastic initial condition $U_0 = U_0(x, \xi)$, $x \in \overline{\mathscr{D}}$ and a corresponding stochastic period $T = T(\xi)$. In general, the trajectories can be computed as a limit of straight forward time integration. However, if the governing equations do not converge towards a stable steady state solution, it is known that a PC expansion can exhibit a convergence breakdown in time, i.e., an increasing order in the PC expansion is required to accurately capture the dynamics of the flow. This is due to a phase-drift of the trajectories in the limit-cycle, since the initial condition U_0 is not uniquely defined. More specifically, for each stochastic realization of the random input ξ, every point on the corresponding deterministic limit-cycle trajectory can serve as another initial condition on that trajectory, requiring a large degree in the PC expansion.

Maintaining high accuracies in time by using low order PC expansions especially for parabolic partial differential equations is still an active field of research. In this work, we address this difficulty by introducing additional constraints to the SNSE, which allow a direct computation of a characteristic tuple (U_0, T) from which the limit-cycle can be reconstructed.

For the remainder of this work we need to assume the following:

Assumption 1 *There exists a solution* (u, p) *to* (5)–(8), $u(\cdot, t, \cdot) \in V \otimes \mathscr{S}$, $p(\cdot, t, \cdot) \in W \otimes \mathscr{S}$ *for* $t \geq 0$, *and a bounded period* $T \in \mathscr{S}$ *with* $\alpha \leq T < \infty$ *a.s. for some* $\alpha > 0$, *such that for* $t \geq 0$:

$$u(x, t + T(\xi); \xi) = u(x, t; \xi), \quad \forall x \in \overline{\mathscr{D}}, \quad a.s.$$

$$p(x, t + T(\xi); \xi) = p(x, t; \xi), \quad \forall x \in \overline{\mathscr{D}}, \quad a.s.$$

The SNSE have to be integrated to a final state at time $t = T(\xi)$, which is unknown. The normalization approach outlined in the following introduces an additional random variable to the governing equations by mapping the uncertain time domain to a deterministic reference interval. Therefore, the trajectories corresponding to different realizations of the random input can be compared to each other on the same deterministic reference interval $[0, 1]$, allowing the use of a deterministic time-stepping methods.

We introduce a scaled time variable $\lambda = \lambda(t, \xi)$:

$$\lambda(t, \xi) := \frac{t}{T(\xi)}, \tag{13}$$

pointwise in Ω and $t \geq 0$. Note that λ is a random process and $\lambda(t, \cdot) \in \mathscr{S}$ for $t \geq 0$ provided that the period T satisfies Assumption 1.

We define:

$$\tilde{u}(x, t/T(\xi), \xi) := u(x, t, \xi), \quad \tilde{p}(x, t/T(\xi), \xi) := T(\xi)p(x, t, \xi).$$

Applying the stochastic Galerkin projection after introducing the mapping $t \to \lambda(t, \cdot)$ into (5)–(8) results in a scaled version of the unsteady stochastic incompressible Navier-Stokes equations for $\lambda > 0$, which we will abbreviate by **S-SNSE**:

$$\begin{aligned}
(\partial_\lambda \tilde{u}_k, \phi) &+ \sum_{i,j=0}^{P} (\tilde{u}_i \cdot \nabla)\tilde{u}_j, \phi) c(T)_{ijk} \\
&+ \sum_{i=0}^{P} (\nabla \tilde{u}_i, \phi)\nu(T)_{ik} - (\tilde{p}_k, \nabla \cdot \phi) = 0, \qquad \forall \phi \in V_0, \qquad (14) \\
&(q, \nabla \cdot \tilde{u}_k) = 0, \qquad \forall q \in W, \qquad (15) \\
&\tilde{u}_k = \langle g, \psi_k \rangle, \qquad \text{on } \Gamma, \qquad (16) \\
&\tilde{u}_k = \langle U_0, \psi_k \rangle, \qquad \text{in } \mathscr{D}, \qquad (17)
\end{aligned}$$

for $\lambda > 0$ and $k = 0, \ldots, P$. The $3rd$ and $2nd$ order tensors $c(T)$ and $\nu(T)$ are defined by:

$$c(T)_{ijk} := \sum_{l=0}^{Q} T_i \langle \psi_i \psi_j \psi_l, \psi_k \rangle, \quad \nu(T)_{ik} := \sum_{l=0}^{Q} T_i \langle \nu \psi_l \psi_i, \psi_k \rangle,$$

for $i, j, k = 0, \ldots, P$. The period T is approximated by means of a qth order PC expansion, such that

$$T(\xi) = \sum_{i=0}^{Q} T_i \psi_i(\xi), \quad Q + 1 = \frac{(q + M)!}{q!M!}.$$

We denote consistently $\mathscr{S}^Q$ the corresponding subspace of $\mathscr{S}$.

In the definition of $c(T)$ we need to evaluate multiple integrals of products of four polynomials. However, since many of these products are equal to zero, the number of coupling terms in (14) is not that large as the size of the tensor suggests.

In the following the tilde notation for $\tilde{u}$ and $\tilde{p}$ will be dropped for notational convenience.

2.3 Characterization of a Limit-Cycle

We define the operator $\mathscr{U}$ to track the velocity as a function of the scaled time variable $\lambda \geq 0$ subject to some period T and initial condition U_0:

$$\mathscr{U}(U_0, T, \lambda) := U_0 + \int_0^{\lambda} \partial_\lambda u(\lambda = \sigma)\, d\sigma, \tag{18}$$

where u satisfies (14)–(17). $\mathscr{U}$ represents the velocity at time λ starting from the initial condition U_0.

The problem of computing limit-cycles of the S-SNSE can be stated in the following way:

Find some $U_0 \in \mathscr{S}^P \otimes V$, whose PC coefficients satisfy (15) and (16), and a corresponding T as in Assumption 1, such that

$$\|\mathscr{U}(U_0, T, 1) - U_0\|^2 = 0, \tag{19}$$

where here and for the rest of this work $\|\cdot\| := \|\cdot\|_{\mathscr{S}\otimes W}$.

3 A Single Shooting Method

In the following we re-interpret an iterative Newton-Galerkin method in the context of single shooting methods, which originally was introduced in our preliminary work [20].

For $\lambda \geq 0$ and some initial condition U_0 we define a distance vector D by:

$$D(U_0, T, \lambda) := U_0 - \mathscr{U}(U_0, T, \lambda).$$

Therefore, $\|D\|$ measures the distance between an initial condition U_0 and its resulting velocity $\mathscr{U}(U_0, T, \lambda)$ at time λ. The goal of the iterative method is to obtain convergence such that:

$$\|D(U_0^k, T^k, \lambda)\| \to 0 \quad \text{as } k \to \infty.$$

3.1 A Simplified Algorithm

We start by choosing some appropriate initial guesses $U_0^0 = \sum_{i=0}^{P} U_{0,i}^0 \psi_i$ and $T^0 = \sum_{i=0}^{Q} T_i^0 \psi_i$ for the initial condition and period, respectively. As an initial guess for U_0^0 and corresponding T^0 we suggest to use a fully developed deterministic flow, i.e., $U_{0,i}^0 = 0$ and $T_i^0 = 0$ for $i > 0$, which can be computed a priori by integration of the deterministic Navier-Stokes equations parametrized by the mean of the random input. Afterwards, for $\epsilon > 0$ we perform iterations in the following sense (k denotes the iteration index):

1. Compute a new period iterate T^{k+1} (see Sect. 3.2).
2. Compute distance $\delta := \|D(U_0^k, T^{k+1}, 1)\|$.
3. If $\delta < \epsilon$ or $k > k_{\max}$ STOP, else set
   $u^{k+1} := \mathscr{U}(U_0^k, T^{k+1}, 1),\ k \leftarrow k+1$ and GOTO 1.

Note that for simplicity, the terminal condition $\mathscr{U}(U_0^k, T^{k+1}, 1)$ is taken as a new iterate for the initial condition. In case that the periodic orbit is stable, it is known that this approach converges as it corresponds to integrating the governing equations in time. However, to speed up convergence, the update of the initial condition in step no. 3 can be performed by employing Newton's method, which is equivalent to a single shooting approach with adapted period values in each Newton iteration. In Sect. 3.3 we provide a description of the single shooting variant.

3.2 Computing the Period Update

We define an optimization problem, which ensures that the distance between the initial condition and its corresponding terminal state remains minimal in an L^2-sense. Therefore, given the *kth* iterates U_0^k and T^k, we correct the period T^k by:

$$T^{k+1} = (1 + d\lambda)T^k \approx \sum_{m=0}^{Q} T_m^{k+1} \psi_m, \tag{20}$$

$$T_m^{k+1} = T_m^k + \sum_{i=0}^{Q} \sum_{j=0}^{Q} d\lambda_i T_j^k \langle \psi_i \psi_j, \psi_m \rangle, \quad m = 0, \ldots, Q, \tag{21}$$

resulting from a Galerkin projection using the PC expansion of $d\lambda \in \mathscr{S}^Q$. Here, $d\lambda$ is a solution of the minimization problem:

$$\min_{d\lambda \in \mathscr{S}^Q} \|U_0^k - \mathscr{U}(U_0^k, T^k, 1 + d\lambda)\|^2. \tag{22}$$

To simplify problem (22), we approximate $\mathscr{U}(U_0^k, T^k, 1 + d\lambda)$ by its first order Taylor series representation around $\lambda = 1$, i.e.

$$\mathscr{U}(U_0^k, T^k, 1 + d\lambda) \approx \mathscr{U}(U_0^k, T^k, 1) + \partial_\lambda \mathscr{U}(U_0^k, T^k, 1) d\lambda. \tag{23}$$

Inserting (23) in (22) results in a linearized stochastic optimization problem for the correction term $d\lambda$:

$$\min_{d\lambda \in \mathscr{S}} \|D(U_0^k, T^k, 1) - \partial_\lambda \mathscr{U}(U_0^k, T^k, 1) d\lambda\|^2. \tag{24}$$

The corresponding optimality condition reads:

$$2\left\langle \left(D(U_0^k, T^k, 1) - \partial_\lambda \mathscr{U}(U_0^k, T^k, 1) d\lambda, \partial_\lambda \mathscr{U}(U_0^k, T^k, 1) d\mu\right)\right\rangle = 0, \quad \forall d\mu \in \mathscr{S}^Q. \tag{25}$$

Introducing the PC expansions of the various stochastic quantities into (25) we arrive at the discrete optimality condition for the PC modes $\mathbf{d\lambda} = [d\lambda_0, \ldots, d\lambda_Q]^t \in \mathbb{R}^{Q+1}$:

$$A\, \mathbf{d\lambda} = b, \tag{26}$$

where $A \in \mathbb{R}^{Q+1,Q+1}$ and $b \in \mathbb{R}^{Q+1}$ are defined by:

$$A_{ml} := \sum_{i,j=0}^{P} \langle \psi_i \psi_j \psi_l, \psi_m \rangle (\partial_\lambda \mathscr{U}_i^k, \partial_\lambda \mathscr{U}_j^k), \; b_m := \sum_{i,j=0}^{P} \langle \psi_i \psi_j, \psi_m \rangle (D_i^k, \partial_\lambda \mathscr{U}_j^k),$$

for $m, l = 0, \ldots, Q$. The corresponding Polynomial Chaos coefficients of $\partial_\lambda \mathscr{U}$ and D are given by:

$$\partial_\lambda \mathscr{U}_i^k = \langle \partial_\lambda \mathscr{U}(U_0^k, T^k, 1), \psi_i \rangle, \quad D_i^k = \langle D(U_0^k, T^k, 1), \psi_i \rangle,$$

for $i = 0, \ldots, P$. In [20] we have shown that this minimization problem is convex.

3.3 A Single Shooting Approach

In this section we employ Newton's method for updating the initial condition iterate. To this end, recapitulate the problem definition:

Find a tuple (U_0, T) such that

$$D(U_0, T, 1) := U_0 - \mathscr{U}(U_0, T, 1) = 0. \tag{27}$$

Clearly, this system is underdetermined due to the two unknowns U_0 and T. However, using the constraint on the period computation as introduced in Sect. 3.2 we can apply Newton's method with respect to the initial condition U_0:

$$\begin{aligned} U_0^{k+1} &= U_0^k + du^k \\ -J^k du^k &= D(U_0^k, T^{k+1}, 1). \end{aligned}$$

with iteration index $k = 0, 1, 2, \ldots$. Here, J^k denotes the Jacobian of D with respect to U_0 at (U_0^k, T^{k+1}).

This results in the following iterative scheme:

1. Compute period T^{k+1} by (21).
2. Compute Newton correction du^k by: $-J^k du^k = D(U_0^k, T^{k+1}, 1)$.
3. Update initial condition $U_0^{k+1} = U_0^k + du^k$.
4. GOTO 1. until convergence.

As an initial guess for U_0^0 and T^0 we suggest to use a fully developed deterministic flow with $U_{0,i}^0 = 0$ and $T_i^0 = 0$ for $i > 0$.

3.3.1 Solving the Newton Step

Note that by definition of D the application of the Jacobian J^k to some du^k can be simplified:

$$J^k du^k = du^k - J_{NS}^k du^k,$$

where J_{NS}^k denotes the Jacobian of $\mathcal{U}(U_0^k, T^{k+1}, 1)$. In [20] we have shown, that computing $J_{NS}^k du^k$ corresponds to a time integration of the linearized Navier-Stokes equations up to the time $\lambda = 1$:

$$\begin{aligned} \partial_\lambda v + T^{k+1}(v \cdot \nabla)u^k + T^{k+1}(u^k \cdot \nabla)v & \\ -\nu T^{k+1}\Delta v + \nabla q = 0, & \quad \text{in } \mathcal{D}, \qquad (28) \\ \nabla \cdot v = 0, & \quad \text{in } \mathcal{D}, \qquad (29) \\ v(x, \lambda, \xi) = 0, & \quad \text{on } \Gamma, \qquad (30) \\ v(x, \lambda = 0, \xi) = du^k(x, \xi), & \quad \text{in } \mathcal{D}, \qquad (31) \end{aligned}$$

such that $[v, q] = J_{NS}^k du^k$. Note, that for computing a solution to the Jacobian in the Newton step usually some iterative solver can be applied. Each iteration of this solver requires an evaluation of $J_{NS}^k du^k$ at a given iterate du^k and therefore the integration of the linearized equations. Alternatively, it is also possible to approximate the Jacobian in terms of Finite-Differences in each solver iteration, i.e.,

$$J_{NS}^k du^k \approx \frac{\mathcal{U}(U_0^k + \epsilon du^k, T^k, 1) - \mathcal{U}(U_0^k, T^k, 1)}{\epsilon},$$

for some $\epsilon > 0$. However, this approach requires the solution of the full nonlinear equations.

3.3.2 Additional Phase Control

In general, the initial condition U_0 is not uniquely defined with respect to the stochastic time scaling through λ. Indeed, if U_0 is a valid initial condition then $\mathscr{U}(U_0, T(\xi), \beta(\xi))$ is another valid initial condition, $\forall \beta(\xi) > 0$. This can cause a phase-drift of the trajectories associated to each realization of ξ, which needs to be controlled in an appropriate way (cf. Fig. 1).

The phase-drift can be measured as an angle between the time tangential of a reference trajectory $\partial_\lambda \hat{u}$ and its distance to the stochastic velocity field u at time λ, i.e.,

$$\Sigma(\lambda, \xi) := \frac{(u - \hat{u}, \partial_\lambda \hat{u})}{(\partial_\lambda \hat{u}, \partial_\lambda \hat{u})^{1/2}}.$$

As a reference, we take $\hat{u}$ to be the stochastic velocity evaluated at the mean of ξ, i.e.,

$$\hat{u}(x, \lambda) := u(x, \lambda, \xi)|_{\xi = \mathbb{E}(\xi)}.$$

Note that $(\partial_\lambda \hat{u}, \partial_\lambda \hat{u}) > 0$ in an almost surely unsteady flow.

We introduce a stochastic rescaling of λ by:

$$\tau(t, \xi) := \frac{\lambda(t, \xi)}{\sigma_t(\xi)},$$

where we choose σ_t to be constant in each time step satisfying

$$\sigma_t(\xi) < 1, \quad \text{if } \Sigma > 0, \quad \text{(slow down)},$$
$$\sigma_t(\xi) > 1, \quad \text{if } \Sigma < 0, \quad \text{(speed up)}.$$

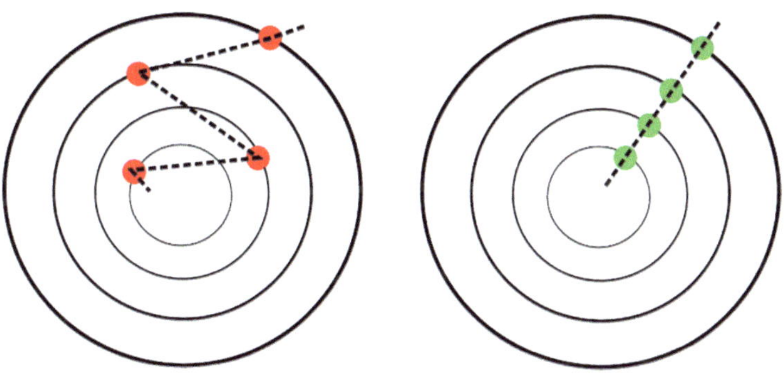

Fig. 1 Illustration of a phase-drift. Fixing an initial condition on each trajectory results in some interpolation between these *points*, which degree of nonlinearity depends their choice

As a simple heuristic we define in each time step

$$\sigma_t(\xi) := (1 + \theta \frac{1}{\Delta\lambda}(\partial_\lambda u, \partial_\lambda u)\Sigma(\lambda, \xi)),$$

where $\theta > 0$ is some prescribed control parameter and $\Delta\lambda > 0$ denotes the time stepping size. If the trajectories are in-phase σ equals to 1 by definition. In the phase-control step, the governing equations (14)–(17) are adapted to the new time scale τ. The resulting rescaled initial condition U_0 can afterwards be reused in the overall single shooting algorithm:

1. Compute flow trajectories starting from U_0^{k-1} with phase-drift control τ. Choose new initial condition U_0^k with minimal drift along the computed trajectories.
2. Compute period T^{k+1} by (21) and (22).
3. Compute Newton correction du^k by: $-J^k du^k = D(U_0^k, T^{k+1}, 1)$.
4. Update initial condition $U_0^{k+1} = U_0^k + du^k$.
5. GOTO 1. until convergence.

For the control parameter θ we suggest to use $\theta := \|D(U_0^k, T^k, 1)\|^2$. A small distance suggests that the trajectories should be close to in-phase requiring no additional phase control.

4 Time Domain Decomposition

4.1 *Multiple Shooting*

In the following we investigate the possibility of extending the single shooting method with respect to a decomposition of the time scale for λ. To this end, we follow the general procedure of defining a multiple shooting approach for which we assume the period to be given. We define a partitioning of the time interval [0, 1] by using $N + 1$ deterministic $\lambda_i \neq \lambda_i(\xi)$ such that $0 = \lambda_0 < \lambda_1 < \ldots < \lambda_{N-1} < \lambda_N = 1$. Furthermore, by $\mathcal{U}_n(U_n, T, \lambda)$ we denote the velocity at time λ obtained by time integration of the S-SNSE starting from the initial condition U_n.

Given a period T, the problem formulation of computing periodic orbits can be stated as follows:

Find $U_0, U_1, \ldots, U_N$ such that:

$$\begin{aligned} U_1 - \mathcal{U}_0(U_0, T, \lambda_1) &= 0 \\ U_2 - \mathcal{U}_1(U_1, T, \lambda_2) &= 0 \\ &\vdots \\ U_N - \mathcal{U}_{N-1}(U_{N-1}, T, \lambda_N) &= 0 \\ U_N - U_0 &= 0. \end{aligned}$$

This nonlinear system, compactly denoted by $F(U_0, U_1, \ldots, U_N, T) = 0$, can be solved by Newton's method resulting in the following iterative procedure:

For $k = 0, \ldots, k_{\max}$ do

$$-J^k C^k = F^k \tag{32}$$

$$U^{k+1} = U^k + C^k, \tag{33}$$

where we define

$$F^k := F(U_0^k, U_1^k, \ldots, U_N^k, T),$$
$$u^k := [U_0^k, U_1^k, \ldots, U_N^k]^t,$$
$$c^k := [C_0^k, C_1^k, \ldots, C_N^k]^t,$$

where J^k denotes the Jacobian of F^k. The resulting system of equations in each Newton step can be written as:

$$-\begin{bmatrix} -\frac{\partial \mathscr{U}_0^k}{\partial U_0} & I & 0 & \cdots & 0 \\ 0 & -\frac{\partial \mathscr{U}_1^k}{\partial U_1} & I & 0 & 0 \\ \vdots & \ddots & \ddots & \ddots & \vdots \\ 0 & \cdots & \cdots & -\frac{\partial \mathscr{U}_{N-1}^k}{\partial U_{N-1}} & I \\ -I & 0 & \cdots & 0 & I \end{bmatrix} \begin{bmatrix} C_0^k \\ C_1^k \\ \vdots \\ C_{N-1}^k \\ C_N^k \end{bmatrix} = \begin{bmatrix} U_1^k - \mathscr{U}_0^k \\ U_2^k - \mathscr{U}_1^k \\ \vdots \\ U_N^k - \mathscr{U}_{N-1}^k \\ U_N^k - U_0^k \end{bmatrix}, \tag{34}$$

where I denotes the identity operator. Usually the assembly of the complete Jacobian matrix is numerically too expensive. Therefore, our goal is to adapt the framework of parareal time integration [5, 16] with respect to the stochastic space, which will be elaborated on in the following section.

4.2 Parareal Time Algorithm

In its standard definition the parareal algorithm approximates the derivatives $\frac{\partial \mathscr{U}_n}{\partial U_n}$, $n = 0, \ldots, N-1$ by a coarse time grid. We transfer this idea towards the stochastic space, for which we will employ lower order PC expansions as a "coarse" grid.

First, observe that the system in (34) can be denoted by:

$$C_n^k = U_n^{k+1} - U_n^k$$

$$-C_{n+1}^k + \frac{\partial \mathscr{U}_n^k}{\partial U_n} C_n^k = U_{n+1}^k - \mathscr{U}_n^k$$

$$C_0^k - C_N^k = U_N^k - U_0^k,$$

for $n = 0, \ldots, N-1$. Rearranging the equations leads to the following iterative scheme:

$$U_0^{k+1} = \mathscr{U}_{N-1}^k + \frac{\partial \mathscr{U}_{N-1}^k}{\partial U_{N-1}}(U_{N-1}^{k+1} - U_{N-1}^k), \tag{35}$$

$$U_{n+1}^{k+1} = \mathscr{U}_n^k + \frac{\partial \mathscr{U}_n^k}{\partial U_n}(U_n^{k+1} - U_n^k), \quad n = 0, \ldots, N-2. \tag{36}$$

In [5] Gander et al. introduced a relaxation of Eq. (35) neglecting the computation of the derivative, which results in

$$U_0^{k+1} = U_N^k, \tag{37}$$

$$U_{n+1}^{k+1} = \mathscr{U}_n^k + \frac{\partial \mathscr{U}_n^k}{\partial U_n}(U_n^{k+1} - U_n^k), \quad n = 0, \ldots, N-1. \tag{38}$$

Computing the derivative $\frac{\partial \mathscr{U}_n^k}{\partial U_n}$, $n = 0, \ldots, N-1$ can be very expensive from a computational point of view. The classical parareal algorithm approximates it on a coarse time grid by:

$$\frac{\partial \mathscr{U}_n^k}{\partial U_n}(U_n^{k+1} - U_n^k) \approx G(U_n^{k+1}) - G(U_n^k),$$

where G denotes some coarse grid approximation to $\mathscr{U}$, e.g., by using lower order in time or larger time steps or a combination of both. The corresponding iterative scheme reads:

$$U_0^{k+1} = U_N^k, \tag{39}$$

$$U_{n+1}^{k+1} = \mathscr{U}_n^k + G(U_n^{k+1}) - G(U_n^k), \quad n = 0, \ldots, N-1. \tag{40}$$

Although a coarse time approximation could accelerate the computation of stochastic periodic orbits as well, in this work we only consider evaluating a coarse approximation with respect to the stochastic space. Specifically, we take a PC expansion of degree p_{fine} for the fine propagator $\mathscr{U}$ resulting in $P_{fine} + 1 = (p_{fine} + M)!/(p_{fine}!M!)$ PC modes. Furthermore, we define $G(U_n) =: v_n$, $n = 0, \ldots, N-1$ as the solution v of the S-SNSE at time $\lambda = \lambda_{n+1}$ starting from the initial condition U_n and using a lower order PC expansion with degree $p_{low} \leq p_{fine}$:

$$(\partial_\lambda v_k, \phi) + \sum_{i,j=0}^{P} (v_i \cdot \nabla) v_j, \phi) c(T)_{ijk}$$

$$+ \sum_{i=0}^{P} (\nabla v_i, \phi) \nu(T)_{ik} - (p_k, \nabla \cdot \phi) = 0, \qquad \forall \phi \in V_0, \tag{41}$$

$$(q, \nabla \cdot v_k) = 0, \qquad \forall q \in W, \tag{42}$$

$$v_k = \langle g, \psi_k \rangle, \qquad \text{on } \Gamma, \tag{43}$$

$$v_k = \langle U_n, \psi_k \rangle, \qquad \text{in } \mathscr{D}, \tag{44}$$

for $\lambda \in [\lambda_n, \lambda_{n+1}]$ and $k = 0, \ldots, P_{low}$. Here, $P_{low} + 1 = (p_{low} + M)!/(p_{low}!M!)$.

Note that we assume the same time and spatial grid both for $\mathscr{U}$ and G. The benefit of this approach is that the fine propagator $\mathscr{U}$ can be computed in parallel, while only a lower order PC expansion needs to be computed sequentially in time.

The parareal iteration (39) and (40) is understood by means of a stochastic Galerkin projection:

$$\langle U_0^{k+1}, \psi_i \rangle = \langle U_N^k, \psi_i \rangle, \tag{45}$$

$$\langle U_{n+1}^{k+1}, \psi_i \rangle = \langle \mathscr{U}_n^k, \psi_i \rangle + \langle (G(U_n^{k+1}) - G(U_n^k)), \psi_i \rangle, \quad n = 0, \ldots, N-1, \tag{46}$$

for $i = 0, \ldots, P_{fine}$.

At the beginning of this section, we assumed the period T to be given. In general, T is not known such that an additional constraint needs to be defined for its determination. In a similar way as described in Sect. 3.2, the period iterate T^{k+1} can be computed for given $U_0^k, U_1^k, \ldots, U_N^k$. However, the period approximation in each iteration will introduce an increase in phase-drift, which necessitates appropriate control similar as in Sect. 3.3.2. This difficulty will be investigated in detail in future work.

5 Numerical Results

5.1 Benchmark Problem

We consider a flow in a channel around a circular domain in two spatial dimensions (cf. Fig. 2) as a benchmark problem. In the following, we use a vector notation $\mathbf{u} := (u, v)$ for the velocity with components u and v in x- and y- direction, respectively.

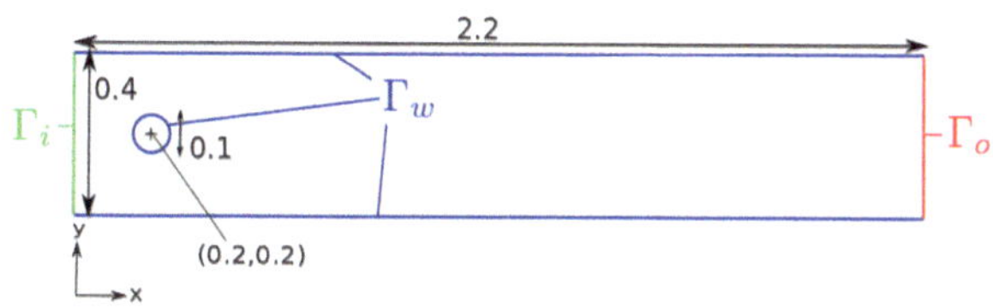

Fig. 2 Geometry of benchmark problem. Flow is considered from left to right

We impose no-slip boundary conditions at the boundary Γ_w, i.e., $\mathbf{u} = 0$. At the inflow Γ_i we use a parabolic flow profile with uncertain magnitude, i.e.,

$$u(x = 0, y, \xi) := 4u_m(\xi)y(H - y)/H^2,$$
$$v(x = 0, y, \xi) := 0,$$

for $y \in [0, 0.4]$ where $u_m(\xi) := 1.5 + 0.15\xi$ with a uniformly distributed $\xi \sim U(-1, 1)$ and $H = 0.4$ denotes the channel width. At the outflow Γ_o we impose natural "do-nothing" boundary conditions, i.e.,

$$\langle \int_{\Gamma_o} \nabla \mathbf{u} \cdot \mathbf{n} - p\mathbf{n} \, dy, \psi_i \rangle = 0, \quad i = 0, 1, 2, 3, \ldots$$

where $\mathbf{n}$ denotes the outward unit normal vector on Γ_o. We assume a deterministic kinematic viscosity $\nu := 0.001$. The Reynoldsnumber depends on ξ and is defined as:

$$Re(\xi) := \frac{2u_m(\xi)D}{3\nu},$$

where $D = 0.1$ denotes the diameter of the circular domain. For this problem setup the mean of the Reynoldsnumber corresponds to $\mathbb{E}(Re) = 100$. In this regime, a laminar time-periodic solution (periodic vortex shedding after circular domain) can be expected.

The spatial part is discretized using the Finite-Element-Method with continuous Lagrange elements of degree 2 for the velocity and degree 1 for the pressure variable (stable Taylor-Hood elements [21]). The mesh (not shown) consists of 3215 triangles resulting in 15045 degrees of freedom for each mode in the PC expansion. For time discretization we employ the Crank-Nicolson scheme (implicit, order 2) with homogeneous time step size $\Delta t = 0.01$. The corresponding time step size $\Delta\lambda > 0$ for the scaled time variable λ is defined by:

$$\Delta\lambda := \frac{\Delta t}{T(\xi = 0)}.$$

5.2 *Single Shooting*

We employ the single shooting algorithm as described in Sect. 3.3. For the PC expansion of the velocity, pressure and period variables we employ a third order Legendre Chaos expansion due to the Uniform distribution of ξ. As initial guesses

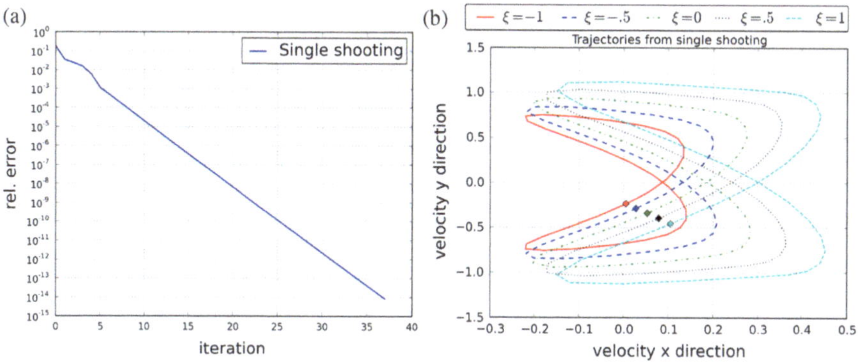

Fig. 3 Convergence properties of the singe shooting approach. (**a**) Evolution of relative distance between terminal state and initial condition. (**b**) Converged velocity trajectories evaluated from the PC expansion a posteriori for different values of ξ. 34 time steps with markers at $\lambda = 0$

for the shooting method we use a fully developed purely deterministic flow, which was obtained by time integration of the deterministic Navier-Stokes equations parametrized by the mean inflow and a corresponding rough deterministic estimate of the period (obtained from time integration of the deterministic equations).

Figure 3 depicts the convergence properties of the single shooting approach. Figure 3a shows the decay of the relative error with respect to the number of iterations up to machine precision. The relative error is measured by evaluating the L^2—norm of the distance between terminal and initial state, i.e.,

$$\text{rel. error}(U_0, T) := \frac{\|U_0 - \mathscr{U}(U_0, T, 1)\|}{\|\mathscr{U}(U_0, T, 1)\|}.$$

In Fig. 3b the PC expansion of the velocity is evaluated a posteriori at the coordinates $(0.35, 0.2)$ (middle of channel, one diameter downstream after circular domain) for different values of ξ. These results have been verified by deterministic sample computations (not shown here, we refer the reader to our preliminary work [20] for more details). Overall the algorithm is capable of approximating the limit-cycle with high accuracy. The single shooting approach, however, can only be computed in a sequential way, although parallel solvers for the S-SNSE (e.g. domain decomposition) can be used to speed up the computation of solutions to the S-SNSE in each iteration.

5.3 Parareal Approach

We evaluate the feasibility of the parareal algorithm as described in Sect. 4.2 for computing the same limit-cycle as in the previous section. Therefore, we again use a purely deterministic fully developed flow as initial guess for U_0. We take the period from the previous section, since we do not consider period corrections in this algorithm. As a fine grid propagator we rely on the third order Legendre expansion of the velocity and pressure variable (as in the previous section). For the coarse propagator we solve the S-SNSE by means of lower order expansions using polynomial degrees of $p = 0, 1, 2$. Since we do not employ a period correction, we compare the obtained results to a straight forward time integration of the S-SNSE (case $p = 3$).

Due to the time step size $\Delta t = 0.01$ it took 34 time steps to complete one cycle. Figure 4 depicts the evolution of the relative distance error with respect to the iteration number. As can be observed, all errors begin to stagnate after some iteration. Indeed, they are decreasing very slowly since the initial condition is not updated by a Newton step. The coarse propagator with degree $p = 0$ performs poorly, its error stagnates at the order of $O(10^{-3})$. It also decreases eventually, but it takes a large number of iterations (not shown here). The other coarse propagators exhibit a more robust convergence behaviour. Figure 5 shows the evolution of their corresponding trajectories at iterations 0, 5 and 10. They look quite similar, but exhibit deviations especially at the "turn-around" points. After 10 iterations all trajectories for $p = 1, 2, 3$ reflect the trajectories of the single shooting approach

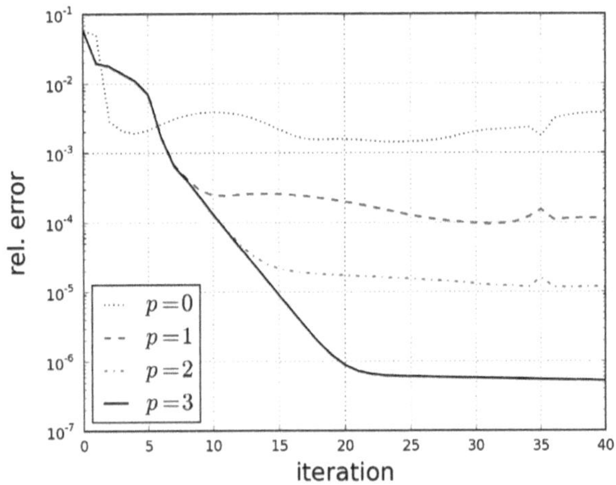

Fig. 4 Evolution of relative distance error for different coarse propagator degrees

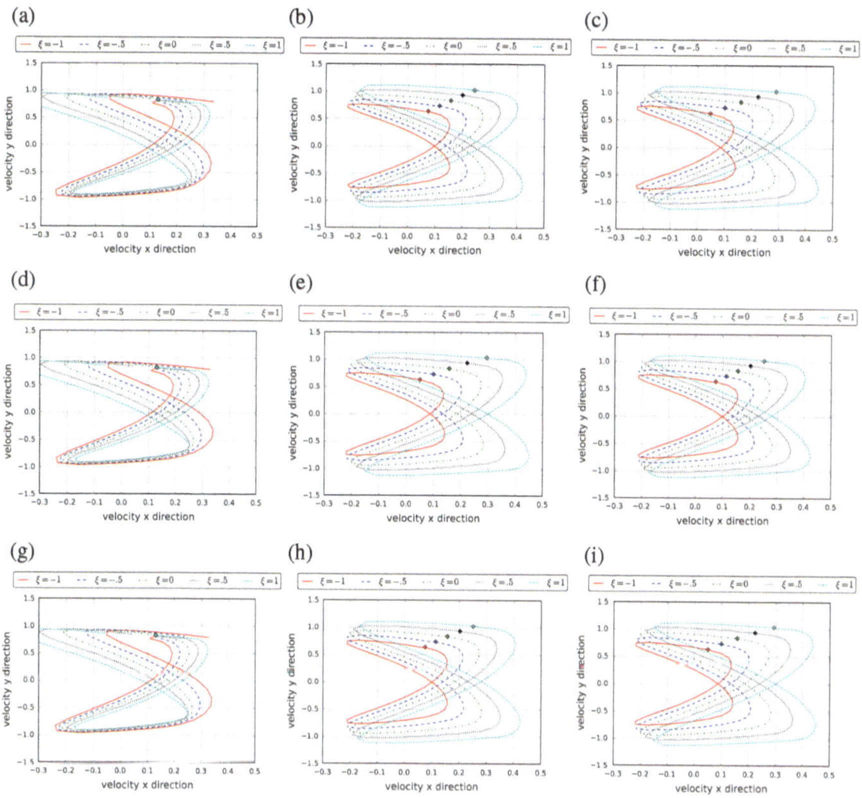

Fig. 5 Evolution of velocity trajectories for different iterations (*left* to *right*) and different coarse grid propagators (*top* to *bottom*). Markers at $\lambda = 0$. (**a**) Iteration 0, $p = 1$. (**b**) Iteration 5, $p = 1$. (**c**) Iteration 10, $p = 1$. (**d**) Iteration 0, $p = 2$. (**e**) Iteration 5, $p = 2$. (**f**) Iteration 10, $p = 2$. (**g**) Iteration 0, $p = 3$. (**h**) Iteration 5, $p = 3$. (**i**) Iteration 10, $p = 3$

(cf. Fig. 3b) with high accuracy. This suggests, that the relative error in the distance between initial condition and terminal state does not need to be computed up to machine precision to reach high accuracy in approximating the limit-cycle. For $p = 0$ the behaviour is different. It takes significantly more iterations (about 150) to reach a comparable accuracy as depicted in Fig. 6. However, the numerical cost of computing the coarse propagator equals the cost of solving purely deterministic equations.

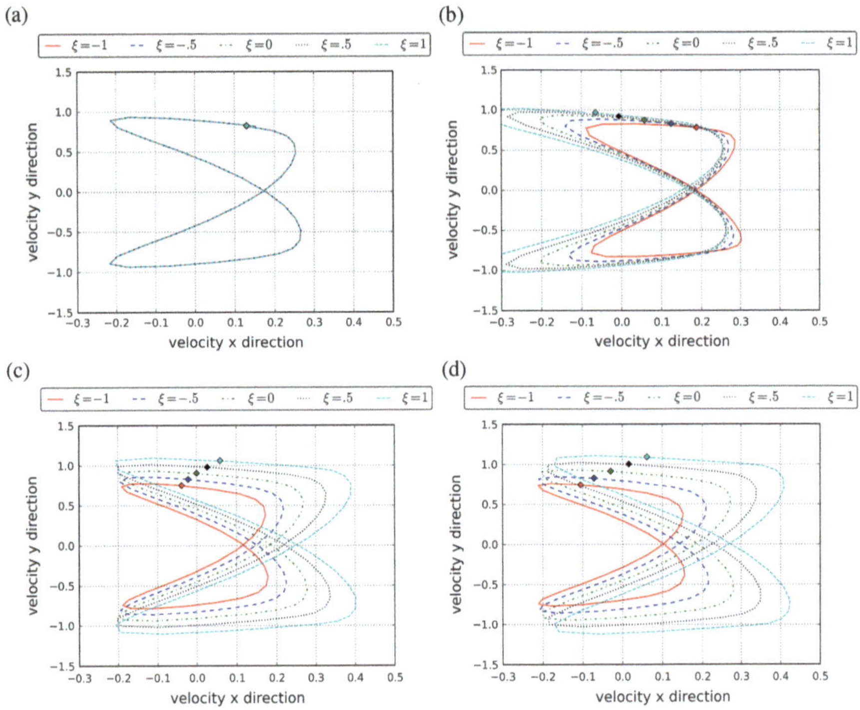

Fig. 6 Evolution of velocity trajectories for different iterations and the coarse grid propagator with polynomial degree $p = 0$. Markers at $\lambda = 0$. (**a**) Iteration 0, $p = 0$. (**b**) Iteration 50, $p = 0$. (**c**) Iteration 100, $p = 0$. (**d**) Iteration 150, $p = 0$

6 Conclusions

Polynomial Chaos expansions are known to exhibit a convergence breakdown in case of for example oscillatory dynamics. Therefore, in time a growing polynomial degree is required to maintain accuracy. In this work, we introduced two types of shooting methods, which are able to overcome this difficulty for the incompressible Navier-Stokes equations with random parameters based on the Spectral-Stochastic-Finite-Element-Method. We re-interpreted a Newton-Galerkin approach originally introduced in [20] as a single shooting method and extended its definition with respect to multiple shooting. Specifically, we adapted the parareal time integration method (resulting from multiple shooting) to compute a stochastic initial condition corresponding to a given stochastic period. Its attractive feature is that solutions corresponding to a fine propagator can be computed in parallel. We evaluated numerically its convergence properties on a benchmark problem computing a time periodic vortex shedding flow.

We observed that coarse propagators using low order PC expansions (down to degree p equal to 0) were able to capture the trajectories of the flow accurately.

Except for $p = 0$, the trajectories converged in few iterations. The case $p = 0$ converged eventually, however, only after a significantly larger number of iterations. The convergence measured by evaluating the norm of the distance of initial condition and corresponding terminal state after one cycle shows a very slow error decay. This is mainly due to the fact that the initial condition is not updated by a Newton method, instead using a relaxation, the terminal state of the previous parareal iterate is used as a new iterate for the initial condition.

Current work is focused on extending the parareal approach to allow for an adaptive period computation, similar as in the single shooting context. In addition, more sophisticated update schemes for the initial condition will be analysed and numerically evaluated on more complex benchmark problems involving larger number of random variables.

Acknowledgements The author wants to thank the Heidelberg Institute for Theoretical Studies for providing financial support for carrying out this work. Furthermore, the author wishes to thank Olivier P. Le Maître, LIMSI-CNRS, France and Duke University, U.S.A. as well as Vincent Heuveline, Heidelberg University, Germany for many fruitful discussions regarding our preliminary work in [20].

References

1. Babuska, I., Nobile, F., Tempone, R.: A stochastic collocation method for elliptic partial differential equations with random input data. SIAM J. Numer. Anal. **45**, 1005–1034 (2007)
2. Beran, P., Pettit, C., Millman, D.: Uncertainty quantification of limit-cycle oscillations. J. Comput. Phys. **217**, 217–247 (2006)
3. Duguet, Y., Pringle, C.C.T., Kerswell, R.R.: Relative periodic orbits in transitional pipe flow. Phys. Fluids **20**(11), 114102 (2008)
4. Fletcher, C.A.J.: Computational Techniques for Fluid Dynamics, vols. I and II. Computational Physics. Springer, Berlin (1988)
5. Gander, M.J., Jiang, Y.L., Song, B., Zhang, H.: Analysis of two parareal algorithms for time-periodic problems. SIAM J. Sci. Comput. **35**, A2393–A2415 (2013)
6. Ghanem, R., Spanos, P.D.: Stochastic Finite Elements: A Spectral Approach. Springer, New York, NY (1991)
7. Guermond, J., Minev, P., Shen, J.: An overview of projection methods for incompressible flows. Comput. Methods Appl. Mech. Eng. **195**, 6011–6045 (2006)
8. Heuveline, V., Schick, M.: A hybrid generalized polynomial chaos method for stochastic dynamical systems. Int. J. Uncertain. Quantif. **4**(1), 37–61 (2014)
9. Knio, O., Le Maître, O.: Uncertainty propagation in CFD using polynomial chaos decomposition. Fluid Dyn. Res. **38**, 616–640 (2006)
10. Le Maître, O., Knio, O., Najm, H., Ghanem, R.: A stochastic projection method for fluid flow. I. Basic formulation. J. Comput. Phys. **173**, 481–511 (2001)
11. Le Maître, O., Reagan, M., Najm, H., Ghanem, R., Knio, O.: A stochastic projection method for fluid flow. II. random process. J. Comput. Phys. **181**, 9–44 (2002)
12. Le Maître, O., Knio, O., Najm, H., Ghanem, R.: Uncertainty propagation using Wiener-Haar expansions. J. Comput. Phys. **197**(1), 28–57 (2004)
13. Le Maître, O.P., Najm, H., Ghanem, R., Knio, O.: Multi-resolution analysis of Wiener-type uncertainty propagation schemes. J. Comput. Phys. **197**(2), 502–531 (2004)

14. Le Maître, O.P., Knio, O.M.: Spectral methods for Uncertainty Quantification: with Applications to Computational Fluid Dynamics. Springer Science+Business Media B.V, Dordrecht. Springer, Berlin (2010)
15. Le Maître, O.P., Mathelin, L., Knio, O.M., Hussaini, M.Y.: Asynchronous time integration for polynomial chaos expansion of uncertain periodic dynamics. Discret. Contin. Dyn. Syst. **28**(1), 199–226 (2010)
16. Lions, J.L., Maday, Y., Turinici, G.: Résolution d'EDP par un schéma en temps pararéel. C. R. Acad. Sci. I Math. **332**(7), 661–668 (2001)
17. Liu, J.S.: Monte Carlo Strategies in Scientific Computing, 1. softcover print edn. Springer Series in Statistic. Springer, New York NY (2008)
18. Powell, C.E., Silvester, D.J.: Preconditioning steady-state Navier–Stokes equations with random data. SIAM J. Sci. Comput. **34**, 2482–2506 (2012)
19. Powell, C.E., Ullmann, E.: Preconditioning stochastic Galerkin saddle point systems. SIAM J. Matrix Anal. Appl. **31**, 2813–2840 (2010)
20. Schick, M., Heuveline, V., Le Maître, O.: A Newton–Galerkin method for fluid flow exhibiting uncertain periodic dynamics. SIAM/ASA J. Uncertain. Quantif. **2**, 153–173 (2014)
21. Taylor, C., Hood, P.: A numerical solution of the Navier-Stokes equations using the finite element technique. Comput. Fluids **1**(1), 73–100 (1973)
22. Turek, S.: Efficient Solvers for Incompressible Flow Problems: an Algorithmic and Computational Approach. Lecture Notes in Computational Science and Engineering, vol. 6. Springer, Berlin (1999)
23. Wiener, N.: The homogeneous chaos. Am. J. Math. **60**(4), 897–936 (1938)
24. Xiu, D., Karniadakis, G.E.: Modeling uncertainty in flow simulations via generalized polynomial chaos. J. Comput. Phys. **187**(1), 137–167 (2003)

MIX
Papier aus verantwortungsvollen Quellen
Paper from responsible sources
FSC® C105338

If you have any concerns about our products,
you can contact us on
ProductSafety@springernature.com

In case Publisher is established outside the EU,
the EU authorized representative is:
Springer Nature Customer Service Center GmbH
Europaplatz 3, 69115 Heidelberg, Germany

Printed by Libri Plureos GmbH
in Hamburg, Germany